SOLID–STATE
LASERS
AND
APPLICATIONS

OPTICAL SCIENCE AND ENGINEERING

Founding Editor
Brian J. Thompson
University of Rochester
Rochester, New York

SOLID-STATE LASERS AND APPLICATIONS

edited by
Alphan Sennaroglu

CRC Press
Taylor & Francis Group
Boca Raton London New York

CRC Press is an imprint of the
Taylor & Francis Group, an **informa** business

CRC Press
Taylor & Francis Group
6000 Broken Sound Parkway NW, Suite 300
Boca Raton, FL 33487-2742

First issued in paperback 2019

ISBN-13: 978-0-8493-3589-1 (hbk)
ISBN-13: 978-0-367-38987-1 (pbk)

Library of Congress Cataloging-in-Publication Data

Solid-state lasers and applications / Alphan Sennaroglu (editor).
    p. cm. -- (Optical science and engineering ; 119)
  Includes bibliographical references and index.
  ISBN-13: 978-0-8493-3589-1 (alk. paper)
    1. Solid-state lasers. I. Sennaroglu, Alphan. II. Series: Optical science and engineering (Boca Raton, Fla.) ; 119.

TA1705.S6748 2007
621.36'61--dc22   2006017534

Preface

Since the invention of the first ruby laser in 1960, rapid progress has taken place in the development of solid-state lasers. In this class of lasers, optical amplification is produced by using insulating crystals or glasses doped with rare-earth or transition-metal ions. Many favorable characteristics such as chemical stability, mechanical durability, and long operational lifetime have put these laser systems among the most preferred candidates for a wide range of applications in science and technology, including spectroscopy, atmospheric monitoring, micromachining, and precision metrology, among others.

The field of solid-state lasers has become so diverse over the last nearly five decades that it is impossible to provide a complete, in-depth review of all of the developments. We will discuss some of the recent trends and the major directions in which active research continues. First, applications requiring lasers that operate at a specific wavelength prompted much interest in the development of new gain media. This has led to the emergence of many novel solid-state materials that produce laser light in different parts of the electromagnetic spectrum from the ultraviolet to the mid infrared. Second, studies have been directed toward achieving compact, cost-effective designs so that solid-state lasers operated in different regimes can be integrated into complete measurement and characterization systems. Examples include microchip solid-state lasers and multipass-cavity femtosecond lasers. Lastly, significant achievements have taken place in the generation and amplification of ultrashort optical pulses from the femtosecond to nanosecond time scales.

Solid-State Lasers and Applications aims at providing an in-depth account of the major advances that have taken place in the field with an emphasis on the most recent trends. For example, Chapter 2 to Chapter 5 discuss the most recent developments and applications of new solid-state gain media in different wavelength regions. Examples include cerium-doped lasers in the ultraviolet, ytterbium lasers near 1 μm, rare-earth ion-doped lasers in the eye-safe region, and tunable Cr^{2+}:ZnSe lasers in the mid infrared. Other chapters focus on specific modes of operation of solid-state laser systems: pulsed microchip lasers (Chapter 1), high-power neodymium lasers (Chapter 6), ultrafast solid-state lasers (Chapter 7 to Chapter 10), amplification of femtosecond pulses with optical parametric amplifiers (Chapter 11), and noise characteristics of solid-state lasers (Chapter 12). A brief overview of each chapter is provided below. Comprising 12 contributed chapters, the handbook targets researchers, graduate students, and engineers who either work in the design of solid-state lasers or who use such systems in applications.

The handbook starts with the chapter on passively Q-switched microchip lasers. An in-depth overview of the basic characteristics of solid-state lasers in general and microchip lasers in particular is first given. Rate equations, energy-level structure of optical amplifiers, and output characteristics of solid-state lasers are discussed. Useful analytical formulas for output pulse energy, peak power, and pulsewidth are derived for passively Q-switched solid-state lasers. The chapter then describes practical demonstrations of passively Q-switched microchip lasers and examines some state-of-the-art applications such as ranging and laser-induced breakdown spectroscopy.

In recent years, ytterbium-doped materials have emerged as important tunable solid-state gain media near 1 μm, owing to the development of high-power InGaAs-based pump diodes. Chapter 2 reviews the general characteristics of ytterbium-doped solid-state lasers and materials. Topics include crystal field effects, basic spectroscopic properties of the Yb^{3+} ion, structural effects, lasing efficiency, variation of quantum efficiency in different hosts, and the role of thermal loading. Finally, recent work in the development of Yb-based femtosecond lasers near 1 μm is discussed.

Chapter 3 provides a review of the recent work aimed at the development of Cr^{2+}:ZnSe lasers operating in the mid infrared between 2 and 3 μm. A brief historical review of tunable solid-state lasers is first given. Synthesis techniques are then discussed with a focus on diffusion doping. After a discussion of the absorption and emission spectroscopy of Cr^{2+}:ZnSe, pulsed, continuous-wave, and mode-locked operations of Cr^{2+}:ZnSe lasers are described. Data showing the dependence of the passive losses, fluorescence lifetime, fluorescence quantum efficiency, and power performance on active ion concentration are presented. Finally, an intracavity-pumped Cr^{2+}:ZnSe laser is described with an ultrabroad tuning range between 1880 and 3100 nm.

Tunable lasers in the ultraviolet have important applications in remote sensing, combustion diagnostics, spectroscopy of wide band-gap semiconductors, and medicine. Chapter 4 discusses all-solid-state cerium lasers which operate in the ultraviolet around 300 nm. Physical properties and tradeoffs of different hosts such as Ce:LLF, Ce:YLF, Ce:LiCAF, and others are discussed. Techniques of short pulse generation from Ce^{3+}-doped fluoride lasers are described. Finally, a Ce:LiCAF-based chirped-pulse amplification system is described, capable of producing femtosecond pulses with 30-GW peak power near 300 nm.

Wavelengths longer than about 1.4 μm are strongly attenuated in the cornea and the vitreous humor of the eye before reaching the retina. Lasers operating above 1.4 μm hence provide enhanced eye safety and are commonly referred to as *eyesafe lasers*. Chapter 5 focuses on the characteristics of a particular class of eyesafe lasers based on bulk solid-state gain media doped with the rare-earth ions Er^{3+}, Tm^{3+}, and Ho^{3+}. Lasing performance of various Er^{3+}, Tm^{3+}, and Ho^{3+} lasers operating around the respective wavelengths of 1.6, 2.0, and 2.1 μm are reviewed. Energy transfer mechanisms via codoping, the quasi-three-level energy structure of the laser-active ions, and the role of reabsorption are discussed. In addition, methods of diode pumping, activator-sensitizer pumping, and direct upper-state pumping are described.

High-power neodymium(Nd)-doped lasers are becoming widely used in several industrial applications such as welding. Chapter 6 reviews the state of the art in high-power Nd-based solid-state lasers. Important neodymium hosts such as YAG, vanadate, and YLF are compared. Possible crystal geometries (for example, rod and slab geometries) and pumping configurations are described. Recent developments in continuous-wave and pulsed Nd lasers at 1064 nm are reviewed. Lastly, wavelength conversion from 1064 nm to the visible and ultraviolet by using high-power Nd lasers is discussed.

Chapter 7 provides an in-depth review of ultrafast solid-state lasers. After a historical review of mode locking, progress in the development of pulsed solid-state lasers over the last 10 years is outlined. Requirements on the gain medium for ultrashort pulse generation, effects of dispersion and nonlinearity, and key design rules are discussed. Different mode-locking techniques are described with a particular emphasis on mode locking with semiconductor saturable absorber mirrors (SESAMs). Several examples of solid-state ultrafast laser systems are examined.

Chapter 8 provides a detailed account of the design rules of multipass cavities (MPC) and their application to pulse energy scaling in femtosecond solid-state lasers. Q-preserving multipass cavities are introduced and their analytical design rules are derived. In the case of MPCs with notched mirrors, use of compensating optics to restore the q-preserving nature of the multipass cavity is also discussed. Variation of the pulse repetition rates for different q-preserving configurations is investigated. Finally, experimental realization of MPCs in compact and/or high-energy mode-locked oscillators with different gain media is discussed.

An important route to femtosecond pulse amplification uses the technique of cavity dumping. Chapter 9 first provides a detailed theoretical description of cavity-dumped femtosecond lasers and investigates the output characteristics in various parameter ranges. Experimental realization of a femtosecond cavity-dumped Yb:KYW laser is then described. A very important emerging application of amplified femtosecond pulses is in the microfabrication of photonic devices. The second part of Chapter 9 discusses in detail the interaction of femtosecond pulses with transparent media through multiphoton absorption and other nonlinear processes. Then, application of the cavity-dumped Yb:KYW laser in waveguide writing and fabrication of waveguide lasers at the telecommunication wavelengths is described.

Chapter 10 is devoted to a discussion of octave-spanning Ti:sapphire lasers and their application in precision metrology. In particular, it is shown that besides their well-known importance for ultrafast time-domain spectroscopy, few-cycle Ti:sapphire lasers yield the highest-quality frequency combs for optical frequency metrology. Carrier-envelope phase-stabilization schemes, noise analysis, and pulse formation dynamics are described. The main technical challenges that must be overcome to enable the generation of ultrabroadband spectra are further discussed. An intriguing alternative to optical clockworks based on difference frequency generation in the infrared spectral region is also presented.

Chapter 11 provides an in-depth review of the state of the art in femtosecond optical parametric amplifiers (OPA). The theory of optical parametric amplification is first reviewed and key design criteria are outlined. OPA systems operating in the visible, near-infrared, and mid-infrared ranges are then described. Use of OPAs in generating pulses shorter than the pump pulses and also their capability in generating pulses with constant carrier-envelope phase are discussed. Finally, an alternative route to petawatt-level peak power generation based on OPAs is presented.

The final chapter of the handbook reviews the subject of noise in solid-state lasers. The authors start with the essential mathematical basics and explain some details of the notation, which often cause confusion in the literature. Then, the noise characteristics of single-frequency lasers are examined first, in order to facilitate the understanding of more complicated systems such as mode-locked lasers. Noise in multimode continuous-wave lasers, Q-switched lasers, and mode-locked lasers is then discussed. Each section begins with a discussion of the basic physics, and most of them end with an overview of experimental methods for measurement.

Many people contributed to the realization of this project. First, I would like to express my gratitude to all of the contributing authors without whose expertise and diligent work, none of this would have been possible. I would also like to thank the acquisitions editor, Taisuke Soda, from Taylor & Francis, for initiating the project. My thanks also go to project coordinator Jill Jurgensen, editorial assistant Jacqueline Callahan, and project editor Jay Margolis, all from Taylor & Francis, for carefully organizing and running the production process. During the preparation of the handbook, I greatly benefited from the support of the Turkish Academy of Sciences in the framework of the Young Scientist Award Program AS/TUBA-GEBIP/2001-1-11 and the NSF-Tubitak travel grant (TBAG-U/110-104T247). Finally, I am indebted to my wife, Figen Sennaroglu, and children, Canan and Özalp Demir, for their patience and never-ending support over the last 2 years.

Alphan Sennaroglu
Koç University, Istanbul, Turkey

Editor

Alphan Sennaroglu was born on November 10, 1966 in Nicosia, Cyprus. He received the B.S., M.S., and Ph.D. degrees in Electrical Engineering from Cornell University in 1988, 1990, and 1994, respectively. In 1994, he joined Koç University in Istanbul, Turkey, where he is currently a Professor in the Departments of Physics and Electrical-Electronics Engineering. In 1994, he established the Laser Research Laboratory at Koç University. His research focuses on the development and modeling of solid-state lasers, ultrafast  optics, spectroscopic investigation of novel laser and amplifier media, and nonlinear optics. He has authored or coauthored more than 80 journal and conference articles in these fields. He was a visiting researcher at the Massachusetts Institute of Technology during the 2002–2003 academic year and in the summers of 2005 and 2006. In 2004 and 2006, he served as the program chair for the conference on Solid-State Lasers and Amplifiers held as part of the Photonics Europe Meeting in Strasbourg, France.

In December 2005, Alphan Sennaroglu was elected as associate member to the Turkish Academy of Sciences (TUBA). He is a Senior Member of IEEE (Institute of Electrical and Electronics Engineers) and a member of OSA (Optical Society of America), SPIE (The International Society for Optical Engineering), and Optics Committee of Turkey. In 1999, he founded the Leos Turkish Chapter of the IEEE Lasers and Electro-Optics Society and served as the Chapter president between 1999 and 2003. He received the 2002 ICTP/ ICO Award (ICTP: International Center for Theoretical Physics, ICO: International Commission for Optics), 2001 Werner-von-Siemens Award (Koç University), 2001 TUBA Young Scientist Award, 1998 Tubitak (Scientific and Technical Research Council of Turkey) Young Scientist Award, Sage Graduate Fellowship (Cornell University, 1989-1990), Sibley Prize of Electrical Engineering (Cornell University, 1988), and AMIDEAST (America-Mideast Educational and Training Services, Inc.) undergraduate scholarship (1984-1988). He married Figen Ecer Sennaroglu in 1995 and has two children, Canan and Özalp Demir.

Contributors

François Balembois
Laboratoire Charles Fabry
de l'Institut d'Optique
Orsay Cedex, France

Giulio Cerullo
Department of Physics
Politecnico di Milano
Milano, Italy

Sebastien Chénais
Laboratoire Charles Fabry
de l'Institut d'Optique
Orsay Cedex, France

Umit Demirbas
Laser Research Laboratory
Department of Physics
Koç University
Istanbul, Turkey

Frédéric Druon
Laboratoire Charles Fabry
de l'Institut d'Optique
Orsay Cedex, France

James G. Fugimoto
Department of Electrical
Engineering and Computer
Science
Research Laboratory of Electronics
Massachusetts Institute of
Technology
Cambridge, Massachusetts, U.S.A.

Romain Gaumé
Laboratoire de Chimie de la Matière
Condensée de Paris
Paris, France

Patrick Georges
Laboratoire Charles Fabry
de l'Institut d'Optique
Orsay Cedex, France

Philippe Goldner
Laboratoire de Chimie de la Matière
Condensée de Paris
Paris, France

Mathieu Jacquemet
Laboratoire Charles Fabry
de l'Institut d'Optique
Orsay Cedex, France

Franz X. Kärtner
Department of Electrical
Engineering and Computer
Science
Research Laboratory of Electronics,
Massachusetts Institute of
Technology
Cambridge, Massachusetts, U.S.A.

Ursula Keller
ETH Zurich
Physics Department
Institute of Quantum Electronics,
Zurich, Switzerland

Alexander Killi
Max-Planck-Institute for Nuclear
Physics
Heidelberg, Germany

Susumu Konno
Mitsubishi Electric Corporation
Hyogo, Japan

Daniel Kopf
HighQLaser Production
Hohenems, Austria

Max Lederer
HighQLaser Production
Hohenems, Austria

Zhenlin Liu
IMRA America, Inc.
Ann Arbor, Michigan, U.S.A.

Cristian Manzoni
Department of Physics
Politecnico di Milano
Milano, Italy

Lia Matos
Department of Physics, and
    Research Laboratory of
    Electronics
Massachusetts Institute of
    Technology
Cambridge, Massachusetts, U.S.A.

Uwe Morgner
Max-Planck-Institute for Nuclear
    Physics
Heidelberg, Germany
Leibniz University of Hannover
Hannover, Germany

Oliver D. Mücke
Department of Electrical
    Engineering and Computer
    Science, and Research Laboratory
    of Electronics,
Massachusetts Institute of
    Technology
Cambridge, Massachusetts, U.S.A.

Shingo Ono
Nagoya Institute of Technology
Nagoya, Japan

Roberto Osellame
Department of Physics
Politecnico di Milano
Milano, Italy

Rüdiger Paschotta
ETH Zurich
Physics Department
Institute of Quantum Electronics,
Zurich, Switzerland

Johan Petit
Laboratoire de Chimie de la Matière
    Condensée de Paris
Paris, France

Nobuhiko Sarukura
Osaka University
Osaka, Japan

Alphan Sennaroglu
Laser Research Laboratory
Department of Physics
Koç University
Istanbul, Turkey

Robert C. Stoneman
Lockheed Martin Coherent
    Technologies
Louisville, Colorado, U.S.A.

Harald R. Telle
Physikalisch-Technische
    Bundesanstalt (PTB)
Braunschweig, Germany

Bruno Viana
Laboratoire de Chimie de la Matière
    Condensée de Paris
Paris, France

J. J. Zayhowski
Lincoln Laboratory
Massachusetts Institute of Technology
Lexington, Massachusetts, U.S.A.

Table of Contents

1

Passively Q-Switched Microchip Lasers

J.J. Zayhowski

CONTENTS

1.1 Introduction

1.1.1 Motivation

Many applications of lasers require subnanosecond optical pulses with peak powers of several kilowatts and pulse energies of several microjoules, or some combination of those properties. The most common method of producing subnanosecond pulses is to modelock a laser, generating a periodic train of short pulses with an interpulse period equal to the round-trip time of the laser cavity, typically 10 ns. Because of the large number of pulses produced each second, even lasers with high average powers (10 W or greater) do not produce much energy per pulse. Energetic pulses can be produced by Q switching. However, the size of conventional Q-switched lasers, along with their physics, precludes producing subnanosecond pulses (Zayhowski and Kelley, 1991). Extremely short, high-energy pulses can be obtained from Q-switched modelocked lasers or amplified modelocked lasers. Both of these approaches require complicated systems, typically several feet long and consuming several kilowatts of electrical power, and are therefore expensive.

The short cavity lengths of Q-switched microchip lasers allow them to produce pulses with durations comparable to those obtained with modelocked systems. At the same time, they take full advantage of the gain medium's ability to store energy. Actively Q-switched microchip lasers, pumped with a 0.5-W diode laser, have produced pulses as short as 115 ps with peak powers of tens of kilowatts and pulse energies of several microjoules (Zayhowski and Dill, 1995). For proper Q switching, these lasers require high-speed, high-voltage electronics. A passively Q-switched microchip laser does not require any switching electronics, thereby reducing system size and complexity, and improving power efficiency. Pumped with a 1.2-W diode laser, passively Q-switched microchip lasers similar to the one shown in Figure 1.1 produce pulses as short as 218 ps with peak powers in excess of 25 kW at pulse repetition rates greater than 10 kHz (Zayhowski and Dill, 1994). More recently, passively Q-switched microchip lasers have been pumped with high-brightness, high-power diode-laser arrays to produce 150-ps pulses at pulse repetition rates of several kilohertz (Zayhowski and Wilson, 2003a), or peak powers in excess of 560 kW at rates up to 1 kHz (Zayhowski, 1998). All of these devices oscillate in a single, transform-limited longitudinal mode and produce a diffraction-limited, linearly polarized, circularly symmetric Gaussian beam.

1.1.2 What Is a Passively Q-Switched Microchip Laser?

A passively Q-switched microchip laser consists of a gain medium and a saturable absorber in a short, plane-parallel Fabry–Pérot cavity. As the gain

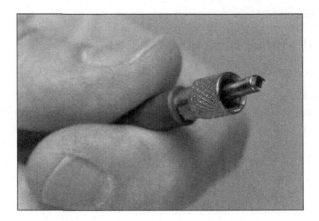

FIGURE 1.1
Photograph of passively Q-switched microchip laser bonded to the ferrule of the fiber used to pump it.

medium is pumped it accumulates stored energy and emits photons. Over many round trips in the resonator, the photon flux sees gain, fixed loss, and saturable loss. If the gain medium saturates before the absorber, the laser will tend to oscillate cw. On the other hand, if the photon flux builds up to a level that saturates, or bleaches, the saturable absorber first, the resonator will see a dramatic reduction in intracavity loss and the laser will Q switch, generating a short, intense pulse of light.

The monolithic, plane-parallel cavity structure of microchip lasers allows for their inexpensive mass production. The planar uniformity of the cavity is broken by the pump beam, which deposits heat as it longitudinally pumps the crystal. Thermal and gain-related effects create a stable optical cavity that defines the transverse dimensions of the laser modes. For many microchip lasers, the fundamental transverse mode is strongly favored and diffraction-limited output beams are generated.

The pulse duration from a Q-switched laser generally decreases with decreasing cavity length. With cavity lengths ranging from a fraction of a millimeter to several millimeters, passively Q-switched microchip lasers generate pulses with durations that range from a few tens of picoseconds to a few nanoseconds. Even at the modest pulse energies produced by such devices, ranging from a fraction of a microjoule to a few hundred microjoules, this can result in peak powers approaching a megawatt.

In addition to simplicity of implementation, other major advantages of passively Q-switched microchip lasers include the generation of pulses with a well-defined energy and duration that are insensitive to pumping conditions. Furthermore, it is much easier to obtain single-frequency operation in a passively Q-switched laser than in an actively Q-switched device.

1.1.3 Organization of Chapter

The technical content of this chapter is organized into three sections: Theory, Demonstrated Device Performance, and Applications. The discussion of theory is divided into eight subsections, starting in Subsection 1.2.1 with a brief discussion of the concepts that are fundamental to understanding the performance of lasers. By including this discussion, Subsection 1.2.1 makes this chapter self-contained and accessible to anyone with a background in calculus and quantum mechanics. Subsection 1.2.1 serves the dual purpose of clarifying the notation used throughout the rest of the chapter. Subsection 1.2.2 introduces models for the gain media of lasers and saturable absorbers. Using these models, Subsection 1.2.3 derives a set of rate equations that can be solved to describe many characteristics of a laser's performance. The solutions to the rate equations are contained in Subsection 1.2.4, followed by a discussion of the stability of the output of passively Q-switched microchip lasers, based on the rate equations, in Subsection 1.2.5. The discussion in Subsections 1.2.4 and 1.2.5 is developed in terms that are most readily applied to bulk saturable absorbers. Semiconductor saturable absorber mirrors represent a second class of saturable absorbers, and are discussed in Subsection 1.2.6. The physical phenomena that determine the transverse modes of microchip lasers are discussed in Subsection 1.2.7. Subsection 1.2.8 concludes the theoretical portion of this chapter with a brief discussion of thermal effects.

Section 1.3, Demonstrated Device Performance, starts, in Subsection 1.3.1, with a description of several of the passively Q-switched microchip lasers that have been reported in the literature. As the required output power of a laser system increases, the need for amplification becomes apparent. For many high-power applications that require subnanosecond pulses, a passively Q-switched microchip laser operating at relatively low power can be used to produce the desired output format, followed by an amplifier to obtain the required pulse energy or power. Compact amplified microchip laser systems are described in Subsection 1.3.2. The discussion includes an amplifier design that scavenges the pump power that is otherwise wasted by the microchip laser and greatly increases the system efficiency. The high peak powers of passively Q-switched microchip lasers make it easy to perform nonlinear frequency generation, with or without amplification. Nonlinear frequency conversion greatly extends the wavelength coverage of these diminutive devices and their utility. Frequency converted devices are discussed in Subsection 1.3.3.

Diode-pumped passively Q-switched microchip lasers are a relatively new family of all-solid-state sources. Nevertheless, numerous applications have already emerged. Section 1.4 describes a few of these in detail, including ranging and three-dimensional imaging in Subsection 1.4.2, laser-induced breakdown spectroscopy in Subsection 1.4.3, and environmental monitoring in Subsection 1.4.4. Each of these applications is enabled by different characteristics of the lasers. Additional applications are briefly described in Subsection 1.4.5.

1.2   Theory

1.2.1   Fundamental Concepts

1.2.1.1   Absorption

Quantum theory shows us that matter exists only in certain allowed energy levels or states. In thermal equilibrium, lower energy states are preferentially occupied, with an occupation probability proportional to $\exp(-E_s / k_B T)$, where E_s is the energy of the state, T is absolute temperature, and k_B is Boltzmann's constant. When light interacts with matter, it is possible for one quantum of optical energy, a photon, to be absorbed while simultaneously exciting the material into a higher-energy state. In this process, the energy difference between the states is equal to the energy of the absorbed photon, and energy is conserved.

Consider a simple material system with only two energy states, and assume that essentially all optically active sites within the material system are in the lowest energy state in thermal equilibrium. The probability that a randomly chosen photon in an optical field of cross section A will be absorbed by a given absorption site with a radiative cross section σ_r as it passes through a material is σ_r / A. If there are g_u identical (degenerate) high-energy states, the probability becomes $g_u \sigma_r / A$. The product $g_u \sigma_r$ is known as the absorption cross section σ_a. When all of the photons in an optical field of intensity I and all of the absorption sites in a material of length dl are accounted for, the intensity of an optical field passing through the material changes by

$$dI = -I \rho_1 \sigma_a dl, \tag{1.1}$$

where ρ_1 is the density of absorption sites in the material. This equation has the solution

$$I(l) = I(0) \exp(-\alpha l), \tag{1.2}$$

where $I(0)$ is the intensity of the optical field as it enters the material at position $z = 0$ and $\alpha = \rho_1 \sigma_a$ is the absorption coefficient of the material.

1.2.1.2   Population Inversion, Stimulated Emission, and Gain

In the material system discussed above, there is the possibility that some sites will be in an upper energy state. Such sites are referred to as inverted. In the presence of an optical field, a transition from an upper state to a lower state can be induced by the radiation, with the simultaneous emission of a photon in phase (coherent) with the stimulating radiation. This stimulated-emission process is the inverse of the absorption process.

The probability that a randomly chosen photon in an optical field of cross section A will stimulate an optical transition with radiative cross section σ_r at a given inverted site as it passes through a material is σ_r/A. The radiative cross section is proportional to the dipole strength of the transition and is the same for absorption and emission. If the low-energy state is degenerate, with a degeneracy of g_l, the probability of stimulated emission becomes $g_l\sigma_r/A$. The product $g_l\sigma_r$ is known as the emission or gain cross section σ_g. When all of the photons in the optical field are accounted for, and absorption and stimulated emission are included, Equation 1.2 becomes

$$I(l) = I(0)\exp[(\sigma_g\rho_u - \sigma_a\rho_l)l], \qquad (1.3)$$

where ρ_u is the density of inverted sites. If the material is forced out of thermal equilibrium (pumped) to a sufficient degree, so that $\sigma_g\rho_u > \sigma_a\rho_l$, stimulated emission occurs at a higher rate than absorption. This leads to the coherent growth or amplification of the optical field. The material is now said to have gain, with a gain coefficient

$$g = \left[\rho_u - \left(\frac{g_u}{g_l}\right)\rho_l\right]\sigma_g. \qquad (1.4)$$

The term $\rho_u - (g_u/g_l)\rho_l$ is referred to as the effective inversion density ρ_{eff}, and $g = \rho_{eff}\sigma_g$. When $\sigma_g\rho_u = \sigma_a\rho_l$ there is no change in the intensity of an optical field as it passes through the material and the material is said to be in a state of transparency.

1.2.1.3 Spontaneous Emission and Lifetime

In the absence of an optical field, materials with an inverted population will evolve toward thermal equilibrium. Transitions from an upper state to a lower state may be facilitated through interactions with the lattice and accompanied by the emission of a phonon (quantum of lattice vibration) or the generation of heat. Alternatively, they can take place through the spontaneous emission of a photon.

The time associated with the spontaneous decay of the upper state is known as the spontaneous lifetime τ and is dominated by spontaneous optical transitions in many laser gain media. For dipole transitions, the radiative lifetime τ_r is related to the gain cross section such that (Siegman, 1986)

$$\int\sigma_g(v)dv = \frac{3^*\lambda_0^2}{8\pi n_g^2\tau_r}, \qquad (1.5)$$

where ν is frequency, λ_0 is the free-space wavelength of the gain peak, n_g is the refractive index of the gain medium, and 3* is a number between 0 and 3, depending on the alignment of the electronic wave functions with the polarization of the optical field. The combination of large gain cross section and large bandwidth implies short lifetime. In general, the spontaneous lifetime is related to the radiative lifetime τ_r and nonradiative lifetime τ_{nr} according to

$$\tau^{-1} = \tau_r^{-1} + \tau_{nr}^{-1}. \tag{1.6}$$

1.2.1.4   Bandwidth

In any material system, the energy levels have a finite spectral (energy) width. This results in a bandwidth for the optical transitions of the system. The bandwidth may be determined by effects that are common to all sites within the system, resulting in a homogeneously broadened transition, or may be determined by local variations in material properties, leading to an inhomogeneously broadened transition. From the above discussion of spontaneous emission and lifetime, the gain cross section σ_g of an optical transition is related to the bandwidth of the transition and the radiative lifetime τ_r. A broad transition generally implies a small gain cross section or a short spontaneous lifetime.

1.2.2   Models of Gain Media and Saturable Absorbers

1.2.2.1   Four-Level Gain Media

The energy-level structure of a laser plays an important role in obtaining inversion. Let us try to understand pumping and relaxation in an "ideal" four-level laser with the aid of Figure 1.2. The pumping process, indicated by the upward arrow, is assumed to excite the system from the lowest energy level, denoted by *g* for ground level, to the highest level, denoted by *e* for excited level. Pumping might occur in a variety of ways. For microchip lasers, it is through radiative excitation using light whose frequency coincides with the transition frequency between *g* and *e*. Level *e* is assumed to relax to the upper laser level *u*. The population of the upper laser level is radiatively transferred, either through spontaneous or stimulated emission, to the lower laser level *l*. Finally, the lower laser level can either relax to the ground level or absorb the laser radiation and repopulate the upper laser level.

Several conclusions concerning optimal operation can be made from this model. First, the relaxation rates from *e* to *u* and from *l* to *g* should be as rapid as possible to maintain the maximum population inversion between *u* and *l*. Second, the pumping rate between *g* and *e* should be sufficiently rapid to overcome the spontaneous emission from *u* to *l*. Third, the thermal equilibrium population of *l* should be as small as possible. Fourth, decay of *e* to any level other than *u* should be as slow as possible (for optical pumping,

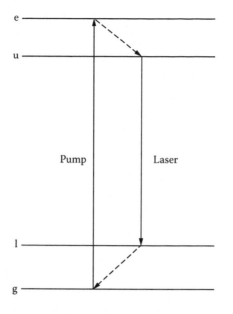

FIGURE 1.2
Schematic representation of four-level system. Population is pumped from *g* to *e* and laser operation occurs on the transition between *u* and *l*.

e can decay radiatively to *g*), and the nonradiative decay of *u* should be slow. For radiative pumping, it is advantageous to have *e* be distinct from *u* and not to have rapid radiative decay of *u* to *g*.

1.2.2.2 Three-Level Gain Media

In a three-level laser, such as ruby, the lower laser level and the ground level are the same (Figure 1.3). Therefore, a large fraction of the ground level must be depopulated to obtain population inversion. For this reason, pumping of a three-level laser requires an extremely high-intensity source.

1.2.2.3 Quasi-Three-Level Gain Media

Energy levels in ionic gain media are grouped into manifolds. Most diode-pumped systems that are referred to as "three-level" lasers have a lower laser level that is slightly above the ground level, typically within the same ground-state manifold. The term quasi-three-level is sometimes used to distinguish these systems from true three-level lasers (e.g., ruby). In quasi-three-level lasers, the lower laser level is partially occupied in thermal equilibrium, and such lasers have properties that are intermediate between true three-level systems and four-level lasers.

Even though the populations of the different energy-level manifolds may be out of equilibrium, there is often rapid thermalization within each manifold. When thermalization within the manifolds occurs on a timescale that

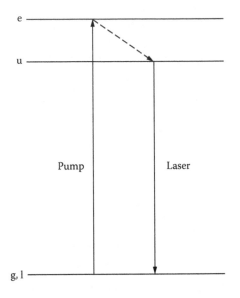

FIGURE 1.3
Schematic representation of three-level system. Note that g and l are now the same.

is short compared to changes in the optical field, it can be modeled as instantaneous. With this approximation, the population of a level within a manifold is given by $g_s f_s N_m$, where g_s is the degeneracy of states in the level, N_m is the total occupation of the manifold, and the Boltzmann coefficients

$$f_s = \frac{\exp(-E_s / k_B T)}{\sum_m g_m \exp(-E_m / k_B T)},$$ (1.7)

with the sum taken over all levels in the manifold. The population of the lower laser level N_l is, therefore, $g_l f_l N_{lm}$, where the subscript l denotes the lower laser level and lm denotes the lower manifold. Similarly, the population of the ground level N_g is $g_g f_g N_{lm}$. To make the model as general as possible, the upper laser level is assumed to be in an upper manifold, with a population $N_u = g_u f_u N_{um}$.

The quasi-three-level model reduces to the three-level model when $g_g f_g = g_l f_l = 1$ and $g_u f_u = 1$. The four-level model is obtained when $g_g f_g = 1$, $g_l f_l = 0$, and $g_u f_u = 1$. In the discussion of the rate-equation model given subsequently (Subsection 1.2.3.2), it is shown that the four-level model is also obtained without the restriction $g_u f_u = 1$ if the rate equations are written using an effective emission cross section $\sigma_{g,eff} = g_u f_u \sigma_g$ instead of the spectroscopic gain cross section σ_g. For similar reasons, $g_u (f_u + f_l)\sigma_g$ is often referred to as the effective cross section for quasi-three-level systems (see Subsection 1.2.3.2). When multiple closely spaced transitions overlap at a given frequency, the effective cross section at that frequency is the sum of $g_u (f_u + f_l)\sigma_g$

over all overlapping transitions. One important example of this is the ~1.064–μm line in room-temperature Nd:YAG, where a pair of transitions centered at 1.06415 and 1.0644 μm both contribute to the effective cross section.

1.2.2.4   Saturable Absorber

Many saturable absorbers can be understood with the four-band model illustrated in Figure 1.4. This model is particularly applicable to one of the most commonly used solid-state saturable absorbers, Cr^{4+}:YAG (Burshtein et al., 1998; Eilers et al., 1992; Il'ichev et al., 1997; Sennaroglu and Yilmaz, 1997). If the saturable absorber is initially in the ground state, absorption of a photon will excite the system from its lowest energy level g to the excited band e. Band e is assumed to relax rapidly to upper level u. From level u the system can relax back to the ground level g or absorb another photon and become excited to the higher energy band h. Band h is assumed to relax rapidly back to upper level u.

1.2.3   Rate-Equation Model

1.2.3.1   Rate Equations

Many of the properties of a laser can be determined from a rate-equation model for the populations of the laser levels and the number of photons in the laser cavity. The rate equations provide a simple and intuitive, yet accurate, picture of the behavior of lasers. In the most simplified form, the increase in photon number within the laser cavity is balanced by the decrease in the

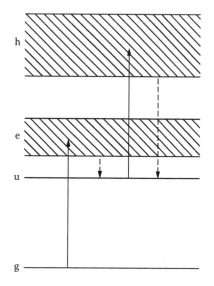

FIGURE 1.4
Schematic representation of four-band saturable absorber.

population difference between the upper and lower laser levels. In addition, the population difference increases on account of pumping, whereas the photon number decreases because of intracavity losses (including absorption and scattering), diffraction of the beam out of the cavity, and transmission through the mirrors.

The rate-equation model can be derived as an approximation to the fundamental equations relating the electromagnetic field, the material polarization, and the quantum-state populations. The validity of the rate equations requires that the material polarization can be accurately approximated by assuming that it instantaneously follows the field; this is a situation that applies to most lasers, including passively Q-switched microchip lasers. To describe the problem in terms of total population and total photon number within a laser cavity, it is necessary that the gain of the laser be small during one pass through the cavity, that the laser operate in a single longitudinal mode, and that the population and optical field be uniform within the cavity. Even when not all of these approximations hold, the resulting equations are still often sufficiently accurate to provide a powerful set of tools both for the design of lasers and for understanding their performance.

1.2.3.2   *Rate Equations without Saturable Absorber*

As stated above, the number of photons q within the laser cavity is affected by two types of events, the emission of a photon by the gain medium ($\dot{q}_e$) and the escape of a photon from the cavity or absorption by unpumped transitions ($\dot{q}_l$). Photon emission can be either stimulated ($\dot{q}_{st}$) or spontaneous ($\dot{q}_{sp}$). Once a laser is above threshold, the stimulated-emission rate is much greater than the spontaneous rate into the oscillating mode and, to first order, spontaneous emission can be ignored. We will return to the issue of spontaneous emission in Subsection 1.2.3.4 when we discuss the buildup of a laser from noise.

The stimulated-emission rate is proportional to the number of photons within the cavity q; the effective population inversion $N_{eff} = \rho_{eff} A_g l_g = N_u - (g_u / g_l) N_l$, where A_g is the cross-sectional area of the laser mode in the gain medium and l_g is the length of the gain medium; and the probability per unit time B that a given photon will interact with a given inverted site. The interaction probability B is the product of the probability that a photon will pass within the gain cross section σ_g of a given inverted site as it traverses the gain medium and the number of times the gain medium is traversed during a round trip within the laser cavity, divided by the round-trip transit time. Mathematically, this reduces to $B = \sigma_g / A_g \times 2^\ddagger / t_{rt}$, where $2^\ddagger$ is the number of times the gain medium is traversed (the notation $2^\ddagger$ is used as a reminder that for the most common type of laser, a standing-wave laser, $2^\ddagger = 2$), $t_{rt} = l_{op,rt} / c$ is the round-trip time of light in the laser cavity, $l_{op,rt} = \oint n dl$ is the round-trip optical length of the cavity, c is the speed of light in vacuum, n is the local refractive index, and the contour integral defining $l_{op,rt}$ is performed along the optical path. Therefore, the stimulated-emission rate $\dot{q}_{st} = 2^{\ddagger}qN_{eff}\sigma_g / t_{rt}A_g$.

For cavities without a saturable absorber or other nonlinear optical element, the escape of photons from the laser cavity and their absorption within the cavity are characterized by the cavity lifetime in the absence of any inversion, $\tau_c = t_{rt} / \gamma_{rt}$, where $\gamma_{rt} = -\ln(1 - \Gamma_{rt})$ is the round-trip loss coefficient and Γ_{rt} is the round-trip loss including transmission through the output coupler. The corresponding decrease in photon number $\dot{q}_l = q\gamma_{rt} / t_{rt}$. Thus, the total rate of change of the number of photons within the laser cavity is

$$\dot{q} = \dot{q}_{st} - \dot{q}_l = \frac{q}{t_{rt}}\left(\frac{2^{\ddagger}N_{eff}\sigma_g}{A_g} - \gamma_{rt}\right). \tag{1.8}$$

We will now derive the rate equations for the population inversion of both a four-level and quasi-three-level laser (see the preceding discussion of models of gain media and saturable absorbers). In both cases, the pump excites the active medium from the ground level to the excited level. It is assumed that the excited level quickly decays to the upper laser level (or manifold), so that the population of the excited level is nearly zero. Lasing occurs between the upper laser level and the lower laser level.

To derive the rate equation for the effective population inversion N_{eff}, we start by considering the population of the upper laser level N_u. The population of the upper level is affected by pumping $\dot{N}_p$, stimulated emission $\dot{N}_{u,st}$, and spontaneous decay $\dot{N}_{u,sp}$. For most pumping schemes, the pump rate is proportional to the number of ions in the ground level and can be written as $\dot{N}_p = W_pN_g$. The stimulated-emission process decreases the population of the upper laser level by one for every photon created, so that $\dot{N}_{u,st} = -\dot{q}_{st}$. Spontaneous decay is characterized by the spontaneous lifetime τ, corresponding to $\dot{N}_{u,sp} = -N_u / \tau$. Thus, the rate equation for the upper-level population is

$$\dot{N}_u = \dot{N}_p + \dot{N}_{u,st} + \dot{N}_{u,sp} = W_pN_g - \frac{2^{\ddagger}qN_{eff}\sigma_g}{t_{rt}A_g} - \frac{N_u}{\tau}. \tag{1.9}$$

In an ideal four-level laser there is a very rapid decay of the lower laser level to the ground level, so that N_l and $\dot{N}_l$ are approximately equal to zero and $N_{eff} \approx N_u$. Because the total number of active ions N_t is constant, $N_t \approx N_g + N_u$. Therefore, the rate equation for the effective population inversion of a four-level laser is

$$\dot{N}_{eff} = W_p(N_t - N_{eff}) - \frac{2^{\ddagger}qN_{eff}\sigma_g}{t_{rt}A_g} - \frac{N_{eff}}{\tau}. \tag{1.10}$$

In a quasi-three-level laser, the population is distributed in the upper and lower manifolds, and $N_t = N_{um} + N_{lm}$. When thermalization of the manifolds is fast relative to changes in the optical field, the population of the upper laser level $N_u = g_u f_u N_{um}$, the population of the lower laser level $N_1 = g_1 f_1 N_{lm}$, and the population of the ground level $N_g = g_g f_g N_{lm}$. As a result, the rate equation for the effective population inversion $N_{eff} = N_u - (g_u/g_1)N_1$ of a quasi-three-level laser reduces to

$$\dot{N}_{eff} = g_g f_g W_p (g_u f_u N_t - N_{eff}) - \frac{2^\ddagger g_u (f_u + f_1) q N_{eff} \sigma_g}{t_{rt} A_g} - \frac{g_u f_1 N_t + N_{eff}}{\tau}. \quad (1.11)$$

Equation 1.11 assumes instantaneous thermalization of the populations within the laser manifolds. Within this approximation, all of the levels in the upper manifold rapidly replenish the upper laser level (on a timescale quicker than changes in the optical field), and a Q-switched pulse can extract energy from the entire upper-manifold population. Likewise, the population of the lower laser level is rapidly depopulated through thermalization with the other levels in the lower manifold.

One of the features of passively Q-switched microchip lasers is that they can produce very short Q-switched pulses and can challenge the instantaneous thermalization approximation. When the thermalization times for the manifolds (they are generally different for the upper and lower manifold) are long compared to the pulse width, repopulation of the upper laser level and depopulation of the lower level through thermalization cannot occur during the pulse. As a result, the effective values of f_u and f_1 approach unity; every laser transition reduces the upper-lever population by one and increases the population of the lower laser level by the same amount. The Boltzmann distribution (Equation 1.7) determines the initial population of the upper and lower levels before the onset of Q switching, but thermalization does not occur during the pulse. In this case, a Q-switched pulse can only extract energy from the upper-laser-level population; it does not have access to the rest of the population in the upper manifold.

When the thermalization times for the manifolds are comparable to the pulse width, the redistribution of the populations within the manifolds will affect the dynamics of pulse formation and more sophisticated analysis is required (Degnan et al., 1998).

In many important laser systems, the lower laser level is essentially empty, and all of the ions are either in the ground level or the upper laser manifold. For such systems, when instantaneous thermalization is a valid approximation, Equation 1.8 and Equation 1.11 can by written as

$$\dot{q} = \frac{q}{t_{rt}} \left(\frac{2^\ddagger N_{um} \sigma_{g,eff}}{A_g} - \gamma_{rt} \right) \quad (1.12)$$

and

$$\dot{N}_{um} = W_p(N_t - N_{um}) - \frac{2^{\ddagger}qN_{um}\sigma_{g,eff}}{t_{rt}A_g} - \frac{N_{um}}{\tau}, \tag{1.13}$$

where $\sigma_{g,eff} = g_u f_u \sigma_g$. These equations are the same as the rate equations for an ideal four-level laser except that the emission cross section has been replaced by an effective emission cross section and the population of the upper laser level has been replaced by the population of the upper-level manifold. The concept of an effective emission cross section has been extended to quasi-three-level lasers with significant population in the lower laser level; $\sigma_{g,eff} = g_u(f_u + f_l)\sigma_g$.

1.2.3.3 Rate Equations with Saturable Absorber

The photon rate equation for a laser containing a saturable absorber must be modified to include the absorption of a photon by a saturable-absorber ion in either the ground state or upper level. Ground-state absorption in the saturable absorber leads to the additional term $\dot{q}_{s,g,ab} = 2^{\ddagger}_s q N_{s,g}\sigma_{s,g} / t_{rt}A_s$ and upper-level absorption leads to the term $\dot{q}_{s,u,ab} = 2^{\ddagger}_s q N_{s,u}\sigma_{s,u} / t_{rt}A_s$, where $N_{s,g}$ is the population of saturable-absorber ions in the ground state, $\sigma_{s,g}$ is the ground-state absorption cross section at the oscillating wavelength, $N_{s,u}$ is the population of saturable-absorber ions in the upper level, $\sigma_{s,u}$ is the upper-level absorption cross section at the oscillating wavelength, A_s is the cross-sectional area of the laser mode in the saturable absorber, and $2^{\ddagger}_s$ is the number of times the same cross-sectional area of the saturable absorber is traversed by light during one round trip within the laser cavity. For microchip lasers, $2^{\ddagger}_s = 2^{\ddagger} = 2$ and $A_s = A_g = A$. If we assume that the decay of the excited and higher energy bands is very rapid, all of the saturable-absorber ions (total population N_s) are either in the ground or upper level, and the new photon rate equation becomes

$$\dot{q} = \frac{q}{t_{rt}} \left[\frac{2^{\ddagger}N_{eff}\sigma_g}{A_g} - \frac{2^{\ddagger}_s N_{s,g}(\sigma_{s,g} - \sigma_{s,u})}{A_s} - \frac{2^{\ddagger}_s N_s \sigma_{s,u}}{A_s} - \gamma_{rt} \right]. \tag{1.14}$$

From this equation, it is clear that the presence of upper-level absorption (excited-state absorption or ESA) decreases the saturable component of loss in the laser and increases the parasitic loss, both undesirable effects.

The rate equation for the gain medium is unaffected by the presence of the saturable absorber. To complete the modeling of a laser containing a saturable absorber, we need a rate equation for the saturable-absorber ground-state population $N_{s,g}$. The population of the saturable-absorber ground state decreases by one for every photon absorbed, resulting in a term $\dot{N}_{s,g,ab} = -\dot{q}_{s,g,ab}$. Decay from the upper level is characterized by the lifetime τ_s, corresponding to $\dot{N}_{s,g,d} = N_{s,u} / \tau_s = (N_s - N_{s,g})/\tau_s$. Many important saturable absorbers have broad absorption bands, and for optically pumped systems where some of the pump light is incident on the saturable absorber we must include a term in the rate equation to account for absorption of the pump light. Consistent with the form of the pump term used for the gain media, $\dot{N}_{s,g,p} = -W_{p,s}N_{sg}$. Thus, the rate equation for the ground state of the saturable absorber is

$$\dot{N}_{s,g} = \dot{N}_{s,g,ab} + \dot{N}_{s,g,d} + \dot{N}_{s,g,p} = -\frac{2_s^{\ddagger}qN_{s,g}\sigma_{s,g}}{t_{rt}A_s} + \frac{(N_s - N_{s,g})}{\tau_s} - W_{p,s}N_{s,g}. \qquad (1.15)$$

1.2.3.4   Buildup from Noise

In the photon rate equations derived above, the term corresponding to spontaneous emission was left out. Laser action is initiated by spontaneous emission, or noise. As a result, these rate equations cannot account for the onset of lasing, as is seen by setting $q = 0$ at time $t = 0$. When spontaneous emission is properly taken into account, the photon rate equation contains the additional term

$$\dot{q}_{sp} = \frac{2^{\ddagger}2^{\dagger}N_u\sigma_g}{2A_gt_{rt}}, \qquad (1.16)$$

where

$$2^{\dagger} \equiv \begin{cases} 1 & \text{for a traveling-wave laser} \\ 2 & \text{for a standing-wave laser} \end{cases}.$$

The difference between a standing-wave laser and a traveling-wave laser can be understood by realizing that the intracavity optical field for a standing-wave laser is the sum of two counter propagating traveling waves, each with its own one-half photon of noise. This one-half photon of noise stimulates optical transitions and initiates lasing. Once lasing has started, this term can usually be ignored.

1.2.4   Solution to Rate Equations

1.2.4.1   Rate Equations for Passively Q-Switched Laser

The output pulses from a passively Q-switched laser are usually much shorter than the spontaneous lifetime of the gain medium, the upper-state lifetime of the saturable absorber, and the pump period prior to the formation of the pulse. This allows us to neglect spontaneous relaxation and pumping during the development of the output pulse, reducing the rate equations for a passively Q-switched laser to

$$\dot{q} = \frac{q}{t_{rt}} \left[\frac{2^{\ddagger} N_{eff} \sigma_g}{A_g} - \frac{2_s^{\ddagger} N_{s,g} (\sigma_{s,g} - \sigma_{s,u})}{A_s} - \frac{2_s^{\ddagger} N_s \sigma_{s,u}}{A_s} - \gamma_{rt} \right], \tag{1.17}$$

$$\dot{N}_{eff} = -\frac{2^* 2^{\ddagger} q N_{eff} \sigma_g}{t_{rt} A_g}, \tag{1.18}$$

and

$$\dot{N}_{s,g} = -\frac{2_s^{\ddagger} q N_{s,g} \sigma_{s,g}}{t_{rt} A_s}, \tag{1.19}$$

where

$$2^* \equiv \begin{cases} 1 & \text{for a four-level laser} \\ g_u(f_{u,eff} + f_{l,eff}) & \text{for a quasi-three-level laser} \\ 2 & \text{for a nondegenerate three-level laser} \end{cases},$$

and $f_{u,eff}$ and $f_{l,eff}$ are the effective values of f_u and f_l, as discussed in Subsection 1.2.3.2.

Through $f_{u,eff}$ and $f_{l,eff}$, the value of 2^* is dependent on the duration of the output pulse from the laser. If the thermalization time for the upper laser manifold is short compared to the pulse width $f_{u,eff} = f_u$; if it is long compared to the pulse width $f_{u,eff} \approx 1$. Likewise, if the thermalization time for the lower laser manifold is short compared to the pulse width $f_{l,eff} = f_l$; if it is long compared to the pulse width $f_l \approx 1$. Similarly, if the lower level of a four-level system has a decay time longer than the output pulse, the system will behave like a three-level system under Q-switched operation (Fan, 1988). The thermalization times and lower-level decay times for even the most commonly used laser gain media are not well known, but there is experimental reason to believe that they may be comparable to or longer than the pulse widths of short-pulse passively Q-switched microchip lasers (several hundred picoseconds) in many material systems, including Nd:YAG (Bibeau et al., 1993; Palombo et al., 1993; Buzelis et al., 1995; Degnan et al., 1998). Fortunately, there are many useful results that can be obtained from the rate equations even without an accurate knowledge of the thermalization times.

It is worth noting that Equation 1.17 to Equation 1.19 are analogous to the set used by Degnan (Degnan, 1995) modified to include the effects of excited-state absorption in the saturable absorber (Xiao and Bass, 1997; Xiao and Bass, 1998) and thermalization of the energy levels in the gain medium (Degnan et al., 1998). There have been several other notable publications on the rate-equation analysis of passive Q switching (Chen et al., 2001; Chen et al., 2002; Liu et al., 2001; Patel and Beach, 2001; Peterson and Gavrielides, 1999; Zhang et al., 1997; Zhang et al., 2000; Zheng et al., 2002). Most cover the whole topic, two focus on estimating the temporal profile of the output pulses (Liu et al.,

2001; Peterson and Gavrielides, 1999), one presents analysis of the quasi-three-level gain medium Yb:YAG (Patel and Beach, 2001), and another presents analysis of passively mode-locked, Q-switched operation (Chen et al., 2002).

Equation 1.17 to Equation 1.19 will tend to overestimate the efficiency of standing-wave Q-switched lasers because they do not include the effects of spatial and spectral hole burning. In addition, many Q-switched lasers operate with a large round-trip gain, so that the optical field within the cavity is not uniform. Although this mitigates the effects of spatial hole burning, it violates one of the assumptions used to reduce the rate equations to such a simple form. The net result is to overestimate the peak power and energy efficiency obtained from the device, and to distort the pulse shape (Stone, 1992). These effects are often small, however, and Equation 1.17 to Equation 1.19 are extremely powerful tools.

1.2.4.2   Initial Conditions

For many systems of practical interest, the upper-state lifetime of the saturable absorber, τ_s, is much longer than the Q-switched output pulse but much shorter than the pump period preceding it. This allows us to solve Equation 1.15 piecewise. Prior to the onset of lasing ($q = \dot{q} = 0$) we are in the quasi-steady state and the population of the ground state of the saturable absorber

$$N_{s,g,0} = \frac{N_s}{1 + W_{p,s} / \tau_s}. \tag{1.20}$$

For optically pumped systems where some of the pump light is incident on the saturable absorber

$$N_{s,g,0} = \int \frac{\rho_s}{1 + I_p / I_{p,sat}} dV, \tag{1.21}$$

where ρ_s is the density of saturable-absorber ions, I_p is the pump intensity in the saturable absorber, $I_{p,sat} = h\nu_p/\sigma_{s,g,p}\tau_s$ is the saturation intensity of the saturable absorber at the pump wavelength, h is Planck's constant, ν_p is the optical frequency of the pump, $\sigma_{s,g,p}$ is the ground-state absorption cross section of the saturable absorber at the pump wavelength, and the integration is over the oscillating-mode volume in the saturable absorber.

For short, longitudinally pumped passively Q-switched microchip lasers in the geometry illustrated in Figure 1.5, the pump light reaching the saturable absorber can approach the saturation intensity $I_{p,sat}$, greatly reducing the number of saturable-absorber ions that can participate in the passive Q-switching process (Welford, 2001; Zayhowski and Wilson, 2003a; Jaspan et al., 2004). This increases the minimum-possible Q-switched pulse width and decreases the pulse energy. If the saturable absorber is separate from the gain medium (as opposed to a single crystal that acts as both the gain medium

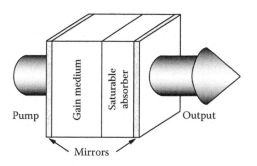

FIGURE 1.5
Simple variation of passively Q-switched microchip laser.

and the saturable absorber), the amount of pump light reaching the saturable absorber can be reduced by any of the following: using a high-concentration gain medium with a large absorption coefficient at the pump wavelength, tuning the pump laser to fall exactly on the absorption peak of the gain medium, narrowing the linewidth of the pump source so that it better overlaps the gain-medium absorption, or putting a dichroic interface between the gain medium and the saturable absorber that reflects the pump light. Alternatively, the saturation intensity could be increased if we could quench the saturable-absorber excited-state lifetime. Finally, in some cases, we can focus the pump light in such a way that it diverges significantly by the time it reaches the saturable absorber, decreasing its intensity.

The population inversion for a passively Q-switched laser at the time the pulse begins to form, N_0, is determined by setting the right-hand side of the photon rate equation (Equation 1.17) equal to zero:

$$N_0 = \frac{A_g}{2^{\ddagger}\sigma_g}\left[\frac{2_s^{\ddagger}N_{s,g,0}(\sigma_{s,g}-\sigma_{s,u})}{A_s}+\frac{2_s^{\ddagger}N_s\sigma_{s,u}}{A_s}+\gamma_{rt}\right]. \qquad (1.22)$$

In contrast to actively Q-switched lasers, the initial population inversion for passively Q-switched lasers is fixed by material parameters and the design of the laser, and, in the absence of pulse bifurcation (see Subsection 1.2.5.3), is identical from pulse to pulse. This can result in extremely stable pulse energies and pulse widths, with measured stabilities of better than 1 part in 10^4. The stability of the pulse energy and pulse width is achieved at the expense of pulse timing stability. Pulse formation will not occur until the population inversion has reached the proper value, and fluctuations in the pump source are the primary cause of pulse-to-pulse timing jitter in many passively Q-switched lasers.

1.2.4.3   Second Threshold

In order to obtain passively Q-switched operation the system must, at some time $t = 0$, go from being absorbing to having net gain. At that transition time the net gain experienced by a photon in the laser cavity is zero, $\dot{q} = 0$, and must be increasing, $\ddot{q} > 0$. This leads to the requirement

$$\alpha_r \equiv \frac{2_s^\ddagger \sigma_{s,g} A_g}{2^* 2^\ddagger \sigma_g A_s} > \frac{2_s^\ddagger N_{s,g,0}(\sigma_{s,g} - \sigma_{s,u}) + 2_s^\ddagger N_s \sigma_{s,u} + A_s \gamma_{tr}}{2_s^\ddagger N_{s,g,0}(\sigma_{s,g} - \sigma_{s,u})}, \tag{1.23}$$

where the right-hand side of the expression is the ratio of the coefficient of total round-trip loss ($\gamma_{rt,t}$) to the coefficient of saturable round-trip loss ($\gamma_{rt,s}$) at $t = 0$. Expression 1.23 is often referred to as the second threshold condition. In systems dominated by saturable loss ($\gamma_{rt,t}/\gamma_{rt,s} \approx 1$), Expression 1.23 reduces to a statement that the loss of the saturable absorber must saturate more quickly than the gain of the gain medium. When Expression 1.23 is satisfied by a large factor, the saturable absorber acts as an ideal, instantaneous Q switch; when it is satisfied by a factor of greater than ten, the instantaneous-Q-switch model is an excellent representation of the system. For a factor of three, the instantaneous-Q-switch model gives results that are within a factor of two of being correct (Szabo and Stein, 1965). When Expression 1.23 is not satisfied, the laser will tend to oscillate continuous-wave.

1.2.4.4   Peak Power

To determine the maximum peak power that can be achieved from a passively Q-switched laser, we start by dividing Equation 1.19 by Equation 1.18 and integrating, to obtain

$$N_{s,g} = N_{s,g,0} \left(\frac{N_{eff}}{N_0} \right)^{\alpha_r}. \tag{1.24}$$

We then divide Equation 1.17 by Equation 1.18 and substitute Equation 1.24 into the result, which we integrate with respect to population inversion to obtain

$$2^*(q - q_0) = N_0 - N_{eff} - \frac{A_g}{2^\ddagger \sigma_g} \left(\frac{2_s^\ddagger N_s \sigma_{s,u}}{A_s} + \gamma_{rt} \right) \ln \left(\frac{N_0}{N_{eff}} \right)$$

$$- N_{s,g,0} \frac{2^*(\sigma_{s,g} - \sigma_{s,u})}{\sigma_{s,g}} \left[1 - \left(\frac{N_{eff}}{N_0} \right)^{\alpha_r} \right], \tag{1.25}$$

where q_0 is the number of photons in the cavity at the time the output pulse begins to develop.

The maximum number of photons in the cavity occurs when $\dot{q} = 0$. Thus, from Equation 1.17 we obtain an implicit expression for the population inversion N_p at the time of peak output power,

$$N_p = \frac{A_g}{2^\ddagger \sigma_g} \left[\frac{2_s^\ddagger N_s \sigma_{s,u}}{A_s} + \gamma_{rt} + \frac{2_s^\ddagger N_{s,g,0}(\sigma_{s,g} - \sigma_{s,u})}{A_s} \left(\frac{N_p}{N_0} \right)^{\alpha_r} \right]. \tag{1.26}$$

Because α_r is much greater than unity for systems of interest, N_p can be obtained by iterative application of this equation using its value for $\alpha_r = \infty$ as a starting point. Neglecting the initial number of photons in the cavity, the peak output power is given by

$$P_{po} = \frac{h\nu_o \gamma_o}{2^* t_{rt}} \left\{ N_0 - N_p - \frac{A_g}{2^\ddagger \sigma_g} \left(\frac{2_s^\ddagger N_s \sigma_{s,u}}{A_s} + \gamma_{rt} \right) \ln\left(\frac{N_0}{N_p} \right) \right.$$

$$\left. - N_{s,g,0} \frac{2^*(\sigma_{s,g} - \sigma_{s,u})}{\sigma_{s,g}} \left[1 - \left(\frac{N_p}{N_0} \right)^{\alpha_r} \right] \right\}, \tag{1.27}$$

where $h\nu_o$ is the energy of a photon at the oscillating wavelength, $\gamma_o = -\ln(1 - T_o)$ is the output-coupling coefficient, and T_o is the transmission of the output coupler.

In the limit of large α_r and negligible absorption from the upper level of the saturable absorber, Equation 1.27 asymptotically approaches the result obtained for instantaneous Q switching (Zayhowski and Kelley, 1991). In that case, the peak power is maximized when

$$\frac{\gamma_{rt} + \gamma_o}{\gamma_{rt}} \ln\left(\frac{N_0}{N_{th}} \right) = \frac{N_0}{N_{th}} - 1, \tag{1.28}$$

where $N_{th} = \gamma_{rt} A_g / 2^\ddagger \sigma_g$ would be the cw threshold of the laser in the absence of the saturable loss. For negligible parasitic loss ($\gamma_{rt} = \gamma_o$), the maximum peak power

$$P_{po,max} = \frac{0.102 N_0^2 h\nu_o \gamma_{rt}}{2^* N_{th} t_{rt}}, \tag{1.29}$$

and is obtained for

$$\gamma_o = 0.28 g_{rt}, \tag{1.30}$$

where $g_{rt} = 2^\ddagger N_0 \sigma_g / A_g$ is the round-trip gain coefficient at the time the pulse begins to develop. The presence of parasitic loss increases the optimal value of γ_o. For smaller values of α_r optimization of the peak power is easily performed numerically, once the material parameters of the system are known.

From Equation 1.27 and Equation 1.29 it is clear that the peak power is dependent on the value of 2* and, therefore, the thermalization times of the laser manifolds. When the output pulse width is longer than the thermalization times of the laser manifolds, thermalization of the upper manifold repopulates the upper laser level as it gets depleted by the optical pulse, and thermalization of the lower manifold keeps the population of the lower laser level as small as possible. This allows the maximum amount of energy extraction from the system and relatively high peak powers. When the pulse width is short compared to the thermalization times only the energy in the upper laser level (not the upper laser manifold) can be extracted by the laser pulse, and the laser will act like a three-level system.

When the pulse width of the laser is comparable to the thermalization times of the laser manifolds it is not possible to accurately calculate the peak output power of the laser pulses without accurately knowing the thermalization times. However, for good saturable absorbers (large α_r), the output coupling that maximizes the peak power (Equation 1.30) is independent of the thermalization times.

1.2.4.5   Pulse Energy

The total pulse energy E_o of a passively Q-switched laser is obtained by integrating the output power P_o using the following technique (Degnan, 1989):

$$E_o = \int_0^\infty P_o \, dt = \frac{h\nu_o\gamma_o}{t_{rt}} \int_0^\infty q \, dt = \frac{h\nu_o A_g \gamma_o}{2^*2^\ddagger\sigma_g} \int_{N_0}^{N_f} \frac{dN_{eff}}{N_{eff}}, \tag{1.31}$$

where the last step used Equation 1.18 and N_f is the population inversion well after the peak of the output pulse. The final integration leads to

$$E_o = \frac{h\nu_o A_g \gamma_o}{2^*2^\ddagger\sigma_g} \ln\left(\frac{N_0}{N_f}\right), \tag{1.32}$$

and the final population inversion is obtained by setting the right-hand side of Equation 1.25 equal to zero, yielding the implicit expression

$$N_0 - N_f - \frac{A_g}{2^\ddagger\sigma_g}\left(\frac{2_s^\ddagger N_s\sigma_{s,u}}{A_s} + \gamma_{rt}\right)\ln\left(\frac{N_0}{N_f}\right) \tag{1.33}$$

$$- N_{s,g,0}\frac{2^*(\sigma_{s,g} - \sigma_{s,u})}{\sigma_{s,g}}\left[1 - \left(\frac{N_f}{N_0}\right)^{\alpha_r}\right] = 0.$$

In the limit of large α_r and negligible absorption from the upper level of the saturable absorber, Equation 1.32 asymptotically approaches the result obtained for instantaneous Q switching, where the maximum value of E_o (Degnan, 1989)

$$E_{o,max} = \frac{N_o h \nu_o \gamma_{rt,p}}{2^* g_{rt}} \left[\frac{g_{rt}}{\gamma_{rt,p}} - 1 - \ln\left(\frac{g_{rt}}{\gamma_{rt,p}}\right) \right],$$ (1.34)

and is obtained for

$$\gamma_o = -\gamma_{rt,p} \left[\frac{(g_{rt}/\gamma_{rt,p}) - 1 - \ln(g_{rt}/\gamma_{rt,p})}{\ln(g_{rt}/\gamma_{rt,p})} \right],$$ (1.35)

where $\gamma_{rt,p} = \gamma_{rt} - \gamma_o$ is the round-trip parasitic loss coefficient. For smaller values of α_r, optimization is easily performed numerically once the material parameters of the system are known.

Like the peak power, and for the same reasons, the pulse energy is dependent on 2^*, but the optimal value of output coupling is not. In a repetitively Q-switched short-pulse laser, some of the energy not extracted from the upper laser manifold by one pulse will be available for a subsequent pulse and the efficiency of the system will improve at high repetition rates.

Bleaching of the saturable absorber consumes stored energy, which can result in lower pump-to-output efficiencies for passively Q-switched lasers than for comparable actively Q-switched lasers. The intrinsic losses of the saturable absorber, including excited-state absorption, also limit the efficiency of a passively Q-switched device. Quite often, these inefficiencies result in a small amount of energy consumption compared to the energy required to drive an active Q switch.

1.2.4.6 Pulse Width

The duration t_w of a passively Q-switched output pulse (full width at half-maximum) is obtained from the energy in the pulse E_o, the peak power of the pulse P_{po}, and the shape of the pulse S_p:

$$t_w = \frac{S_p t_{rt}}{\gamma_{rt} + \frac{2_s^{\ddagger} N_s \sigma_{s,u}}{A_s}}$$

$$\times \left\{ \frac{N_0 - N_f - N_{s,g,0} \frac{2^*(\sigma_{s,g} - \sigma_{s,u})}{\sigma_{s,g}} \left[1 - \left(\frac{N_f}{N_0}\right)^{\alpha_r}\right]}{\left[N_0 - N_P - \frac{A_g}{2^{\ddagger}\sigma_g}\left(\frac{2_s^{\ddagger} N_s \sigma_{s,u}}{A_s} + \gamma_{rt}\right)\ln\left(\frac{N_0}{N_P}\right) - N_{s,g,0}\frac{2^*(\sigma_{s,g} - \sigma_{s,u})}{\sigma_{s,g}}\left[1 - \left(\frac{N_P}{N_0}\right)^{\alpha_r}\right]\right]} \right\}.$$ (1.36)

Single-mode Q-switched pulses have a pulse shape factor $S_p \approx 0.86$ (Zayhowski and Kelley, 1991), which is midway between the pulse shape factor for sech- and sech²-shaped pulses (S_p = 0.84 and 0.88, respectively). The rising edge of the pulse is exponential with a rise time determined by the gain of the cavity; the exponential decay of the trailing edge is determined by the cavity lifetime.

In the limit of large α_r the minimum pulse width (Zayhowski and Kelley, 1991)

$$t_{w,min} = \frac{8.1t_{rt}}{g_{rt}}$$

(1.37)

is obtained for

$$\gamma_o = 0.32g_{rt} - \gamma_{rt,p}.$$

(1.38)

The term 2^* does not appear in Equation 1.37, indicating that the pulse width is independent of the thermalization times of the laser manifolds and whether the gain medium is a three- or four-level system. A more careful analysis shows that the laser will have the same pulse width when the thermalization times are either much longer than or much shorter than the pulse width. However, when the pulse width and thermalization times are comparable, thermalization of the manifolds will affect the dynamics of pulse formation, changing the pulse shape and its duration (Degnan et al., 1998). In a system optimized for minimum pulse width, the rising edge of the laser pulse is sharper than the falling edge. As the pulse starts to decay, thermalization of the laser manifolds will pump the laser, slowing its decay and broadening the pulse. When the thermalization times and the pulse width are nearly equal, this can increase the pulse width of an optimized system by ~50%. This effect may help account for the fact that all published values for the pulse widths of short-pulse Nd:YAG microchip lasers are longer than the values calculated neglecting thermalization (Zayhowski and Wilson, 2003a).

For systems with a large α_r the pulse duration is determined primarily by the ratio of the resonator length to some combination of the fixed and the saturable loss. The pulse energy depends on the same parameters but is also directly proportional to the laser mode volume, which provides the means to scale the pulse energy independent of the pulse duration.

1.2.4.7 CW-Pumped Passively Q-Switched Lasers

For a cw-pumped passively Q-switched laser, the repetition rate of the pulses increases as the pump power is increased above threshold. The average interpulse period is given by (Liu et al., 2003)

$$\tau_p = \tau \ln\left[\frac{P_a - P_i - (P_{th} - P_i)N_f / N_0}{P_a - P_{th}}\right],$$

(1.39)

where P_a is the total pump power absorbed within the lasing mode volume, P_i is the absorbed pump power required to reach inversion ($N_{eff} = 0$), and P_{th} is the absorbed pump power necessary to obtain $N_{eff} = N_0$. This expression does not include the effects of intramanifold thermalization, which could decrease the interpulse period of short-pulse systems at high pulse repetition rates. For low-power devices the upper limit to the pulse repetition rate (the inverse of the interpulse period) is set by the saturable absorber recovery time; at higher pulse rates the laser will deteriorate into unstable pulse generation and ultimately cw operation. (In high-power devices the maximum pulse repetition rate is limited by thermal effects.) Pulse rates as high as 300 kHz have been demonstrated with a Cr^{4+}:YAG saturable absorber (Jaspan et al., 2000; Zheng et al., 2002). The use of semiconductor saturable structures, with short upper-level lifetimes (see Subsection 1.2.6), allows picosecond Q-switched pulses to be generated at megahertz rates (Keller et al., 1996).

1.2.5 Mode Beating, Afterpulsing, and Pulse-to-Pulse Stability

1.2.5.1 Single-Frequency Operation

When Q-switched lasers operate in more than one longitudinal mode the output pulse is intensity modulated as a result of mode beating; single-longitudinal-mode lasers produce pulses with a smoothly varying temporal profile. Smooth temporal profiles are desirable for many applications in high-resolution ranging (Abshire et al., 2000; Afzal et al., 1977; Degnan, 1993; Degnan, 1999), altimetry (Degnan et al., 2001), and light detection and ranging (LIDAR). Traditionally, the methods used to obtain single-frequency operation in a Q-switched laser involve the addition of intracavity frequency-selective optical elements or the injection of an optical signal from a cw single-frequency oscillator (optical seeding).

In an actively Q-switched laser, potential lasing modes start to build up when the cavity Q is switched. All of the modes start from spontaneous emission and the amount of gain seen by each mode determines how quickly it builds up. To derive a criterion for single-frequency operation of an actively Q-switched laser, we will require that the number of photons in the primary mode (mode 1) is at least 100 times greater than the number in any other mode (mode 2) by the time the output pulse forms. (Other authors have suggested that a factor of ten is sufficient [Sooy, 1965]; the number is somewhat arbitrary.) In this case, the primary mode will extract most of the stored energy in the cavity, leaving very little energy for the second mode. From Equation 1.8 it follows that, for the case of instantaneous Q switching, our criterion reduces to

$$\frac{N_{eff}(N_{th,2} - N_{th,1})}{N_{th,2}(N_{eff} - N_{th,1})} > \frac{\ln(100)}{\ln(q_p)} = \frac{4.6}{\ln(q_p)}, \tag{1.40}$$

where $N_{th,1}$ and $N_{th,2}$ are the cw thresholds for modes 1 and 2 ($N_{th} = \gamma_{rt} A_g / 2^{\ddagger} \sigma_g$) when the cavity is in the low-loss state, q_p is the number of photons in the laser cavity at the peak of the output pulse, and we have assumed that both modes build up from noise.

An alternative criterion for single-frequency operation is that the build-up time for two competing modes differ by an amount comparable to or greater than the pulse duration (Isyanova and Welford, 1999). The first mode to lase (mode 1) extracts most of the stored energy in the cavity, slowing the build-up of the second mode (mode 2) and greatly decreasing its intensity. For the case of instantaneous Q switching, this criterion reduces to

$$\frac{N_{\text{eff}}(N_{\text{th},2} - N_{\text{th},1})}{(N_{\text{eff}} - N_{\text{th},2})(N_{\text{eff}} - N_{\text{th},1})} > \frac{2.6t_w}{\ln(q_p)t_{w,\text{min}}}, \tag{1.41}$$

where $t_{w,\text{min}}$ is the minimum obtainable pulse width as given by Equation 1.37. Equation 1.40 and Equation 1.41 give similar results for lasers designed to produce short pulses; both ways of treating the problem are equally valid.

For an instantaneously Q-switched laser designed to produce pulses with the minimum possible duration, with typical values of q_p in the range of 10^{11} to 10^{18}, the thresholds for modes 1 and 2 (as calculated with Equation 1.40 or Equation 1.41) must differ by more than 7% to 12% in order to ensure single-frequency operation. This requirement can be relaxed if the cavity Q is switched more slowly, because the resulting longer pulse build-up times lead to more optical passes through any mode-selective elements in the resonator (Sooy, 1965). Nevertheless, for typical gain linewidths in solid-state media, the gain differential between longitudinal modes is not enough to ensure single-frequency operation for a cavity length greater than ~1 cm and, if single-frequency operation is desired, intracavity frequency-selective loss elements may be required. For shorter cavity lengths the gain differential may be sufficient, if care is taken to keep one of the longitudinal modes near the peak of the gain profile.

In a passively Q-switched laser, the cavity Q is not switched until after a mode starts to build up; it is the light in the first mode that causes Q switching to occur. As a result, when the cavity Q switches, that initial mode already contains a significant photon population and has a head start in the race to form a pulse. As the initial mode has the highest net gain, the number of photons in the first mode at any time during the pulse buildup will exceed the number in any other mode by a factor greater than the number of photons in the first mode at the time the second mode reaches threshold. Because a factor of 100 is more than enough to ensure single-frequency operation, single-frequency pulses are easily achieved in passively Q-switched systems.

Once one mode (mode 1) in a passively Q-switched laser begins to oscillate, the ground-state population of saturable-absorber ions $N_{s,g}$ and inversion N_{eff} begin to change rapidly. By the time a second mode (mode 2) reaches threshold, they have changed by

$$\Delta N_{s,g}(1,2) \approx \frac{2^{\ddagger} N_0 A_s (\sigma_{g,1} - \sigma_{g,2})}{2_s^{\ddagger} A_g (\sigma_{s,g} - \sigma_{s,u})} \left[1 - \frac{N_0 \sigma_{s,g}}{2^* \alpha_r^2 N_{s,g,0}(\sigma_{s,g} - \sigma_{s,u})} \right]^{-1} \tag{1.42}$$

and

$$\Delta N_{eff}(1,2) \approx \frac{2^{\ddagger} N_0^2 A_s (\sigma_{g,1} - \sigma_{g,2})}{2_s^{\ddagger} \alpha_r N_{s,g,0} A_g (\sigma_{s,g} - \sigma_{s,u})} \left[1 - \frac{N_0 \sigma_{s,g}}{2^* \alpha_r^2 N_{s,g,0} (\sigma_{s,g} - \sigma_{s,u})} \right]^{-1}, \quad (1.43)$$

where we have assumed that the difference in thresholds between modes 1 and 2 is the result of a small difference in the gain cross sections $\sigma_{g,1}$ and $\sigma_{g,2}$. (A similar result can be derived if we assume a small difference in cavity loss.) The bracketed term on the right-hand side of these equations approaches unity for large α_r and can often be ignored.

To calculate the photon population of the first mode at the onset of the second mode we use the approximation

$$q(1,2) = q_0 + \Delta N_{eff}(1,2) q'(0) + \frac{[\Delta N_{eff}(1,2)]^2 q''(0)}{2!}, \quad (1.44)$$

where the primes indicate differentiation with respect to N_{eff}, q_0 is assumed to be negligible (1/2 photon per mode), and the last term is included because $q'(0) = 0$. From Equation 1.25,

$$q''(0) = -\frac{\alpha_r^2 N_{s,g,0} (\sigma_{s,g} - \sigma_{s,u})}{N_0^2 \sigma_{s,g}} \left[1 - \frac{\gamma_{rt,t}}{\alpha_r \gamma_{rt,s}} \right], \quad (1.45)$$

where $\gamma_{rt,t}/\gamma_{rt,s}$ is the fraction that appears on the right-hand side of Expression 1.23 and is of order unity. Thus,

$$q(1,2) = \frac{N_0}{2^* 2 \alpha_r} \left(\frac{\sigma_{g,1} - \sigma_{g,2}}{\sigma_{g,1}} \right)^2 \left[1 - \frac{\gamma_{rt,t}}{\alpha_r \gamma_{rt,s}} \right] \left[1 - \frac{N_0 \sigma_{s,g}}{2^* \alpha_r^2 N_{s,g,0} (\sigma_{s,g} - \sigma_{s,u})} \right]^{-2}, \quad (1.46)$$

where the two bracketed terms approach 1 for large α_r and can usually be ignored.

To gain an appreciation for how easy it is to obtain single-frequency operation with a passively Q-switched laser, consider a 1.064-μm Nd:YAG/Cr^{4+}:YAG microchip laser, with $A_g = A_s$ and $\alpha_r = 21$. If we assume an output pulse energy of 100 μJ and an extraction efficiency of 80% ($N_0 = 6.7 \times 10^{14}$), and consider only one possible polarization, $q(1,2) > 100$ for $(\sigma_{g,1} - \sigma_{g,2})/\sigma_{g,1}$ as small as 2.5×10^{-6}. If we make no attempt to control the position of the cavity modes relative to the gain peak, the probability of multimode operation is

$$P_{mm} = \left(\frac{\delta \nu_L}{2 \Delta \nu} \right)^2 \left(\frac{\sigma_{g,1} - \sigma_{g,2}}{\sigma_{g,1}} \right)_{q(1,2)=100}, \quad (1.47)$$

where δv_L is the (Lorentzian) linewidth of the gain medium (160 GHz for Nd:YAG), Δv is the mode spacing of the cavity, and the value of $(\sigma_{g,1} - \sigma_{g,2})/\sigma_{g,1}$ is determined by setting $q(1,2) = 100$ in Equation 1.46. With a cavity length of 1 cm, the probability of the laser producing a multimode pulse is less than 2.5×10^{-4}. Note that this is a conservative estimate of the maximum likelihood of multimode operation, because it does not include the fact that the first mode to oscillate will continue to develop more quickly than the other modes after the second mode reaches threshold, and we are insisting that the energy in the primary mode be at least 100 times greater than the energy in any other mode.

1.2.5.2   *Afterpulsing*

The cavity Q of passively Q-switched lasers cannot be rapidly decreased once the Q-switched output pulse exits the system. The residual gain left by the first output pulse will contribute to the continued development of pulses in competing longitudinal modes, and afterpulsing is difficult to prevent. Because the afterpulse is typically a different longitudinal mode than the primary pulse, afterpulsing is minimized when there is a large difference in the gain of the primary and competing modes. This is most easily accomplished in cavities with a large mode spacing (short cavity length) and when the primary mode is positioned at the peak of the gain profile. The latter can often be accomplished by controlling the pump power to the laser or its temperature, or through stress tuning of the laser mode (Owyoung and Esherick, 1987). As the primary pulse gets closer to the peak of the gain profile the time delay between the primary pulse and the afterpulse increases, and the magnitude of the afterpulse decreases.

Afterpulsing is also minimized by positioning the saturable absorber near the center of a standing-wave passively Q-switched laser cavity. Because of longitudinal spatial hole burning (due to the standing-wave nature of the oscillating mode; see, for example, Zayhowski, 1990a; Zayhowski, 1990b), the first mode to lase does not completely bleach the saturable absorber for the second mode. The differential loss for the adjacent modes is greatest when the saturable absorber is located in the center of the cavity.

For applications where afterpulsing cannot be tolerated, it is often possible to eliminate the afterpulse with spectral filtering (an appropriately tuned etalon) outside of the laser cavity, if the laser is bifurcation free (see next section). In uncontrolled environments, it may be necessary for the filter to actively track the laser, or vice versa.

There is an additional mechanism for afterpulsing when the thermalization times of the laser manifolds are long compared to the Q-switched output pulse. Themalization of the manifolds can rapidly pump the laser transition and lead to gain-switched pulses following the primary Q-switched pulse. These gain-switched pulses can be in the same longitudinal mode as the primary pulse and can be very stable in amplitude and time.

1.2.5.3 Pulse Bifurcation, Pulse-to-Pulse Amplitude Stability

At high pulse repetition rates some of the energy not extracted by one pulse will still be present in the laser cavity when the next pulse forms. As a result of spatial hole burning in the gain medium (Zayhowski, 1990a; Zayhowski, 1990b), the residual gain will be greater at positions where it favors the development of a pulse in a different longitudinal mode than the first pulse. When the gain differential between modes is extremely small, even small amounts of spatial hole burning can cause pulse bifurcation, and the effects can be significant even when the pulse-to-pulse spacing is several times the spontaneous lifetime of the gain medium. The second pulse will preferentially deplete the gain most available to it, creating a situation where the first mode is favored in the subsequent pulse. At higher pulse repetition rates more modes can come into play, creating trains of three or more modes occurring in a regular sequence or, depending on the stability of the system, chaotically. In systems with small mode-gain differentials operating at high repetition rates, time-averaged spectra will show the presence of multiple oscillating modes. This should not be interpreted as proof of multimode pulses; each individual pulse may still be single frequency with a smooth temporal profile, free of mode beating.

Pulse bifurcation in a cw-pumped passively Q-switched laser usually results in alternating strong and weak pulses. The timing interval between the pulses typically varies in accordance with the amplitudes; the period preceding a weak pulse is shorter than the period preceding a strong pulse. At higher pulse repetition rates, as the pulse train subdivides into pulses of more longitudinal modes, there will be a greater variation in pulse amplitudes and pulse-to-pulse timing. Because of gain-related index-guiding effects (see Subsection 1.2.7.3), each of the modes will have slightly different transverse dimensions and divergences.

For some applications, pulse-to-pulse amplitude stability is critical, and pulse bifurcation cannot be tolerated. To maximize the pulse repetition rate at which bifurcation-free operation is achieved, the gain differential between pulses must be maximized. Techniques that reduce afterpulsing also encourage bifurcation-free operation. Thus, observations of the afterpulse can be useful in tuning a passively Q-switched laser for maximum pulse-to-pulse stability.

Because the amplitude and pulse width of a passively Q-switched laser are determined by material characteristics and cavity design, they can be extremely stable. Passively Q-switched microchip lasers operating on a single longitudinal mode and producing pulse energies of 10 μJ have demonstrated pulse-to-pulse amplitude stabilities better than 1 part in 10^4 at pulse repetition rates of 7 kHz.

1.2.5.4 Pulse-to-Pulse Timing Stability

The pulse-to-pulse timing stability of passively Q-switched lasers is generally worse than that of actively Q-switched systems (Mandeville et al., 1996) and, in bifurcation-free systems, is often limited by noise in the pump source to ~1% of the pump duration. The timing stability is improved by using a pulsed pump source, operating at as high a peak power as possible for just long enough to produce an output pulse. When operated in this fashion, the pump pulses are turned on by an external clock and turned off when a Q-switched output pulse is detected. This approach has the added benefit of minimizing thermal effects in passively Q-switched lasers because it minimizes efficiency losses due to spontaneous decay of the gain medium, leading to additional benefits in performance.

In a system that uses a pulsed pump source, the pulse-to-pulse timing jitter of the passively Q-switched laser gets smaller as the pump period decreases. The minimum possible pump duration is limited by the power (or cost) of the available pump diodes. One approach to minimize this limitation is to cw pump the Q-switched laser just below threshold and then quickly raise the pump power to its maximum value just before an output pulse is desired. This technique results in the greatest pulse-to-pulse timing stability.

Other techniques used to reduce the pulse timing jitter include hybrid active/passive loss modulation (Arvidsson et al., 1998) and external optical synchronization (Dascalu et al., 1996). These approaches have resulted in timing jitter as low as 65 ps (Hansson and Arvidsson, 2000), at the expense of added system complexity and some sacrifice in the amplitude and pulse-width stability of a true passively Q-switched system.

1.2.6 Semiconductor Saturable-Absorber Mirrors

All of the formalism developed so far was based on bulk saturable absorbers. Semiconductor saturable-absorber mirrors (SESAMs) are a second class of devices used as passive Q switches in solid-state lasers. SESAMs consist of an antiresonant semiconductor Fabry–Pérot etalon formed by a semiconductor layer grown on top of a highly reflecting semiconductor Bragg mirror and covered by a dielectric reflector. The semiconductor layer typically consists of absorptive quantum-well layers in an otherwise transparent medium. The bandgap of the quantum wells can be engineered to provide saturable absorption at a wide variety of wavelengths.

By their design, SESAMs are high reflectors with a saturable loss γ_s. In addition, there is an undesirable parasitic loss $\gamma_{s,p}$, which typically accounts for ~20% of the unsaturated loss. The effective absorption cross section of the saturable loss, determined by the absorption cross section of the semiconductor quantum wells and the reflectivity of the dielectric reflector, is usually much larger than the gain cross section of the gain medium, leading to a very large α_r in most applications. As a result, SESAMs are, in some ways, ideal saturable absorbers and can often be modeled as instantaneous Q switches. This is almost always the case when they are used in microchip and other miniature lasers, where typical values of α_r are several thousand.

In a cavity containing a SESAM, the SESAM is used as one of the cavity mirrors, with the oscillating light incident on the dielectric reflector. Prior to the onset of lasing the total round-trip cavity loss $\gamma_{rt,t}$ consists of three components:

$$\gamma_{rt,t,t<0} = \gamma_0 + \gamma_{rt,p} + \gamma_s, \tag{1.48}$$

where γ_0 is the output-coupling loss, $\gamma_{rt,p}$ is the parasitic loss (which now includes the unsaturable loss of the SESAM), and γ_s is the saturable loss of the SESAM. When the laser reaches threshold, the intracavity loss rapidly decreases to

$$\gamma_{rt,t,t>0} = \gamma_0 + \gamma_{rt,p}. \tag{1.49}$$

The change in cavity Q, for the very large values of α_r typically associated with SESAMs, is rapid enough to treat the transition as instantaneous for most purposes.

The inversion within the cavity at the time Q switching occurs has the value

$$N_0 = \frac{(\gamma_0 + \gamma_{rt,p} + \gamma_s)A_g}{2^{\ddagger}\sigma_g}, \tag{1.50}$$

and the resulting pulse width is

$$t_w = \frac{S_p t_{rt}}{\gamma_s}\left[\frac{R_s(1+R_s)\eta_e}{R_s - \ln(1+R_s)}\right], \tag{1.51}$$

where $R_s = \gamma_s/(\gamma_{rt,p} + \gamma_0)$ is the ratio of saturable to unsaturable cavity losses. For a given amount of saturable loss, the pulse width asymptotically approaches its minimum value of $t_w = 4S_p t_{rt}/\gamma_s$ in the limit of large unsaturable losses ($\gamma_{rt,p} + \gamma_0 \gg \gamma_s$). However, this minimum pulse width is obtained at the expense of high threshold and low efficiency. In the opposite limit ($\gamma_{rt,p} + \gamma_0 \ll \gamma_s$) the pulse width asymptotically approaches $t_w = S_p t_{rt}/(\gamma_{rt,p} + \gamma_0)$, the threshold of the laser is reduced, and the extraction efficiency can be high. The best compromise between pulse width, threshold, and efficiency will depend on the application of the laser.

Though SESAMs act like instantaneous Q switches in many ways, they still benefit from being saturable absorbers in their single-frequency performance. It takes extremely large values of α_r before this benefit disappears.

SESAMs offer both advantages and limitations when compared to bulk saturable absorbers. The advantages include the fact that they can be engineered to operate at a wide variety of wavelengths, making SESAMs particularly interesting where good gain medium/bulk saturable absorber combinations have not been identified. The effective absorption cross sections are very large, leading to large values of α_r and nearly instantaneous Q switching. The physical length of the saturable-absorber region l_s is small and its contribution to the round-trip time of light in the laser cavity is negligible, resulting in the shortest possible pulses. The dielectric top reflector on the SESAM can be (and often is) designed to be highly reflecting at the wavelength of the pump light so that there is no pump-induced bleaching of the saturable absorber.

One of the main limitations of SESAMs is their relatively low damage threshold. Typical values for the damage fluence of the semiconductor materials are ~10 mJ cm^{-2}. The damage fluence of the SESAMs can be increased beyond this value by using highly reflective dielectric layers. This, however, reduces the saturable loss γ_s, which generally reduces the Q-switched pulse energy and increases the pulse duration. Typical values for the saturable loss are in the range of 10% or less. The largest pulse energy reported for a microchip laser using a SESAM is 4 μJ, with fractions of a microjoule being more typical. Other limitations of SESAMs are that they cannot be used as input or output couplers, and that their thermal expansion coefficients are not matched to the gain media, precluding the possibility of bonding a SESAM to a gain medium in a quasi-monolithic fashion.

Finally, the carrier recombination time in semiconductors (the upper-state lifetime of the SESAM) can be quite short, ranging from ~10 ps to ~1 nsec, depending on the material and the growth conditions. When the upper-state lifetime is comparable to the output pulse width, it can introduce a significant parasitic loss. On the other hand, it effectively closes the Q switch and can prevent the formation of afterpulses. It also allows SESAMs to be operated at extremely high pulse repetition rates, and 7-MHz operation has been reported (Braun et al., 1997).

As a result of their advantages and limitation, SESAMs are most attractive in applications requiring short, low-energy pulses, and where requirements on the system's robustness can be relaxed. In this regime, they can be operated at very high repetition rates and can be engineered to work with gain media at many different wavelengths.

1.2.7 Transverse Mode Definition

1.2.7.1 Issues in Laser Design

A very important issue in the design of many lasers is the extraction of heat from the gain medium. In the process of pumping the gain medium, heat is generated. As the temperature of the gain medium changes so, too, do its physical length and refractive index. Each of these contributes to changes in the optical length and resonant frequencies of the laser cavity. These changes are especially significant in miniature lasers where the active volume of the gain medium constitutes much of the resonator. Nonuniform heating results in a nonuniform refractive index and internal stress. Index gradients lead to thermal lensing, which changes the confocal parameters of the laser cavity and can destabilize an otherwise stable cavity, or vice versa. Internal stress leads to stress birefringence and, eventually, stress-induced fracture.

Other issues that must be considered in high-power lasers are nonlinear optical effects and optical damage. The electrical field within the optical beam of a high-power laser can be large enough to damage optical components. This is particularly important in high-peak-intensity pulsed lasers such as passively Q-switched microchip lasers.

1.2.7.2 Cavity Designs

One common feature of many applications for miniature lasers is that mode quality is at least as important as total power. As a result, miniature lasers are usually designed to operate in the fundamental transverse mode. Most of the results and formulas presented in this section are derived for fundamental-mode operation and may require some modification for lasers operating in multiple transverse modes.

The transverse modes of a laser are determined by the cavity design and the pump-energy deposition profile. The cavities for many miniature solid-state lasers are small versions of larger devices, designed according to the same principles (Hall and Jackson, 1989). However, the use of longitudinal pumping allows additional possibilities for obtaining stable cavity modes.

1.2.7.3 Microchip Fabry–Pérot Cavities

1.2.7.3.1 Thermal Guiding

Efficient lasers can be produced using small, longitudinally pumped standing-wave laser cavities defined by two plane mirrors. The planar uniformity of such a cavity is broken by the pump beam, which deposits heat as it pumps the crystal. The heat diffuses away from the pump beam, generally resulting in a radially symmetric temperature distribution. In materials with a positive change in refractive index n with temperature T ($dn/dT > 0$), such as Nd:YAG, this results in a thermal waveguide. In addition, when the cavity mirrors are deposited on the gain medium, there is some thermally induced curvature of the mirrors as the warmer sections of the gain medium expand or contract. In materials with a positive thermal expansion coefficient α_e this effect also contributes to the stabilization of the transverse mode. In some materials, such as Nd:YLF, this term dominates and can lead to stable transverse-mode operation in an otherwise flat–flat cavity, despite a negative dn/dT. Another effect is strain-induced variation of the refractive index caused by nonuniform heating and expansion of the gain medium. This effect tends to be less important than the others in determining the transverse mode characteristics of the cavity, although it can cause local birefringence. It will be ignored in this section.

When both index change and thermal expansion are considered, the variation in the optical length of a material nl (where l is the physical length of the material) as a function of temperature is given by

$$\frac{\delta(nl)}{nl} = \left(C\alpha_e + \frac{1}{n}\frac{dn}{dT} \right)\delta T, \tag{1.52}$$

where C is a number between 0 and 1, depending on whether the material is constrained ($C = 0$) or free to expand ($C = 1$). For the case of nonuniform heating, the thermal expansion of the warmer sections of the material will be constrained by the cooler regions, and C may be a function of the thermal gradients. If the cavity length is short compared to the confocal parameter, as in the case of microchip lasers, the total change in optical length as a function of transverse cavity position can be modeled as a simple lens between the two flat mirrors or as an axially uniform waveguide with a radially varying index.

To simplify the analysis, we will assume that C is independent of position, so that $\delta(nl) = nl\Delta_{nl,T}\delta T$, where $\Delta_{nl,T} = C\alpha_e + dn/ndT$ is a constant. For a monolithic, longitudinally pumped, short cavity with radial heat flow, the radius of the oscillating mode r_m is given by (Zayhowski, 1991a)

$$r_m^2 = \lambda_0 \left[\frac{(r_0^2 + r_p^2)lk_c}{\pi\kappa P_a n\Delta_{nl,T}} \right]^{1/2}, \tag{1.53}$$

where λ_0 is the free-space wavelength of the oscillating mode, l is the cavity length, k_c is the thermal conductivity of the gain medium, κ is the heat-generating efficiency of the gain medium, P_a is the absorbed pump power, r_p is the average radius of the pump beam within the laser cavity, and r_0 determines the waist size of the oscillating mode for extremely small pump radii. Once $\Delta_{nl,T}$ and r_0 are determined, this equation does an excellent job of describing the pump-power dependence of the oscillating-mode radius for many microchip lasers.

For a thermally guided Fabry–Pérot laser cavity to create a symmetric fundamental transverse mode, parallelism between the cavity mirrors is critical. The maximum angle ψ_{max} that can be tolerated between the mirrors (for symmetric fundamental-mode operation) is given by (Zayhowski, 1991a)

$$\psi_{max} = 1.8\frac{\kappa P_a\Delta_{nl,T}}{4\pi k_c r_p}. \tag{1.54}$$

A linear thermal gradient across the oscillating mode will contribute to the effective wedge between the two mirrors, through both the thermal expansion and the temperature-induced change in index of the material.

1.2.7.3.2 *Aperture Guiding in Three-Level Gain Media*

In three-level or quasi-three-level lasers, there can be significant absorption of the oscillating radiation in unpumped regions of the gain medium. For longitudinally pumped devices, this creates a radially dependent loss (aperture) which can restrict the transverse dimensions of the lasing mode, resulting in smaller mode radii than predicted by Equation 1.53. Aperture guiding may be important in three-level systems such as Yb:YAG microchip lasers (Fan, 1994).

1.2.7.3.3   Gain Guiding

The absence of gain in the unpumped regions of the gain medium is sufficient to define a stable transverse mode (Kogelnik, 1965). This effect is similar to aperture guiding. It is usually insignificant compared to thermal guiding but can be important in low-duty-cycle lasers where thermal effects are minimized. It can also be important in lasers operating near threshold. Near threshold there is little saturation of the population inversion, the round-trip amplification of the mode is highest near the center of the mode, and gain guiding will tend to reduce the mode cross section. Well above threshold the population inversion at the center of the mode is saturated and residual gain in the wings (if there is any) can lead to a slight increase in the mode radius (Kemp et al., 1999).

1.2.7.3.4   Gain-Related Index Guiding

Optical gain provides dispersion. Laser modes that are spectrally detuned from the center of the gain profile will see a refractive index that is a function of their detuning. Modes that fall on the long-wavelength side of the gain profile will see an increased refractive index; modes on the short-wavelength side will see a decreased refractive index.

If we assume a Lorenzian gain profile, the change in refractive index seen by a mode as it moves away from the gain peak is

$$\Delta n(r,z,\lambda) = \frac{\lambda_0(\lambda - \lambda_0)g(r,z,\lambda)}{2\pi\Delta\lambda}, \tag{1.55}$$

where $g(r,z,\lambda)$ is the saturated, spatially and spectrally dependent gain in the presence of laser oscillation. Hence, $\Delta n(r,z,\lambda)$ depends on the mode profile and must be calculated self-consistently. (This can be done iteratively starting with the mode profile due to thermal considerations alone.) Near threshold the gain will have a maximum near the center of the mode; well above threshold the gain will be strongly saturated near the mode center and will have a local minimum on axis. In either case, the deformation of the optical path near the center of the mode, due to gain-related index guiding, can usually be modeled as a spherical (positive or negative) intracavity lens.

As a result of gain-related index guiding, each of the longitudinal modes of a laser has a slightly different spatial profile, a different amount of overlap with the pump, and a different far-field divergence. Gain-related index guiding has been shown to play an important role in Nd:YVO$_4$ microchip lasers

(Kemp et al., 1999), and can lead to interesting effects such as self Q switching without a saturable absorber (Conroy et al., 1998).

1.2.7.3.5   Self-Focusing

The refractive index of a material can be affected by the presence of a high-intensity optical field through the optical Kerr effect:

$$n = n_0 + n_2 I, \tag{1.56}$$

where n_0 is the refractive index in the absence of an optical field and n_2 is the nonlinear Kerr index. Light traveling through a Kerr medium will experience self-focusing. When the Kerr medium is inside a laser cavity, this will cause a reduction in the diameter of the oscillating mode. Self-focusing is usually insignificant for miniature cw lasers but is the driving mechanism in Kerr-lens mode-locked lasers and can play a role in determining the mode diameter in some miniature Q-switched and gain-switched lasers. For the miniature Q-switched and gain-switched lasers demonstrated to date, self-focusing introduces only a small perturbation to the mode diameter that would result in its absence.

1.2.7.3.6   Aperture Guiding in Saturable Absorber

The transverse mode of miniature passively Q-switched lasers, especially microchip lasers, is nearly diffraction limited. In addition to the mode-defining mechanisms already discussed in this section, the saturable absorber acts as a soft aperture and spatially filters the resonator modes, contributing to the definition of the fundamental mode and the suppression of higher-order transverse modes. However, because the saturable absorber bleaches at the center of the beam first, where the optical intensity is greatest, this soft aperture evolves as the pulse forms. As a result, there can be a measurable change in the output beam profile during the pulse (Yao et al., 1994; Ardvisson, 2001).

1.2.7.4   Pump Considerations

To ensure oscillation in the fundamental transverse mode, it is necessary for the fundamental mode to use most of the gain available to the laser. If the radius of the fundamental mode is much less than the radius of the pumped region of the gain medium, higher-order transverse modes will oscillate.

For longitudinally pumped lasers the active length of the gain medium l_g and the desired round-trip gain $g_{rt} = 2^{\ddagger} g l_g$ together determine the required brightness of the pump source B_p, where brightness is defined as intensity per unit solid angle. For an incoherent pump source, the required brightness is given by (Fan and Sanchez, 1990)

$$B_p > \frac{4}{P_p} \left(\frac{h\nu_p g_{rt} l_g}{2^{\ddagger} n_g \sigma_{g,\text{eff}} \tau \eta_p} \right)^2, \tag{1.57}$$

where P_p is the total pump power, $h\nu_p$ is the energy of a pump photon, and η_p is the fraction of the pump power absorbed. For a diffraction-limited pump source, the brightness requirement is reduced by a factor of four (Fan and Sanchez, 1990):

$$B_p > \frac{1}{P_p} \left(\frac{h\nu_p g_{rt} l_g}{2^{\ddagger} n_g \sigma_{g,\text{eff}} \tau \eta_p} \right)^2 . \tag{1.58}$$

For single-transverse-mode operation the pump light should be focused within the volume of the fundamental oscillating mode.

1.2.7.5   Polarization Control

It is often desirable for a laser to oscillate in a single linear polarization. For lasers with isotropic gain media, this can be ensured by including a polarizing element, such as a Brewster plate, within the cavity. In single-frequency devices, the presence of any tilted optical element may be sufficient to polarize a laser because the reflection coefficient of surfaces is often different for *s*- and *p*-polarized waves; only a small amount of modal discrimination is needed to select one of two frequency-degenerate polarizations (Zayhowski, 1990a; Zayhowski, 1990b). If the design of a cavity does not favor a given polarization, as is the case for microchip lasers, the polarization degeneracy of an isotropic gain medium can often be removed by applying uniaxial transverse stress.

If there is very little polarization selectivity within a laser cavity, feedback from external surfaces can determine the polarization of the oscillating mode. Although this effect is usually undesirable, it has been used to controllably switch the polarization of microchip lasers at rates up to 100 kHz (Zayhowski, 1991b).

Finally, in the absence of any strong polarizing mechanism, the polarization of the pump light may determine the polarization of the laser.

1.2.8   Additional Thermal Effects

Temperature affects many of the properties of the materials in a microchip laser. In addition to the refractive index, it affects the population distributions within the laser manifolds of the gain medium, the thermalization times, and the magnitude and spectral position of the gain. Thermal effects are most important in lasers that require a high average inversion density, i.e., short-pulse lasers operated at high repetition rates.

The rate-equation analysis presented in Subsection 1.2.4 shows that the pulse parameters of a passively Q-switched microchip laser are fixed by the material parameters and cavity design. To first order, the pulse width is independent of pulse repetition rate (Zayhowski and Dill, 1994; Shimony et al., 1996). However, short-pulse lasers require aggressive pumping, and the temperature of the active region in the gain medium can increase significantly at high repetition rates. As it does, the population distributions within the laser manifolds and the thermalization times change. In short-pulse Nd:YAG-based lasers, this often results in shorter pulses at higher repetition rates.

Some of the thermal effects in passively Q-switched microchip laser are dependent on changes in temperature across the active area of the laser. Thermal guiding, discussed in Subsection 1.2.7.3, is a very important example. Another such effect, which can limit the performance of these devices, is the temperature dependence of the spectral position of the gain peak. (The 1.064-μm transition in Nd:YAG shifts to longer wavelengths by about 0.46 nm per 100°C variation in temperature.) As the temperature gradients across the active area of the gain medium increase, the spectral position of the gain peak at the center of the lasing mode will begin to deviate from the spectral position of the gain in the wings. The average gain coefficient for the mode therefore decreases, and a larger inversion is required to reach threshold. A larger inversion requires additional pumping, which further increases the temperature gradients, and the process can run away. For passively Q-switched Nd:YAG/Cr^{4+}:YAG microchip lasers, this effect contributes to making it difficult to obtain very short pulses at high repetition rates (Zayhowski and Wilson, 2003a).

There are numerous additional subtleties that can only be captured by a full three-dimensional model of the laser, including the temperature dependences of the relative material parameters at all points in the active volume of the device. Such an analysis is outside of the scope of this chapter. The concepts and analysis presented above, supplemented with a small amount of experimentation, are adequate for the design of passively Q-switched microchip lasers and for understanding most of their characteristics. They are an essential part of any more complete model.

1.3   Demonstrated Device Performance

1.3.1   Demonstrated Passively Q-Switched Lasers

1.3.1.1   Saturable Absorbers

Most of the commonly used solid-state saturable absorbers are vibrationally broadened transition-metal-doped single-crystal materials or, to a lesser degree, glasses. The most commonly used material for passive Q switching of miniature lasers is Cr^{4+}:YAG. The use of Cr^{4+}:YAG dates back to 1991 (Andrauskas and Kennedy, 1991). It has been used to Q switch Nd:YAG microchip lasers operating at 1.064 μm (Zayhowski and Dill, 1994), 946 nm (Zayhowski et al., 1996), and 1.074 μm (Zayhowski et al., 2000); and Nd:YVO$_4$ and Nd:GdVO$_4$ microchip lasers operating at 1.064 and 1.062 μm (Jaspan et al., 2000; and Liu et al., 2003; respectively). This material has been very successful despite the presence of significant excited-state absorption and polarization anisotropy (Eilers et al., 1992; Il'ichev et al., 1997). The latter property does not impose any significant limitations on passively Q-switched laser performance, and can be used to help fix the polarization of passively Q-switched lasers with isotropic gain media, such as Nd:YAG. Some of the relevant material properties of Cr^{4+}:YAG are listed in Table 1.1. Relevant properties of YAG (Aggarwal et al., 2005) and Nd:YAG, the most common gain medium used in passively Q-switched lasers, are included in Table 1.2 and Table 1.3. Table 1.4 lists several gain medium/bulk saturable absorber combinations that have been used at a variety of wavelengths.

Semiconductor saturable-absorber mirrors (SESAMs) (Kärtner et al., 1995; Keller et al., 1996; Spühler et al., 1999) are used to generate Q-switched pulses with durations <200 ps, and have been used with a variety of gain media operating at a wide range of wavelengths. They are, however, typically limited to pulse energies in the nanojoule regime by the onset of optical damage (Braun et al., 1996; Braun et al., 1997).

Bulk semiconductor saturable absorbers (Gu et al., 2000; Kajave and Gaeta, 1996; Tsou et al., 1993) and semiconductor-doped glasses (Bilinsky et al.,

TABLE 1.1

Properties of Cr^{4+}:YAG Saturable Absorber

Property	Symbol	Value	Units	Comments
Ground-state absorption cross section	$\sigma_{s,g}$	7×10^{-18}	cm^2	1.064 μm
		2×10^{-18}	cm^2	808 nm
Upper-level absorption cross section	$\sigma_{s,u}$	2×10^{-18}	cm^2	1.064 μm
Upper-level lifetime	τ_s	4.1×10^{-6}	s	Room temperature

TABLE 1.2

Properties of Yttrium Aluminum Garnet ($Y_3Al_5O_{12}$, YAG)

Property	Symbol	Value	Units	Comments
Crystal symmetry				Cubic
Lattice constant	a_0	12.01	Å	
Refractive index	n	1.818		1.064 μm
Temperature index variation	dn/dT	7.8×10^{-6}	K^{-1}	1.064 μm
Thermal expansion coefficient	α_e	6.1×10^{-6}	K^{-1}	
Thermal conductivity	k_c	0.11	$W\ cm^{-1}\ K^{-1}$	
Specific heat	c_{sh}	0.6	$J\ g^{-1}\ K^{-1}$	
Thermal diffusivity	α_d	0.041	$cm^2\ s^{-1}$	
Mass density	ρ_m	4.56	$g\ cm^{-3}$	
Poisson's ratio		0.25		
Hardness		1215	$kg\ mm^{-2}$	Knoop
		8–8.5		Moh's scale
Tensile strength		2.0×10^6	$g\ cm^{-2}$	
Melting point		1950	°C	

TABLE 1.3

Laser Parameters for Neodymium-Doped YAG (Nd^{3+}:$Y_3Al_5O_{12}$)

Property	Symbol	Value	Units	Comments
Nd^{3+} concentration (1.0 at. %)		1.39×10^{20}	cm^{-3}	
Wavelength at gain peak	λ_0	946	nm	
		1.064	μm	
		1.319	μm	
Linewidth	$\Delta\lambda_g$	0.8	nm	946 nm
		0.6	nm	1.064 μm
		0.6	nm	1.319 μm
Spontaneous lifetime	τ	230	μsec	1.0 at. % Nd
Concentration-quenching parameter	ρ_{cq}	2.63	at. %	
Emission cross section	σ_g	0.4×10^{-19}	cm^2	946 nm
		6.5×10^{-19}	cm^2	1.06415 μm
		1.2×10^{-19}	cm^2	1.0644 μm
		1.7×10^{-19}	cm^2	1.319 μm
Occupation probability	f_u	0.60		946 nm
	f_l	0.008		946 nm
	f_u	0.40		1.06415 μm
	f_l	0		1.06415 μm
	f_u	0.60		1.0644 μm
	f_l	0		1.0644 μm
	f_u	0.40		1.319 μm
	f_l	0		1.319 μm
Energy of lower level	E_l	857	cm^{-1}	946 nm
Effective emission cross section	$\sigma_{g,eff}$	0.3×10^{-19}	cm^2	946 nm
		3.3×10^{-19}	cm^2	Sum of 1.06415 and 1.0644 μm
		0.7×10^{-19}	cm^2	1.319 μm
Pump absorption coefficient	α_p	9.5	cm^{-1}	808.5 nm, 1.0 at. % Nd

1998), although demonstrated, are rarely used for passive Q switching. Although organic dyes may be used as saturable absorbers, they are generally not used in solid-state devices operating in the near-infrared because photothermal degradation leads to impractically short operational lifetimes.

1.3.1.2   *Passively Q-Switched Microchip Lasers*

1.3.1.2.1   *Low-Power Embodiments*

A simple embodiment of a passively Q-switched microchip laser is illustrated in Figure 1.5. In this embodiment, the laser is constructed by diffusion bonding a thin, flat wafer of gain medium to a similar wafer of saturable absorber (Zayhowski and Dill, 1994). The composite structure is polished flat and parallel on the two faces normal to the optic axis, with a typical cavity length between 0.75 and 1.5 mm. The pump-side face of the gain medium is coated dielectrically to transmit the pump light and to be highly reflecting at the oscillating wavelength. The output face is coated to be partially reflecting

TABLE 1.4

Passively Q-Switched Laser Gain Media and Bulk Saturable Absorber Combinations

Wavelength (μm)	Gain Medium	Saturable Absorber	Reference
0.694	Ruby	Cr^{4+}:GSGG	Chen et al., 1993
0.694	Ruby	Cr^{4+}:CaGd$_4$(SiO$_4$)$_3$	Yumashev et al., 1998
0.781–0.806	Cr^{3+}:LiCaAlF$_6$	Cr^{4+}:Y$_2$SiO$_5$	Kuo et al., 1995
0.946*	Nd:YAG	Cr^{4+}:YAG	Zayhowski et al., 1996a
1.03*	Yb:YAG	Cr^{4+}:YAG	Zhou et al., 2003
1.04*	Yb:KGW	V^{3+}:YAG	Lagatsky et al., 2000
1.062	Nd:GdVO$_4$	Cr^{4+}:YAG	Liu et al., 2003
1.064*	Nd:YAG	Cr^{4+}:YAG	Andrauskas and Kennedy, 1991; Zayhowski and Dill, 1994
1.064*	Nd:YVO$_4$	Cr^{4+}:YAG	Bai et al., 1997; Jaspan et al., 2000
1.064	Nd:YAG	Cr:Mg$_2$SiO$_4$	Demchuk et al., 1992
1.064	Nd:YAG	F^{-2}:LiF	Isyanova and Welford, 1993
1.064	Nd:YAG	Cu$_2$Se:glass	Yumashev et al., 2001
1.067	Nd:KGW	Cu$_{2+x}$Se:glass	Yumashev et al., 2001
1.074*	Nd:YAG	Cr^{4+}:YAG	Zayhowski et al., 2000
1.08	Nd:YALO	Cr^{4+}:SrGd$_4$(SiO$_4$)$_3$	Yumashev et al., 1998
1.32	Nd:YAG	Cu$_{2+x}$Se:glass	Yumashev et al., 2001
1.34*	Nd:YAG	V^{3+}:YAG	Jabczynski et al., 2001
1.34*	Nd:YAP	V^{3+}:YAG	Jabczynski et al., 2001
1.34*	Nd:YVO$_4$	V^{3+}:YAG	Jabczynski et al., 2001
1.34	Nd:YAP	Cu$_2$Se:glass	Yumashev et al., 2001
1.34	Nd:YAG	Co^{2+}:MgAl$_2$O$_4$	Wu et al., 2000
1.34	Nd:YALO	Co^{2+}:LaMgAl$_{11}$O$_{19}$	Yumashev et al., 1998
1.44	Nd:KGW	Co^{2+}:MgAl$_2$O$_4$	Wu et al., 2000
1.44*	Nd:YAG	V^{3+}:YAG	Kuleshov et al., 2000
1.5*	Yb,Tm:YLF	Co^{2+}:LaMgAl$_{11}$O$_{19}$	Braud et al., 2000
1.53	Er:glass	U^{2+}:CaF$_2$	Stultz et al., 1995
1.54*	Er:glass	Co^{2+}:LaMgAl$_{11}$O$_{19}$	Thony et al., 1999; Yumashev et al., 1999
1.54*	Er:glass	Co^{2+}:MgAl$_2$O$_4$	Wu et al., 2000
1.54	Er:glass	Co^{2+}:glass	Malyarevich et al., 2001a
1.54	Er:glass	Cu$_{2+x}$Se:glass	Yumashev et al., 2001
2.0	Tm,Cr:YAG	Ho:YLF	Kuo et al., 1994
2.1	Ho:YAG	PbSe:glass	Malyarevich et al., 2001b

Note: An asterisk denotes demonstration in a miniature or microchip laser.

at the lasing wavelength and provides the optical output from the device. The laser is completed by dicing the wafer into small squares, typically 1 to 2 mm on a side. The laser cavity is then mounted on an appropriate heatsink and pumped with the output of a diode laser. Figure 1.1 shows a photograph of a Nd:YAG/Cr^{4+}:YAG passively Q-switched microchip laser epoxied directly to the ferrule of the pump fiber, with the ferrule serving as the heatsink.

Alternative approaches to constructing passively Q-switched microchip lasers include the use of a single material that acts as both the gain medium and the saturable absorber (Zhou et al., 1993; Wang et al., 1995; Zhou et al., 2003), the growth of one material on the other (Fulbert et al., 1995), and the use of semiconductor saturable absorbers (Braun et al., 1996; Braun et al., 1997). In all cases, the simplicity of the passively Q-switched microchip laser and the small amount of material used by it give it the potential for inexpensive mass production; nearly monolithic construction results in robust devices.

The minimum pulse width that can be obtained from any Q-switched laser is given by Equation 1.37. Because microchip lasers are physically very short, they have very short cavity round-trip times and can produce very short output pulses. As an example, a 0.75-mm-long Nd:YAG/Cr^{4+}:YAG microchip laser has a round-trip time of 8.2 ps. Pumped with an incident power of 1 W from a semiconductor diode laser, it is possible to achieve a round-trip gain of 1.4, resulting in a pulse width as short as 200 ps. To put this in perspective, before the work on electro-optically Q-switched microchip lasers (Zayhowski and Dill, 1992; Zayhowski and Dill, 1995), the shortest Q-switched pulse obtained from a solid-state laser was about 1 nsec, and Q-switched Nd:YAG lasers typically produce pulse widths in excess of 5 nsec long.

The Nd:YAG/Cr^{4+}:YAG passively Q-switched microchip laser shown in Figure 1.1 produces an output pulse with a 218-ps pulse width. It operates at 10 kHz with a pulse energy of 4 μJ and peak power in excess of 15 kW. The high peak powers of passively Q-switched microchip lasers are another aspect of the devices that make them very interesting.

In addition to producing extremely short Q-switched pulses with high peak powers, the short length of the microchip cavity promotes single-frequency operation (Subsection 1.2.5), so that the output pulses are free of mode beating. The flat cavity mirrors strongly favor fundamental-transverse-mode operation (Subsection 1.2.7). The result is an output beam with nearly ideal temporal and spatial mode properties. The pulse-to-pulse amplitude fluctuations of the high-performance devices have been measured to be <0.05%. Pulse-to-pulse timing jitter tracks fluctuations in the output of the pump diodes and is typically 1% of the pump period (Subsection 1.2.5.4).

Composite-cavity microchip lasers containing a Nd:YAG gain medium diffusion bonded to a Cr^{4+}:YAG saturable absorber have demonstrated pulses as short as 200 ps when pumped with a 1-W diode laser (Zayhowski and Dill, 1994; Zayhowski et al., 1995). They can produce pulse energies in excess of 14 μJ, peak powers in excess of 30 kW, and time-averaged powers up to 120 mW. They typically operate at repetition rates between 8 and 15 kHz. Repetition rates as high as 70 kHz have been demonstrated (Zayhowski and Dill, 1994), but at the expense of pulse width and peak power. Specifications and performance parameters for several diffusion-bonded Nd:YAG/Cr^{4+}:YAG microchip lasers are given elsewhere (Zayhowski, 1998), and devices are commercially available. Monolithic passively Q-switched lasers using Cr^{4+},Nd^{3+}:YAG as both the gain medium and saturable absorber have demonstrated similar performance (Wang et al., 1995).

Pulses as short as 180 ps have been obtained by Q switching Nd:$LaSc_3(BO_3)_4$ microchip lasers with semiconductor antiresonant Fabry–Pérot saturable absorbers (Braun et al., 1996). Pulses as short as 37 ps were obtained using the same method in a Nd:YVO$_4$ laser (Spühler et al., 1999). Devices based on semiconductor saturable absorbers operate at lower pulse energies and peak powers than Cr^{4+}:YAG devices but have been pulsed at repetition rates up to 7 MHz (Braun et al., 1997).

All of the passively Q-switched microchip lasers mentioned above operate on the ~1.06-μm Nd^{3+} transition. Passively Q-switched microchip lasers operating on the 1.32-μm (Keller, 1996), 0.946-μm (Zayhowski et al., 1996, Zayhowski, 1997c), and 1.074-μm (Zayhowski et al., 2000) Nd^{3+}:YAG transitions have also been demonstrated. Other passively Q-switched microchip lasers use the 1.34-μm transition in Nd:YVO$_4$ (Fluck et al., 1997) and the 1.55-μm transition in Er:glass (Thony et al., 1999). A slightly larger miniature passively Q-switched laser used the 1.5-μm Yb:Tm:YLF transition (Braud et al., 2000).

1.3.1.2.2 High-Power Embodiments

The Q-switched microchip lasers discussed above were pumped with ~1-W of optical power. Several devices have been designed to be pumped with the higher-power output of fiber-coupled diode-laser arrays (Zayhowski, 1998; Zayhowski, 1997a; Zayhowski, 1997b; Zayhowski et al., 1999).

In addition to the gain medium and saturable absorber, the higher-power devices typically have undoped YAG endcaps, as shown in Figure 1.6. The endcaps perform several functions. One thing they do is lengthen the cavity. The higher-power devices typically have a cavity length of between 2 and 24 mm. Longer cavities produce oscillating modes with a larger diameter, which can more efficiently use the power deposited by the pump diodes. Increasing the cavity length also increases the output pulse length, keeping the peak intensity of these higher-power devices below the damage thresholds of the materials used. In addition, a longer pulse width may be desirable if the laser is to be used to pump optical parametric oscillators (OPOs) (Zayhowski, 1997c; Zayhowski, 1997b). In this case, longer pulses result in a lower thresholds for the OPOs (Byer, 1975).

The undoped YAG endcaps also remove heat from the active region of the gain medium, reducing thermal gradient and allowing higher-repetition-rate operation of short-pulse devices. Finally, they increase the damage threshold of the microchip lasers by reducing the thermal stress on the dielectric mirrors and by removing the active materials from the air interfaces.

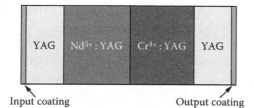

FIGURE 1.6
Illustration of high-power passively Q-switched microchip laser.

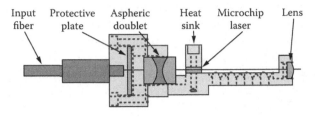

Input   Protective   Aspheric   Heat   Microchip   Lens
fiber   plate   doublet   sink   laser

FIGURE 1.7
Schematic of high-power passively Q-switched microchip laser head.

The increased heat load of the high-power microchip lasers forces us to pay more attention to heat sinking, and these devices can no longer be bonded to the ferrule of the pump fiber. High-power lasers are typically heatsunk on the top and bottom. The direction of the heat flow and the stress induced by mounting the heat sink breaks the symmetry of the isotropic gain medium (in the case of Nd:YAG) and defines the eigenpolarizations of the system. A typical high-power microchip laser head is shown schematically in Figure 1.7.

High-power devices pumped with high-brightness 10- to 25-W fiber-coupled diode-laser arrays have demonstrated pulse widths as short as 150 ps, pulse energies as high as 250 μJ, peak powers up to 565 kW, and time-averaged powers up to 1.2 W. They typically operate at pulse repetition rates between 1 and 20 kHz. Like the low-power devices, the high-power passively Q-switched microchip lasers have nearly ideal mode properties and excellent pulse-to-pulse stability. Specifications and performance parameters for several high-power Nd:YAG/Cr^{4+}:YAG microchip lasers are given elsewhere (Zayhowski, 1998; Zayhowski et al., 1999; Zayhowski and Wilson, 2003a), and high-power devices are commercially available.

1.3.2 Amplification

Passively Q-switched microlasers excel in generating subnanosecond, single-frequency pulses, but often at the expense of pulse energy, because the small volume of the gain medium limits the stored energy capacity. The highest pulse energy demonstrated with a passively Q-switched microchip laser using flat cavity mirrors is 250 μJ; the highest average power demonstrated with such a device is 1.2 W (Zayhowski, 1998; Zayhowski et al., 1999). Using an unstable resonator with a much larger mode, pulse energies as high as 2.15 mJ have been obtained at low repetition rates (Liu et al., 1998).

As the required output power of the laser system increases, the need for amplification becomes apparent. For a 1.064-μm Nd:YAG/Cr^{4+}:YAG passively Q-switched microchip laser either Nd:YAG (Druon et al., 1999; Zayhowski and Wilson, 2002; Zayhowski and Wilson, 2003b) or Nd:YVO$_4$ (Isyanova et al., 2001; Zayhowski and Wilson, 2004) can be used for the amplifier. For microchip lasers with low pulse energies, the need to drive the amplifier into saturation for efficient energy extraction makes Nd:YVO$_4$ the preferred gain

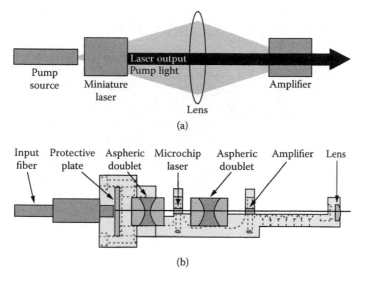

FIGURE 1.8
Energy-scavenging amplifier: (a) illustration of concept; (b) schematic of current embodiment.

medium because of its larger emission cross section and lower saturation fluence. For higher-energy microchip lasers, where reaching saturation in the amplifier is not an issue, the mechanical and thermal properties of Nd:YAG make it the preferred gain medium.

A modest amount of amplification can be achieved by scavenging the unused pump light from the microchip laser to pump an in-line amplifier. Longitudinally pumped microchip lasers tend to be inefficient, in part due to inefficient absorption of the pump light owing to the short length of the gain medium. This becomes more pronounced in lasers designed for very short pulse widths. By scavenging the unabsorbed pump light to pump an in-line Nd:YVO$_4$ amplifier, as shown in Figure 1.8 and Figure 1.9, the efficiency of short-pulse, high-power passively Q-switched microchip laser systems can be more than tripled, with minimal added cost, size, or complexity (Zayhowski and Wilson, 2004). With the use of energy-scavenging amplifiers, short-pulse microchip-laser systems operate with optical-to-optical efficiencies up to 15% at 1.064 μm. Such systems can be used anywhere that simple microchip lasers could be used.

To obtain higher output power, a double-pass Nd:YAG microchip amplifier can be introduced into the system, as shown in Figure 1.10 (Zayhowski and Wilson, 2002; Zayhowski and Wilson, 2003b). At 10 kHz, the microchip amplifier boosts the output of the passively Q-switched laser from 75 μJ (typically) to 210 μJ. At repetition rates below 2.5 kHz, 500 μJ/pulse have been obtained, with peak powers in excess of 1 MW. The output of the amplifier, at the highest output powers, is better than 1.2 times diffraction limited. In its current embodiment, the oscillator/amplifier system occupies a volume of <0.25 liters and can put out >2 W of time-averaged power, and there is considerable room for further size reduction (Figure 1.9).

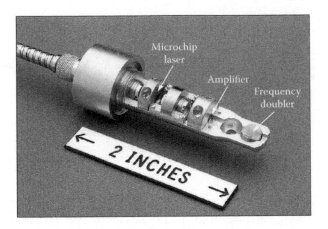

FIGURE 1.9
Photograph of miniature amplified, frequency-doubled microchip laser.

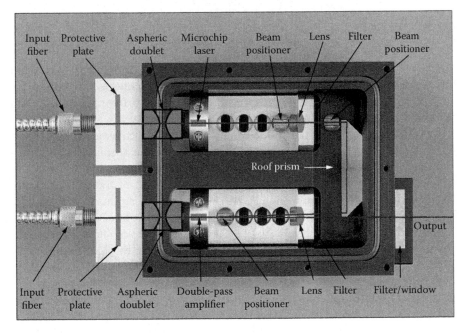

FIGURE 1.10
Photograph of amplified passively Q-switched microchip laser system. Optical components are labeled. Hidden optical components and the beam paths are drawn on top of the photograph.

Multipass amplifiers have been used to increase the output power of low-power microchip lasers operating at several tens of kilohertz (Druon et al., 1999). Systems based on this approach are commercially available. Power amplifiers (Isyanova et al., 2001) have been used to increase the output pulse energy of passively Q-switched microchip laser systems beyond 1 mJ. There are ongoing efforts, driven by applications, to reach much higher energies.

1.3.3 Frequency Conversion

1.3.3.1 Nonlinear Frequency Conversion

The high peak powers obtained from passively Q-switched microchip lasers enable a variety of miniature nonlinear optical devices. Harmonic generation, frequency mixing, parametric conversion, and stimulated Raman scattering have been used with microchip lasers for frequency conversion to wavelengths covering the entire spectrum from 213 nm to 4.3 μm in extremely compact optical systems.

1.3.3.1.1 Harmonic Conversion

Because of its high peak intensity, the output of even the low-power Nd:YAG/Cr^{4+}:YAG passively Q-switched microchip lasers can be efficiently frequency-doubled by simply placing a 5-mm-long piece of KTiOPO$_4$ (KTP) near the output facet of the laser, with no intervening optics. With 1-W-pumped devices, doubling efficiencies as high as 70% have been demonstrated (Zayhowski and Dill, 1999), although more typical numbers are between 45% and 60%. Because there is still high peak intensity and good mode quality in the green, it is possible to frequency-convert the green radiation into the UV by placing the appropriate nonlinear material (β-BaB$_2$O$_4$, BBO) adjacent to the output facet of the KTP, as shown in Figure 1.11. In all cases, the crystals are polished flat on the faces normal to the optic axis and proximity-coupled to each other without any intervening optics, allowing for very simple, and potentially inexpensive, fabrication. With this approach, and a 1-W-pumped Nd:YAG/Cr^{4+}:YAG passively Q-switched microchip laser, up to 7 μJ of green, 1.5 μJ of third-harmonic, 1.5 μJ of fourth-harmonic, and 50 nJ of 213-nm (fifth-harmonic) light have been demonstrated at a typical pulse repetition rate of 10 kHz (Zayhowski et al., 1995; Zayhowski, 1997c; Zayhowski, 1996a; and Zayhowski, 1996b). In each case, the optical head of the device, including the passively Q-switched microchip laser and the nonlinear crystals, has been packaged in a 1-cm-diameter $\times$ 2.5-cm-long stainless-steel housing, as shown in Figure 1.12. The electronics required to operate the system fit in an 11 $\times$ 17 $\times$ 4-cm box that consumes ~8 W of power at room temperature. The wavelength diversity offered by harmonic conversion opens up numerous applications. To

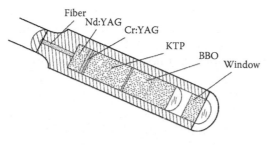

FIGURE 1.11
Schematic of UV harmonically converted passively Q-switched microchip laser.

FIGURE 1.12
Optical head of low-power UV microchip-laser system packaged in a 1-cm-diameter × 2.5-cm-long stainless-steel housing.

address those applications, microchip laser systems with characteristics similar to those described are commercially available at 532, 355, and 266 nm.

The high-power passively Q-switched microchip lasers can be harmonically converted to produce high-power UV output. Devices producing over 19 μJ of 355-nm output, or 12 μJ of 266-nm output, at 5 kHz, have been packaged in robust, 2.5-cm-diameter × 8-cm-long housings (Zayhowski, 1998; Zayhowski et al., 1999). A high-power 1.074-μm Nd:YAG/Cr^{4+}:YAG passively Q-switched microchip laser has been frequency-converted to generate more than 1 μJ of 215-nm light, its fifth harmonic, in the same size package (Zayhowski et al., 2000).

Microchip oscillator/amplifiers have been used in high-power green laser systems that produce 100 μJ of 532-nm radiation in a 460-ps pulse at pulse rates up to 10 kHz, in an output beam that is 1.4 times diffraction limited (Zayhowski and Wilson, 2002). A high-energy UV system, based on a microchip oscillator/amplifier, produces 110 μJ of 355-nm radiation in a 700-ps pulse, with near-diffraction-limited output, at pulse repetition rates up to 500 Hz (Zayhowski and Wilson, 2003b). The optical heads for each of these systems contain two equal-size compartments; the first compartment houses the oscillator/amplifier (similar to that shown in Figure 1.10), the second contains an optical breadboard that can accommodate a variety of nonlinear frequency-conversion schemes, including those just mentioned. The volume of the optical heads is <0.5 liters. A photograph of one of the optical heads is shown in Figure 1.13.

An interesting variation of the harmonically converted passively Q-switched microchip laser uses a self-frequency-doubling material, Nd^{3+}:GdCa$_4$O(BO$_3$)$_3$, as the gain medium (Zhang et al., 2001). Systems using a single material as the gain medium, saturable absorber, and nonlinear crystal have been proposed, but not yet attained in practice.

1.3.3.1.2 Optical Parametric Conversion

The high-power passively Q-switched microchip lasers can also be used to pump optical parametric amplifiers (OPAs), oscillators (OPOs), and generators (OPGs). The unfocused 1.064-μm output of a microchip laser has been used to drive several periodically poled lithium niobate (PPLN) OPGs, covering the

FIGURE 1.13
Photograph of optical head used for nonlinear optical systems and gain-switched lasers pumped with a microchip oscillator/amplifier.

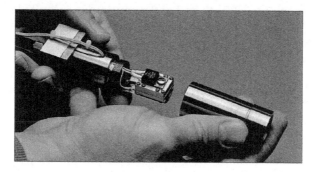

FIGURE 1.14
Photograph of passively Q-switched microchip laser/OPG system.

spectral range from 1.4 to 4.3 μm (Zayhowski, 1997a; Zayhowski and Wilson, 2000). In these experiments, nearly 100% conversion of the pump radiation was observed at the peak of the pulse. A photograph of the OPG system is shown in Figure 1.14. The microchip laser used to pump the OPG is contained in the first, covered section of the housing. An oven containing the OPG is in a second section and is shown uncovered in the photo.

A similar range of wavelengths has been accessed with microchip-laser-pumped OPOs (Capmany et al., 2001). KTP OPOs have been pumped with the second harmonic of a high-power microchip laser and operated at signal and idler wavelengths between 700 and 2000 nm (Zayhowski, 1997b).

1.3.3.1.3 Raman Conversion

Intracavity Raman frequency conversion has been demonstrated in Nd:YAG/Cr^{4+}:YAG and Nd:LSB/Cr^{4+}:YAG passively Q-switched microchip lasers using Ba(NO$_3$)$_2$ as the Raman medium (Demidovich et al., 2003). Up to 1.2 μJ of energy was obtained at the Stokes wavelength of 1.196 μm, corresponding to 8% conversion efficiency. The pulse duration, at 1.196 μm, was as short as 118 ps.

Raman combined with other nonlinear effects in fibers can be used to create extremely broadband, spatially coherent optical continua. Because the output of microchip lasers is diffraction limited, it is possible to focus the beam into a single-mode optical fiber. The intensities in the fiber core can reach 100 GW/cm². These intensities lead to very efficient stimulated Raman scattering (Agrawal, 1995; Zayhowski, 1997c; Zayhowski, 1999; Zayhowski, 2000; Johnson et al., 1999). The input wavelength efficiently generates the first Stokes line, which is red-shifted and broadened. This line generates a second Stokes line, which is further shifted and further broadened. The second line generates a third, and so on. Once Raman shifting generates light on both sides of the fiber's zero-dispersion wavelength, four-wave mixing broadens the spectrum beyond what is expected from Raman shifting alone and generates a smoother continuum. As shown in Figure 1.15a, a relatively featureless output spectrum extending from 850 nm to 2.25 μm is obtained by pumping a 100-m length of 10-μm-core fiber with the output from a 1.064-μm passively Q-switched microchip laser. The long-wavelength end of the continuum is limited by absorption in the fiber. The spectrum obtained from a 100-m length of 4.6-μm-core fiber pumped with 532-nm second-harmonic light from a passively Q-switched microchip laser extends from the green to 950 nm, as shown in Figure 1.15b. In this case, all of the light is on the short-wavelength side of the fiber's zero dispersion point, frequency shifting is dominated by the Raman effect, and many of the individual Raman lines are easily identified.

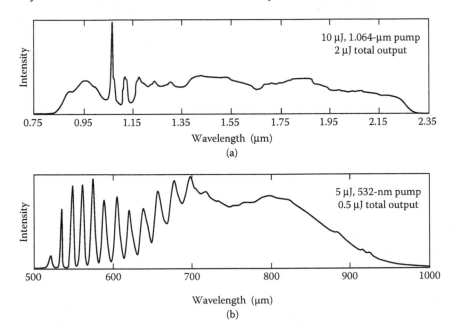

FIGURE 1.15
Output spectra obtained through cascaded stimulated Raman scattering in: (a) 100-m length of 10-μm-core fiber pumped with 1.064-μm light; (b) 100-m length of 4.6-μm-core fiber pumped with 532-nm light.

Microstructured fibers can have their zero-dispersion wavelength any-where throughout the visible or near-IR, making it possible to obtain smooth, broadband continua covering the entire transmission window of the fiber (Ranka et al., 2000; Provino et al., 2001; Wadsworth et al., 2004; Champert et al., 2004). Microstructured fibers can also have much larger mode cross sections and result in higher-power white-light systems (Champert et al., 2002).

1.3.3.2   Gain-Switched Lasers

Gain-switched lasers with spectrally broad gain bandwidths offer an alternative to optical parametric devices as tunable short-pulse sources. As the pump-pulse durations get shorter, the thresholds of optical parametric oscillators (OPOs) increase because of the cavity build-up time. This is a result of the fact that the gain of the system is only present during the pump pulse. Because the cavity build-up time increases with cavity length, there is a large penalty (in threshold) for including passive elements within the OPO cavity (Byer, 1975; Zayhowski, 1997b). Gain-switched lasers do not suffer the same penalty, because energy (hence gain) is stored in the gain medium until it is extracted. In tunable systems, or when narrow linewidths are desired, it is often necessary to use intracavity frequency-selective elements. For short-pulse applications where tunability or narrow linewidth is required, gain-switched lasers can have a lower threshold than OPOs and provide more robust operation.

The equations that describe a Q-switched pulse also describe a gain-switched pulse. The difference is that the operation of the laser is controlled by gain rather than loss. Once we know the inversion density at the beginning of the output pulse, all of the formalism developed in Section 1.2 applies. Essentially, given the temporal pump profile, we must calculate the pulse build-up time and integrate the pump rate up to that time to obtain N_0. For a given pump energy, the minimum pulse width is obtained when the inversion density in the gain-switched laser is about three times above threshold at the time the pulse forms. If the goal is to have a single output pulse extract the maximum amount of energy from each pump pulse, the duration of the pump pulse should typically be less than three to ten times the output pulse width. Longer pump pulses can lead to multiple output pulses.

Miniature gain-switched lasers, pumped by passively Q-switched microchip lasers or their harmonics, can be broadly tunable (Ti:sapphire is tunable from 650 to 1100 nm) and are an attractive alternative to parametric devices for many short-pulse applications (Zayhowski et al., 2001). Figure 1.16 shows a schematic of a miniature gain-switched Ti:sapphire laser; a photograph is shown in Figure 1.17. The input mirror of the laser is coated directly onto the Ti:sapphire ($Ti:Al_2O_3$) crystal. To control the bandwidth of the laser output, intracavity frequency-selective elements are used, including a birefringent filter and quartz etalons. The Ti:sapphire laser was packaged with a frequency-doubled high-power Nd:YAG/Cr^{4+}:YAG passively Q-switched microchip-laser pump in an

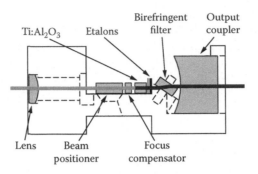

FIGURE 1.16
Schematic of miniature gain-switched Ti:sapphire laser.

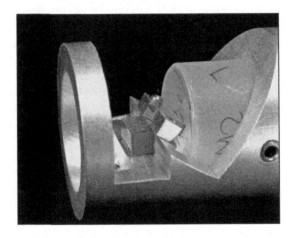

FIGURE 1.17
Photograph of miniature gain-switched Ti:sapphire laser.

optical head that occupies a volume of <0.2 liters, as shown in Figure 1.18 (Zayhowski and Wilson, 2002). It operates at a center wavelength of 780 nm, producing 5.5 μJ in a 1-nsec pulse with a 0.3-nm bandwidth. The amplitude stability is better than 2% at pulse repetition rates from 0 to 10 kHz, and the output beam is TEM_{00}, diffraction limited.

A high-power gain-switched Ti:sapphire laser, based on a microchip oscillator/amplifier, has also been built (Zayhowski and Wilson, 2002), The high-power Ti:sapphire laser generates 18.4 μJ of 780-nm radiation in a 700-ps pulse with a 0.17-nm bandwidth at pulse rates up to 10 kHz, in a diffraction-limited, TEM_{00} output beam. The optical head is similar to that shown in Figure 1.13, with a volume <0.5 liters. This is one of the most complex microchip-laser-based sources built. Yet, it is an environmentally sealed, turnkey system designed to survive temperature excursions from –40° to 71°C, shocks in excess of 20 G, and the vibrations encountered in demanding flight scenarios. This system was built to meet the demands of military use, industrial use, and flight.

FIGURE 1.18
Photograph of passively *Q*-switched microchip laser/miniature gain-switched Ti:sapphire laser system.

Several other miniature Ti:sapphire gain-switched lasers were built and tested. Pulse widths as short as 350 ps were demonstrated, bandwidths as narrow as 0.05 nm were measured, and tuning was demonstrated over the full free-spectral range of the intracavity etalons (Zayhowski et al., 2001).

Miniature gain-switched lasers using Cr^{4+}:YAG as a gain medium were also demonstrated. Pumped with the 1.064-μm output of a passively *Q*-switched microchip laser, they produced 2-μJ pulses at 1.44 μm, with a 4-nsec pulse width (Zayhowski et al., 2001). The broad gain bandwidth of Cr^{4+}:YAG should allow this laser to be operated at wavelengths from 1.3 to 1.6 μm.

It should also be possible to pump Ce-doped-fluoride crystals with the fourth harmonic of a Nd:YAG/Cr^{4+}:YAG passively *Q*-switched microchip laser to obtain broadly tunable miniature gain-switched lasers operating in the near-UV (Marshall et al., 1994).

1.4   Applications

1.4.1   Overview

Passively *Q*-switched microchip lasers, and compact solid-state laser systems based on them, are attractive for a wide range of applications. The short pulses are useful for high-precision ranging using time-of-flight techniques, with applications in three-dimensional imaging, target identification, and robotics. The short pulse durations and ideal mode properties can also be used to advantage in the characterization of materials. The high peak powers of microchip lasers can be used to photoablate materials, leading to applications in laser-induced breakdown spectroscopy, micromachining, and marking. As has been discussed, the high peak powers also enable the construction of extremely compact nonlinear optical systems. The ultraviolet systems, in particular, have already been used to perform UV fluorescence spectroscopy for a variety of applications, including environmental monitoring and the detection of biological particles.

Here, I will only discuss microchip laser applications that we have worked on at MIT Lincoln Laboratory. Three of these applications will be discussed in detail, including: three-dimensional imaging, where the short pulse width is the key feature of the laser; laser-induced breakdown spectroscopy, where the extremely high peak powers are the enabling laser characteristic; and environmental monitoring using a cone penetrometer, where the small size of the UV head and the ability to fiber pump it at IR wavelengths are essential. A brief review of other applications will follow.

1.4.2 Ranging and Imaging

1.4.2.1 Scanning Three-Dimensional Imaging Systems

Time-of-flight optical ranging is an important application for passively Q-switched microchip lasers. The resolution of a time-of-flight ranging system is one half of the speed of light multiplied by the pulse width. A 200-ps optical pulse can provide a range resolution (minimum separation between two resolvable objects) of 3 cm. When the shape of the optical pulse is repeatable, as is the case for the microchip laser, the accuracy of the system can be much better. At MIT Lincoln Laboratory, we developed a compact time-of-flight optical transceiver using a low-power frequency-doubled green microchip laser attenuated to the Class-II eye-safe level of 0.2 μJ per pulse. With it, we were able to range to objects, including black felt, with a single-pulse range accuracy of 1 mm at distances up to 50 m. The system used a commercial 1-GHz detector and 5-cm-diameter collection optics. Coupled to a two-dimensional scanning system, the high repetition rate of the laser made it possible to obtain high-resolution three-dimensional images in minutes. This system, originally called Cyrax™, was developed in collaboration with Cyra Technologies, Inc., which has since been acquired by Leica Geosystems. Leica Geosystems offers a commercial, tripod-mounted, battery-operated version of this system, with software that can quickly convert the captured images to three-dimensional CAD models. Applications include automated production, civil engineering, construction, and architecture.

One of the sites used to beta test Cyrax was the Norfolk Naval Shipyard, where the U.S. Navy was interested in creating a virtual model of an aircraft carrier (Zayhowski, 1999). The ship used in the beta tests was the USS Tarawa. The image in the left of Figure 1.19 shows the Cyrax imaging system set up on the deck of the ship. From that position, it collected a three-dimensional cloud of points representing the ship's surfaces. The image of the mast, in the center of Figure 1.19, was captured by Cyrax at a distance of 60 m, and the two-dimensional drawing shown in the right side of Figure 1.19 was generated from it. Measurements taken from this drawing are accurate to within 6 mm.

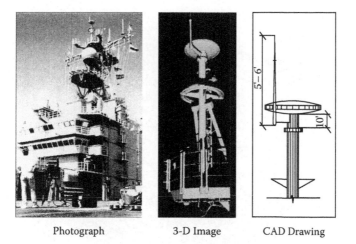

Photograph   3-D Image   CAD Drawing

FIGURE 1.19
Photograph of Cyrax imaging system set up on the deck of the USS Tarawa (left); (center) three-dimensional image of ship's mast captured by Cyrax at a distance of 60 m; (right) two-dimensional CAD drawing generated from three-dimensional image, accurate to 6 mm.

The system just described uses one of the least powerful of the Nd:YAG/Cr^{4+}:YAG microchip lasers discussed in this chapter and attenuates its frequency-doubled output. Higher-power devices have the capability of ranging to much greater distances. The unamplified output of high-power passively Q-switched microchip lasers can be used to perform earth-to-satellite ranging with centimeter accuracy (Degnan and McGarry, 1997; Degnan and Zayhowski, 1999). The output of microchip-laser-pumped OPAs, OPOs, and OPGs; Yb,Er passively Q-switched lasers; or Cr^{4+} gain-switched lasers can be used to perform ranging at eye-safe wavelengths.

1.4.2.2   Flash Three-Dimensional Imaging Systems

What if in addition to having more transmitted power we also had the ability to detect single photons? The impulse response of the detector used in Cyrax had a noise-equivalent energy of ~50 photons. What if we had two-dimensional arrays of detectors with single-photon sensitivity? At MIT Lincoln Laboratory, we have been working on such systems, and have recently demonstrated the world's first flash three-dimensional ladar (laser detection and ranging) system (Heinrichs et al., 2001; Albota et al., 2002a; Albota et al., 2002b; Aull et al., 2004). The detector in the system, built at MIT Lincoln Laboratory, is a 32 × 32 array of Geiger-mode avalanche photodiodes (APDs) bonded to a commensurate array of CMOS timing circuits with 500-ps accuracy (Aull et al., 2004). The hardware is shown in Figure 1.20. An outgoing light pulse starts the clocks for all of the timing circuits. A single return photon incident on one of the APDs stops the clock for that element. After a predetermined amount of time, the timing information is read out digitally from the full

FIGURE 1.20
Prototype flash ladar system.

array and used to create a three-dimensional image from a single laser pulse. Spectral filtering is used to reduce the effects of background light; range gating is used to reduce the effects of electrical noise, or dark counts.

With some of the flash ladar systems we built, the object being imaged is flood illuminated, and the return light is concentrated on the active portions of the APD array using an array of microlenses. In other systems, the transmitted light passes through a holographic element to produce a 32×32 array of spots at the object. The spot array is imaged onto the active portions of the detector array, so that background light outside of the illuminated spots is not seen.

Early trials of the flash ladar system used strong laser illumination, resulting in greater than one return photon per detector element. They demonstrated the ability to create a three-dimensional image from a single laser pulse and proved that a system with single-photon sensitivity could operate in full sunlight.

In some cases it is preferable to operate the Geiger-mode imager with an average return of less than one photon per pixel per laser pulse. Under such conditions, it is possible to see through semitransparent screens, foliage, or camouflage netting. The probability of the return photon coming from the "hidden" object depends on the transmission of the obscurant. A test of the system operating under these conditions was performed by looking down at a heavily wooded area from a tower at Redstone Arsenal. The ladar system was mounted at three different positions on the tower, and the three-dimensional images from the three locations were registered to form a single image. For each position of the ladar system, a small number of transmitted photons passed though gaps in the foliage and struck the objects below. Each of the three vantage points provided different gaps, and different portions of the hidden objects were illuminated. As a result, it was possible to image picnic tables, a gazebo, a jeep, a P.T. Cruiser, and a pile of sticks, despite heavy obscuration by the foliage, as shown in Figure 1.21. (A single two-dimensional black and white image does not do the three-dimensional image justice.) More recently, the flash ladar system has been made rugged and mounted on a helicopter. Field tests using the helicopter resulted in even more impressive images, because there were many more vantage points and more opportunities to penetrate the foliage at different angles.

1.4.3 Laser-Induced Breakdown Spectroscopy

Laser-induced breakdown spectroscopy (LIBS) is an example of a less developed, but very promising, application of the passively Q-switched microchip laser. Although the per pulse energy is relatively low compared to lasers conventionally used in this application, the diffraction-limited beam of the passively Q-switched microchip laser can be focused to a spot size as small as 1 μm in diameter. Even the low-power microchip lasers can be focused to intensities in excess of 1 TW/cm^2. This is sufficient to break down metals and many other solids (Zayhowski, 1996a; Zayhowski and Johnson, 1996; Bloch et al., 1998). In the resulting plasma, there are highly excited ions, atoms, and molecules; each emits a unique spectrum as it recombines. By examining the recombination spectra, it is possible to determine the elemental composition of the material. We have demonstrated the detection of various metals in soil using a low-power microchip laser and a compact diode-array-based spectrometer (Bloch et al., 1998). Because the optical pulse width is short and the resulting plasma volume is small, the plasma continuum radiation decays rapidly (in ~15 nsec). This has allowed us to measure concentrations of metals as low as 100 ppm without any of the standard temporal or spatial gating that conventional LIBS systems employ (Radziemski and Cremers, 1989; Theriault and Lieberman, 1995; Castle et al., 1997). Potential applications include the identification of heavy-metal contaminants such as Pb, Hg, Cd, Cr, and Zn (Bell, 1994).

As the power of the microchip laser increases, so too does the sensitivity of the resulting LIBS system and the variety of materials that can be examined. Midpower devices, pumped with 3-W diode-laser arrays, easily break down transparent media, including glasses and water; the focused output of the highest-power devices can break down clean air, as shown in Figure 1.22, with potential applications in the monitoring of effluents and closed-loop process control.

1.4.4 Environmental Monitoring

Laser-based techniques are highly sensitive methods for determining concentrations of chemical species, including pollutants. For many applications,

FIGURE 1.21
View of foliage as seen by ladar system mounted on tower at Redstone Arsenal (top); (middle) photograph of objects obscured by foliage; (bottom) three-dimensional image of objects obtained through foliage.

FIGURE 1.22
Focused output of high-power passively Q-switched microchip laser creating breakdown in clean air.

the optimal measurement wavelengths lie in the UV. In recent years, remote detection has been performed with UV light delivered to the remote area with an optical fiber. Unfortunately, optical fibers transmit UV poorly. Thus, powerful lasers are required to provide sufficient energy at the fiber's distal end to ensure adequate detection sensitivity. In addition, sensitivity is critically dependent on the fiber length. These limitations can be overcome by using a multimode fiber to deliver easily transmitted near-IR diode-laser pump radiation to a remote head containing a UV frequency-converted passively Q-switched microchip laser (Zayhowski, 1996a; Zayhowski and Johnson, 1996; Bloch et al., 1998).

At MIT Lincoln Laboratory, we constructed a sensor head that is 2.5-cm diameter × 7-cm long for use in a cone penetrometer (Lieberman et al., 1991; Chudyk et al., 1989; Schade and Bublitz, 1996; Bujewski and Rutherford, 1997) to characterize subsurface soil contamination at depths up to 50 m. The sensor head contains a frequency-quadrupled low-power passively Q-switched Nd:YAG/Cr⁴⁺:YAG microchip laser and collection optics, as shown in Figure 1.23. The laser output is filtered to remove the IR and visible light before the UV light is focused outside through a sapphire window. Fluorescence from material contacting the window is collected and focused into a 500-μm-core return fiber for spectral analysis. The short duration of the excitation pulse facilitates accurate measurements of the decay times of even the short-lived benzene, toluene, ethylbenzene, and xylene (BTEX) compounds (decay times from 2 to 60 nsec). Measurements of fluorescence decay times offer greater chemical selectivity than that of spectra alone (St. Germain and Gillispie, 1992). In field tests, from spectra including those shown in Figure 1.24, we identified BTEX compounds as well as heavier aromatic hydrocarbons that resulted from contamination due to aviation and heating fuels (Bloch et al., 1998). These tests demonstrated that the microchip-laser-based probe offers the potential for in situ, real-time characterization of soils and groundwater in a robust, compact, inexpensive package.

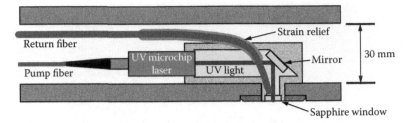

FIGURE 1.23
Cross section of cone penetrometer containing a sensor comprising a frequency-quadrupled passively *Q*-switched microchip laser and collection optics. When deployed, a cone-shaped end is attached to one end of this section and several additional lengths of pipe are added to the other.

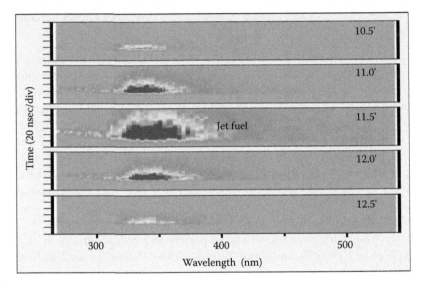

FIGURE 1.24
Wavelength-time spectra obtained with cone penetrometer at depths of 10.5, 11.0, 11.5, 12.0, and 12.5 ft. The intensity of the fluorescence signal, indicated in false color/intensity, is proportional to the concentration of the contaminant; the spectral and temporal details identify the contaminant.

For compounds that do not fluoresce appreciably, such as chlorinated solvents, we have measured Raman spectra, using the time domain to distinguish the Raman lines from fluorescence due to interferents in the same spectral region (Bloch et al., 1998). UV excitation spectroscopy, using the frequency-quintupled output of a 1.074-µm passively *Q*-switched microchip laser, has been used to detect NO_x, for the detection of energetic contaminants such as the explosives TNT, RDX, and HMX (Wormhoudt et al., 2000a; Wormhoudt et al., 2000b; Zayhowski et al., 2000).

A related application of the passively *Q*-switched microchip laser is the detection of biological warfare agents. In this application, UV laser-induced fluorescence is used for the detection and classification of airborne biological

FIGURE 1.25
Prototype biological agent warning sensor.

particles (Jeys, 1998; Primmerman, 2000). The small size of the UV laser has allowed us to reduce the size of the sensor system to make it man-portable. The turnkey nature of the laser, and its robustness, make it possible to put these sensors in the hands of combat soldiers. The relatively low cost makes it practical to distribute enough of them to do some good. Systems like the one shown in Figure 1.25 have already seen use in several parts of the world.

1.4.5   Other Applications

Passively Q-switched microchip lasers are attractive devices for a wide range of applications, beyond those discussed in detail above. It should be apparent from the discussion of laser-induced breakdown spectroscopy that the Q-switched output of a microchip laser can photoablate most materials, including metals, semiconductors, glasses, and biological tissues. We have used a low-power Nd:YAG/Cr^{4+}:YAG passively Q-switched microchip laser at both 1.064 µm and 532 nm to cut clean 5-µm-wide lines in the metallization on semiconductor wafers and to drill holes through the substrate. Higher-power devices have been used to bore holes in glass, scribe alumina, etc. Applications ranging from marking and labeling to microsurgery are being pursued.

Active Impulse Systems, since acquired by Philips Analytical, developed the Impulse 300™, an instrument designed to measure the mechanical and thermal properties of thin films, around a low-power passively Q-switched microchip laser. In this application, the output of a frequency-doubled microchip laser is split into two beams that are recombined interferometrically on the surface of a thin-film material. Using the technique of impulsive stimulated thermal scattering (Duggal et al., 1992; Banet et al., 1998; Rogers et al., 2000), the Impulse 300 (and the later instruments derived from it) can make nondestructive measurements of thin-film thickness to an accuracy of 5 nm, with a transverse resolution of 10 μm. It can also measure the anisotropic elastic moduli and thermal diffusivity, as well as determine whether or not there is delamination of the film, all at about 1000 measurements per second. In addition to semiconductor manufacturing, potential applications of this technology include checking for defects in painted or laminated surfaces, and monitoring the curing of epoxies and resins.

SPARTA, Inc. uses the green and UV harmonics from a single low-power passively Q-switched microchip laser in their second-generation nanAlign™ interferometer system to remove measurement error that results from single-wavelength interferometric measurements over a turbulent air path. By measuring the amount of spectral dispersion in the path, this system improves the ability to position the stage used to hold semiconductor wafers during the lithography process.

High-power passively Q-switched microchip lasers, with their ability to break down air when focused, have potential applications in combustion diagnostics, effluent monitoring, and closed-loop process control. In addition to fluorescence spectroscopy, the harmonically converted UV microchip lasers can be used for flow cytometry, photo-ionization spectroscopy (Kunz et al., 1999), matrix-assisted laser desorption and ionization (MALDI), and stereolithography. Microchip-laser-pumped optical parametric devices and tunable gain-switched lasers are robust, compact tools for IR spectroscopy. The broad spectrum generated by Raman scattering and four-wave mixing in fibers has applications in time-resolved absorption, reflection, and excitation spectroscopy; active three-dimensional hyperspectral imaging (Johnson et al., 1999); and white-light interferometry (optical coherence tomography). The list goes on, and new applications continue to emerge as this technology becomes more readily available.

1.5   Conclusion

Passively Q-switched microchip lasers are a new and important family of high-performance devices. These lasers, and the systems based on them, are small, efficient, robust, and low cost. They have the proven potential to take what once were complicated laser-based experiments out of the laboratory and into the field, enabling applications in diverse areas. The applications discussed above account for only a few of the many passively Q-switched microchip laser programs we have had at MIT Lincoln Laboratory since we started working on the technology a decade ago; a much wider range of applications is being developed worldwide.

This technology is still in its infancy; passively Q-switched microchip lasers are just now becoming commercially available. Further work can be expected to lead to shorter pulses, higher peak powers, increased pulse energies, and new wavelengths of operation. With advances in the technology, and increased availability, the applications of passively Q-switched microchip lasers will continue to expand.

Acknowledgments

Large portions of this chapter were taken from J.J. Zayhowski, D. Welford, and J. Harrison, "Miniature solid-state lasers," *CRC Handbook of Photonics*, 2nd ed., M.C. Gupta and J. Ballato, Editors-in-Chief (CRC Press, Boca Raton, FL), to be published; its predecessor, J.J. Zayhowski and J. Harrison, "Miniature solid-state lasers," *CRC Handbook of Photonics*, M.C. Gupta, Editor-in-Chief (CRC Press, Boca Raton, FL, 1997), pp. 326–392; and J.J. Zayhowski, "Compact solid-state sources and their applications," *Proceedings of SPIE* **5620**, *Solid State Laser Technologies and Femtosecond Phenomena*, J.A.C. Terry and W.A. Clarkson, Eds. (SPIE Press, Bellingham, WA, 2004), pp. 155–169.

This work was sponsored by the Department of the Air Force under Air Force Contract F19628-00-C-0002. Opinions, interpretations, conclusions, and recommendations are those of the author and are not necessarily endorsed by the United States Government.

References

Abshire, J., Ketchum, E., Afzal, R., Millar, P., and Sun, X., 2000, The geoscience laser altimeter system (GLAS) for the ICEsat mission, *Conf. Lasers and Electro-Optics 2000 Tech. Dig.*, 602.

Afzal, R.S., Yu, A.W., Zayhowski, J.J., and Fan, T.Y., 1997, Single-mode, high-peak-power passively Q-switched diode-pumped Nd:YAG laser, *Opt. Lett.* **22**, 1314.

Aggarwal, R.L., Ripin, D.J., Ochoa, J.R., and Fan, T.Y., 2005, Measurement of thermo-optic properties of $Y_3Al_5O_{12}$, $Lu_3Al_5O_{12}$, $YAlO_3$, $LiYF_4$, $LiLuF_4$, BaY_2F_8, $KGd(WO_4)_2$, and $KY(WO_4)_2$ laser crystals in the 80–300 K temperature range, *J. Appl. Phys.* **98**, 103514.

Agrawal, G.P., 1995, *Nonlinear Fiber Optics*, 2nd ed., Academic Press, San Diego, CA, chap. 8.

Albota, M.A., Heinrichs, R.M., Kocher, D.G., Fouche, D.G., Player, B.E., O'Brien, M.E., Aull, B.F., Zayhowski, J.J., Mooney, J., Willard, B.C., and Carlson, R.R., 2002a, Three-dimensional imaging laser radar with a photon-counting avalanche photodiode array and microchip laser, *Appl. Opt.* **41**, 7671.

Albota, M.A., Aull, B.F., Fouche, D.G., Heinrichs, R.M., Kocher, D.G., Marino, R.M., Mooney, J., O'Brien, M., Player, B.E., Willard, B.C., and Zayhowski, J.J., 2002b, Three-dimensional imaging laser radar using Geiger-mode avalanche photodiode arrays and short-pulse microchip lasers, *Lincoln Lab. J.* **13**, 351.

Andrauskas, D.M. and Kennedy, C., 1991, Tetravalent chromium solid-state passive Q switch for Nd:YAG laser systems, *OSA TOPS* **10**, *Advanced Solid-State Lasers*, Dubé, G. and Chase, L., Eds., Optical Society of America, Washington, D.C., 393.

Arvidsson, M., 2001, Far-field timing effects with passively Q-switched lasers, *Opt. Lett.* **26**, 196.

Arvidsson, M., Hansson, B., Holmgren, M., and Lindstrom, C., 1998, A combined actively and passively Q-switched microchip laser, *SPIE* **3265**, 106.

Aull, B.F., Loomis, A.H., Young, D.J., Stern, A., Felton, B.J., Daniels, P.J., Landers, D.J., Retherford, L., Rathman, D.D., Heinrichs, R.M., Marino, R.M., Fouche, D.G., Albota, M.A., Hatch, R.E., Rowe, G.S., Kocher, D.G., Mooney, J.G., O'Brien, M.E., Player, B.E., Willard, B.C., Liau, Z.L., and Zayhowski, J.J., 2004, Three-dimensional imaging with arrays of Geiger-mode avalanche photodiodes, *SPIE* **5353**, 105.

Bai, Y., Wu, N., Zhang, J., Li, J., Li, S., Xu, J., and Deng, P., 1997, Passively Q-switched Nd: YVO_4 laser with a Cr^{4+}:YAG crystal saturable absorber, *Appl. Opt.* **36**, 2468.

Banet, M.J., Fuchs, M., Rogers, J.A., Reinold, J.H., Knecht, J.M., Rothschild, M., Logan, R., Maznev, A.A., and Nelson, K.A., 1998, High-precision film thickness determination using a laser-based ultrasonic technique, *Appl. Phys. Lett.* **73**, 169.

Bell, T.E., October 1994, A pyrotechnical test for smokestack pollution, *IEEE Spectrum*, 14.

Bibeau, C., Payen, S.A., and Powell, H.T., 1993, Evaluation of the terminal level lifetime in sixteen Neodymium-doped crystals and glasses, *OSA Proc. Adv. Solid-State Lasers* **15**, Pinto, A.A. and Fan, T.Y., Eds., Optical Society of America, Washington, D.C., p. 74.

Bilinsky, I.P., Fujimoto, J.G., Walpole, J.N., and Missaggia, L.J., 1998, Semiconductor-doped-silica saturable-absorber films for solid-state laser mode locking, *Opt. Lett.* **23**, 1766.

Bloch, J., Johnson, B., Newbury, N., Germaine, J., Hemond, H., and Sinfield, J., 1998, Field test of a novel microlaser-based probe for *in situ* fluorescence sensing of soil contamination, *Appl. Spectrosc.* **52**, 1299.

Braud, A., Girard, S., Doualan, J.L., and Moncorge, R., 2000, Wavelength tunability and passive Q-switching of a (Yb,Tm):YLF laser operating around 1.5 μm, *Conf. Lasers and Electro-Optics, Tech. Dig.*, 463.

Braun, B., Kärtner, F.X., Keller, U., Meyn, J.-P., and Huber, G., 1996, Passively Q-switched 180-psec Nd:LaSc$_3$(BO$_3$)$_4$ microchip laser, *Opt. Lett.* **24**, 405.

Braun, B., Kärtner, F.X., Zhang, G., Moser, M., and Keller, U., 1997, 56-psec passively Q-switched diode-pumped microchip laser, *Opt. Lett.* **22**, 381.

Bujewski, G. and Rutherford, B., 1997, The rapid optical screening tool (ROST) laser-induced fluorescence (LIF) system for screening of petroleum hydrocarbons in subsurface soils, USA EPA Report # EPA/600/R-97/020I, U.S. Environmental Protection Agency, Washington, D.C.

Burshtein, Z., Blau, P., Kalisky, Y., Shimony, Y., and Kokta, M.R., 1998, Excited-state absorption studies of Cr^{4+} ions in several garnet host crystals, *IEEE J. Quantum Electron.* **34**, 292.

Buzelis, R., Dement'ev, A.S., Kosenko, E.K., and Murauskas, E., 1995, Stimulated-Brillouin-scattering compression of pulses from an Nd:YAG laser with a short cavity and measurement of the nonradiative relaxation time of the lower active level, *Quantum Electron.* **25**, 540.

Byer, R.L., 1975, Optical parametric oscillators, *Quantum Electronics: A Treatise*, Rabin, H. and Tang, C.L., Eds., Academic Press, New York, p. 587.

Capmany, J., Bermudez, V., Callejo, D., and Dieguez, E., June 2001, Microchip OPOs operate in the infrared, *Laser Focus World*, p. 143.

Castle, B.C., Visser, K., Smith, B.W., and Winefordner, J.D., 1997, Spatial and temporal dependence of lead emission in laser-induced breakdown spectroscopy, *Appl. Spectrosc.* **51**, 1017.

Champert, P.A., Popov, S.V., and Taylor, J.R., 2002, Generation of multiwatt, broadband continua in holey fibers, *Opt. Lett.* **27**, 122.

Champert, P.A., Couderc, V., Leproux, P., Février, S., Tombelaine, V., Labonté, L., Roy, P., and Froehly, C., 2004, White-light supercontinuum generation in normally dispersive optical fiber using original multi-wavelength pumping system, *Opt. Express* **12**, 4366.

Chen, W., Spariosu, K., Stultz, R., Kuo, Y.K., Birnbaum, M., and Shestakov, A.V., 1993, Cr^{4+}:GSGG saturable absorber Q-switch for the ruby laser, *Opt. Commun.* **104**, 71.

Chen, Y.F. and Tsai, S.W., 2001, Simultaneous Q-switching and mode-locking in a diode-pumped Nd:YVO_4-Cr^{4+}:YAG laser, *IEEE J. Quantum Electron.* **37**, 580.

Chen, Y.F., Lan, Y.P., and Chang, H.L., 2001, Analytical model for design criteria of passively Q-switched lasers, *IEEE J. Quantum Electron.* **37**, 462.

Chen, Y.F., Lee, J.L., Hsieh, H.D., and Tsai, S.W., 2002, Analysis of passively Q-switched lasers with simultaneous mode locking, *IEEE J. Quantum Electron.* **38**, 312.

Chudyk, W., Pohlig, P., Wolf, L., and Fordiani, R., 1989, Field determination of ground water contamination using laser fluorescence and fiber optics, *SPIE* **1172**, 123.

Conroy, R.S., Lake, T., Friel, G.J., Kemp, A.J., and Sinclair, B.D., 1998, Self-Q-switched Nd:YVO_4 microchip laser, *Opt. Lett.* **23**, 457.

Dascalu, T., Pavel, N., Lupei, V., Philipps, G., Beck T., and Weber, H., 1996, Investigation of a passive Q-switched, externally controlled, quasicontinuous or continuous pumped Nd:YAG laser, *Opt. Eng.* **35**, 1247.

Degnan, J.J., 1989, Theory of the optimally coupled Q-switched laser, *IEEE J. Quantum Electron.* **25**, 214.

Degnan, J.J., 1993, Millimeter accuracy satellite laser ranging: a review, *Contributions of Space Geodesy to Geodynamics: Technology*, Vol. 25, AGU Geodynamics Series, p. 133.

Degnan, J.J., 1995, Optimization of passively Q-switched lasers, *IEEE J. Quantum Electron.* **31**, 1890.

Degnan, J.J., 1999, Engineering progress on the fully automated photon-counting SLR2000 satellite laser ranging station, *SPIE* **3865**, 76.

Degnan, J.J. and McGarry, J.F., 1997, SLR2000: eye-safe and autonomous single-photoelectron satellite laser ranging at kilohertz rates, *SPIE* **3218**, 63.

Degnan, J.J. and Zayhowski, J.J., 1999, SLR2000 microlaser performance: theory vs. experiment, *Proceedings of the 11th International Workshop on Laser Ranging*, Vol. 2, Deggendorf, Germany, September 20–25, 1998, Schlüter, W., Schreiber, U., and Dassing, R., Eds., Verlag des Bundesamtes für Kartographie und Geodäsie, Frankfurt am Main, p. 458.

Degnan, J.J., Coyle, D.B., and Kay, R.B., 1998, Effects of thermalization on Q-switched laser properties, *IEEE J. Quantum Electron.* **34**, 887.

Degnan, J.J., McGarry, J., Zagwodzki, T., Dabney, P., Geiger, J., Chabot, R., Steggerda, C., Marzuok, J., and Chu, A., 2001, Design and performance of an airborne multikilohertz, photon-counting microlaser altimeter, *Int. Arch. Photogramm. Remote Sensing* **XXXIV-3/W4**, 9.

Demchuk, M.I., Mikhailov, V.P., Zhavoronkov, N.I., Kuleshov, N.V., Prokoshin, P.V., Yumashev, K.V., Livshits, M.G., and Minkov, B.I., 1992, Chromium-doped fosterite as a solid-state saturable absorber, *Opt. Lett.* **17**, 929.

Demidovich, A.A., Apanasevich, P.A., Batay, L.E., Grabtchikov, A.S., Kuzmin, A.N., Lisinetskii, V.A., Orlovich, V.A., Kuzmin, O.V., Hait, V.L., Kiefer, W., and Danailov, M.B., 2003, Sub-nanosecond microchip laser with intracavity Raman conversion, *Appl. Phys. B* **76**, 509.

Druon, F., Balembois, F., Georges, P., and Brun, A., 1999, High-repetition-rate 300 ps pulsed ultraviolet source with a passively Q-switched microchip laser and a multipass amplifier, *Opt. Lett.* **24**, 499.

Duggal, A.R., Rogers, J.A., and Nelson, K.A., 1992, Real-time optical characterization of surface acoustic modes of polyimide thin-film coatings, *J. Appl. Phys.* **72**, 2823.

Eilers, H., Hoffman, K.R., Dennis, W.M., Jacobsen, S.M., and Yen, W.M., 1992, Saturation of 1.064 μm absorption in $Cr,Ca:Y_3Al_5O_{12}$ crystals, *Appl. Phys. Lett.* **61**, 2958.

Fan, T.Y., 1988. Effect of finite lower level lifetime on Q-switched lasers, *IEEE J. Quantum Electron.* **24**, 2345.

Fan, T.Y., 1994. Aperture guiding in quasi-three-level lasers, *Opt. Lett.* **19**, 554.

Fan, T.Y. and Sanchez, A., 1990, Pump source requirements for end-pumped lasers, *IEEE J. Quantum Electron.* **26**, 311.

Fluck, R., Braun, B., Gini, E., Melchoir, H., and Keller, U., 1997, Passively Q-switched 1.34 μm $Nd:YVO_4$ microchip laser with semiconductor saturable-absorber mirrors, *Opt. Lett.* **22**, 991.

Fulbert, L., Marty, J., Ferrand, B., and Molva, E., 1995, Passively Q-switched monolithic microchip laser, *Conf. Lasers Electro-Optics, Tech. Dig.* **15**, 176.

Gu, J., Tam, S.C., Lam, Y.L., Chen, Y., Kam, C.H., Tan, W., Xie, W.J., Zhao, G., and Yang, H., 2000, Novel use of GaAs as a passive Q-switch as well as an output coupler for diode-infrared solid state lasers, *SPIE* **3929**, 222.

Hall, D.R. and Jackson, P.E., Eds., 1989, *The Physics and Technology of Laser Resonators*, Adam Hilger, Bristol, U.K.

Hansson, B. and Arvidsson, M., 2000, Q-switched microchip laser with 65 ps timing jitter, *Electron. Lett.* **36**, 1123.

Heinrichs, R.M., Aull, B.F., Marino, R.M., Fouche, D.G., McIntosh, A.K., Zayhowski, J.J., Stephens, T., O'Brien, M.E., and Albota, M.A., 2001, Three-dimensional laser radar with APD arrays, *SPIE*, **4377**, 106.

Il'ichev, N.N., Kir'yanov, A.V., Gulyamova, E.S., and Pashinin, P.P., 1997, An influence of passive shutter Cr^{4+}:YAG latent anisotropy on output energy and polarisation characteristics of neodymium laser at passive Q-switching, *OSA TOPS* **10**, *Advanced Solid-State Lasers*, Pollock, C.R. and Bosenberg, W.R., Eds., Optical Society of America, Washington D.C., p. 137.

Isyanova, Y. and Welford, D., 1993, 2.4-nsec pulse generation in a solid-state, passively Q-switched, laser-diode-pumped Nd:YAG laser, *OSA Proc. Adv. Solid-State Lasers* **15**, Pinto, A.A. and Fan, T.Y., Eds., Optical Society of America, Washington, D.C., p. 20.

Isyanova, Y. and Welford, D., 1999, Temporal criterion for single-frequency operation of passively Q-switched lasers, *Opt. Lett.* **24**, 1035.

Isyanova, Y., Manni, J.G., Welford, D., Jaspers, M., and Russell, J.A., 2001, High-power, passively Q-switched microlaser-power amplifier system, *OSA TOPS* **50**, *Advanced Solid-State Lasers*, Marshall, C., Ed., Optical Society of America, Washington, D.C., p. 186.

Jabczynski, J.K., Kopczynski, K., Mierczyk, Z., Agnesi, A., Guandalini, A., and Reali, G.C., 2001, Application of V^{3+}:YAG crystals for Q-switching and mode-locking of 1.3-μm diode-pumped neodymium lasers, *Opt. Eng.* **40**, 2802.

Jaspan, M.A., Welford, D., Xiao, G., and Bass, M., 2000, Atypical behavior of Cr:YAG passively Q-switched Nd:YVO$_4$ microlasers at high-pumping rates, *Conf. Lasers and Electro-Optics, Tech. Dig.*, 454.

Jaspan, M.A., Welford, D., and Russell, J.A., 2004, Passively Q-switched microlaser performance in the presence of pump-induced bleaching of the saturable absorber, *Appl. Opt.* **43**, 1.

Jeys, T.H., 1998, Bioaerosol fluorescence sensor, *M.I.T. Lincoln Lab., Solid State Res., Q. Tech. Rep.*, 1998:2, ESC-TR-97-136, 1.

Johnson, B., Joseph, R., Nischan, M., Newbury, A., Kerekes, J.P., Barclay, H.T., Willard, B., and Zayhowski, J.J., 1999, A compact, active hyperspectral imaging system for the detection of concealed targets, *SPIE* **3710**, 144.

Kajave, T.T. and Gaeta, A.L., 1996, Q switching of a diode-pumped Nd:YAG laser with GaAs, *Opt. Lett.* **21**, 1244.

Kärtner, F.X., Brovelli, L.R., Kopf, D., Kamp, M., Calasso, I., and Keller, U., 1995, Control of solid state laser dynamics by semiconductor devices, *Opt. Eng.* **34**, 2024.

Keller, U., 1996, Modelocked and Q-switched solid-state lasers using semiconductor saturable absorbers, *Conf. Proc. of 9th Annual Meeting IEEE Lasers and Electro-Optics Society* **1**, 50.

Keller, U., Weingarten, K.J., Kärtner, F.X., Kopf, D., Braun, B., Jung, I.D., Fluck, R., Hönninger, C., Matuschek, N., and Aus der Au, J., 1996, Semiconductor saturable absorber mirrors (SESAMs) for femtosecond to nanosecond pulse generation in solid-state lasers, *IEEE J. Sel. Top. Quantum Electron.* **2**, 435.

Kemp, A.J., Conroy, R.S., Friel, G.J., and Sinclair, B.D., 1999, Guiding effects in Nd:YVO$_4$ microchip lasers operating well above threshold, *IEEE J. Quantum Electron.* **35**, 675.

Kogelnik, H., 1965, On the propagation of Gaussian beams of light through lenslike media including those with a loss or gain variation, *Appl. Opt.* **4**, 1562.

Kuleshov, N.V., Podlipensky, A.V., Yumashev, K.V., Kretschmann, H.M., and Huber, G., 2000, V:YAG saturable absorber as a Q-switch for diode-pumped Nd:YAG-lasers at 1.44 μm and 1.34 μm, *Conf. Lasers and Electro-Optics, Tech. Dig.*, 228.

Kuo, Y.K., Birnbaum, M., and Chen, W., 1994, Ho:YLiF$_4$ saturable absorber Q-switch for the 2 μm Tm,Cr:Y$_3$Al$_5$O$_2$ laser, *IEEE J. Quantum Electron.* **31**, 657.

Kuo, Y.K., Huang, M.F., and Birnbaum, M., 1995, Tunable Cr^{4+}:YSO Q-switched Cr:LiCAF laser, *Appl. Phys. Lett.* **65**, 3060.

Kunz, R.R., Zayhowski, J.J., Becotte-Haigh, P., and McGann, W.J., 1999, Detection of contraband via laser ionization-chemical ionization ion mobility spectroscopy using a 30-kilowatt microchip UV laser, *Proc. 1999 ONDCP Int. Tech. Symp.*, 9-1.

Lagatsky, A.A., Abdolvand, A., and Kuleshov, N.V., 2000, Passive Q switching and self frequency Raman conversion in a diode-pumped Yb:KGd(WO$_4$)$_2$ laser, *Opt. Lett.* **25**, 616.

Lieberman, S.H., Theriault, G.A., Cooper, S.S., Malone, P.G., Olsen, R.S., and Lurk, P.W., 1991, Rapid, subsurface, *in situ* field screening of petroleum hydrocarbon contamination using laser induced fluorescence over optical fibers, *2nd International Symposium on Field Screening Methods for Hazardous Wastes and Toxic Chemicals*, Air and Waste Management Association, Sewickley, PA, p. 57.

Liu, H., Zhou, S.H., and Chen, Y.C., 1998, High-power monolithic unstable-resonator solid-state laser, *Opt. Lett.* **23**, 451.

Liu, J., Shen, D., Tam, S.C., and Lam, T.L., 2001, Modeling pulse shape of Q-switched lasers, *IEEE J. Quantum Electron.* **37**, 888.

Liu, J., Ozygus, B., Yang, S., Erhard, J., Seelig, U., Ding, A., Weber, H., Meng, X., Zhu, L., Qin, L., Du, C., Xu, X., and Shao, Z., 2003, Efficient passive Q-switching operation of a diode-pumped Nd:GdVO$_4$ laser with a Cr^{4+}:YAG saturable absorber, *J. Opt. Soc. Am. B* **20**, 652.

Lucas-Leclin, G., Augé, F., Auzanneau, S.C., Balembois, F., Georges, P., Brun, A., Mougel, F., Aka, G., and Vivien, D., 2000, Diode-pumped self-frequency-doubling Nd:GdCa$_4$O(BO$_3$)$_3$ lasers: towards green microchip lasers, *J. Opt. Soc. Am. B* **17**, 1526.

Malyarevich, A.M., Denisov, I.A., Yumashev, K.V., Dymshits, O.S., and Zhilin, A.A., 2001a, Co^{2+}-doped glass ceramic as saturable absorber Q switch for 1.54 μm Er-glass laser, *OSA TOPS* **50**, *Advanced Solid-State Lasers*, Marshall, C., Ed., Optical Society of America, Washington, D.C., p. 241.

Malyarevich, A.M., Savitski, V.G., Prokoshin, P.V., Yumashev, K.V., and Lipovskii, A.A., 2001b, Passive Q-switch operation of PbSe-doped glass at 2.1 μm, *SPIE* **4350**, 32.

Mandeville, W., Dindorf, K.M., and Champigny, N.E., 1996, Characterization of passively Q-switched microchip lasers for laser radar, *SPIE* **2748**, 358.

Marshall, C.D., Speth, J.A., Payne, S.A., Krupke, W.F., Quarles, G.J., Castillo, V., and Chai, B.H.T., 1994, Ultraviolet laser emission properties of Ce^{3+}-doped LiSrAlF$_6$ and LiCaAlF$_6$, *J. Opt. Soc. Am. B* **11**, 2054.

Owyoung, A. and Esherick, P., 1987, Stress-induced tuning of a diode-laser-excited monolithic Nd:YAG laser, *Opt. Lett.* **12**, 999.

Palombo, F.K., Matthews, S., Sheldrake, S., and Kapps, D., 1993, Determination of the effective lower level lifetime for Nd:YLF and Nd:YAG through experimental measurements and computer modeling, *OSA Proc. Adv. Solid-State Lasers* **15**, Pinto, A.A. and Fan, T.Y., Eds., Optical Society of America, Washington, D.C., 78.

Patel, F.D. and Beach, R.J., 2001, New formalism for the analysis of passively Q-switched laser systems, *IEEE J. Quantum Electron.* **37**, 707.

Peterson, P. and Gavrielides, A., 1999, Pulse train characteristics of a passively Q-switched microchip laser, *Opt. Express* **5**, 149.

Primmerman, C.A., 2000, Detection of biological agents, *Lincoln Lab. J.* **12**, 3.

Provino, L., Dudley, J.M., Maillotte, H., Grossard, N., Windeler, R.S., and Eggleton, B.J., 2001, Compact broadband continuum source based on microchip laser-pumped microstructured fiber, *Electron. Lett.* **37**, 558.

Radziemski, R.J. and Cremers, D.A., 1989, *Laser-Induced Plasmas and Applications*, Marcel Dekker, New York.

Ranka, J.K., Windeler, R.S., and Stentz, A.J., 2000, Visible continuum generation in air-silica microstructure optical fibers with anomalous dispersion at 800 nm, *Opt. Lett.* **25**, 25.

Rogers, J.A., Maznev, A.A., Banet, M.J., and Nelson, K.A., 2000, Optical generation and characterization of acoustic waves in thin films: fundamentals and applications, *Annu. Rev. Mater. Sci.* **30**, 117.

Schade, W. and Bublitz, J., 1996, On-site laser probe for the detection of petroleum products in water and soil, *Environ. Sci. Technol.* **30**, 1451.

Sennaroglu, A. and Yilmaz, M.B., 1997. Experimental and theoretical study of thermal loading in chromium-doped YAG saturable absorbers, *OSA TOPS* **10**, *Advanced Solid-State Lasers*, Pollock, C.R. and Bosenberg, W.R., Eds., Optical Society of America, Washington, D.C., p. 132.

Shimony, Y., Burshtein, Z., Baranga, B.A., Kalisky, Y., and Strauss, M., 1996, Repetitive Q-switching of a cw Nd:YAG laser using Cr^{4+}:YAG saturable absorbers, *IEEE J. Quantum Electron.* **32**, 305.

Siegman, A.J., 1986, *Lasers*, University Science Books, Mill Valley, CA.

Sooy, W.R., 1965, The natural selection of modes in a passive Q-switched laser, *Appl. Phys. Lett.* **7**, 66.

Spühler, G.J., Paschotta, R., Fluck, R., Braun, B., Moser, M., Zhang, G., Gini, E., and Keller, U., 1999, Experimentally confirmed design guidelines for passively Q-switched microchip lasers using semiconductor saturable absorbers, *J. Opt. Soc. Am. B* **16**, 376.

St. Germain, R.W. and Gillispie, G.D., 1992, *In-situ* tunable laser fluorescence analysis of hydrocarbons, *SPIE* **1637**, 159.

Stultz, R.D., Camargo, M.B., and Birnbaum, M., 1995, Passive Q-switching at 1.53 μm using divalent uranium ions in calcium fluoride, *J. Appl. Phys.* **78**, 2959.

Stone, D.H., 1992, Effects of axial nonuniformity in modeling Q-switched lasers, *IEEE J. Quantum Electron.* **28**, 1970.

Szabo, A. and Stein, R.A., 1965, Theory of laser giant pulsing by saturable absorber, *J. Appl. Phys.* **36**, 1562.

Theriault, G.A. and Lieberman, S.H., 1995, Remote in-situ detection of heavy metal contamination in soils using a fiber optic laser-induced breakdown spectroscopy (FOLIBS) system, *SPIE* **2504**, 75.

Thony, P., Fulbert, L., Besesty, P., and Ferrand, B., 1999, Laser radar using a 1.55-μm passively Q-switched microchip laser, *SPIE* **3707**, 616.

Tsou, Y., Garmire, E., Chen, W., Birnbaum, M., and Asthana, R., 1993, Passive Q switching of Nd:YAG lasers by use of bulk semiconductors, *Opt. Lett.* **18**, 1514.

Wadsworth, W.J., Joly, N., Knight, J.C., Birks, T.A., Biancalana, F., and Russell, P.St.J., 2004, Supercontinuum and four-wave mixing with Q-switched pulses in endlessly single-mode photonic crystal fibers, *Opt. Express* **12**, 299.

Wang, P., Zhou, S.-H., Lee, K.K., and Chen, Y.C., 1995, Picosecond laser pulse generation in a monolithic self-Q-switched solid-state laser, *Opt. Commun.* **114**, 439.

Welford, D., 2001, Passively Q-switched lasers: Short pulse duration, single frequency sources, LEOS Annual Meeting, San Diego, CA, Paper MP1.

Wormhoudt, J., Shorter, J.H., and Zayhowski, J.J., 2000a, Diode-pumped 214.8-nm Nd:YAG/Cr^{4+}:YAG microchip-laser system for the detection of NO, *OSA TOPS* **36**, *Laser Applications to Chemical and Environmental Analysis*, Optical Society of America, Washington, D.C., p. 33.

Wormhoudt, J., Shorter, J. H., Cook, C.C., and Zayhowski, J.J., 2000b, Diode-pumped 214.8-nm Nd:YAG/Cr^{4+}: YAG microchip laser system for the detection of NO, *Appl. Opt.* **39**, 4418.

Wu, R., Myers, J.D., Myers, M.J., Denker, B.I., Galagan, B.I., Sverchkov, S.E., Hutchinson, J.A., and Trussel, W., 2000, Co^{2+}:$MgAl_2O_4$ crystal passive Q-switch performance at 1.34, 1.44, and 1.54 micron, *SPIE* **3929**, 42.

Xiao, G. and Bass, M., 1997, A generalized model for passively Q-switched lasers including excited state absorption in the saturable absorber, *IEEE J. Quantum Electron.* **33**, 41.

Xiao, G. and Bass, M., 1998. Additional experimental confirmation of the predictions of a model to optimize passively Q-switched lasers, *IEEE J. Quantum Electron.* **34**, 1142.

Yao, G., Lee, K.K., Chen, Y.C., and Zhou, S., 1994, Characteristics of transverse mode of diode-pumped self-Q-switched microchip laser, *OSA Proc. Adv. Solid-State Lasers* **20**, Fan, T.Y. and Chai, B.H.T., Eds., Optical Society of America, Washington, D.C., p. 28.

Yumashev, K.V., Psonov, N.N., Denisov, I.A., Mikhailov, V.P., and Moncorge, R., 1998, Nonlinear spectroscopy and passive Q-switching operation of Cr^{4+}-doped $SrGd_4(SiO_4)_3O$ and $CaGd_4(SiO_4)_3O$ crystals, *J. Opt. Soc. Am. B* **15**, 1707.

Yumashev, K.V., Denisov, I.A., Posnov, N.N., Mikhailov, V.P., Moncorge, R., Vivien, D., Ferrand, B., and Guyot, Y., 1999, Nonlinear spectroscopy and passive Q-switching operation of a $Co^{2+}:LaMgAl_{11}O_{19}$ crystal, *J. Opt. Soc. Am. B* **16**, 2189.

Yumashev, K.V., Prokoshin, P.V., Zolotovskaya, S.A., Gurin, V.S., Prokopenko, V.B., and Alexeenko, A.A., 2001, Copper selenide-doped glass saturable absorbers for solid-state lasers of 1.0-1.5 μm region, *OSA TOPS* **50**, *Advanced Solid-State Lasers*, Marshall, C., Ed., Optical Society of America, Washington, D.C., p. 77.

Zayhowski, J.J., 1990a, Limits imposed by spatial hole burning on the single-mode operation of standing-wave laser cavities, *Opt. Lett.* **15**, 431.

Zayhowski, J.J., 1990b, The effects of spatial hole burning and energy diffusion on the single-mode operation of standing-wave lasers, *IEEE J. Quantum Electron.* **26**, 2052.

Zayhowski, J.J., 1991a, Thermal guiding in microchip lasers, *OSA Proc. Adv. Solid-State Lasers* **6**, Jenssen, H.P. and Dubé, G., Eds., Optical Society of America, Washington, D.C., p. 9.

Zayhowski, J.J., 1991b, Polarization-switchable microchip lasers, *Appl. Phys. Lett.* **58**, 2746.

Zayhowski, J.J., April 1996a, Microchip lasers create light in small places, *Laser focus world*, Penwell Publishing Co., Tulsa, Oklahoma, p. 73.

Zayhowski, J.J., 1996b, Ultraviolet generation with passively Q-switched microchip lasers, *Opt. Lett.* **21**, 588; Ultraviolet generation with passively Q-switched microchip lasers: errata, *Opt. Lett.* **21**, 1618.

Zayhowski, J.J., 1997a, Periodically poled lithium niobate optical parametric amplifiers pumped by high-power passively Q-switched microchip lasers, *Opt. Lett.* **22**, 169.

Zayhowski, J.J., 1997b, Microchip optical parametric oscillators, *IEEE Photon. Technol. Lett.* **9**, 925.

Zayhowski, J.J., 1997c, Covering the spectrum with passively Q-switched picosecond microchip laser systems, *Conf. Lasers Electro-Optics, Tech. Dig.* **11**, 463.

Zayhowski, J.J., 1998, Passively Q-switched microchip lasers and applications, *Rev. Laser Eng.* **26**, 841.

Zayhowski, J.J., August 1999, Q-switched microchip lasers find real-world application, *Laser Focus World,* Penwell Publishing Co., Tulsa, Oklahoma, p. 129.

Zayhowski, J.J., 2000, Passively Q-switched Nd:YAG microchip lasers and applications, *J. Alloys Compd.* **303–304**, 393.

Zayhowski, J.J., 2004, Compact solid-state sources and their applications, *SPIE* **5620**, 155.

Zayhowski, J.J. and Dill, C., III, 1992, Diode-pumped microchip lasers electro-optically Q switched at high pulse repetition rates, *Opt. Lett.* **17**, 1201.

Zayhowski, J.J. and Dill, C., III, 1994, Diode-pumped passively Q-switched picosecond microchip lasers, *Opt. Lett.* **19**, 1427.

Zayhowski, J.J. and Dill, C., III, 1995, Coupled-cavity electro-optically Q-switched Nd:YVO$_4$ microchip lasers, *Opt. Lett.* **20**, 716.

Zayhowski, J.J. and Harrison, J., 1996, Miniature solid-state lasers, *CRC Handbook of Photonics*, Gupta, M.C., Ed., CRC Press, Boca Raton, FL, chap. 8.

Zayhowski, J.J. and Johnson, B., 1996, Passively Q-switched microchip lasers for environmental monitoring, *Laser Applications to Chemical, Biological and Environmental Analysis, Tech. Dig. Series* **3**, Optical Society of America, Washington, D.C., p. 37.

Zayhowski, J.J. and Kelley, P.L., 1991, Optimization of Q-switched lasers, *IEEE J. Quantum Electron.* **27**, 2220; Corrections to optimization of Q-switched lasers, *IEEE J. Quantum Electron.* **29**, 1239.

Zayhowski, J.J. and Wilson, A.L., 2000, Miniature sources of subnanosecond 1.4 – 4.0-μm pulses with high peak power, *OSA TOPS* **34**, *Advanced Solid-State Lasers*, Keller, U., Injeyan, H., and Marshall, C., Eds., Optical Society of America, Washington, D.C., p. 308.

Zayhowski, J.J. and Wilson, A.L., Jr., 2002, Miniature, pulsed Ti:sapphire laser system, *IEEE J. Quantum Electron.* **38**, 1449.

Zayhowski, J.J. and Wilson, A.L., Jr., 2003a, Pump-induced bleaching of the saturable absorber in short-pulse Nd:YAG/Cr^{4+}:YAG passively Q-switched microchip lasers, *IEEE J. Quantum Electron.* **39**, 1588.

Zayhowski, J.J. and Wilson, A.L., Jr., 2003b, Miniature, high-power 355-nm laser system, *OSA TOPS* **83**, *Advanced Solid-State Photonics*, Zayhowski, J.J., Ed., Optical Society of America, Washington, D.C., p. 357.

Zayhowski, J.J. and Wilson, A.L., Jr., 2004, Energy-scavenging amplifiers for miniature solid-state lasers, *Opt. Lett.* **29**, 1218.

Zayhowski, J.J., Ochoa, J., and Dill, C., III, 1995, UV generation with passively Q-switched picosecond microchip lasers, *Conf. Lasers and Electro-Optics, Tech. Dig.* **15**, 139.

Zayhowski, J.J., Fan, T.Y., Cook, C., and Daneu, J.L., 1996, 946-nm passively Q-switched microlasers, *M.I.T. Lincoln Lab., Solid State Res., Q. Tech. Rep., 1996:3*, ESC-TR-96-096, 5.

Zayhowski, J.J., Dill, C., III, Cook, C., and Daneu, J.L., 1999, Mid- and high-power passively Q-switched microchip lasers, *OSA TOPS* **XXVI**, *Advanced Solid-State Lasers*, Fejer, M.M., Injeyan, H., and Keller, U., Eds., Optical Society of America, Washington, D.C., p. 178.

Zayhowski, J.J., Cook, C.C., Wormhoudt, J., and Shorter, J.H., 2000, Passively Q-switched 214.8-nm Nd:YAG/Cr^{4+}:YAG microchip-laser system for the detection of NO, *OSA TOPS* **34**, *Advanced Solid-State Lasers*, Keller, U., Injeyan, H., and Marshall, C., Eds., Optical Society of America, Washington, D.C., p. 409.

Zayhowski, J.J., Buchter, S.C., and Wilson, A.L., 2001, Miniature gain-switched lasers, *OSA TOPS* **50**, *Advanced Solid-State Lasers*, Marshall, C., Ed., Optical Society of America, Washington, D.C., p. 462.

Zayhowski, J.J., Welford, D., and Harrison, J., to be published, Miniature solid-state lasers, *CRC Handbook of Photonics*, 2nd ed., Gupta, M.C. and Ballato, J., Eds., CRC Press, Boca Raton, FL.

Zhang, X., Zhao, S., Wang, Q., Zhang, Q., Sun, L., and Zhang, S., 1997, Optimization of Cr^{4+}-doped saturable absorber Q-switched lasers, *IEEE J. Quantum Electron.* **33**, 2286.

Zhang, X., Zhao, S., Wang, Q., Ozygus, B., and Weber, H., 2000, Modeling of passively Q-switched lasers, *J. Opt. Soc. Am. B* **17**, 1166.

Zhang, X., Zhao, S., Wang, Q., Zhang, S., Sun, L., Liu, X., Zhang, S., and Chen, H., 2001, Passively Q-switched self-frequency-doubled Nd^{3+}:GdCa$_4$O(BO$_3$)$_3$ laser, *J. Opt. Soc. Am. B* **18**, 770.

Zheng, J., Zhao, S., and Chen, L., 2002, Laser-diode end-pumped passively Q-switched Nd:YVO$_4$ laser with Cr^{4+}:YAG saturable absorber, *Opt. Eng.* **41**, 2271.

Zhou, S., Lee, K.K., Chen, Y.C., and Li, S., 1993, Monolithic self-Q-switched Cr,Nd:YAG laser, *Opt. Lett.* **18**, 511.

Zhou, Y., Thai, Q., Chen, Y.C., and Zhou, S., 2003, Monolithic Q-switched Cr,Yb:YAG laser, *Opt. Commun.* **219**, 365.

List of Symbols

2^*	$\equiv \begin{cases} 1 & \text{for a four-level laser} \\ g_u(f_{u,\text{eff}} + f_{l,\text{eff}}) & \text{for a quasi-three-level laser} \\ 2 & \text{for a nondegenerate three-level laser} \end{cases}$	

$2^\dagger \equiv \begin{cases} 1 & \text{for a traveling-wave laser} \\ 2 & \text{for a standing-wave laser} \end{cases}$

$2^\ddagger$	Number of times the same cross-sectional area of the gain medium is traversed by light during one round trip within the laser cavity
$2_s^\ddagger$	Number of times the same cross-sectional area of the saturable absorber is traversed by light during one round trip within the laser cavity
3^*	Number between 0 and 3, depending on the alignment of the electronic wave functions with the polarization of the optical field
A	Cross-sectional area of the optical field (m^2)
A_g	Cross-sectional area of the oscillating mode in the gain medium (m^2)
A_s	Cross-sectional area of the oscillating mode in the saturable absorber (m^2)
B	Probability per unit time that a photon will interact with a given inverted site (sec^{-1})
B_p	Pump brightness (W m^{-2} st^{-1})
c	Speed of light in vacuum (m sec^{-1})
C	Number between 0 and 1 that characterizes mechanical boundary conditions
E_o	Output pulse energy (J)
$E_{o,\text{max}}$	Maximum obtainable output pulse energy (J)
E_s	Energy of a state (J)
f_g	Thermal occupation probability of states in the ground level
f_l	Thermal occupation probability of states in the lower laser level
$f_{l,\text{eff}}$	Effective value of f_l used in simplified rate equations
f_s	Thermal occupation probability of a state within a manifold
f_u	Thermal occupation probability of states in the upper laser level
$f_{u,\text{eff}}$	Effective value of f_u used in simplified rate equations
g	Gain coefficient (m^{-1})
$g(r,z,\lambda)$	Saturated, spatially and spectrally dependent gain coefficient (m^{-1})
g_g	Degeneracy of states in the ground level
g_l	Degeneracy of states in the lower laser level
g_{rt}	Round-trip gain coefficient at the time a Q-switched pulse begins to develop
g_s	Degeneracy of a state
g_u	Degeneracy of states in the upper laser level
h	Planck's constant (J sec)
I	Circulating optical intensity (W m^{-2})
I_p	Pump intensity in the saturable absorber (W m^{-2})

$I_{p,sat}$	Saturation intensity of the saturable absorber at the pump wavelength (W m^{-2})
k_B	Boltzmann's constant (J K^{-1})
k_c	Thermal conductivity (W m^{-1} K^{-1})
l	Length of the material (m)
l_g	Length of the gain medium (m)
$l_{op,rt}$	Round-trip optical length of the laser cavity (m)
l_s	Length of the saturable absorber (m)
n	Refractive index of the material
n_0	Refractive index in the absence of an optical field
n_2	Nonlinear Kerr index (m^2 W^{-1})
n_g	Refractive index of the gain medium
$\Delta n(r,z,\lambda)$	Spatially and spectrally dependent change in the refractive index seen by an oscillating mode as it moves away from the gain peak
N_0	Effective population inversion immediately before Q switching
N_{eff}	Effective population inversion
N_f	Effective population inversion well after the peak of a Q-switched output pulse
N_g	Population of the ground level
N_l	Population of the lower laser level
N_{lm}	Population of the lower laser manifold
N_m	Population of a manifold
N_p	Effective population inversion at the time of peak output power
N_s	Population of saturable absorber ions
$N_{s,g}$	Population of saturable absorber ions in the ground state
$N_{s,g,0}$	Population of saturable absorber ions in the ground state at the time an optical pulse begins to form
$N_{s,u}$	Population of saturable absorber ions in the upper level
N_t	Population of active ions in the gain medium
N_{th}	Effective population inversion at threshold
$N_{th,1}$	Effective population inversion at threshold for the primary mode of a Q-switched laser in the low-loss state
$N_{th,2}$	Effective population inversion at threshold for the second mode of a Q-switched laser in the low-loss state
N_u	Population of the upper laser level
N_{um}	Population of the upper laser manifold
$\Delta N_{eff}(1,2)$	Change in the effective population inversion between the times when the first and second modes of a Q-switched laser begin to oscillate
$\Delta N_{s,g}(1,2)$	Change in the population of saturable absorber ions in the ground state between the times when the first and second modes of a Q-switched laser begin to oscillate
$\dot{N}_p$	Pump rate (sec^{-1})
$\dot{N}_{u,sp}$	Spontaneous decay rate of the upper-laser-level population (sec^{-1})
$\dot{N}_{u,st}$	Stimulated decay rate of the upper-laser-level population (sec^{-1})
$\dot{N}_{s,g,ab}$	Rate of change of the saturable-absorber ground-state population due to absorption of photons at the oscillating wavelength (sec^{-1})
$\dot{N}_{s,g,d}$	Rate of change of the saturable-absorber ground-state population due to spontaneous decay of the upper level (sec^{-1})
$\dot{N}_{s,g,p}$	Rate of change of the saturable-absorber ground-state population due to absorption of photons at the pump wavelength (sec^{-1})
$\dot{N}_{s,u,p}$	Rate of change of the saturable-absorber upper-level population due to absorption of photons at the pump wavelength (sec^{-1})
P_a	Absorbed pump power (W)
P_i	Absorbed pump power required to obtain inversion (W)

P_{mm}	Probability of multimode operation in a passively Q-switched laser
P_o	Output power (W)
P_p	Pump power (W)
P_{po}	Peak output power (W)
$P_{po,max}$	Maximum obtainable peak output power (W)
P_{th}	Pump power at threshold (W)
q	Number of photons in the laser cavity
$q(1,2)$	Number of photons in the first mode at the time the second mode reaches threshold
q_0	Number of photons in the laser cavity at the time a Q-switched pulse begins to form
q_p	Number of photons in the laser cavity at the peak of the output pulse
$\dot{q}_e$	Photon emission rate (sec^{-1})
$\dot{q}_l$	Photon loss rate (sec^{-1})
$\dot{q}_{s,g,ab}$	Absorption rate of photons by saturable-absorber ions in the ground state (sec^{-1})
$\dot{q}_{s,u,ab}$	Absorption rate of photons by saturable-absorber ions in the upper level (sec^{-1})
$\dot{q}_{sp}$	Spontaneous-emission rate of photons (sec^{-1})
$\dot{q}_{st}$	Stimulated-emission rate of photons (sec^{-1})
Q	"Quality" of the optical cavity
r	Radial coordinate (m)
r_0	Characteristic, thermally defined radius of the oscillating mode (m)
r_m	Radius of the oscillating mode (m)
r_p	Radius of the pump beam (m)
R_s	Ratio of saturable to unsaturable cavity losses
S_p	Pulse shape factor
t	Time (sec)
t_{rt}	Cavity round-trip time (sec)
t_w	Pulse width (full width at half-maximum) (sec)
$t_{w,min}$	Minimum obtainable pulse width (full width at half-maximum) (sec)
T	Temperature (K)
T_o	Transmission of the output coupler of a laser
W_p	Pump rate factor (sec^{-1})
$W_{p,s}$	Pump rate factor for the saturable absorber (sec^{-1})
z	Coordinate along the cavity axis (m)
α	Absorption coefficient of a material (m^{-1})
α_e	Thermal expansion coefficient (K^{-1})
α_r	Ratio of the optical power required to saturate the gain medium to the optical power required to saturate the saturable absorber
γ_o	Output-coupling loss coefficient
γ_{rt}	Round-trip loss coefficient, excluding loss due to a saturable absorber
$\gamma_{rt,p}$	Round-trip parasitic loss coefficient
$\gamma_{rt,s}$	Round-trip saturable loss coefficient
$\gamma_{rt,t}$	Total round-trip loss coefficient, including loss due to the saturable absorber
$\gamma_{rt,t,t<0}$	Total round-trip loss coefficient prior to the onset of lasing
$\gamma_{rt,t,t>0}$	Total round-trip loss coefficient after the laser has Q switched
γ_s	Saturable loss coefficient of a SESAM
$\gamma_{s,p}$	Unsaturable, parasitic loss coefficient of a SESAM
Γ_{rt}	Round-trip loss
$\Delta_{nl,T}$	Fractional thermally induced change in optical length (K^{-1})
η_p	Pump absorption efficiency
κ	Heat-generating efficiency of the pump
λ	Free-space wavelength (measured in vacuum) = c/ν (m)

λ_0	Free-space wavelength of the gain peak (m)
λ_o	Free-space wavelength of the oscillating mode (m)
ν	Optical frequency (sec^{-1})
ν_o	Optical frequency of the oscillating mode (sec^{-1})
ν_p	Optical frequency of the pump (sec^{-1})
$\delta\nu_L$	Lorentzian linewidth (sec^{-1})
$\Delta\nu$	Frequency difference between adjacent longitudinal modes of a cavity (sec^{-1})
ρ_{eff}	Effective inversion density (m^{-3})
ρ_l	Density of sites in the lower laser level (m^{-3})
ρ_m	Mass density (kg m^{-3})
ρ_s	Density of saturable-absorber ions (m^{-3})
ρ_u	Density of sites in the upper laser level (m^{-3})
σ_a	Absorption cross section (m^2)
σ_g	Stimulated-emission cross section (m^2)
$\sigma_{g,1}$	Stimulated-emission cross section of the first mode to lase in a Q-switched laser (m^2)
$\sigma_{g,2}$	Stimulated-emission cross section of the second mode to lase in a Q-switched laser (m^2)
$\sigma_{g,eff}$	Effective stimulated-emission cross section of the gain medium (m^2)
σ_r	Radiative cross section of a transition (m^2)
$\sigma_{s,g}$	Ground-state absorption cross section of the saturable absorber at the oscillating wavelength (m^2)
$\sigma_{s,g,p}$	Ground-state absorption cross section of the saturable absorber at the pump wavelength (m^2)
$\sigma_{s,u}$	Upper-level absorption cross section of the saturable absorber at the oscillating wavelength (m^2)
τ	Spontaneous lifetime of the upper-state population in the gain medium (sec)
τ_c	Cavity lifetime (sec)
τ_{nr}	Nonradiative lifetime of the upper-state population (sec)
τ_p	Temporal separation between pulses (sec)
τ_r	Radiative lifetime of the upper-state population in the gain medium (sec)
τ_s	Radiative lifetime of the upper-state population in the saturable absorber (sec)
ψ_{max}	Maximum angle that can be tolerated between the mirrors of a Fabry–Pérot cavity

2

Yb-Doped Solid-State Lasers and Materials

Bruno Viana, Johan Petit, Romain Gaumé, Philippe Goldner,
Mathieu Jacquemet, Frédéric Druon, Sebastien Chénais,
François Balembois, and Patrick Georges

CONTENTS

2.1   Introduction

Thanks to the development of high power InGaAs diodes emitting around
975 nm [1] and specific pumping schemes [2], ytterbium (Yb)-doped materials have turned out in the early 1990s to be very relevant competitors of
neodymium (Nd)-doped materials. The trivalent Yb^{3+} ion possesses only two
electronic states that are optically accessible: the $^2F_{7/2}$ ground state and the
$^2F_{5/2}$ excited state, separated by approximately 10,000 cm^{-1}. The splitting of
these electronic levels under the action of the crystal field enables laser action
with a quasi-three-level scheme [3,4]. Consequently, the spectroscopic characteristics and laser properties of ytterbium are particularly host-dependent
[5]. The first reason is the sensitivity of the energy levels distribution towards
crystal field. In order to limit thermal population of the terminal laser level
and to tend toward a four-level laser operating scheme, a ground state
splitting as large as possible is desirable [6,7]. It implies that the crystal field
experienced by the Yb^{3+} ions has to be strong. After this first introductory
part, Section 2.2 describes the crystal field effect and the basic spectroscopic
properties of the Yb^{3+} cations in solid state laser hosts.

Owing to the quasi-three-level scheme of ytterbium, high pumping intensities are required to deplete the ground state and prevent reabsorption [8,9].
With commercially available high power diode arrays, the accessible pump
intensities incident onto the crystal now commonly exceed several tens of
kW/cm^2 [10]. In contrast, the minimum pump intensity at pump wavelength
λ_p needed to reach transparency at laser wavelength is commonly in the
range of several kW/cm^2 only [11]. This means that excellent efficiencies can
be obtained in end-pumped geometries (up to 50% diode-to-laser conversion
efficiency has been reported in Yb: $Y_3Al_5O_{12}$ (YAG), together with C.W. output
powers in the kW range). Section 2.3 presents laser parameters, saturation
effect as well as the expected laser efficiency in the case of the Yb^{3+} lasers.

For these lasers, the quantum defect — that is, the fraction of the pump
photons not converted into laser photons — is small, typically less than 10%,
whereas it reaches 30% for Nd^{3+} lasers. If the quantum defect is to be easily
calculated, more attention has to be paid to the quantum efficiency, and we
have focused on the analysis of the quantum efficiency in Section 4. Thermal
lensing measurements have recently been used for the determination of the
quantum defect [11,12]. The comparison between thermal lensing under
lasing and nonlasing conditions is a simple way to infer the radiative quantum efficiency and constitutes an all-optical original alternative to photometric, calorimetric, and spectroscopic-lifetime methods. Several authors have
demonstrated that in Yb:YAG, the quantum efficiency is far below unity
[13,14]. We are presenting in Section 2.4, quantum efficiency results obtained
in several hosts with the help of thermal lensing measurements.

The performance of ytterbium lasers is therefore not fundamentally limited
by the quasi-three-level nature of the laser scheme but more classically by

thermal issues [15–17]. Indeed, the fraction of absorbed power deposited as heat generates a temperature gradient inside the crystal, which in turn causes the crystal to be under stress, and ultimately could lead to fracture. One way to limit thermal load inside the crystal is to obtain a laser oscillation at a very short wavelength in order to reduce the energy difference between the pump and the laser photons and therefore reduce the quantum defect. Laser hosts with broad emission bands represent one possible solution to obtain short laser wavelengths. This point is developed in Section 2.5. Another application of this property is that the materials presenting broad emission spectra are able to generate ultrashort pulses. A review of the state of art of the Yb-doped femtosecond lasers is presented in Section 2.5 [18]. Ultrafast laser technology involving pulses of light with durations of typically a few tens of femtoseconds (10^{-15} sec) has become one of the most active fields in physics.

One different way to considerably decrease the impact of pump heating is to choose a gain medium that could well manage heat inside its structure [19]. In Section 2.6 are presented the structural/thermomechanical relationships leading to materials with high thermomechanical properties.

In order to develop laser materials well adapted to diode pumping, attention has to be paid to the overlap between diode emission and material absorption. Indeed, a broad absorption band can better fit the wavelength thermal shift of the pumping source, which needs to be controlled through an accurate temperature monitoring of the diode. As materials with broad emission bands (and, correspondingly, broad absorption bands) are presented adequately in this chapter, this requirement is fulfilled and is not further developed here.

2.2 Crystal-Field Effects on Yb^{3+} Energy Level Distributions

In the case of ytterbium lasers, the fundamental and terminal laser levels belong to the same manifold. Therefore, the spectroscopic characteristics and laser properties of ytterbium are particularly host-dependent. This arises from the sensitivity of the energy levels distribution on the crystal field. For instance, the $^2F_{7/2}$ manifold overall splitting is only of 455 cm^{-1} in double tungstate [20], whereas it reaches 1190 cm^{-1} in $Sr_5(PO_4)_3F$ [21] (SFAP), which is one of the highest known values. No global survey of crystal field effect in various Yb-doped hosts has been carried out. The reason is twofold. The first explanation is related to the very simple electronic structure of Yb^{3+}: only two manifolds $^2F_{7/2}$ and $^2F_{5/2}$ composed of four and three Stark levels, respectively. With the experimental determination of these 7 energy levels, it is impossible to univocally derive a phenomenological free-ion Hamiltonian and crystal field potential described, according to the site symmetry, by up to 27 parameters. All available crystal field descriptions thus concern hosts that have been studied with several lanthanide doping ions [22,23], the result being extrapolated to ytterbium. The second reason for the lack of crystal field data is the difficulty to derive unambiguously the energy level diagram from the low-temperature absorption and emission spectra. This is mainly due to the strong interaction of Yb^{3+} ions with the lattice vibrations [24]. These effects can give rise to strong vibronic sidebands [25] or to deformation or splitting of electronic lines in the case of resonant coupling, which significantly complicates the low-temperature spectra. Two useful tools have been derived to help in determining energy-level diagrams of ytterbium from low-temperature spectra and are described in the following section.

2.2.1 Effect on the Spin–Orbit Splitting: the "Barycenters Plot"

The separation between the $^2F_{7/2}$ (ground state) and $^2F_{5/2}$ (excited state) manifolds results from the spin–orbit interaction. For an ytterbium ion embedded in a host lattice, two distinct effects could alter this energy separation. The first one would be a modification in the spin–orbit coupling constant ζ. However theory predicts that this variation should be very weak for lanthanide ions. The second possible effect is the J-mixing, which corresponds to the mixing by crystal field of states belonging to the $^2F_{7/2}$ and $^2F_{5/2}$ levels. This second-order interaction should increase the energy separation between the two manifolds. In the case of ytterbium, this energy difference (about 10200 cm^{-1}) is large enough to neglect the J-mixing to a good approximation [26].

Hence, it is reasonable to consider that the energy separation between Yb^{3+} ground and excited states is constant whatever the host, and equal to the spin–orbit splitting measured for the free ion. More precisely, when taking the lowest Stark level as origin of energies and plotting the $^2F_{5/2}$ manifold energy barycenter vs. the $^2F_{7/2}$ energy, the representative points should align on a straight line whose slope is equal to one. In Figure 2.1, we present this plot for several hosts for which reliable data are available. As expected, data obtained from the literature are aligned on the constant energy separation line with an exception concerning the CFAP and SFAP matrices, for which vibronic lines may have been assigned to electronic transitions. This "barycenters plot" is then a useful tool to derive energy level distribution from the low-temperature spectroscopic data and has been extensively used recently [27,28].

2.2.2 Crystal Field Strength and Yb³⁺ Manifolds Splitting

A second useful concept is the crystal field strength, introduced in the early 1980s. A scalar parameter calculated from the crystal field parameters B_{kq} is defined to evaluate the effect of crystal field interaction on a $^{2S+1}L_J$ level, as follows:

$$N_J = \left[\sum_{\substack{k=2,4,6 \\ k \leq 2J}} \sum_{q=-k}^{k} (B_{kq})^2 \frac{4\pi}{2k+1} \right]^{1/2} \tag{2.1}$$

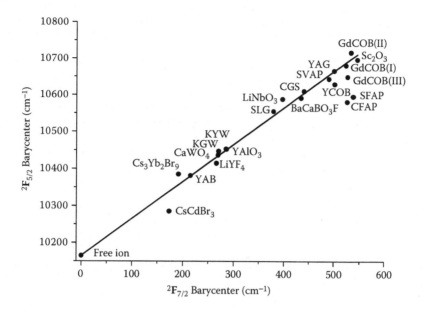

FIGURE 2.1
$^2F_{5/2}$ level barycenter as a function of $^2F_{7/2}$ level barycenter for Yb^{3+} ion.

A major interest in this parameter is the fact that it can be related to the crystal field splitting of the $^{2S+1}L_J$ level through:

$$\Delta E_J = \left(\frac{3g_a^2}{8(g_a+2)(g_a+1)\pi} \right)^{1/2} \left| \prod_{k=2,4,6} \langle 3\|C_k\|3\rangle\langle LSJ\|U_k\|LSJ\rangle \right|^{1/3} N_J \quad (2.2)$$

where g_a is the degeneracy lifted by the crystal field and g is the total degeneracy of the $^{2S+1}L_J$ manifold. Hence, the splitting of some spectroscopic levels and in particular $^2F_{7/2}$ Yb-ground state is proportional to crystal field strength. The reduced matrix elements of tensor operators as described in [29,30], and the tensor operator C_k are presented in the same references using the definition of the spherical harmonics. In the case of the Yb^{3+} $^2F_{7/2}$ ground state, using Equation 2.2, one obtains [31]:

$$\Delta E(^2F_{7/2}) = 0.261\, N(^2F_{7/2}) \quad (2.3)$$

With the knowledge of the lanthanide contraction, it is possible to establish a relationship between the N_J parameter for neodymium and ytterbium ions in a given host [32].

$$N(^2F_{7/2}) = 0.88\, N(^4I_{9/2}) \quad (2.4)$$

where the $^4I_{9/2}$ and $^2F_{7/2}$ are the fundamental energy levels of Yb^{3+} and Nd^{3+}, respectively. Thus, by transitivity of the previous linear relations, it appears that the overall splitting of the ytterbium ground state in a given host is proportional to the one obtained for the Nd^{3+} $^4I_{9/2}$ manifold in the same host:

$$\Delta E(Yb^2F_{7/2}) \approx 1.6\ \Delta E(Nd^4I_{9/2}) \qquad (2.5)$$

In contrast to Yb-activated materials, Nd-doped compounds have been extensively studied [22] and numerous data are reported in the literature. The use of Equation 2.5 could then be an efficient way to detect attractive Yb^{3+} doped hosts as far as splitting of the ground state level is considered.

Figure 2.2 presents the optical spectral — emission ($\beta = 1$) and absorption ($\beta = 0$) features — of $Yb:Ca_5(PO_4)_3F$ (CFAP), Yb:YAG, and $Yb:Y_2SiO_5$ (YSO) hosts showing the crystal field effect. Indeed for these oxides and fluoroapatite hosts, very different behavior is observed. Large crystal field splitting is observed for the Yb:CFAP and for one site of the Yb:YSO (corresponding to the shortest Yb-O distances; see Figure 2.7 for the description of the two ytterbium sites in the YSO structure). The Yb:YAG laser host presents an intermediate situation. The general trend is that high covalency and short Yb-O distances leads to high N_J and, therefore, important energy level splitting. From the analysis of the structure of the laser hosts and the crystal field study, the potential of the new laser could be estimated. This was extensively performed in the last few years, which have seen a significant increase of publications on the presentation and spectroscopic analysis of Yb^{3+} doped laser hosts (see, for instance, [28,5,31]). The next section presents laser modeling in the particular case of the Yb^{3+} energy level scheme.

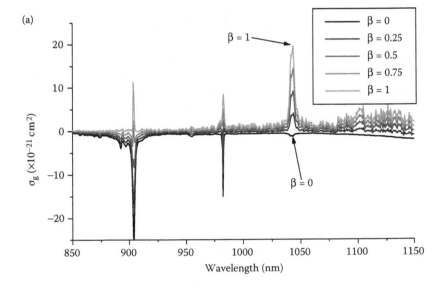

FIGURE 2.2
Gain cross-section $\sigma_{gain}(\lambda)=\beta\sigma_e(\lambda)-(1-\beta)\sigma_{abs}(\lambda)$ for (a) Yb:CFAP at room temperature.

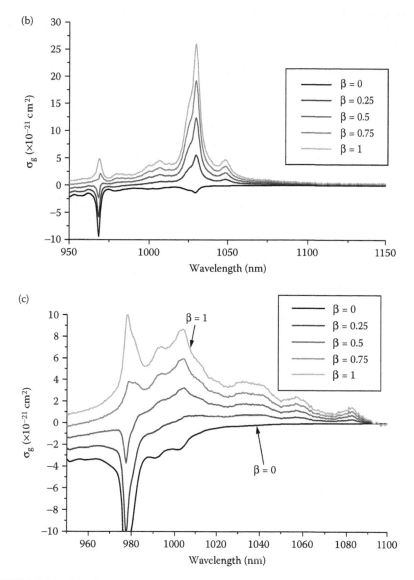

FIGURE 2.2 (continued)
Gain cross-section $\sigma_{gain}(\lambda)=\beta\sigma_e(\lambda)-(1-\beta)\sigma_{abs}(\lambda)$ for (b) Yb:YAG, and (c) Yb:YSO at room temperature.

2.3   Laser Modeling

2.3.1   Quasi-Thermal Equilibrium

Figure 2.3 shows the energy level diagram for the quasi-three-level laser when the crystal field splitting is taken into account. The laser transition could occur between Stark levels. The thermal population of the fundamental laser level leads to residual absorption (reabsorption). According to Boltzmann law, the population within the sublevels is expressed as:

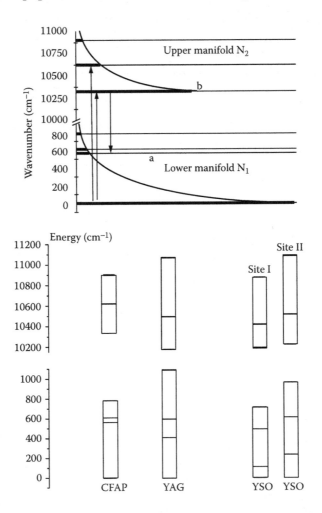

FIGURE 2.3
Energy level diagram of Yb:YAG, Yb:CFAP, and Yb:YSO and on the top scheme of the energy level diagram in the case of Yb:YAG. The population corresponds to the length of the energy level ($\beta = 0.35$).

$$f_a = \frac{\exp(-E_{la} \, / \, kT)}{Z_l} \tag{2.6}$$

where f_a (f_b respectively) represents the fraction of population in the *a* sublevel (*b*, respectively) (Figure 2.3). E_{la} is the energy of the level a, Z_l is the partition function of the lower manifold;

$$Z_l = \sum_i \exp(-E_{li} \, / \, kT) \, .$$

The fraction of the population in the excited level for the population inversion $f_b N_2 \geq f_a N_1$ corresponds to the following β coefficient:

$$\beta \geq \frac{f}{1+f} \text{ with } f = \frac{f_a}{f_b} \quad \text{and} \quad \beta = \frac{N_2}{N_{total}} \tag{2.7}$$

The lowest β value is also called β_{min}. For a given Yb^{3+} host, lower β_{min} values correspond to a "four-level" laser tendency. One can notice here that these lasers are called quasi-three or quasi-four level according to the authors [3]. For the Yb:YAG presented in Figure 2.3, the laser transition at 1.03 μm corresponds to f = 0.066. Clearly for this transition, only a relatively small fraction of the population needs to be in the upper manifold in order to reach population inversion. The main gain line for the Yb:YAG at 1.03 μm corresponds to a transition between the lowest lying crystal field splitting of the upper manifold and a level situated at 612 cm^{-1} in the ground state manifold. Considering the absorption, two transition lines are possible at 968 nm and at 943 nm with approximately the same absorption strength. The band at the lower wavelength was initially chosen because of the broader absorption feature well adapted to laser diode, even if this corresponds to a much higher thermal load in the crystal. At the present time, pumping in the so-called zero line at 968 nm has been considered in order to limit the quantum defect. One can notice that the fluorescence lifetime in the material is about 1.16 ms. Measuring the experimental fluorescence lifetime in the case of Yb^{3+} cations is not an easy task, as the reabsorption effect could considerably vary the measured lifetime [33,34]. In order to have good information on the lifetime values, this measurement has to be performed on powder with a very low dopant concentration. More accurate measurements can be performed of fine powder in an immersion liquid as proposed in [34] providing values very close to the calculated radiative lifetime in the case of double tungstate hosts.

2.3.2 Transparency Intensity I_{min}

In ytterbium lasers, as opposed to four-level lasers, the unpumped medium is absorbing at laser wavelength; consequently, the pump intensity I_{min} required to reach threshold in a perfect lossless cavity is not zero. $I_{min} (\lambda_p, \lambda_l)$ is also called transparency intensity because it represents the amount of pump intensity (at λ_p) necessary to make the medium transparent at the laser wavelength λ_l.

In the most general case, the rate equation for a quasi-three-level system is [35]:

$$
\frac{dN_1}{dt} = \frac{N_2}{\tau} + \sigma_{em}(\lambda_l) \frac{\lambda_l I}{hc} N_2 + \sigma_{em}(\lambda_p) \frac{\lambda_p I_p}{hc} N_2 - \sigma_{abs}(\lambda_l) \frac{\lambda_l I}{hc} N_1
$$
$$
- \sigma_{abs}(\lambda_p) \frac{\lambda_p I_p}{hc} N_1
$$

(2.8)

where σ_{abs} and σ_{em} are the absorption and stimulated emission cross sections, I is the intracavity laser intensity (W.cm^{-2}), I_p is the pump intensity, τ is the lifetime defined as

$$
\left(\tau_{rad}^{-1} + \tau_{NR}^{-1} \right)^{-1}
$$

where τ_{rad} is the radiative lifetime and τ_{NR} the nonradiative lifetime (s), λ_p and λ_l are the pump and laser wavelengths, respectively. At steady state, considering that the fraction of total population in the excited manifold is:

$$
\beta = \frac{N_2}{N} = \frac{\sigma_{abs}(\lambda_p) \frac{\lambda_p I_p}{hc} + \sigma_{abs}(\lambda_l) \frac{\lambda_l I}{hc}}{\left(\sigma_{abs}(\lambda_p) + \sigma_{em}(\lambda_p) \right) \frac{\lambda_p I_p}{hc} + \left(\sigma_{abs}(\lambda_l) + \sigma_{em}(\lambda_l) \right) \frac{\lambda_l I}{hc} + \frac{1}{\tau}}
$$

(2.9)

The medium linear gain at laser wavelength is defined by:

$$
g(\lambda_l) = \sigma_{em}(\lambda_l) N_2 - \sigma_{abs}(\lambda_l) N_1
$$

(2.10)

The transparency is attained when g = 0, or equivalently when $\beta = \beta_{min}$ where

$$
\beta_{min}(\lambda_l) = \frac{\sigma_{abs}(\lambda_l)}{\sigma_{abs}(\lambda_l) + \sigma_{em}(\lambda_l)}
$$

(2.11)

Writing I = 0 and $I_p = I_{min}$ leads to the expression of the transparency intensity:

$$
I_{min}(\lambda_l, \lambda_p) = \frac{hc}{\lambda_p \left[\sigma_{abs}(\lambda_p) \times \frac{\sigma_{em}(\lambda_l)}{\sigma_{abs}(\lambda_l)} - \sigma_{em}(\lambda_p) \right] \tau}
$$

(2.12)

This expression is rigorously true whatever the pump or laser wavelengths. DeLoach et al. [36] use a slightly different expression of I_{min}, defined by :

$$I^*_{min} = \beta_{min} I_{sat} \qquad (2.13)$$

where the pump saturation intensity is defined by:

$$I_{sat} = \frac{hc}{\lambda_p \sigma_{abs}(\lambda_p)\tau} \qquad (2.14)$$

The definition Equation 2.13 is then more restrictive than Equation 2.12: the two expressions are consistent, provided that two conditions are met: i) $\sigma_{em}(\lambda_p) \ll \sigma_{abs}(\lambda_p)$ which means that stimulated emission is ignored at pump wavelength: this is true at "short pump wavelengths" (typically 900 or 940 nm for Yb-doped materials) but not at the zero-line pump wavelength (around 980 nm) ; and ii) $\sigma_{abs}(\lambda_l) \ll \sigma_{em}(\lambda_l)$ which is reasonable, but strictly speaking only true for "long laser wavelengths," that is, for wavelengths that do not experience significant reabsorption [35,37].

2.3.3 Laser Extraction Efficiency

The laser extraction efficiency η_l is defined as the photons that relax through stimulated emission [38]. An expression of the laser extraction efficiency can be derived by expressing the stimulated emission rate Q_{stim} (number of stimulated photons emitted per second and per unit volume), the spontaneous emission rate Q_{spont}, and nonradiative relaxation rate Q_{nr}:

$$
\begin{aligned}
Q_{stim} &= \frac{\lambda_l}{hc}\left[\sigma_{em}(\lambda_l)N_2 I - \sigma_{abs}(\lambda_l)N_1 I\right] \\
&= \frac{\lambda_l I}{hc}\left(\left[\sigma_{abs}(\lambda_l)+\sigma_{em}(\lambda_l)\right]N_2 - \sigma_{abs}(\lambda_l)N\right) \\
Q_{spont} &= \frac{N_2}{\tau_{rad}} \\
Q_{nr} &= \frac{N_2}{\tau_{nr}}
\end{aligned}
\qquad (2.15)
$$

Then the laser extraction efficiency is given by:

$$\eta_l = \frac{Q_{stim}}{Q_{stim} + Q_{spont} + Q_{nr}} \qquad (2.16)$$

By writing the rate equations, as done in [39] it is possible to obtain an exact formulation of η_l. However, in most cases one can neglect reabsorption at laser wavelength (that is, $\sigma_{em}(\lambda_l)N_2 \gg \sigma_{abs}(\lambda_l)N_1$, or following the standard notations $\beta(\lambda_l) \gg \beta_{min}(\lambda_l)$). In the latter case we obtain the simplified expression:

$$\eta_l \approx \frac{\sigma_{em}(\lambda_l)I}{\sigma_{em}(\lambda_l)I + \dfrac{1}{\eta_r \tau_{rad}}} \tag{2.17}$$

with

$$\eta_r = \frac{Q_{spont}}{Q_{spont} + Q_{nr}} = \frac{\tau_{nr}}{\tau_{nr} + \tau_{rad}} \tag{2.18}$$

As shown by Equation 2.17, the laser extraction efficiency tends towards unity for intracavity laser intensities that surpass the laser saturation intensity. Generally, C.W. oscillators based on Yb-doped materials work with high reflectivity output couplers: as a consequence, the intracavity laser intensity is very high, at least one order of magnitude higher than the laser saturation intensity, so that η_l is typically close to unity in an operating Yb-laser. In this case, as seen from Equation 2.19, the thermal load becomes nearly independent of the radiative quantum efficiency and is only governed by the quantum defect. Nevertheless, the quantum efficiency directly affects the excited state population and, hence, has crucial importance for Q-switched lasers or low repetition rate amplifiers. Incidentally, Equation 2.17 also illustrates that the performance of an Yb-based C.W. oscillator becomes nearly independent of the emission cross section at laser wavelength, provided that the pump intensity is far higher than the pump saturation intensity.

2.3.4 Thermal Loading

Considering the heat sources in an ytterbium-doped material [40], the fractional thermal load η_h (that is, the fraction of the absorbed pump power converted into heat) can be written as:

$$\eta_h = 1 - \eta_p \left[\left(1 - \eta_l\right)\eta_r \frac{\lambda_p}{\lambda_f} + \eta_l \frac{\lambda_p}{\lambda_l} \right] \tag{2.19}$$

Here η_p is the pump quantum efficiency, which is the fraction of absorbed pump photons contributing to inversion. Nonunity pump quantum efficiency accounts for residual absorption of the undoped crystal or can be related to the presence of nonradiative sites. η_r denotes the radiative quantum efficiency for the upper manifold: it represents the fraction of excited atoms that decay by a radiative path (in absence of stimulated emission). Nonunity radiative quantum efficiency can be related to multiphonon relaxation (although it is very unlikely because a large number of phonons are necessary to bridge the 10,000 cm^{-1} gap separating the excited and ground manifolds) or more probably to concentration quenching. The latter phenomenon corresponds to the trapping of the energy by a color center, an impurity, or a lattice defect (Yb^{2+}, rare-earth impurities or hydroxyl groups have been evoked) [41–43] after several transfers of excitation between neighboring ions. Consideration and measurements of the quantum efficiency are presented in the following sections of this chapter.

Furthermore, we have shown that it is possible to include pump absorption saturation effects. The pump saturation intensity is defined by [35,40]:

$$I_{p_{sat}} = \frac{hc}{\lambda_p \left[\sigma_{abs}\left(\lambda_p\right) + \sigma_{em}\left(\lambda_p\right) \right] \tau} \qquad (2.20)$$

The low brightness of the diode pump beam in comparison with the brightness of the laser beam makes the effective Raleigh distance of the pump beam considerably shorter than the crystal length. For this reason, the divergence of the pump beam inside the crystal must also be considered, in order to correctly account for saturation issues.

Figure 2.4 shows a 3D view of temperature distribution without saturation (Figure 2.4a) and in presence of strong saturation (Figure 2.4b). It appears in the latter case that the region where the pump density is the strongest (near the pump beam waist) is not the region where the temperature is the highest (near the faces of the crystal) because of absorption saturation. Then, pump beam divergence appears to be an important parameter, as for instance, the temperature could be higher at the exit face than at the entrance face of the crystal. It appears that fracture risks may be paradoxically reduced when the pump beam is focused on a smaller spot, because in this case the absorption saturation is made stronger.

In presence of strong laser extraction, the pump intensity evolution through the crystal could be approximated by [12]:

$$\frac{dI_p}{dz} = -\alpha_{NS}^p \, I_p \qquad (2.21)$$

where α_{NS}^p is the linear absorption coefficient at the pump wavelength. This means that ground manifold is repopulated so that absorption is not saturated anymore under lasing conditions. Because the saturation of absorption is reduced under lasing conditions, pump absorption is higher. This can lead to higher temperature and then fracture is more likely to happen, which is not commonly observed in four-level laser systems.

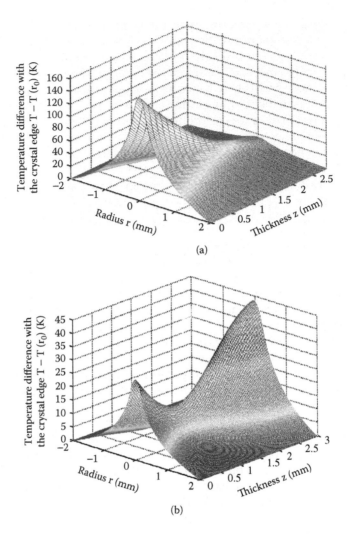

FIGURE 2.4
Temperature distribution under nonlasing condition. (a) The saturation of absorption is ignored; (b) pump absorption saturation is taken into account ($I_{P_{sat}} = 4.1$ kW/cm², corresponding to Yb:GdCOB parameters).

2.4   Evaluation of Quantum Efficiency

It has been generally accepted that the quantum efficiency should approach unity in ytterbium-doped materials. In a paper dedicated to heat generation in Nd:YAG and Yb:YAG, T.Y. Fan [44] predicted that the thermal load under lasing or nonlasing conditions should be substantially identical in Yb:YAG. This prediction was based upon the assumption that the quantum efficiency

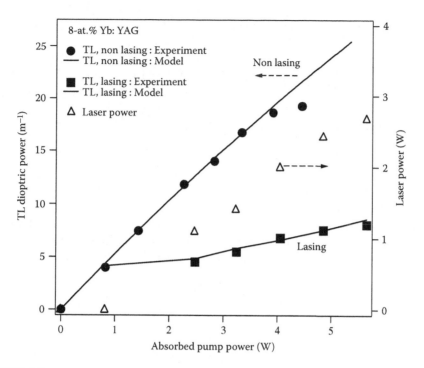

FIGURE 2.5
Thermal lensing dioptric power (left) and laser power (right) in Yb:YAG.

was high enough to compensate for the difference between the energy of the laser photon and the mean energy of the fluorescence photons. However, recent studies have shown that the quantum efficiency was less than unity in Yb:YAG [13,45]. In the meantime, some authors have proved the existence of quenching mechanisms. Thermal lensing measurements appear as an important tool [46,11,12] for the determination of the quantum defect. This is now presented in the following section. As a consequence of the thermal load in the laser material, thermal lens can be defined by its focal length, or dioptric power. We have studied the evolution of the thermal lensing dioptric power vs. the absorbed pump power, under nonlasing condition and lasing conditions, before and after threshold. The results in the case of Yb:YAG crystal are presented in Figure 2.5. When the pump power exceeds the laser threshold, thermal lensing dioptric power under laser effect experiences a decrease and then increases following a nearly straight line. The clear difference between the dioptric power measured under lasing and nonlasing conditions, and for an identical absorbed power, proves the existence of significant nonradiative effects, which turn out to be "short-circuited" by stimulated emission when laser occurs. The behavior with lasing and non-lasing effect is presented in Figure 2.6.

Assuming that the absorption saturation presented before is negligible under laser condition, the thermal lensing dioptric power can be written as

$$D_{th} = A \times P_{abs} \times \eta_h \tag{2.22}$$

$$\left(D_{th}\right)_{after\ threshold} = A \times P_{abs} \times \left(1 - \eta_p \left[\left(1 - \eta_l\right)\eta_r \frac{\lambda_p}{\lambda_f} + \eta_l \frac{\lambda_p}{\lambda_l}\right]\right) \tag{2.23}$$

$$and\ \left(D_{th}\right)_{before\ threshold} = A \times P_{abs} \times \left(1 - \eta_p\,\eta_r \frac{\lambda_p}{\lambda_F}\right)$$

where A is a constant which depends on the thermo-optic coefficient and on the pump diameter, and this leads [12] to the determination of the η_r value. For the Yb:YAG host a value of $\eta_r = 0.7$ has been determined [12]. The nonradiative relaxations are significant just before threshold ($\eta_l = 0$) and become negligible (because short-circuited by laser emission) when η_l tends towards unity as previously seen in Subsection 2.3.3. Far above laser threshold:

$$\left(D_{th}\right)_{far\ above\ threshold} \approx A \times P_{abs} \times \left(1 - \frac{\lambda_p}{\lambda_l}\right) \tag{2.24}$$

(here $\eta_p = 1$), which means that the unique heat source, far above threshold, is the quantum defect. Using the same experimental procedure for the Yb:Ca$_4$YB$_3$O$_{10}$ (YCOB) (15% Yb corresponding to 6.7 × 10^{20} atoms/cm^3), Yb:Y$_3$Al$_5$O$_{12}$ (YAG) (8% Yb, 11 × 10^{20} atoms/cm^3), Yb:Gd$_3$Ga$_5$O$_{12}$ (GGG) (5.7% Yb, 7.2 × 10^{20} atoms/cm^3), and Yb:KGd(WO$_4$)$_2$ (KGW) (5% Yb, 2.2 × 10^{20} atoms/cm^3), the fluorescence quantum efficiencies are 0.9, 0.7, 0.9 and 0.96, respectively. Other methods presented in the literature indicate that nonradiative effects occur in all these crystals [13,14,44,45]. However, among the various Yb-doped laser materials, the Yb:KGW crystal exhibits a different behavior. Thermal lensing dioptric power is here higher under lasing action than under nonlasing conditions [12]. This can be interpreted as illustrated in Figure 2.6b: the average fluorescence wavelength is particularly low in this material (993 nm, which has justified its interest for radiation-balanced lasers in particular [47,48]). The quantum defect with laser oscillation at 1030 nm is thus higher than without oscillation. This explanation implies of course that the quantum efficiency is high in this sample: indeed, the procedure described above yields $\eta_r = 0.96$.

Then thermal lensing measurement appears for the determination of the quantum efficiency as a very good alternative [49] to photometric and calorimetric methods rather difficult to manage. The lifetime method, also reported for quantum efficiency calculation (consisting in comparing the fluorescence and radiative lifetime), is particularly intricate in Yb-doped materials because of the radiation trapping process shortly presented in Subsection 2.3.1.

In all the Yb-doped crystals under investigation in this study, nonradiative relaxations have been evident. Up to now, to the best of our knowledge,

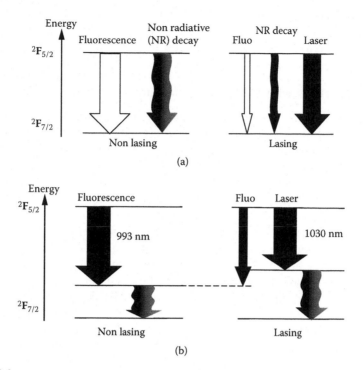

FIGURE 2.6

(a) Qualitative interpretation of the behavior observed in all the Yb-doped materials. The arrow's thickness stands for the corresponding rate; (b) qualitative interpretation of the behavior observed in Yb:KGW. The nonradiative relaxations are considered negligible here for simplification.

less-than-unity quantum efficiencies have been reported in Yb:YAl$_3$(BO$_3$)$_4$ (YAB) [46], in Yb:MgO:LiNbO$_3$ [50] and in Yb:YAG. For the Yb:YAG crystals, Barnes et al. [13] used a photometric and a calorimetric method, and measured a quantum efficiency of 0.898 and 0.932, respectively, with the two methods in a 1% atoms-doped YAG samples. Notice that our measurement on Yb:YAG has been performed on a 8% atoms-doped crystal. Other authors reported values between 0.85 and 0.97 with temperature and lifetime measurements [45]. In highly-doped Yb:YAG samples nonradiative relaxations were attributed to cooperative processes between two Yb^{3+} ions towards Yb^{2+} impurities [41]. Owing to the intrinsic nature of concentration quenching and the major role played by impurities, it is clear that the radiative quantum efficiency is a parameter that pertains to a single given sample, characterized by its doping concentration, the growth technique, and, of course, the degree of purity of the compounds. This could explain the dispersion in the literature, even in such a well known laser host like Yb:YAG.

In conclusion, we have first presented in this chapter spectroscopic and laser parameters, as well as laser formalism, in the case of trivalent ytterbium. The Yb^{3+} host strongly influences these properties, and next in this chapter are presented various laser hosts with broad emission, which leads to a low quantum defect and could allow oscillation in femtosecond regime.

2.5 Structural/Optical Properties Relationships for Femtosecond Lasers

2.5.1 Yb³⁺ Hosts with Broad Emission Bands

The influence of the crystal host on the broadening of the spectral lines for a given doping ion is fundamental. We have already described the crystal field effect on the energy level splitting. On the other hand, Yb^{3+} exhibits a strong electron–phonon coupling in crystals. Some authors explain this qualitatively by the antagonist effects of the lanthanide contraction and of the shielding of the 4f electrons [51]. Consequently, strong electron–phonon coupling homogeneously broadens the electronic lines. The addition of these two phenomena, splitting of the energy levels and electron–phonon coupling, both strongly dependent on the crystal, are the main contributions that lead to very broad and smooth emission spectra [25].

The broad emission could also be obtained in the laser hosts where the ytterbium cation could occupy several cationic sites [52,53]. This is well expressed in the $Yb:Y_2SiO_5$ (YSO) and $Yb:Lu_2SiO_5$ (LSO) presented in Figure 2.7. Yb^{3+} ions equally share the two Yb^{3+} crystallographic sites. The broad and smooth emission band is due to the localization of the Yb^{3+} dopant in the two cationic sites.

The broadening could also be due to a disorder around one type of substitution site. This is the case for the $Yb:CaGdAlO_4$ (CAlGO), which will be presented in more detail in the following section. In that case, Yb^{3+} cations could occupy one type of site shared either by the divalent calcium or by the trivalent gadolinium [54]. This is also the case in the tetragonal double tungstate and molybdate [55–57] where there is an almost random distribution of Na^+ and trivalent rare-earth cations over two lattice sites. Several crystalline borate hosts present very similar effect [58].

This property leads to locally variable crystal field around the dopant ions and the linewidths of the electronic transitions for the ytterbium cations are found to be broader in these so called "disordered" crystals than in "ordered" laser hosts. Such "disordered" hosts present intermediate behavior between ordered crystal and glass systems. Some examples of the broad emission obtained for some new laser hosts are presented in Figure 2.8 in emission cross section units. Crystals could indeed present broader emission bands than the glass system.

2.5.2 Performances of Some Yb³⁺ Doped Laser Materials in C.W. Diode Pumping

Limiting the quantum defect by pumping on the zero-line around 980 nm is the first way to decrease the thermal load into Yb-doped crystals. We have performed experiments using as pump source a 15-W fiber-coupled laser diode emitting around 978 nm, and a laser cavity setup allowing us to

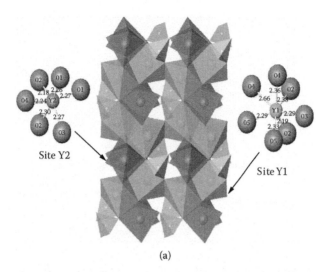

(a)

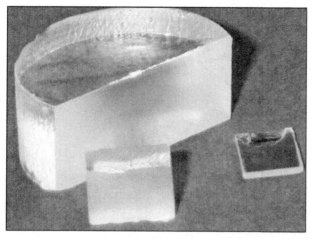

(b)

FIGURE 2.7
(a) M_2SiO_5 (M= Y, Lu) structure with the two sites. The distances are indicated for the YSO compound; (b) example of laser crystals (Yb:YSO elaborated at the LETI-CENG by the Czochralski process).

investigate the tuning range (by inserting a Lyot filter into the collimated arm of the cavity). The laser results in the case of Yb:LSO and Yb:YSO are presented in Figure 2.9 [59]. Laser effect extends from 1025 nm to 1091 nm for Yb:YSO and from 1030 nm to 1095 nm for Yb:LSO. Yb:YSO provides more than 4 W over a bandwidth of 60 nm with a relatively flat tuning curve, whereas Yb:LSO exhibits more than 4 W over more than 50 nm, but with more pronounced emission peak around 1080, 1068, and 1057 nm. In that case, the quantum defect is about 4.4%, but much lower values have been recently obtained under off-axis configuration, because, in this configuration, the tuning range

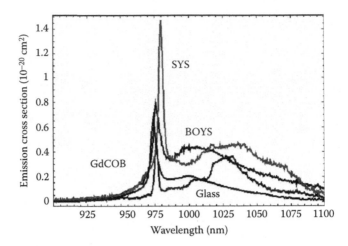

FIGURE 2.8
Broad emission bands of Yb-doped crystalline hosts in comparison with Yb-doped glass.

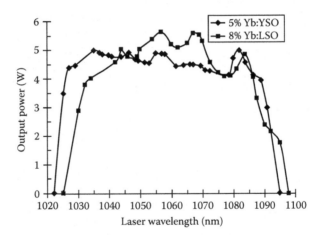

FIGURE 2.9
Tuning curves obtained with a Lyot filter for Yb:YSO and Yb:LSO for 14 W of pump power.

is not limited on the short wavelengths by the dichroic coating of both input and folding mirrors. In the case of Yb:YSO value as low as 2.5% has been obtained (laser wavelength at 1003 nm) under diode pumping at 980 nm [60] and, in the case of Yb:CaGdAlO$_4$ (Yb:CAlGO), the quantum defect could be even lower (0.8%), as laser effect at 987.6 nm has been recently obtained under pumping at 980 nm under titanium–sapphire pumping [61]. The emission features of the Yb:CAlGO are presented in Figure 2.10 for different β values. Indeed, this material presents an outstanding broad and flat emission in the near infrared range.

Let us now notice that maximum output powers of 7.7 W and 7.3 W were obtained, respectively, for Yb:YSO and Yb:LSO, corresponding to optical-to-optical efficiencies of around 53% (and 50%) and to a large value of the slope

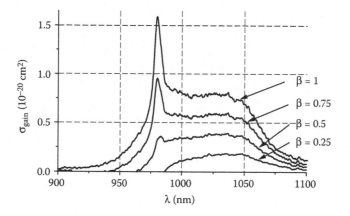

FIGURE 2.10
Gain cross section ($\sigma_{gain}(\lambda)=\beta\sigma_e(\lambda)-(1-\beta)\sigma_{abs}(\lambda)$) for the Yb:CAlGO laser material in sigma polarization. Only the positive values are presented in this spectrum.

efficiencies (67% and 62%, respectively, for the two hosts). One can also indicate that, under lasing conditions and at maximum power, both crystals absorbed about 90% of the pump power, corresponding approximately to the unsaturated absorption regime. Indeed, when the laser effect occurs it induces predominant flow from the excited $^2F_{5/2}$ laser level to the lower $^2F_{7/2}$ level of the laser transition as previously indicated. In particular, we show here that these crystals exhibit comparable optical conversion efficiencies to Yb:KYW in more complicated pumping scheme (thin disk configuration) [62]. Of course, the slope efficiency of Yb^{3+} laser could be much higher under titanium–sapphire pumping which correspond to a Gaussian profile. In C.W. regime, slope efficiency as high as 78% in Yb:KYW has been obtained [63], and the laser slope efficiency could even reach 86.8% in pulsed titanium–sapphire pumping (laser pumping of 981 nm and laser at 1025 nm) [63]. In that latter case, the slope efficiency is rather close to the maximal value of 95% obtained from the difference between absorption and emission wavelengths. This shows that "quasi-three level" lasers can, indeed, be very efficient, and this power slope efficiency is higher than any 808 nm-pumped 1.06 μm Nd^{3+} laser demonstrated up to date. We have then demonstrated two main interests of Yb^{3+} lasers (i) for high optical/optical conversion in C.W. regime and (ii) for tunable emission and therefore low quantum defect. One immediate application of the broad emission bands of these lasers is their capabilities to lead to ultrashort pulses. This is now developed in the following section.

2.5.3   Performances of Some Yb^{3+} Doped Laser Materials in Fs Regime

Direct diode pumping and high optical-to-optical efficiency of the Yb^{3+}-lasers represent the main advantages for new device development. In the development of ultrafast lasers, Ti^{3+}:Al$_2$O$_3$ (titanium–sapphire) indeed leads to

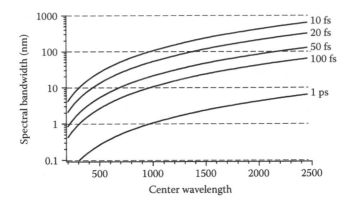

FIGURE 2.11
Required bandwidth for the production of femtosecond pulses (10 fs, 20 fs, 50 fs, 100 fs, and 1000 fs) versus the central wavelength (in nm) of the laser.

very short pulses (< 5 fs) but as these lasers could not be directly diode pumped, the efficiency and cost of these products is quite high. With Yb^{3+} system, direct diode pumping is possible, and that unique property explains the rapid expansion on the industrial market of such systems.

If one wants to generate ultrashort pulses, the laser crystal has to be able to allow amplification of very broad spectra (conservation of the time–bandwidth product). The emission bandwidth of the crystal has to be larger than the spectral bandwidth of the pulse. For example, producing around 100 fs pulses requires at least a spectral bandwidth of 10 nm for a central wavelength around 1050 nm (Figure 2.11).

However, the Yb-doped crystals exhibit some disadvantages. First, their emission cross sections are in general relatively low and only the narrowest emission band materials (such as Yb:YAG or Yb:KYW, Yb:KGW) have a relatively high gain. Choosing the ideal Yb-doped material for femtosecond laser development strongly depends on the application: either favoring a crystal matrix host with high disorder (broad emission band but poor thermal conductivity and low emission cross section) suitable for ultrashort oscillators or favoring a crystal matrix host with low disorder and narrow emission band (high emission cross section) more suitable for amplifiers. Another drawback of Yb-doped crystals is the quasi-three-level structure of these laser transitions; in fact, as shown in Figure 2.2, there is a strong overlap between the emission and the absorption bands which leads to strong reabsorption effects and to a reduction of the effective emission bandwidth. The final disadvantage of Yb-doped crystals is related to the very long fluorescence lifetime which implies some difficulties in the modelocking process, because the laser gain medium tends to store too much energy and to favor the Q-switched regime instead of the modelocked regime [64].

During the past few years many interesting studies and results have been reported in literature. Table 2.1 summarizes the properties of the most

TABLE 2.1

Main Experimental Results of Yb-Doped Laser Materials Developed in Diode-Pumped Fs Mode-Locked Laser Oscillators

Material	Experimental Duration	Central Laser Emission	Minimum Theoretical Duration	Average Power	Type of Saturable Absorber	Dispersion Compensation System	Year
Yb:YAG	340 fs	1031 nm	124 fs	110 mW	SESAM	Prisms	1999
Yb^{3+}:Y$_3$Al$_5$O$_{12}$	810 fs	1030 nm		60 W	SESAM	Chirp mirrors	2003
Yb^{3+}:glass	58 fs	1020 nm	31 fs	65 mW	SESAM	Prisms	1998
Yb:GdCOB	89 fs	1044 nm	26 fs	40 mW	SESAM	Prisms	2000
Yb^{3+}:Ca$_4$GdO(BO$_3$)$_3$							
Yb:BOYS	69 fs	1062 nm	18 fs	80 mW	SESAM	Prisms	2002
Yb^{3+}:Sr$_3$Y(BO$_3$)$_3$							
Yb:KGW	112 fs	1045 nm	44 fs	200 mW	SESAM	Prisms	2000
Yb^{3+}:KGd(WO$_4$)$_2$	240 fs	1028 nm	44 fs	22 W	SESAM	Chirp mirrors	2002
Yb:KYW	71 fs	1025 nm	46 fs	120 mW	KLM	Prisms	2001
Yb^{3+}: KY(WO$_4$)$_2$							
Yb:SYS	94 fs	1070 nm		110 mW			2002
Yb^{3+}:SrY$_4$(SiO$_4$)$_3$O	110 fs	1070 nm	16 fs	420 mW	SESAM	Prisms	2003
	100 fs	1070 nm		1 W			2004
Yb^{3+}:Sc$_2$O$_3$	230 fs	1044 nm	95 fs	540 mW	SESAM	Prisms	2003
Yb^{3+}:CaF$_2$	150 fs	1033 nm	30 fs	880 mW	SESAM	Prisms	2004
	220 fs	1033 nm		1.4 W			
Yb^{3+}:Y$_2$O$_3$ ceramic	615 fs	1076 nm	75 fs	420 mW	SESAM	Prisms	2004
Yb^{3+}:YVO$_4$	120 fs	1021 nm	/	185 mW	SESAM	Chirp mirrors	2004
Yb:SYS//YAG	130 fs	1072 nm	40 fs	1 W	SESAM	Chirp mirrors	2005
Yb:LSO	260 fs	1059 nm	75 fs	2.6 W	SESAM	Chirp mirrors	2005
Yb^{3+}: Lu$_2$SiO$_5$							
Yb :CALGO	47 fs	1050 nm	15 fs	300 mW	SESAM	Chirp mirrors	2006
Yb^{3+}:CaGdAlO$_4$							

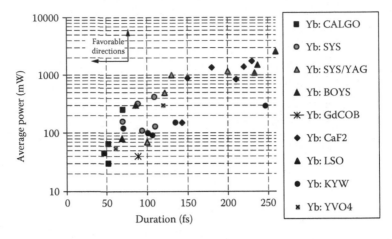

FIGURE 2.12
Recent results of ultrashort pulse generation obtained with Yb^{3+}-doped crystals.

interesting Yb-doped crystals for the femtosecond laser developments: Yb:YAG [65], Yb: $Sr_3Y(BO_3)_3$ (BOYS) [66], Yb: $SrY_4(SiO_4)_3O$ (SYS) [67], Yb:KGW, Yb:KYW [68], Yb:GdCOB [58], Yb:$CaGdAlO_4$ [69], and Yb:Y_2O_3 [70]. For instance, Figure 2.8 shows the absorption and emission spectra of several Yb^{3+} doped laser hosts developed for the fs lasers. The recent results of ultrashort pulses generation obtained with Yb^{3+}-doped crystals are presented in Figure 2.12.

To obtain ultrashort laser pulses, the first element that needed to be added into the laser cavity is the modelocking generator which favors the pulse production vs. the C.W. regime. Two ways are generally used: the Kerr effect (the optical Kerr effect is based on the refraction index variation under intense laser beam) and a semiconductor saturable absorber (SESAM), which is made of a few quantum wells of InGaAs semiconductor grown on semiconductor Bragg mirrors [71].

With the Yb^{3+} doped materials, the shortest pulses achieved so far 58 fs in glass [72], 47 fs in crystals [69] and 615 fs in ceramics [73] — (this latter value should be much lower with high quality ceramics now commercially available) are longer than the 5 fs obtained in Ti:sapphire lasers. Indeed, in the duration range smaller than 200 fs, several crystals can be proposed. It should also be noticed that recent interests arise in fluorides such as Yb:$LiYF_4$ (YLF) and Yb:CaF_2 [74], which could be grown in the form of large and good optical quality crystals. In addition undoped CaF_2 presents relatively high thermal conductivity value. This property is important for all the laser materials investigated for such purpose (see details in the last part of this chapter).

For laser efficiency, it is also crucial to consider the emission cross section peak σ_{em} and the lifetime of the upper state τ of the laser transition. One observes that the higher the product $\sigma_{em}.\tau$ is, the lower the laser threshold will be. The crystal well-adapted for the development of an efficient ultrashort pulsed laser will require both high peak and very broad emission band. However, these two criteria are often in contradiction and a compromise has to be found between short duration and efficiency.

2.6 Structural/Thermomechanical Effects in High Power Yb-Doped Lasers

We have previously discussed the fact that for the development of femtosecond lasers, the evaluation of the spectral emission properties of a crystal is important but not sufficient. Many other criteria have to be taken into account and, in particular, thermal properties have to be favorable. Indeed, for developing efficient, reliable, and high power oscillators, the thermal properties play a key part. In fact, if the crystal tends to store too much heat during the pumping process, some problems of efficiency, stability, or even fracture of the crystal may occur. When the temperature increase in the crystal is moderate, the induced thermal lens can be easily compensated by carefully adjusting the cavity mirrors because the thermal lens does not exhibit any geometrical aberrations (focus coefficient is the only significant term in the analysis of the wavefront). At higher crystal temperature, high order aberrations in the thermal lens start to appear, and it is no longer possible to compensate these wavefront distortions. Higher temperature can lead to the fracture of the crystal. A good thermal conductivity is thus an essential condition in the strategy of choosing a laser crystal. This section concerns the thermomechanical properties of the Yb^{3+}-doped laser hosts.

2.6.1 Thermal Shock Resistance

To obtain high average laser output power, materials with high thermal shock resistance capabilities are required. One defines the thermal shock resistance parameter R'_T as [75]:

$$R'_T = \frac{(1-\upsilon)\,K_c\,\kappa}{E\,\alpha}$$

(2.25)

where υ is Poisson's constant, E the Young's modulus of the material and K_c the material toughness, which represents the resistance to crack propagation. κ and α are the thermal conductivity and thermal expansion, respectively. The higher this parameter, the better it is, as it corresponds to the resistance of the material under thermal solicitations. The evaluation of this thermomechanical figure of merit may, however, be simplified if one assumes that whatever the considered solid host, the Poisson's coefficient is close to 0.25 and the ratio K_c/E is almost constant for crystalline materials. Therefore, the thermal shock resistance coefficient R'_T is $\approx \kappa/\alpha$ [19]. One can notice here that in the case of YAG ceramics, K_c, the material toughness, is estimated to be higher [76] as the crack propagation could be limited in such polycrystalline hosts. Further work is indeed required to check this behavior in various transparent ceramics.

The maximum of the temperature difference depends also on Yb^{3+} quantum defect in the host. Introducing the notion of quantum defect in the maximal admissible temperature difference that a laser crystal can withstand, we have expressed (according to the first Fourier law) that ΔT_{max} is proportional to the maximal heat flow inside the crystal, namely φ_{max} as:

$$\Delta T_{max} \propto \frac{\varphi_{max}}{\kappa} \tag{2.26}$$

It is therefore possible to define a laser power resistance parameter R_P, taking into account the α and κ thermomechanical coefficients and the energy difference between the absorption and the laser wavelength [61]:

$$R_P \propto \frac{\kappa^2}{\alpha\eta} \tag{2.27}$$

To obtain materials with high power resistance, this latter parameter has to be high, and indeed Yb^{3+} laser matrices should present low quantum defect.

2.6.2 Estimation of Thermal Expansion and Thermal Conductivity Parameters

In a preceding section, we have proposed [19,77] a relation between the thermal expansion parameter, the melting point temperature T_{mp}, and the covalency of the material. The α value increases with the ionic character of the material [19], assuming that only the acoustic phonon modes participate in the heat conduction process in insulators. The thermal conductivity can be expressed as a function of a limited set of physical parameters such as the melting point temperature, the molar mass, the number of atoms per formula unit, and the reduced mass μ (harmonic average of all ion masses of the formula unit) [77].

If only the acoustic phonon modes participate in the heat conduction process, one obtains [78]:

$$\kappa = \frac{1}{T}\frac{\rho v^4}{\gamma^2\omega_D} \tag{2.28}$$

where T is the temperature, ρ is the density, v is the sound velocity, γ is the Gruneisen anharmonicity parameter, and ω_D is the Debye frequency.

Then, one can assume that first, the material is isotropic and second, that at the melting point temperature, the interatomic distance a is increased by a factor ε $(a = \varepsilon a_0)$ [79]. That permits us to express ω_D as a function of the melting point temperature. The Debye frequency is the cut-off frequency for a vibration propagating along an atomic chain. Thus, ω_D is approximately given by:

$$\omega_D = \sqrt{\frac{6\,N\,k_B\,T_{mp}}{\mu\,a_0^2\,\varepsilon^2}} \tag{2.29}$$

Here, N and k_B are the Avogadro's and Boltzmann's constants, and the sound velocity is given by:

$$v = \frac{a_0\,\omega_D}{\pi} \tag{2.30}$$

where a_0 is an average interatomic distance :

$$a_0 = 2\,\sqrt[1/3]{\frac{3}{4\pi}\frac{M}{N\,n\,\rho}} \tag{2.31}$$

M is the molar mass and n is the number of atoms per formula unit. Thus, one obtains:

$$\kappa = \frac{1}{T}\frac{A}{\gamma^2\,\varepsilon^3}\,T_{mp}^{3/2}\,\rho^{2/3}\,M^{1/3}\,n^{-1/3}\,\mu^{-3/2} \tag{2.32}$$

where A is a numerical constant independent of the physical properties of the crystal. Consequently, considering that the factor $A/(T\gamma^2\varepsilon^3)$ is a constant, the thermal conductivity of an insulating material is a simple function of its melting temperature, its density, its molar mass, its number of atom per formula unit, and the harmonic average of all the ion masses. However, this model could be considered as more qualitative than quantitative because the error bars are of the order of 50% [77].

Figure 2.13 presents the $\kappa \cdot T$ product at room temperature for different materials reported in the literature, as a function of $T_{mp}^{3/2} \cdot \rho^{2/3} \cdot M^{1/3} \cdot n^{-1/3} \cdot \mu^{-3/2}$. Covalent compounds corresponding to low ε values and presenting a very high thermal conductivity are also indicated for the sake of comparison. This model leads to good results even in the case of rather complex oxides such as, for instance, garnet compounds like $Y_3Al_5O_{12}$ (YAG). In order to obtain a large κ value, the following requirements must be fulfilled: low molar mass M; high melting point T_{mp} (high strength atomic bonds); compact crystal structures, and covalent character of the host.

We have also investigated the influence of doping ion on thermal conductivity. The method is based on Klemens method [80] simplifying Debye frequency cut-off into $\omega_D \sim \pi v/a_0$. One obtains [77]:

$$\kappa = \frac{1}{\pi a_0} \sqrt{\frac{2 k_B v \kappa_0}{\delta}} \, A \tan\left(\pi a_0 \sqrt{\frac{\kappa_0 \delta}{2 k_B v}} \right) \qquad (2.33)$$

$$\text{with: } \delta = \sum_i c_i \left(\frac{M_i - M}{M} \right)^2 \text{ and } M = \sum_i c_i M_i \qquad (2.34)$$

δ represents the mass variance of the lattice substitution sites of average mass M that have an occupation probability c_i to be occupied with ions i with mass M_i; κ_0 is the undoped material thermal conductivity value.

Thus, if the mass difference between the doping ion and the substituted ion is important, as in the case where Y^{3+} (M = 88.9 g/mol) is replaced by Nd^{3+} (M = 144.2 g/mol) or Yb^{3+} (M = 173.0 g/mol), the thermal conductivity decrease is important, even for low doping rates. This is, for instance, presented in Yb-doped garnet and in $Yb:CaF_2$ in Figure 2.14, which are two laser hosts recently investigated.

Further, we have compared the thermomechanical properties of some laser hosts. Among several hundreds of hosts investigated in the case of neodymium

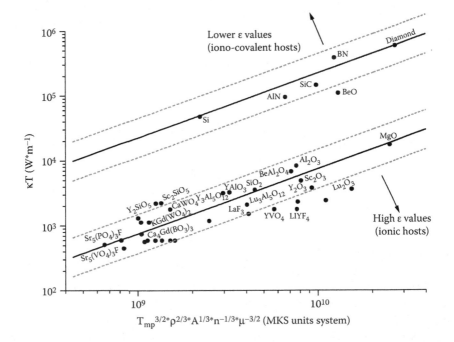

FIGURE 2.13
κT vs $T_{mp}^{3/2} \cdot \rho^{2/3} \cdot M^{1/3} \cdot n^{-1/3} \cdot \mu^{-3/2}$ for different insulating crystals. Dashed curves correspond to 50% variation from the predicted values.

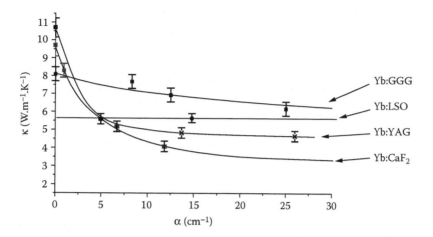

FIGURE 2.14
Thermal conductivity values of the Yb^{3+} doped $Y_3Al_5O_{12}$ (YAG) and $Gd_3Ga_5O_{12}$ (GGG), $Yb:Lu_2SiO_5$ (LSO) and $Yb:CaF_2$ for different absorption coefficient values (in cm^{-1}) at the maximum of the absorption band around 980 nm (notice here that comparison with Yb content (in percentage) is strongly dependant of the host structure).

doping — and more than 85 for the ytterbium ion — we have selected some matrices of particular interest for high power applications. Table 2.2 collates the thermomechanical properties of several important laser hosts.

One can observe both from literature and from calculations made under the assumptions presented in this chapter that the sesquioxides Ln_2O_3 appear as the most appropriate hosts to reach high power laser applications. The thermal shock resistance parameter reaches 9 $W \cdot m^{-1/2}$, which is one order of magnitude larger than for the CFAP material for instance. Unfortunately, these sesquioxide hosts (Gd_2O_3, Lu_2O_3 and Sc_2O_3) are rather difficult to grow in the single crystal form because of their high melting points [81,82]. Silicate hosts like Yb:LSO could be an alternative to Yb:YAG, as the thermal conductivity is comparable to that of Yb:YAG and because the synthesis of this material is perfectly controlled in the case of Ce^{3+}-doped LSO [83]. In the field of scintillators, several hundreds of Czochralski boules of this material are grown every year by several companies all over the world [84]. The Yb:LSO host has a number of advantages such as low quantum defect, pump wavelength at 980 nm in a broad absorption band, high laser efficiency, and generation of ultrashort pulses with a 2.6 W average power.

But other materials are also worth developing. Alternative matrices deduced from this work may be vanadate [85], aluminate (see for instance the outstanding properties of $CaGdAlO_4$ in [69]), and perovskite materials.

TABLE 2.2

Thermomechanical Parameters of Laser Hosts

Crystal Hosts	Acronyms	κ (Wm^{-1}K^{-1})	α 10^{-6} K^{-1}	R$'_T$ (W·m$^{-1/2}$)	dn/dT (10^{-6} K^{-1})
Y$_3$Al$_5$O$_{12}$	YAG	10.7	20.1	7.9	9.1
Gd$_3$Sc$_2$Ga$_3$O$_{12}$	GSGG	5.8	22.5	4.4	10.5
Gd$_3$Ga$_5$O$_{12}$	GGG	8.2	24.0	5.4	/
Lu$_3$Al$_5$O$_{12}$	LuAG	6.6[a]	/	/	/
Y$_2$SiO$_5$	YSO	5.0	18.0	3.3	7.2
Lu$_2$SiO$_5$	LSO	5.6[a]	/	/	/
LiYF$_4$	YLF	5.0	42.0	1.3	–2.8
Sc$_2$O$_3$	/	16.5	27.0	9.2	/
Y$_2$O$_3$	/	12.8	25.5	8.4	7.0
Lu$_2$O$_3$	/	12.2	24.0	8.5	/
Ca$_5$(PO$_4$)$_3$F	CFAP	2.0	28.8	0.8	–9.3
Sr$_5$(PO$_4$)$_3$F	SFAP	2.0	27.3	0.6	–10.0
Sr$_5$(VO$_4$)$_3$F	SVAP	1.7	30.0	0.5	/
CaF$_2$	/	10.0	56.6	1.2	–10.6
SrF$_2$	/	8.0	60.0	2.1	/
Ca$_4$Gd(BO$_3$)$_3$O	GdCOB	2.0	33.5	0.7	2.0
KGd(WO$_4$)$_2$	KGW	3.8	16.1	2.8	0.4
Gd$_2$O$_3$	/	(9.0)	(16.2)	(7.1)	9.1
YAlO$_3$	YAP	8.5[a]	24.6	> 5.5	/
LuAlO$_3$	LuAP	(9.8)	(21.3)	(8.6)	/
GdAlO$_3$	GAP	(10.4)	26.0	(6.9)	/
CaGdAlO$_4$	CAlGO	6.9[a]	35.0	> 4.5	/
ZrO$_2$	/	(8.2)	(12.8)	(8.7)	/
ThO$_2$	/	(8.0)	(10.9)	(9.9)	/
YVO$_4$	/	(7.0)	11.2	(8.0)	6.7
GdVO$_4$	/	8.1[a]	22.7	> 5.3	/

Note: κ is the thermal conductivity of undoped materials, α is the volumic thermal expansion coefficient, and R$'_T$ is the thermal shock resistance parameter. Values in brackets are calculated values from the presented models.

[a] κ measured for absorption coefficient around 10 cm^{-1} in Yb^{3+}-doped samples.

2.7 Conclusions

In this chapter, we first presented the singular aspects of the Yb^{3+} spectroscopy in solid state laser hosts. Because of the three-level scheme, crystal field and electron–phonon coupling effects play an important role for this trivalent ion, which is indeed very attractive for applications such as high power laser and ultrashort pulses lasers. Considerations are made in this chapter, taking into account the materials-science view rather than laser physics. We have tried to correlate basic optical properties, thermomechanical properties, and laser parameters. Of course, when considering the final laser system, not only the laser host has to be considered but the laser cavity plays a key role.

For instance, a thin disc can be helpful to remove the heat, and crystal bonding of doped and undoped material is also very efficient.

An effort is made in this chapter to present several aspects that are singular to thermal effects in Yb-doped materials working as quasi-three-level system. Particular attention is paid to the quantum efficiency and the quantum defects which are considered as the main thermal sources. Thermal lensing measurements with end-pumping are used to calculate the quantum efficiency of different systems. Comparison between experiments and models yields the radiative quantum efficiency of several Yb-doped samples. They are in the range of 0.7–0.96. Pump absorption saturation effects and pump beam divergence inside the crystal are also considered. This provides a formulation of the temperature distribution in end-pumped crystals, under nonlasing conditions. Because the saturation of absorption is reduced under lasing conditions, fracture is more likely to happen in this latter case, which is not observed commonly in four-level lasers.

For developing efficient, reliable and high power lasers, the thermal properties play a key part. If the crystal tends to store too much heat during pumping process, some problems of efficiency, stability, or even fracture of the crystal may occur. We have shown in this chapter how the estimation of these parameters in the case of insulating crystals might be carried out. To obtain host materials with a large thermal conductivity value, the following requirements must be fulfilled: low molar mass M; high melting point (high strength atomic bonds); compact crystal structures and covalent character of the chemical bonds. Consequently, we can derive materials with high thermal shock resistance parameters $R'_T = \kappa/\alpha$. For instance $Yb:Sc_3O_3$; $Yb:Y_2O_3$, $Yb:YAG$, and $Yb:CAlGO$ present very high R'_T values.

Furthermore, we have also emphasized in this chapter other very important features such as dopant effects on the thermal conductivity and structural effects on the thermo-optical properties. These effects can dramatically affect the gain media performance for the generation of high power laser beams.

Finally, in this chapter, the state of the art of femtosecond oscillators based on the Yb-doped materials is presented. The development of new crystal hosts is still very important in order to extend to new application domains by finding crystals with better laser properties.

Acknowledgments

We wish to acknowledge Gerard Aka and Daniel Vivien from the Laboratoire de Chimie de la Matière Condensée de Paris (Ecole Nationale Supérieure de Chimie Paris, France) for fruitful discussions, Bernard Ferrand from CEA-LETI (Grenoble, France) for growing some of the materials presented here. We wish to thank the CNRS, (Groupement de Recherches,

GdR Matériaux Laser, and CMDO) as well as the Délégation Générale de l'Armement (DGA) for funding the Ph.D. thesis of Sébastien Chénais, Johan Petit, and Romain Gaumé.

References

1. T.Y. Fan and R.L. Byer, Diode pumped solid state lasers, *IEEE J. Quantum Electron.* 24, 895, 1988.
2. A. Giesen, H. Gudel, A. Voss, K. Witting, U. Brauch, H. Opower, Scalable concept for diode-pumped high power solid state laser, *Appl. Phys. B* 58, 365–372, 1994.
3. T.Y. Fan, Quasi-three level lasers, in *Solid State Lasers News Developments and Applications*, Eds., M. Inguscio and R. Wallenstein, Series B, 317, 1993, p. 189.
4. W. Koechner, *Solid State Laser Engineering*, 2nd ed., Springer-Verlag, Berlin, 1988.
5. W.F. Krupke, Ytterbium solid state lasers: the first decade, *IEEE J. Sel. Top. in Quantum Electron.* 6(6), 1287–1296, 2000.
6. R. Gaumé, P.H. Haumesser, E. Antic-Fidancev, P. Porcher, B. Viana, D. Vivien, Crystal field calculations of Yb^{3+}-doped double borate crystals for laser applications, *J. Alloys Compd.* 341, 160–164, 2002.
7. P.H. Haumesser, R. Gaumé, B. Viana, D. Vivien, Determination of laser parameters of ytterbium-doped oxide crystalline materials, *J. Opt. Soc. Am. B*, 19(10), 2365–2375, 2002.
8. P. Lacovara, H. Choi, C. Wang, R. Aggarwal, T. Fan, Room-temperature diode-pumped Yb:YAG laser, *Opt. Lett.* 16(14), 1089–1091, 1991.
9. D. Sumida, A. Betin, H. Bruesselbach, R. Byren, S. Matthews, R. Reeder, M. Mangir, Diode-pumped Yb:YAG catches up with Nd:YAG, *Laser Focus World*, June 1999, pp. 63–70.
10. Jeff Hecht, in *Laser Focus World*, 40(12), December 2004, p. 81.
11. S. Chénais, F. Druon, F. Balembois, G. Lucas-Leclin, Y. Fichot, P. Georges, R. Gaumé, B. Viana, G.P. Aka, D. Vivien, Thermal lensing measurements in diode-pumped Yb-doped GdCOB, YCOB, YSO, YAG and KGW, *Opt. Mater.* 22, 129–137, 2003.
12. S. Chénais, F. Balembois, F. Druon, G. Lucas-Leclin, P. Georges, Thermal lensing in diode-pumped ytterbium laser — Part I: Theoretical analysis and wavefront measurements, *IEEE J. Quantum Electron.* 40(9), 1217–1234, 2004.
13. N. Barnes, B. Walsh, Quantum efficiency measurements of Nd:YAG, Yb:YAG, and Tm:YAG, *OSA TOPS Adv. Solid State Lasers* 68, 284–287, 2002.
14. M. Larionov, K. Schuhmann, J. Speiser, C. Stolzenburg, A. Giesen, Nonlinear decay of the excited state in Yb:YAG in *Advanced Solid State Photonics*, OSA, Washington, D.C., 2005, presentation number TuB49.
15. W.A. Clarkson, Thermal effects and their mitigation in end-pumped solid-state lasers, *J. Phys. D: Appl. Phys.* 34, 2381–2395, 2001.
16. D.C. Hanna, W. Clarkson, A review of diode-pumped lasers, in *Advances in Lasers and Applications: Proc. 52nd Scottish Universities Summer School in Physics*, St. Andrews, Scotland, Eds., D. Finlayson and B. Sinclair, SUSSP Publications and Institute of Physics Publishing, 1999.

17. E. Wyss, M. Roth, T. Graf, H. Weber, Thermooptical compensation methods for high-power lasers, *IEEE J. Quantum Electron.* 38(12), 1620–1628, 2002.
18. B. Viana, F. Druon, F. Balembois, P. Georges, Laser crystals for the production of ultra-short laser pulses, *Ann. Chim. Sci. Mater.* 28, 47–72, 2003.
19. R. Gaumé, B. Viana, Optical and thermo-mechanical properties of solid state laser materials, *Ann. Chim. Sci. Mater.* 28, 89–102, 2003.
20. L.D. Merkle, M. Dubinskii, B. Zandi, J.B. Gruber, D.K. Sardar, E.P. Kokanyan, V.G. Babajanyan, G.G. Demirkhanyan, R.B. Kostanyan, Spectroscopy of potential laser material Yb^{3+} ($4f^{13}$) in NaBi $(WO_4)_2$, *Opt. Mater.* 27, 343–349, 2004.
21. S.A. Payne, L.D. DeLoach, L.K. Smith, W.L. Kway, J.B. Tassano, W.F. Krupke, B.H.T. Chai, G. Loutts, Ytterbium-doped apatite-structure crystals: a new class of laser materials, *J. Appl. Phys.* 76, 497, 1994.
22. A.A. Kaminskii, *Laser Crystals: Their Physics and Properties*, Vol. 14, 2nd ed., Springer series in Optical Sciences, Springer-Verlag, Berlin, 1990.
23. E. Antic-Fidancev, What kind of information is possible to be obtained from ^{2S+1}L terms of ^4fN configurations, in *Physics of Laser Crystals*, Eds., J.C. Krupa and N. Kulagin, 126, 75, 2003.
24. A. Ellens, H. Andres, M. L. H. der Heerdt, R. T. Wegh, A. Meijerink, G. Blasse, The variation of the electron-phonon coupling strength through the trivalent lanthanide ion series, *J. Luminescence*, 66–67, 240–243, 1995.
25. A. Lupei, G.P. Aka, E. Antic-Fidancev, B. Viana, D. Vivien, Phonon effects in Yb^{3+} and Nd^{3+} spectra of GdCOB, *J. Luminescence* 94–95, 691–694, 2001.
26. E. Antic-Fidancev, Simple way to test the validity of $^{2S+1}L_J$ barycenters of rare earth ions, *J. Alloys Compd.* 300–301, 2, 2000.
27. P.-H. Haumesser, R. Gaumé, B. Viana, E. Antic-Fidancev, D. Vivien, Spectroscopic and crystal field analysis of new Yb-doped laser materials, *J. Phys. Cond. Matter* 13(23), 5427–5447, 2001.
28. H. Cañibano, G. Boulon, L. Palatella, Y. Guyot, A. Brenier, M. Voda, R. Balda, J. Fernandez, Spectroscopic properties of new Yb^{3+}-doped $K_5Bi(MoO_4)_4$ crystals, *J. Luminescence* 102–103, 318–326, 2003.
29. A.R. Edmonds, *Angular Momentum in Quantum Mechanics*, Princeton University Press, Princeton, New York, 1960.
30. B.R. Judd, *Operator Technique in Atomic Spectroscopy*, McGraw-Hill, New York, 1963.
31. F. Auzel, On the maximum splitting of the $^2F_{7/2}$ ground state in Yb^{3+} doped solid state laser materials, *J. Luminescence* 93, 129, 2001.
32. F. Auzel, A relationship for crystal field strength along the lanthanide series; application to the prediction of the $^2F_{7/2}$ Yb^{3+} maximum splitting, *Opt. Mater.* 19, 89, 2002.
33. K. Petermann, D. Fagundes-Peters, J. Johannsen, M. Mond, V. Peters, J.J. Romero, S. Kutovoi, J. Speiser, A. Giesen, Highly Yb-doped oxides for thin-disc lasers, *J. Cryst. Growth* Vol. 275, 1–2, 135, 2005.
34. V. Kisel, A. Troshin, V. Schcherbitshy, N. Kuleshov, Luminescence Lifetime Measurements in Yb^{3+} doped KYW and KGW, Conference on Advanced Solid State Photonics Santa Fe, U.S., Technical Paper WB7, 2004.
35. S. Chénais, F. Balembois, F. Druon, G. Lucas-Leclin, P. Georges, Thermal lensing in diode-pumped ytterbium laser — Part II Evaluation of quantum efficiencies and thermo-optic coefficients, *IEEE J. Quantum Electron.* 40(9), 1235, 2004.
36. L. DeLoach, S. Payne, L. Chase, L. Smith, W. Kway, W. Krupke, Evaluation of absorption and emission properties of Yb^{3+}-doped crystals for laser applications, *IEEE J. Quantum Electron.* 29, 1179, 1993.

37. R. Gaumé, P.-H. Haumesser, B. Viana, D. Vivien, G. Aka, B. Ferrand, Optical and laser properties of Yb:Y$_2$SiO$_5$ single crystals and discussion of the figure of merit relevant to compare ytterbium-doped laser materials, *Opt. Mater.* 19(1), 81–88, 2002.

38. W. Koechner, in *Solid State Laser Engineering*, 5th ed., Springer-Verlag, Berlin, 1999.

39. F. Augé, F. Druon, F. Balembois, P. Georges, A. Brun, F. Mougel, G.P. Aka, D. Vivien, Theoretical and experimental investigations of a diode-pumped quasi-three-level laser : the Yb^{3+}-doped Ca$_4$GdO(BO$_3$)$_3$ (Yb:GdCOB) laser, *IEEE J. Quantum Electron.* 36, 5, 598, 2000.

40. T.Y. Fan, Diode-pumped Solid-State Lasers", in Laser Sources and Applications, *Proc. 47th Scottish Universities Summer School in Physics*, Eds., A. Miller and D. Finlayson, SUSSP Publications and Institute of Physics Publishing, St. Andrews, Scotland, 1996.

41. D.F. de Sousa, N. Martynyuk, V. Peters, K. Lunstedt, K. Rademaker, K. Petermann, S. Basun, Quenching behaviour of highly-doped Yb:YAG and YbAG, Conference on Lasers and Electro-Optics Europe (CLEO Europe 2003), Technical digest, Conference edition (Optical Society of America, 2003), paper CG1-3.

42. P. Yang, P. Deng, Z. Yin, Concentration quenching in Yb:YAG, *J. Luminescence* 97, 51–54, 2002.

43. H. Yin, P. Deng, F. Gan, Defects in Yb:YAG, *J. Appl. Phys.* 83(8), 3825–3828, 1998.

44. T.Y. Fan, Heat generation in Nd:YAG and Yb:YAG, *IEEE J. Quantum Electron.* 29, 1457. 1993.

45. F. Patel, E. Honea, J. Speth, S. Payne, R. Hutcheson, R. Equall, Laser demonstration of Yb$_3$Al$_5$O$_{12}$ (YbAG) and materials properties of highly doped Yb:YAG, *IEEE. J. Quantum. Electron.* 37(1), 135–144, 2001.

46. J.L. Blows, P. Dekker, P. Wg, J.M. Dawes, T. Omatsu, Thermal lensing measurements and thermal conductivity of Yb:YAB, *Appl. Phys. B*, 76(3), 289–292, 2003.

47. S. Bowman, Lasers without internal heat generation, *IEEE J. Quantum Electron.* 35(1), 115–122, 1999.

48. S. Bowman, C. Mungan, New materials for optical cooling, *Appl. Phys. B*, 71, 807–811, 2000.

49. V. Pilla, T. Catunda, J. Jenssen, A. Cassanho, Fluorescence quantum efficiency measurements in the presence of Auger upconversion using the thermal lens method, *Opt. Lett.* 28(4), 239–241, 2003.

50. M.O. Ramirez, D. Jaque, L.E. Bausa, J.A. Sanz Garcia, J. Garcia Solé, Thermal loading in highly efficient diode pumped Ytterbium doped lithium niobate lasers, Conference on Lasers and Electro-Optics Europe (CLEO Europe), Europhysics Conference Abstracts Vol. 27E CP1-3-THU. 2003.

51. S. Hufner, in *Optical Spectra of Transparent Rare Earth Compounds*, Academic Press, New York, 1978, p 107.

52. S. Campos, S. Jandl, B. Viana, D. Vivien, P. Loiseau, B. Ferrand, Spectroscopic studies of Yb^{3+}-doped rare earth orthosilicate crystals, *J. Phys. Condens. Matter* 16, 25, 4579–4590, 2004.

53. S. Campos, J. Petit, B. Viana, S. Jandl, D. Vivien, B. Ferrand, Spectroscopic investigation of the laser materials Yb^{3+}:RE$_2$SiO$_5$; RE=Y, Sc, Lu, *SPIE Solid State Lasers and Amplifiers* 5460, 323–334, 2004.

54. A.A. Lagatskii, N.V. Kuleshov, V.G. Shcherbiskii, V.F. Kleptsyn, V.P. Mikhailov, V.G. Ostroumov, G. Huber, Lasing characteristics of a diode-pumped Nd^{3+}:CaGdAlO$_4$ crystal, *IEEE J. Quantum Electron.* 27(1), 15–17, 1997.

55. N.V. Kuleshov, A.A. Lagatsky, V.G. Shcherbitsky, V.P. Mickhailov, E. Heumann, T. Jansen, A. Diening, G. Huber, CW laser performance of Yb and Er/Yb doped tungstates, *Appl. Phys. B* 64, 409, 1997.

56. X. Mateos, V. Petrov, M. Aguilo, R.M. Solé, J. Gavalda, J. Massons, F. Diaz, U. Griebner, CW laser oscillation of Yb^{3+} in monoclinic $KLu(WO_4)_2$, *IEEE J. Quantum Electron.* 40, 1056, 2004.

57. J. Liu, J.M. CanoTorres, C. Cascales, F. Esteban-Betegon, M.D. Serrano, V. Volkov, C. Zaldo, M. Rico, U. Griebner, V. Petrov, Growth and CW laser operation of disordered crystals of $Yb:NaLa(WO_4)_2$ and $Yb:NaLa(MoO_4)_2$, *Phys. Status Solidi* 202(4), R29, 2005.

58. F. Druon, F. Balembois, P. Georges, A. Brun, A. Courjaud, C. Honninger, F. Salin, A. Aron, F. Mougel, G. Aka, D. Vivien, Generation of 90-fsec pulses from a mode-locked diode-pumped $Yb^{3+}:Ca_4GdO(BO_3)_3$ laser, *Opt. Lett.*, 25, 423, 2000.

59. M. Jacquemet, C. Jacquemet, N. Janel, F. Druon, F. Balembois, P. Georges, J. Petit, B. Viana, D. Vivien, B. Ferrand, Efficient laser action of Yb:LSO and Yb:YSO oxy-orthosilicatees crystals under high-power diode pumping, *Appl. Phys. B: Laser and Optics* 80(2), 171–176, 2005.

60. M. Jacquemet, F. Balembois, S. Chénais, F. Druon, P. Georges, R. Gaumé, B. Ferrand, First diode pumped solid state laser continuously tunable between 1000 and 1010 nm, *Appl. Phys. B* 78, 13, 2003.

61. J. Petit, P. Goldner, B. Viana, Laser emission with low quantum defect in $Yb:CaGdAlO_4$, *Opt. Lett.* 30, 1345–1347, 2005.

62. C. Stewen, M. Larionov, A. Giesen, K. Kontag, Yb:YAG thin disk laser with 1 kW output power, *OSA Trends in Optics and Photonics*, Vol. 34, Advanced Solid State Lasers, p. 35, 2000.

63. N.V. Kuleshov, A.A. Lagatsky, A.V. Podlipensky, V.P. Mikhailov, E. Heumann, A. Diening, G. Huber, Highly efficient CW and pulsed lasing of doped tungstate, *OSA Trends in Optics and Photonics*, Eds., C.R. Pollock, W.R. Bosenberg, OSA, Washington, D.C., 10, 415, 1997.

64. C. Honninger, R. Paschotta, F. Morier-Genoud, M. Moser, and U. Keller, Q-switching stability limits of continuous-wave passive mode locking, *J. Opt. Soc. Am. B* 16, 46, 1999.

65. C. Honninger, R. Paschotta, M. Graf, F. Morier-Genoud, G. Zhang, M. Moser, S. Biswal, J. Nees, A. Braun, G. Mourou, I. Johannsen, A. Giesen, W. Seeber, U. Keller, Ultrafast ytterbium-doped bulk laser amplifiers, *Appl. Phys. B* 69, 3, 1999.

66. F. Druon, S. Chénais, P. Raybaut, F. Balembois, P. Georges, R. Gaumé, G. Aka, B. Viana, S. Mohr, D. Kopf, Diode-pumped $Yb:Sr_3Y(BO_3)_3$ femtosecond laser, *Opt. Lett.* 27 197, 2002.

67. F. Druon, S. Chénais, P. Raybaut, F. Balembois, P. Georges, R. Gaumé, P. H. Haumesser, B. Viana, D. Vivien, S. Dhellemmes, V. Ortiz, C. Larat, Apatite-structure crystal, $Yb:SrY_4(SiO_4)_3O$, for the development of diode-pumped femtosecond lasers, *Opt. Lett.* 27, 1914, 2002.

68. F. Brunner, G.J. Spülher, J. Aus der Au, L. Krainer, F. Morier-Genoud, R. Paschotta, N. Lichtenstein, S. Weiss, C. Harder, A.A. Lagatsky, A. Abdolvand, N.V. Kuleshov, U. Keller, Diode-pumped femtosecond $Yb:KGd(WO_4)_2$ laser with 1.1-W average power, *Opt. Lett.* 25(15), 1119–1121, 2000.

69. Y. Zaouter, J. Didierjean, F. Balembois, G. Lucas Leclin, F. Druon, P. Georges, J. Petit, P. Goldner, B. Viana, 47 fsec diode-pumped $Yb^{3+}:CaGdAlO_4$ laser, *Opt. Lett.* 31(1), 119–121, 2006.

70. P. Klopp, V. Petrov, U. Griebner, K. Petermann, V. Peters, G. Erbert, Highly efficient mode locked Yb:Sc$_2$O$_3$ laser, *Opt. Lett.* 29, 391–393, 2004.
71. D. Strickland, G. Mourou, Compression of amplified chirped optical pulses, *Opt. Commun.* 55(6), 447, 1985.
72. C. Honninger, F. Morier-Genoud, M. Moser, U. Keller, L. R. Brovelli, C. Harder, Efficient and tunable diode-pumped femtosecond Yb:glass lasers, *Opt. Lett.* 23(2), 126–128, 1998.
73. A. Shirakawa, K. Takaichi, H. Yagi, J.F. Bisson, J. Lu, M. Musha, K. Ueda, T. Yanagitani, T.S. Petrov, A.A. Kaminskii, Diode pumped mode locked Yb:Y$_2$O$_3$ ceramic laser, *Opt. Express* 11(22), 2911–2916, 2003.
74. A. Lucca, G. Debourg, M. Jacquemet, F. Druon, F. Balembois, P. Georges, P. Camy, J.L. Doualan, R. Moncorgé, High power diode-pumped Yb^{3+}:CaF$_2$ femtosecond laser, *Opt. Lett.* 29, 23, 2767, 2004.
75. W.F. Krupke, M.D. Shinn, J.E Marion, J.A. Caird, S.E. Stokowski, Spectroscopic, optical and thermomechanical properties of neodymium and chromium doped gadolinium scandium gallium garnet, *J. Opt. Soc. Am. B* 3(1), 102–113, 1986.
76. A.A. Kaminskii, Modern developments in the physics of crystalline laser materials, *Phys. Stat. Sol. (a)*, 200, 215, 2003.
77. R. Gaumé, B. Viana, D. Vivien, J.P. Roger, D. Fournier, Simple model for the prediction of thermal conductivity in pure and doped insulating crystals, *Appl. Phys. Lett.* 83(7), 1355–1357, 2003.
78. A. Akhieser, On absorption of sound in solids, *J. Phys. (Moscow)*, 1, 277, 1939.
79. G.A. Slack and D.W. Oliver, Thermal conductivity of garnets and phonon scattering by rare-earth ions, *Phys. Rev. B* 4, 592–609, 1971.
80. G. Klemens, Thermal resistance due to point defects at high temperature, *Phys. Rev.* 119(2), 507, 1960.
81. V. Peters, A. Bolz, K. Petermann, G. Huber, Growth of high-melting sesquioxides by the heat exchanger method, *J. Cryst. Growth* 237–239, 879, 2002.
82. K. Petermann, L. Fornasiero, E. Mix, and V. Peters, High melting sesquioxides: crystal growth, spectroscopy, and laser experiments, *Opt. Mater.* 19, 67, 2002.
83. C. van Eijk, Perspectives on the future development of new scintillators, *Nuclear Instruments and Methods in Physics Research Section A: Accelerators, Spectrometers Detectors and Associated Equipment*, 537 1–2, 6, 2005.
84. CTI Momecular Imagine Inc. (http://www.ctimi.com); Saint Gobain Crystals (http://www.detectors.saint-gobain.com).
85. J. Petit, B. Viana, P. Goldner, D. Vivien, P. Loiseau, B. Ferrand, Oscillation with low quantum defect in Yb:GdVO$_4$, a crystal with high thermal conductivity, *Opt. Lett.* 29(8), 833–835, 2004.

3

Tunable Cr²⁺:ZnSe Lasers in the Mid-Infrared

Alphan Sennaroglu and Umit Demirbas

CONTENTS

3.1   Introduction

3.1.1   Overview of Cr^{2+}:ZnSe Lasers

Undoped bulk II-VI semiconductors (also referred to as chalcogenides) such as CdSe, ZnSe, and ZnTe have a wide transparency window in the near-infrared and mid-infrared regions of the electromagnetic spectrum. If divalent transition metal ions (TM^{2+}) such as Co^{2+}, Cr^{2+}, and Fe^{2+} are introduced into these materials, crystal field splitting and strong electron–phonon coupling lead to the formation of broad absorption and emission bands in the mid-infrared. As an example, let us consider the ZnSe medium which has a band-gap of 2.67 eV [1] and phonon energy of 250 cm^{-1} [2]. In contrast to pure bulk ZnSe whose transparency window extends from 0.5 to 22 μm [3], chromium-doped ZnSe (Cr^{2+}:ZnSe) possesses a strong absorption band centered near 1800 nm. Furthermore, optical excitation of the substitutional Cr^{2+} ions with a pump emitting in the 1500–2100 nm region produces a broad, strong emission band in the mid infrared between 2 and 3 μm. As a matter of fact, these spectroscopic properties of Cr^{2+}:ZnSe in particular and of other TM^{2+}-doped chalcogenides in general have long been known [4–9]. For example, infrared absorption spectra of chromium-containing II-VI compounds were studied in detail by researchers since early 1960s [9–11]. In addition, absorption and emission characteristics of other TM^{2+} ions such as Fe^{2+} were thoroughly investigated [6–8,12,13]. Interestingly, despite this considerable interest in the spectroscopy of TM^{2+}-doped II-VI compounds, lasing action was not reported until mid 1990s. In the pioneering work reported by DeLoach et al., absorption and emission characteristics of various zinc chalcogenides doped with Cr^{2+}, Co^{2+}, Ni^{2+}, and Fe^{2+} were studied to evaluate their potential as active media in the mid infrared, and room-temperature lasing was demonstrated with Cr^{2+}:ZnSe and Cr^{2+}:ZnS near 2.4 μm [14,15]. Since then, tunable room-temperature laser action has also been successfully demonstrated with other chalcogenide hosts doped with Cr^{2+}, including CdSe [16–18], CdTe [19], ZnS_xSe_{1-x} [20], $Cd_xMn_{1-x}Te$ [21–25], CdZnTe [26], and others.

Cr^{2+}:ZnSe remains the most extensively studied member of this class of lasers and, over the last decade, has emerged as a versatile source of broadly tunable laser radiation in the mid infrared region between 2 and 3 μm. This gain medium possesses many favorable spectroscopic characteristics that enable efficient lasing. These include a four-level energy structure and absence of excited-state absorption which allow low-threshold continuous-wave (cw) operation, a broad absorption band that overlaps with the operating wavelength of many laser systems for optical pumping, a phonon-broadened emission band that gives rise to wide tunability, and near-unity fluorescence quantum efficiency at room temperature. To date, gain-switched [14,15], cw [27], diode-pumped [28], mode-locked [29], random-lasing [30], and single-frequency [31] operations have been demonstrated. Broad tunability in the 2000–3100 nm wavelength range was also reported [2]. Among the potential applications of Cr²⁺:ZnSe lasers, we can list atmospheric imaging [32–34], vibrational spectroscopy [35–40], self-difference frequency mixing [41], and optical pumping of mid infrared lasers as well as optical parametric oscillators [42].

In this chapter, we provide an extensive review of the recent studies aimed at the development of Cr²⁺:ZnSe lasers. Since Cr²⁺:ZnSe belongs to the family of tunable solid-state lasers, we start with a brief historical review of this field in Subsection 3.1.2. The key milestones are outlined and various examples of tunable solid-state lasers are mentioned. In Section 3.2, we describe various techniques that are commonly used for the synthesis and spectroscopic characterization of Cr²⁺:ZnSe. In particular, we concentrate on diffusion doping and discuss how the diffusion time and diffusion temperature influence the strength of the absorption bands. Different techniques for the determination of the diffusion constant are then described. In Section 3.3, we discuss the fluorescence properties of Cr²⁺:ZnSe. Experimental data showing the dependence of the fluorescence efficiency and fluorescence lifetime on chromium ion concentration are presented. The temperature dependence of the fluorescence lifetime is discussed. In Subsection 3.3.4, we further describe techniques with which the absorption cross section can be determined from cw and pulsed absorption saturation data. Pulsed operation of Cr²⁺:ZnSe lasers is covered in Section 3.4. After a review of the earlier work, the dependence of the power performance on active ion concentration is discussed. In tuning experiments, we describe a pulsed, intracavity-pumped Cr²⁺:ZnSe laser with an ultrabroad tuning range extending from 1880 to 3100 nm. Continuous-wave operation of Cr²⁺:ZnSe lasers is next discussed in Section 3.5. Here, after a review of the work on cw Cr²⁺:ZnSe lasers, we describe the operation of a cw 1800-nm-pumped Cr²⁺:ZnSe laser. The power performance of several Cr²⁺:ZnSe samples were also investigated in order to determine the optimum chromium concentration for cw operation. Finally, Section 3.6 describes the mode-locked operation of Cr²⁺:ZnSe lasers. Experiments employing both active and passive mode-locking techniques are reviewed.

3.1.2 Historical Review of Tunable Solid-State Lasers

The Cr²⁺:ZnSe laser belongs to the family of tunable solid-state lasers, also referred to as vibronic lasers due to the role of electron–phonon coupling in the broadening of the absorption and emission bands. In tunable solid-state lasers, the output wavelength can be tuned over a substantial fraction of the central emission wavelength. As such, the fractional tuning range, defined as $\Delta\lambda/\lambda_0$ ($\Delta\lambda$ = full width of the tuning range and λ_0 = the central emission wavelength) can be several tens of percent. Table 3.1 shows some examples of tunable solid-state gain media and their reported fractional tuning range at room temperature. Tunable solid-state lasers have a long history dating back to the early days of lasers. In this section, we briefly outline some of the important historical developments in this field. For a comprehensive review on their physical properties and operational characteristics, we refer the reader to several excellent review articles on this subject [2,30,43–49].

Work on tunable solid-state lasers started shortly after the invention of the ruby laser in 1960 [50]. First experimental demonstration of tunable laser action was reported by Johnson et al. in flashlamp-pumped Ni^{2+}-doped magnesium fluoride (MgF_2) [51]. This was followed by the development of other tunable sources based on several crystal hosts doped with divalent ions such as Ni^{2+}, Co^{2+}, and V^{2+} [52–54]. These lasers could produce radiation in the near infrared between 1.1 and 1.8 μm but due to their low luminescence efficiency at elevated temperatures, they needed to be operated at cryogenic temperatures. In the meantime, a rigorous theoretical treatment of phonon-terminated lasers was also given by McCumber [55–58]. For over a decade between mid 1960s and late 1970s, limited activity continued in search for new tunable solid-state gain media and dye and color-center lasers were instead used widely as tunable sources. Toward the end of 1970s, interest in tunable solid-state lasers was revived. This was driven by two important developments. First, laser end pumping was introduced by Moulton et al. for the optical excitation of solid-state lasers [59]. Since the nearly diffraction-limited output of already existing commercial lasers could be focused to small volumes, lasing thresholds were considerably lowered. Second, as a result of improvements in crystal growth techniques, new high-quality tunable laser crystals became commercially available. One of the early examples of tunable solid-state media that emerged at this time was the alexandrite

TABLE 3.1

Room-Temperature Tuning Range of Several Transition Metal Ion-Doped Solid-State Gain Media

Gain Medium	Tuning Range (nm)	$\Delta\lambda/\lambda_0$	Reference
$Ti^{3+}:Al_2O_3$	660–1180	0.57	140
$Cr^{3+}:BeAl_2O_4$	701–818	0.15	48
$Cr^{4+}:Mg_2SiO_4$	1130–1367	0.19	141
$Cr^{4+}:Y_3Al_5O_{12}$	1309–1596	0.20	142
$Co^{2+}:MgF_2$	1750–2500	0.35	143
$Fe^{2+}:ZnSe$	3900–4800	0.21	68
$Cr^{2+}:ZnS$	2110–2840	0.29	144
$Cr^{2+}:CdSe$	2300–2900	0.23	17
$Cr^{2+}:ZnSe$	1880–3100	0.49	This work

laser (Cr^{3+}:$BeAl_2O_4$). Since their invention in 1979 [60], there has been an ever growing interest in the spectroscopic characterization and development of new tunable solid-state lasers. To date, lasing action has been reported from many vibronic gain media including Ti^{3+}:sapphire [61], various chromium (Cr)-doped lasers [37–45], Fe^{2+}:ZnSe [46], and others. By using different ion–host combinations, the wavelength range between 665 and 4500 nm can now be covered.

In the search for new tunable gain media, one of the most extensively explored transition metal ions has been chromium. This is primarily due to the chemical stability of the different ionic charge states in the lattice and the existence of broad pump bands. Depending on the charge state of the substitutional chromium ion introduced into the host, chromium-doped laser sources can be classified into three categories. The first group consists of Cr^{3+}-doped solid-state lasers such as alexandrite [60], Cr^{3+}-doped garnets [62,63], Cr^{3+}:LiCAF [64], Cr^{3+}:LiSAF [65], and others. Some of these such as alexandrite and Cr^{3+}-doped garnets are among the earliest room-temperature tunable solid-state lasers. The tuning range is host-dependent and extends approximately from 700 nm to 1000 nm. The second category consists of the near-infrared Cr^{4+} lasers such as Cr^{4+}:forsterite and Cr^{4+}:YAG which have been extensively reviewed in Reference 44. The third and the most recently explored group consists of divalent chromium-doped hosts such as Cr^{2+}:ZnSe which is the main theme of this chapter.

We note in passing that another important TM^{2+}-doped chalcogenide laser is Fe^{2+}:ZnSe in which lasing action was reported in the 4–4.5 μm wavelength range [66]. There is a lot of ongoing activity in the investigation of solid-state lasers and saturable absorbers based on Fe^{2+}:ZnSe. We refer the reader to references 66 to 72 for further discussion on these studies.

3.2 Synthesis and Absorption Spectroscopy of Cr²⁺:ZnSe

3.2.1 Synthesis Methods for Cr²⁺:ZnSe

Various methods can be used to introduce the laser-active TM^{2+} ions into the chalcogenide hosts either during the growth of the host or after the growth process. Techniques that incorporate chromium ions during the growth of the host include melt growing [10], vapor growing [73,74], pulsed laser deposition [75], molecular beam epitaxy [76], and temperature solution growth [77]. The most widely used postgrowth technique is thermal diffusion [78]. In this section, we provide a brief comparative review of these methods. In Subsection 3.2.2, thermal diffusion doping is discussed in detail.

Melt growing (solidification from melt [79]) of chromium doped II-VI materials has been one of the extensively used methods [10,14,19,21,23,25,78,80–83]. Here, a given amount of the dopant is first mixed with the host, and doped samples are then pulled from the molten mixture by using a modified vertical Bridgman growth technique. Previous studies showed that melt-grown samples suffer from being coarsely polycrystalline and contain defects such as inclusions and voids [14,80]. Besides, controlling the dopant concentration is difficult since the actual dopant concentration in the sample is not necessarily the same as the starting concentration [23]. Hence, concentration gradients were observed in samples prepared with this method [14,80]. Possibility of uncontrolled contamination during the melting process, leading to unwanted passive losses, has also been pointed out [78,81].

Vapor growth of Cr^{2+}:ZnSe using physical vapor transport (PVT) has been shown to produce samples with higher crystal and optical quality [15,73]. However, with the PVT technique, obtaining samples with uniform chromium concentration is difficult [73,78,80,84,85]. Besides, previous studies also showed that chromium diffusion efficiency in this technique is relatively low for CdSe [86,87] and CdS_xSe_{1-x} [88]. In a comparative study, it was shown that postgrowth thermal diffusion doping is more effective in introducing chromium in CdS_xSe_{1-x} in comparison with the PVT technique [88].

Pulsed laser deposition of Cr^{2+}:ZnS thin films on silicon substrate has also been demonstrated [75]. In this technique, precise control of chromium dopant concentration was possible [75], but there is very little work in the literature on the optical quality of the samples prepared with this method. In some studies, pulsed laser deposition was also used to coat the host surfaces with chromium films, and then, thermal diffusion doping was employed in the second step to add the chromium ions into the host [89,90]. This is discussed further in Subsection 3.2.2.

Another alternative method for introducing chromium into the chalcogenide host during the growth is molecular beam epitaxy [76]. This method allows for the fabrication of complex heterostructures and the adjustment of chromium concentration within the sample, enabling the growth of integrated structures [76]. On the other hand, the growth rates in this technique are very slow (~1 μm/h) for obtaining millimeter-sized bulk samples [91]. In addition, surface segregation of chromium during crystal growth may degrade the optical quality of the samples [76].

Thermal diffusion doping is the widely used postgrowth technique for the preparation of Cr^{2+}:ZnSe samples [92]. In comparison with the above techniques which require sophisticated instrumentation, thermal diffusion doping offers a simple, efficient, and cost-effective alternative. This method has been known for decades, and in the earliest studies on chromium doped II-VI semiconductors, samples prepared with this technique were used [9]. Chromium doping level of the samples can be adjusted easily by varying the diffusion temperature and/or the diffusion time. However, spatial inhomogeneities in chromium concentration may result due to the nature of the diffusion process. Furthermore, slightly higher passive losses have been observed in diffusion-doped samples in comparison with those obtained by using other methods such as PVT [93].

3.2.2 Thermal Diffusion Doping

The thermal diffusion doping method uses commercially available ZnSe samples in crystalline, polycrystalline or ceramic [94] form as the starting host material. Polycrystalline samples have been widely used since their laser and spectroscopic properties are comparable with those of single-crystal Cr^{2+}:ZnSe samples [95]. As the dopant, sputtered film of metallic chromium deposited on the ZnSe host surface [9,10,78,81,96] or powders of CrSe (or Cr) [74,79,97,98] can be used. The deposition of metallic chromium on ZnSe surface was accomplished by using magnetron sputtering systems [96] or pulsed laser deposition [89,90]. One important issue that needs to be addressed is that both the dopant and host must be extremely pure in order to obtain laser-quality samples (purity better than 99.99%). Sometimes, this may require additional purification steps before diffusion. Even trace amounts of impurities will cause unwanted losses in the doped material. As an example, Fe^{2+} in ZnSe has a strong absorption band between 2.2 and 5 μm, overlapping with the emission band of Cr^{2+}:ZnSe [66,92]. Hence, the presence of Fe^{2+} may greatly reduce the lasing efficiency [3,67,87].

In our studies, polycrystalline ZnSe samples were used for diffusion doping. Highly pure powders of CrSe or Cr were used as dopant. Figure 3.1 shows a sketch of the silica ampoules used for diffusion doping. The dopant (Cr or CrSe powder) and the host (ZnSe) are placed in different compartments so that the deposition of the dopant on the ZnSe occurs only via the gas phase. As a result, contamination due to less volatile impurities such as metal oxides could be minimized. Besides, preventing a direct contact between the CrSe powder and ZnSe gives rise to a more uniform distribution of the dopant inside the sample and prevents the formation of hot spots on the sample surface [78]. Diffusion was activated thermally by heating the system to temperatures between 800 and 1100°C. The ampoule is also kept under high vacuum during the diffusion process ($P < 10^{-5}$ mbar). The diffusion time varies from several hours to tens of days. By adjusting the diffusion temperature or the diffusion time, one can control the average chromium concentration in the host as will be delineated in Subsection 3.2.3.

Cr^{2+}:ZnSe samples prepared by thermal diffusion doping require polishing after the synthesis process due to two reasons. First, the surface roughness increases during diffusion and polishing improves the optical quality of the doped samples. Second, polishing also removes the heavily chromium-doped layer near the surface. As will be further noted in the following sections, high chromium concentration causes several unwanted effects such as enhanced

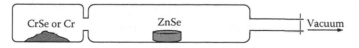

FIGURE 3.1
Schematic of the silica ampoule used in the preparation of Cr^{2+}:ZnSe by thermal diffusion doping.

nonradiative decay and increased passive losses. Hence, removal of the highly doped layer can improve the laser performance of the samples.

3.2.3 Modeling of Thermal Diffusion Doping

The concentration of the chromium ions inside the host varies with position, diffusion time, and diffusion temperature. Here, we discuss a fairly accurate model of the thermal diffusion doping process based on the solution of the diffusion equation in cylindrical coordinates [92]. In this model, a constant dopant vapor concentration is assumed on all sides of the sample during diffusion. The concentration $N(\vec{r},t)$ of the chromium ions inside the host can be calculated by using the well-known diffusion equation

$$\frac{\partial N(\vec{r},t)}{\partial t} = D\nabla^2 N(\vec{r},t), \tag{3.1}$$

where t is the diffusion time, $\vec{r}$ gives the position inside the ZnSe sample, and D is the diffusion coefficient. In our case, the samples were cylindrical in shape, with radius R and thickness L. In this case, Equation 3.1 can be solved exactly in cylindrical coordinates by using the following set of initial/boundary conditions:

$$N(r,z,t=0) = 0$$

$$N(r=R,z,t>0) = n_o \tag{3.2}$$

$$N\left(r,z=\pm\frac{L}{2},t>0\right) = n_o.$$

Above, n_0 is the constant chromium ion vapor concentration on all sides of the ZnSe sample. The solution of the above boundary value problem can be expressed as

$$N(r,z,t) = n_0 - \frac{8n_0}{\pi R}\sum_{m=1}^{\infty}\sum_{n=0}^{\infty}\frac{(-1)^n J_0(r\alpha_m)}{(2n+1)\alpha_m J_1(R\alpha_m)}\cos\left(\frac{(2n+1)\pi}{L}z\right)e^{\left(-\left(\frac{(2n+1)^2\pi^2}{L}+\alpha_n^2\right)Dt\right)}, \tag{3.3}$$

where n and m are summation indices, $J_0(x)$ is the Bessel function of the first kind of order zero, $R\alpha_m$ is the mth positive root of $J_0(x) = 0$ and $J_1(x)$ is the Bessel function of the first kind of order one [99].

3.2.4 Room-Temperature Absorption Spectra of Cr²⁺:ZnSe

Chromium ions enter the ZnSe host as substitutional impurities and occupy the Zn^{2+} cation sites in the lattice [11]. Electron spin resonance experiments revealed that chromium ion in ZnSe can have the charge states of Cr^{1+}, Cr^{2+}, and Cr^{3+} [10,74]. However, it was shown that Cr^{2+} is the stable charge state in ZnSe, and more than 95% of the chromium have a valence of two [10]. In support of this, recent deep-level spectra in Cr^{2+}:ZnTe have shown that the concentration of Cr^{1+} ions is about 10^{-5} of the total added chromium [91]. Figure 3.2 shows a simplified energy level diagram of the Cr^{2+} ($3d^4$) ion in the ZnSe host [2,100]. The ground level splits into the $^5T_2$ and ^5E energy levels under the influence of tetrahedral crystal field of the ZnSe host. Further splitting of these energy levels occurs as a result of the Jahn–Teller effect. The energy levels all lie within the forbidden bandgap of the ZnSe semiconductor.

Figure 3.3 shows the room-temperature absorption spectra of undoped and chromium-doped ZnSe samples in the 300–3100 nm spectral range. The three chromium doped samples shown in the figure (Samples 1–3) were prepared using thermal diffusion doping at 1000°C with diffusion times of 1 (s1), 4 (s2), and 5 (s3) days, respectively. The undoped material is transparent at wavelengths above 470 nm, corresponding to the band-gap energy of 2.67 eV for ZnSe [1]. Diffusion doping of chromium modifies the absorption spectra of ZnSe in several ways. First, a strong absorption band of the laser-active Cr^{2+} ion forms in the near-infrared region centered at 1775 nm. This absorption band is due to the optical transitions between the $^5T_2$ and ^5E levels of the Cr^{2+} ions (Figure 3.2). The differential absorption coefficient $\alpha(r,z,t)$ of this band can be expressed in terms of the chromium ion concentration $N(r,z,t)$ and the absorption cross section $\sigma_a(\lambda)$ as

$$\alpha(r,z,t) = \sigma_a(\lambda)N(r,z,t). \tag{3.4}$$

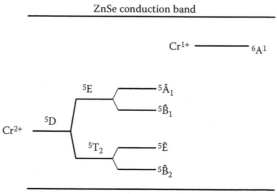

FIGURE 3.2
Energy level diagram of Cr^{2+} in ZnSe. ^5D energy level of the Cr^{2+} ion splits into ^5E and $^5T_2$ energy levels under the influence of the crystal field. These levels are further split due to the Jahn–Teller effect. All these energy levels lie inside the forbidden bandgap of ZnSe.

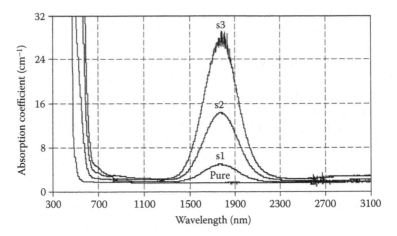

FIGURE 3.3
Room-temperature absorption spectra of pure and chromium-doped ZnSe samples. Cr^{2+}:ZnSe Samples 1–3 (s1 through s3 in the figure) were subjected to diffusion at 1000°C for a period of 1, 4, and 5 days, respectively. Each sample had a thickness of about 2 mm.

This band has a Gaussian shape, and the wavelength dependence of absorption cross section can be accurately represented by the following empirical formula:

$$\sigma_a(\lambda) = \sigma_o \exp\left[-4\left(\frac{\lambda - \lambda_o}{\Delta\lambda \Big/ \sqrt{\ln(2)}}\right)^2\right]. \tag{3.5}$$

In Equation 3.5, σ_o is the peak absorption cross section, λ is the wavelength, λ_o and $\Delta\lambda$ are the central wavelength and the full-width at half-maximum (FWHM) of the absorption band, respectively [78]. Experimental absorption measurements taken with Cr^{2+}:ZnSe samples show that λ_o and $\Delta\lambda$ do not vary significantly with concentration. Best fit between experimental data and Equation 3.5, gave λ_o = 1775 nm and $\Delta\lambda$ = 365 nm for Cr^{2+}:ZnSe. This transition is very wide ($\Delta\lambda$ = 365 nm), mainly due to vibronic coupling [2], enabling the use of numerous different pump sources to excite the Cr^{2+}:ZnSe gain medium. Equation 3.4 and Equation 3.5 can be used to determine the Cr^{2+} concentration from absorption measurements. Most studies have used the σ_o-value of 11.5×10^{-19} cm^2 reported by Vallin et al. [10].

The second effect of chromium ion on the ZnSe absorption spectra is the shift of the short-wavelength absorption edge of ZnSe (~470 nm) to longer wavelengths. This shift is due to an absorption band around 500 nm originating from the $Cr^{2+} + e_{VB} \rightarrow Cr^{1+} + h_{VB}$ transition (where e_{VB} is a valance band electron and h_{VB} is a valance band hole) [101–103]. This transition is one of the ways of generating rarely found Cr^{1+} charge state of chromium in ZnSe

[10]. Previous studies showed that the strength of this band scales with the Cr^{2+} ion concentration. The amount of the short-wavelength absorption edge shift is also related to the dopant concentration. In our studies, we observed a maximum shift of 160 nm (band edge ~640 nm) for a sample with an active Cr^{2+} ion concentration of 66×10^{18} ions/cm³. There are two more transitions near the band gap edge of ZnSe, one at 610 nm and another at 680 nm which are hardly visible in the room-temperature absorption spectra shown in Figure 3.3 [79]. The origin of these transitions is not clear and more detailed information can be found in Reference 79. Another important band of Cr^{2+} ion in ZnSe is centered around 6.5 μm, extending up to 14 μm on the long wavelength side [79]. The importance of this band lies in the fact that its short wavelength side also overlaps with the lasing wavelength range of Cr^{2+}:ZnSe, causing unwanted absorption losses [79].

3.2.5 Dependence of Passive Laser Losses on Chromium Concentration

Earlier studies with Cr^{2+}:ZnSe gain medium showed that increasing chromium concentration leads to increased passive losses at the lasing wavelengths [28]. We used the measured absorption spectra of Cr^{2+}:ZnSe samples with different chromium concentrations to investigate this effect in detail. Comparison of the measured absorption spectra of the Cr^{2+}:ZnSe samples with those of the undoped ZnSe allows the estimation of the passive losses at the lasing wavelengths (around 2500 nm, see Figure 3.3). Our measurements showed that, despite slight variations from sample to sample, passive losses of Cr^{2+}:ZnSe increase linearly with increasing Cr^{2+} concentration. This can be clearly seen in the absorption spectra of the three Cr^{2+}:ZnSe samples shown in Figure 3.3. For example, whereas sample 1 with a Cr^{2+} concentration of 2.7×10^{18} ions/cm³ has an estimated single-pass loss of about 2% at 2500 nm, the loss increases to above 10% for sample 3 which has a concentration of 23.2×10^{18} ions/cm³. Based on the absorption spectra of more than 10 polycrystalline Cr^{2+}:ZnSe samples, we obtained the following empirical equation for the differential loss coefficient at 2500 nm (α_{loss}, in the units of cm⁻¹) as a function of the average Cr^{2+} concentration N_{Cr} (cm⁻³):

$$\alpha_{loss} \approx (0.04 \pm 0.02) + \left\{ (0.02 \pm 0.01) \times 10^{-18} \right\} N_{Cr}. \quad (3.6)$$

The constant in Equation 3.6 (0.04 ± 0.02 cm⁻¹) is probably due to scattering losses.

3.2.6 Determination of the Diffusion Coefficient

Knowledge of the diffusion coefficient for Cr^{2+} ions in the ZnSe host facilitates the preparation of Cr^{2+}:ZnSe samples with a desired active ion concentration. In previous studies, Ndap et al. studied the preparation of Cr^{2+}:ZnSe samples by thermal diffusion doping between 800 and 950°C to determine the diffusion coefficient. Both single-crystal and polycrystalline hosts were used. It was shown that diffusion is faster in polycrystalline samples than in single crystals below 910°C probably due to the enhancement of diffusion along grain boundaries [78]. For temperatures above 910°C, diffusion speed was similar in both cases due to the grain growth process observed in polycrystalline samples at temperatures above ~900°C [78]. Both sputtered films and CrSe powders were used as the dopant source, and diffusion was observed to be faster with CrSe powder than with sputtered metallic chromium for temperatures higher than 850°C [78]. In a recent work, thermal diffusion process in ZnSe single crystals was studied between 800 and 1000°C by using sputtered metallic chromium layers [104].

In our work, we used the model discussed in Subsection 3.2.3, to determine the diffusion coefficient for Cr^{2+} ions (from CrSe powder source) in the polycrystalline ZnSe host between 900 and 1100°C [92]. Two different methods were employed. In the first method, the average absorption coefficient at the center of the cylindrical Cr^{2+}:ZnSe samples was measured as a function of the diffusion time. Sixteen samples prepared at 1000°C for different diffusion times were used. Figure 3.4 shows the measured variation of the average absorption coefficient (α_{av}) at 1775 nm as a function of the diffusion time. Least-squares fit between the experimental data and the diffusion model gave the best value of 6×10^{-10} cm^2/sec for D.

We also used a second method which allows the determination of the diffusion coefficient with one sample only. In this case, a Cr:YAG laser operating at 1510 nm was used to measure the spatial variation of the absorption

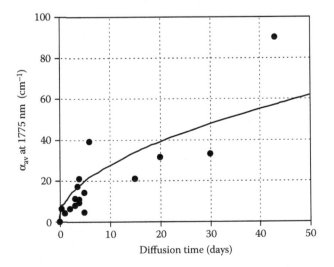

FIGURE 3.4
Measured variation of the peak absorption coefficient at the sample center as a function of the diffusion time for several Cr^{2+}:ZnSe samples. All the samples were prepared at 1000°C. The solid curve is the best theoretical fit based on the diffusion model.

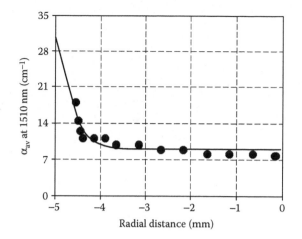

FIGURE 3.5

Measured and calculated variation of the absorption coefficient as a function of the radial distance in the Cr²⁺:ZnSe sample. The sample was subjected to diffusion for 10 days at 1000°C. The thickness and radius are 2 mm and 5 mm, respectively.

coefficient inside the sample. In these experiments, the output beam of the Cr:YAG laser was focused to a 20-μm spot and scanned across the sample cross section. Figure 3.5 shows the measured and calculated variation of the average absorption coefficient (α_{av}) as a function of the radial distance for a Cr²⁺:ZnSe sample subjected to diffusion at 1000°C for 10 days. Note that this measurement also gives information on the spatial uniformity of the Cr²⁺:ZnSe samples prepared by thermal diffusion doping. For the specific sample in Figure 3.5, spatially uniform section of the sample extends over 65% of the total sample cross section. Using this method, best-fit between experiment and theory was obtained for a D value of 5.45×10^{-10} cm²/sec, in good agreement with the result of 6×10^{-10} cm²/sec obtained above. The average diffusion coefficient determined at 1000°C (5.7×10^{-10} cm²/sec) in our study is lower than the previously reported values [78,104], possibly due the variation in the properties of the host samples. We also used Cr²⁺:ZnSe samples prepared at 900 and 1100°C to investigate the temperature dependence of the diffusion coefficient. Figure 3.6 shows the variation of the best-fit value of the diffusion coefficient as a function of diffusion temperature between 900 and 1100°C.

3.3 Fluorescence Spectroscopy of Cr²⁺:ZnSe

In this section, we review the fluorescence properties of the Cr²⁺:ZnSe medium. In particular, we present data that show how the lifetime and the fluorescence efficiency vary with the active ion concentration. An empirical

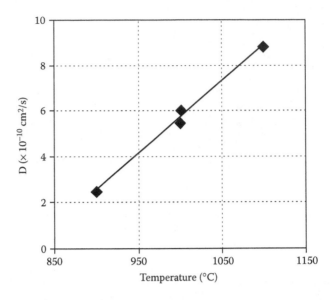

FIGURE 3.6
Temperature dependence of the diffusion coefficient D for the diffusion of Cr^{2+} into polycrystalline ZnSe between 900 and 1100°C.

fit is also obtained for the concentration dependence of the lifetime. We further discuss the models used for the determination of the absorption cross section from cw and pulsed absorption saturation measurements. Since lifetime data are needed in the cw analysis of saturation, we included this discussion on the modeling of saturation after Subsection 3.3.3.

3.3.1 Emission Spectrum of Cr^{2+}:ZnSe

When the Cr^{2+}:ZnSe gain medium is excited by using optical sources operating between 1500 and 2100 nm ($^5T_2 \rightarrow {}^5E$ excitation), efficient, Stokes-shifted fluorescence can be obtained in the 2–3 μm wavelength range. Figure 3.7 shows the emission spectrum of Cr^{2+}:ZnSe samples excited at 1510 nm by a cw Cr^{4+}:YAG laser. The Cr^{2+}:ZnSe samples have the concentrations of 3×10^{18} ion/cm³ (sample a) and 14×10^{18} ion/cm³ (sample b) with peak absorption coefficients of 3.4 cm⁻¹ and 16 cm⁻¹, respectively, at 1775 nm. Samples showed wide emission between 1.7 and 3.1 μm. Note that due to the self-absorption of the medium between 1.5 and 2.1 μm, emission intensity on the low-wavelength side of the spectrum is reduced for sample b. This suggests that samples with relatively low chromium concentration are more suitable to obtain lasing in the low-wavelength end of the emission band. Experiments aimed at operating a gain-switched Cr^{2+}:ZnSe laser below 2000 nm are further discussed in Subsection 3.4.3. We also note in passing that even ultraviolet (355 nm) and visible (532 nm) pump sources can be used to obtain mid infrared emission from Cr^{2+}:ZnSe [105]. The

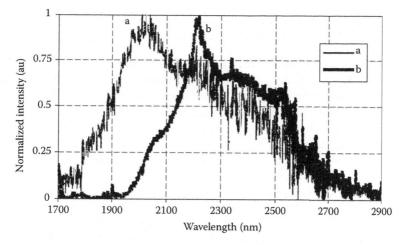

FIGURE 3.7

Measured emission spectra of two Cr^{2+}:ZnSe samples with concentrations of a) 3×10^{18} ion/cm³ and b) 14×10^{18} ion/cm³. The samples were excited at 1510 nm using a cw Cr^{4+}:YAG laser.

details of this indirect excitation mechanism and a discussion of lasing at 2.4 μm under 532-nm excitation can be found in Reference 105.

The stimulated emission cross section σ_e is an important laser parameter which is related to the strength of the emission band and the amount of optical amplification that can be obtained from the gain medium. σ_e can be determined from the fluorescence spectrum by using the Fuchtbauer–Ladenburg formula [18,81,106]:

$$\sigma_{em}(\lambda) = \frac{\lambda^5}{8\pi c n^2 \tau_{rad} \int_{band} \lambda I_e(\lambda) d\lambda} I_e(\lambda). \tag{3.7}$$

Here, c is the speed of light, n is the index of refraction, τ_{rad} is the radiative lifetime, and $I_e(\lambda)$ is the spectral distribution of the fluorescence intensity. The emission data for the sample with lower chromium concentration (sample a) can be used to determine $I_e(\lambda)$. In our analysis the refractive index n and the radiative lifetime τ_{rad} were taken as 2.45 and 5.5 μsec for Cr^{2+}:ZnSe, respectively [107,108], yielding a peak emission cross section of 13×10^{-19} cm² at 2050 nm. This is in good agreement with the previously reported values [14,43,106,109,110].

3.3.2 Temperature and Concentration Dependence of the Fluorescence Lifetime

In general, the total fluorescence decay rate of an energy level is the sum of the radiative (W_{rad}) and nonradiative ($W_{non-rad}$) decay rates. The experimentally measured fluorescence lifetime (τ_f) is related to W_{rad} and $W_{non-rad}$ through

$$\frac{1}{\tau_f} = \frac{1}{\tau_{rad}} + \frac{1}{\tau_{non\text{-}rad}} = W_{rad} + W_{non\text{-}rad}, \tag{3.8}$$

where τ_{rad} is the radiative lifetime and $\tau_{non\text{-}rad}$ is the inverse of the nonradiative decay rate $W_{non\text{-}rad}$. As $W_{non\text{-}rad}$ increases, the fluorescence lifetime of the sample decreases and a larger fraction of the input pump energy is converted into heat. An important parameter that influences the power efficiency of a solid-state laser is the fluorescence quantum efficiency, which is the ratio of the fluorescence lifetime (τ_f) to the radiative lifetime (τ_{rad}). This parameter (τ_f/τ_{rad}) measures the fraction of the input pump energy that can be converted to laser emission. As noted earlier, Cr^{2+}:ZnSe attracted a lot of attention because at low doping densities, it has a near-unity fluorescence quantum efficiency near room temperature. This is quite rare in TM-doped tunable solid-state lasers. Near-unity quantum efficiency at room temperature further suggests that both the multiphonon relaxation and thermally activated nonradiative decay processes are negligible in Cr^{2+}:ZnSe at room temperature [30,43,46,111,112]. Here, we also note that the reduced role of multiphonon relaxation is due to the low phonon cutoff energy (250 cm^{-1} in ZnSe) of the medium [2].

3.3.2.1　Concentration Dependence of the Fluorescence Lifetime

As the doping concentration is increased, ion-ion interactions can also increase the strength of nonradiative decay processes, leading to a reduction in fluorescence quantum efficiency [111]. In this section, we discuss the experiments performed in our group to investigate the concentration dependence of the fluorescence lifetime.

Figure 3.8 shows the room-temperature time-dependent fluorescence decay curve of a polycrystalline Cr^{2+}:ZnSe sample with a Cr^{2+} ion concentration of 9.5×10^{18} cm^{-3}. The corresponding peak absorption coefficient is 10.7 cm^{-1} at 1775 nm. A pulsed optical parametric oscillator (OPO) operating at 1570 nm and producing 65-ns pulses at a repetition rate of 1 kHz was used to excite the samples. The inset in Figure 3.8 shows the variation of the natural logarithm of the fluorescence intensity. We note that the fluorescence signal shows a single-exponential decay. By doing a single-exponential fit to the experimental data, the fluorescence lifetime was determined to be 4.6 ± 0.2 μsec for this particular sample.

The above measurement was repeated for various Cr^{2+}:ZnSe samples doped with different amounts of chromium. Figure 3.9 shows the measured variation of the fluorescence lifetime at room temperature as a function of the average chromium concentration. As can be seen, the fluorescence lifetime shows a rapid decay with Cr^{2+} concentration due to concentration quenching. Similar decrease in the fluorescence lifetime with Cr^{2+} concentration was also observed in previous studies [74,113,114]. The concentration dependence of the fluorescence lifetime τ_f can be described by an empirical formula of the form

$$\tau_F = \frac{\tau_{F0}}{1+\left(\dfrac{N_{Cr}}{N_0}\right)^2},\qquad (3.9)$$

where N_{Cr} is the chromium concentration, τ_{F0} is the low-concentration value of the fluorescence lifetime, and N_0 is the concentration where τ_F is reduced

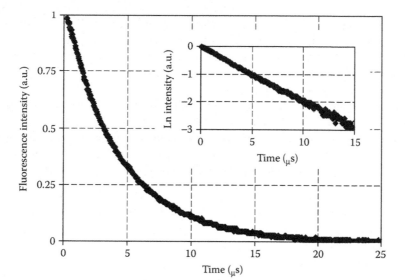

FIGURE 3.8
Room temperature time-dependent fluorescence decay of a Cr²⁺:ZnSe sample with a concentration of 9.5×10^{18} cm⁻³. The sample was excited with 65-ns-long pulses at 1570 nm. The inset shows the variation of the natural logarithm of the fluorescence intensity. The fluorescence lifetime was determined to be 4.6 ± 0.2 μsec.

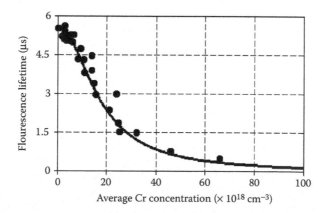

FIGURE 3.9
Measured variation of the fluorescence lifetime of polycrystalline Cr²⁺:ZnSe as a function of the chromium concentration at room temperature. The solid line is the empirical fit given by Equation 3.9 in the text.

to $\tau_{F0}/2$. The solid line in Figure 3.9 shows the least-squares fit to the experimental lifetime data by using Equation 3.9. The best-fit values of τ_{F0} and N_0 were determined to be 5.56 μsec and 17×10^{18} ions/cm^3, respectively. $N_0 = 17 \times 10^{18}$ ions/cm^3 corresponds to a peak absorption coefficient of 19.5 cm^{-1} at 1775 nm. τ_{F0} values reported by other groups are in the 5–7 μsec range [74,106,108,114]. The observed variation may be due to the difference in the structural properties of the ZnSe host. The value of N_0 measured in our study (17×10^{18} ions/cm^3) is also lower than those obtained in other studies (~25×10^{18} ions/cm^3 [108,113], ~26×10^{18} ions/cm^3 [114], ~30×10^{18} ions/cm^3 [74]). Despite the variation in the reported values of N_0 and τ_{F0}, we see clearly that the concentration dependence of the lifetime is relatively weak for concentrations lower than 10×10^{18} ions/cm^3 (see Figure 3.9 and References 74, 108, 113). This value of concentration (~10×10^{18} ions/cm^3) appears as the critical concentration where ion–ion interactions and nonradiative decay processes start to become effective. We can hence conclude, based on the lifetime measurements, that samples with chromium concentrations lower than about 10×10^{18} ions/cm^3 (peak absorption coefficient = 11.5 cm^{-1} at 1775 nm) are more suitable for laser applications, especially in the cw regime.

3.3.2.2 *Temperature Dependence of the Fluorescence Lifetime*

Data on the temperature dependence of the fluorescence lifetime gives very useful information about the strength of multiphonon relaxation processes at different temperatures. In Cr^{2+}:ZnSe, the temperature dependence of the fluorescence lifetime was investigated by several groups [14,74,81,106,113]. In some of the previous measurements, a slight increase in fluorescence lifetime was observed as the temperature was increased from cryogenic temperatures to room temperature [14,43,106]. The reason for this increase remained unclear and several explanations were offered [14,106]. In a recent work, Kisel et al. suggested that the observed increase in the fluorescence lifetime may be attributed to reabsorption in the bulk of the samples [108]. In support of this, the temperature dependence of the fluorescence lifetime was measured by using the luminescence from the surface and the bulk of Cr^{2+}:ZnSe samples. Insignificant increase in the fluorescence lifetime was observed with temperature when the fluorescence signal was collected from sample surface, in comparison with what was taken from the bulk of the samples.

In our studies, we characterized the temperature dependence of the fluorescence lifetime at elevated temperatures between 0 and 160°C. Such data provide useful information for the design of power-scalable Cr^{2+}:ZnSe lasers since the unused pump power can lead to thermal gradients inside the gain medium, and reduce the population inversion due to the temperature dependence of the fluorescence lifetime. Figure 3.10 shows the experimentally measured variation of the fluorescence lifetime as a function of temperature between 0 and 160°C for a Cr^{2+}:ZnSe sample with a Cr^{2+} concentration of 5.7×10^{18} ions/cm^3 (peak absorption coefficient = 6.5 cm^{-1} at 1775 nm). A

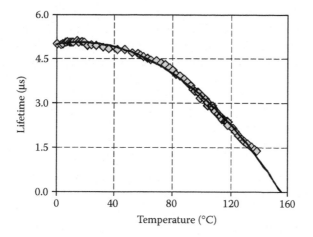

FIGURE 3.10
Measured variation of the fluorescence lifetime for a Cr^{2+}:ZnSe sample with a Cr^{2+} concentration of 5.7×10^{18} ions/cm³ between 0 and 150°C.

sharp decrease in the fluorescence lifetime was observed above 60°C similar to the results of other studies [14,74,81,106,113]. An empirical polynomial fit to the data shows that the fluorescence lifetime $\tau_F(T)$ can be estimated from

$$\tau_F(T) = \tau_{RT}\left(1 + 0.00166T - 0.000052T^2\right), \tag{3.10}$$

where τ_{RT} is the fluorescence lifetime of the sample at room temperature and T is the temperature in degrees Celsius. The solid line in Figure 3.10 is the best fit to the experimental data based on Equation 3.10 with $\tau_{RT} = 5$ μsec.

3.3.3 Dependence of the Fluorescence Efficiency on Chromium Concentration

A direct consequence of the concentration dependence of the fluorescence lifetime is a reduction in the fluorescence efficiency in Cr^{2+}:ZnSe with increasing active ion concentration. As a result, the population inversion of the upper laser level for a given pump power decreases and the cw lasing threshold increases. In our studies, we have directly measured the effect of the ion concentration on the fluorescence quantum efficiency in Cr^{2+}:ZnSe. Figure 3.11 shows the measured variation of fluorescence efficiency (η_F) as a function of Cr^{2+} concentration at 2400 nm. Fluorescence efficiency η_F at 2400 nm is defined as

$$\eta_F = \frac{I_{2400}}{P_{abs}}, \tag{3.11}$$

where I_{2400} is the measured fluorescence intensity at 2400 nm and P_{abs} is the absorbed pump power at the excitation wavelength. Measurements were performed by using a Cr:YAG laser at 1510 nm and a Tm-fiber laser at 1800 nm. Results showed a monotonic decrease in the fluorescence efficiency with increasing Cr^{2+} concentration. The observed decrease in η_F (Figure 3.11) is sharper than that for the fluorescence lifetime (Figure 3.9). This may be attributed to the presence of higher passive losses at the lasing wavelength in more heavily doped samples, as explained in Subsection 3.2.5.

Based on the fluorescence efficiency data, it is possible to estimate the optimum chromium concentration where the emitted photons at the lasing wavelength will be maximized. Note that η_F gives the fraction of the absorbed pump power that is converted to photons at the emission wavelength. By multiplying η_F with the total absorption A of the sample, we obtain a fraction which is proportional to the total number of photons at the emission wavelength. Note that the total absorption $A = 1 - \exp(-\sigma_a N_{Cr} \ell)$ (σ_a = absorption cross section, N_{Cr} = chromium concentration, and ℓ = sample length) increases with ion concentration N_{Cr} and tends to unity at large N_{Cr}. Figure 3.12 shows the calculated variation of $A\eta_F$ for a 2-mm thick Cr^{2+}:ZnSe sample by taking $\sigma_a = 11.5 \times 10^{-19}$ cm^2 [10]. As can be seen from Figure 3.12, this simple analysis shows that the fluorescence intensity for Cr^{2+}:ZnSe will be maximized in a sample with an average chromium concentration of about 6×10^{18} ions/cm^3. We note here that the optimum chromium concentration will also depend on the chosen sample length. The result obtained here is for the specific case of $\ell = 2$ mm.

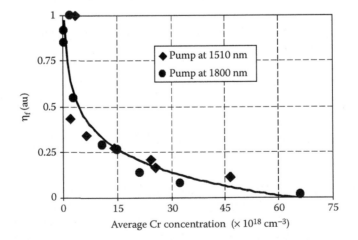

FIGURE 3.11
Variation of the fluorescence efficiency in Cr^{2+}:ZnSe samples at 2400 nm as a function of the chromium concentration. Two different pump sources (a Cr:YAG laser at 1510 nm and a Tm-fiber laser at 1800 nm) were used.

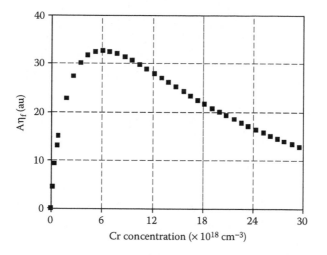

FIGURE 3.12
Estimated variation of $A\eta_f$ (A = absorption, η_f = fluorescence efficiency) for a 2-mm thick Cr²⁺:ZnSe gain medium as a function of chromium concentration.

3.3.4 Determination of Absorption Cross Sections

3.3.4.1 Model Equations

If the active ion density N_{Cr} inside the Cr²⁺:ZnSe medium is accurately known, the absorption cross section $\sigma_a(\lambda)$ at the wavelength λ can be readily determined from

$$\sigma_a(\lambda) = \frac{\alpha_{p0}(\lambda)}{N_{Cr}}, \tag{3.12}$$

where the small-signal absorption coefficient $\alpha_{p0}(\lambda)$ is typically measured with a spectrophotometer as described before in Subsection 3.2.4. If N_{Cr} is not accurately known, absorption saturation measurements can instead be employed in the determination of $\sigma_a(\lambda)$. In this case, a laser whose operating wavelength λ overlaps with the absorption band of the medium is required. The transmission of the sample is measured as a function of the input power or input energy depending on whether a cw or a pulsed source is used, respectively. Alternatively, the input power or energy is kept fixed and the transmission is measured as a function of the pump beam waist location (z-scan method). In the following, we employ the rate-equation formalism [115] to analyze the saturation characteristics of an absorber for the cw and pulsed cases and describe how $\sigma_a(\lambda)$ can be determined from the measured data. We model the Cr²⁺:ZnSe medium as a modified four-level system with the possibility of excited-state absorption. Transverse intensity variations across

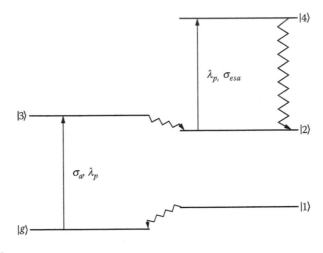

FIGURE 3.13
Energy level diagram for Cr²⁺:ZnSe medium, with the possibility of excited state absorption at the pump wavelength.

the beam cross section are also taken into account. As we will point out, knowledge of the fluorescence lifetime is necessary in order to determine $\sigma_a(\lambda)$ from the cw saturation measurements.

3.3.4.2   *Continuous-Wave Case*

In the case of cw saturation, we derive the differential equation satisfied by the beam power during propagation through the saturable absorber. A schematic of the energy-level diagram is shown in Figure 3.13. Spectroscopic designations of the energy levels are not used in order to keep the analysis general. Let us suppose that pump photons at the wavelength of λ_p are incident on the absorber. If λ_p overlaps with the absorption band, absorber ions will be raised from the ground state $|g\rangle$ to the first excited-state $|3\rangle$. We assume that fast nonradiative decay of the excited ions then takes place to the upper level $|2\rangle$. In the case of Cr²⁺:ZnSe, level $|2\rangle$ is the upper laser level from which ions decaying via stimulated emission produce laser radiation around 2.5 μm. In the analysis of the saturable absorber, we neglect stimulated emission and only consider spontaneous emission to the lower level $|1\rangle$. Finally, fast nonradiative decay brings the ions back to the ground state $|g\rangle$. Typically, the nonradiative decay rates are extremely fast compared to spontaneous decay and populations of levels $|1\rangle$ and $|3\rangle$ may be neglected. We further assume that excited-state absorption may occur from level $|2\rangle$ to a higher lying level $|4\rangle$ at the pump wavelength of λ_p and that once ions are excited to level $|4\rangle$, rapid nonradiative decay takes them back to level $|2\rangle$. In the following analysis, we drop the spatial and temporal dependence of the populations and intensities to simplify the notation. The population density N_2 of the upper level obeys the differential equation

$$\frac{\partial N_2}{\partial t} = \frac{\sigma_a \lambda_p I_p}{hc} N_g - \frac{N_2}{\tau_f}, \tag{3.13}$$

where h is Planck's constant, c is the speed of light, I_p is the intensity of the pump radiation at λ_p, σ_a is the ground state absorption cross section at λ_p, N_g is the population density of the ground state, and τ_f is the fluorescence lifetime.

As the pump beam propagates through a thin slab of the saturable absorber with cross-sectional area A and thickness dz, conservation of energy yields

$$\frac{\partial I_p}{\partial z} + \frac{1}{v_g}\frac{\partial I_p}{\partial t} = -\sigma_a I_p \left(N_g + f_p N_2 \right). \tag{3.14}$$

Here, v_g is the group velocity of the pump beam, and f_p is the normalized strength of excited-state absorption at λ_p ($f_p = \sigma_{esa}/\sigma_a$, σ_{esa} = excited-state absorption cross section). Note that here, $N_2 + N_g = N_{Cr}$. In steady state, N_2 becomes

$$N_2 = N_{Cr} \frac{I_p/I_{sa}}{1 + I_p/I_{sa}}, \tag{3.15}$$

where I_{sa} is the absorption saturation intensity given by

$$I_{sa} = \frac{hc}{\sigma_a \lambda_p \tau_f}. \tag{3.16}$$

Note that the smaller the value of I_{sa}, the easier it will be to saturate the absorption. By using the steady-state expressions for N_2 and N_g, it can be shown that I_p will satisfy the differential equation

$$\frac{\partial I_p}{\partial z} = -\alpha_{p0} I_p \left[\frac{1 + f_p \dfrac{I_p}{I_{sa}}}{1 + \dfrac{I_p}{I_{sa}}} \right]. \tag{3.17}$$

Here, α_{p0} is the small-signal differential pump absorption coefficient defined in Equation 3.12 in the preceding section. If excited-state absorption is present, note from Equation 3.17 that it is not possible to fully saturate the absorber even at very high pump intensities.

If the normalized transverse distribution Φ_p of the intensity is assumed to be cylindrically symmetric, then the beam power $P_p(z)$ at the location z satisfies

$$\frac{dP_p}{dz} = -\alpha_{p0}P_p \int\limits_0^\infty dr 2\pi r \Phi_p \left[\frac{1 + f_p \dfrac{P_p \Phi_p}{I_{sa}}}{1 + \dfrac{P_p \Phi_p}{I_{sa}}} \right],\qquad(3.18)$$

where r is the radial coordinate. As an example, Φ_p for a Gaussian beam is given by

$$\Phi_p = \frac{2}{\pi \omega_p^2} \exp\left(-\frac{2r^2}{\omega_p^2} \right),\qquad(3.19)$$

where ω_p is the position-dependent pump spot-size distribution. The transmitted power will depend on the input power and the waist location of the pump spot-size function. Best fits can be made to experimental data by varying f_p and I_{sa}. From the values of the best-fit parameters, σ_a and σ_{esa} can be determined. To find σ_a, the fluorescence lifetime τ_f of the sample is also needed (see Equation 3.16).

3.3.4.3 Pulsed Case

As an alternative to the cw saturation measurements, a train of pulses can also be sent through the medium to measure the dependence of the transmission on input pulse energy and beam focusing. Here, we analyze the behavior of the saturable absorber by assuming that the duration of the pulses is much shorter than the fluorescence lifetime. We also assume that the absorber completely recovers before the next pulse comes. The pulse intensity $I_p(t)$ and the integrated energy density $\bar{E}_p$ are related through

$$\int\limits_{-\infty}^{\tau} dt I_p(t) = \bar{E}_p.\qquad(3.20)$$

Here, τ is much greater than the pulsewidth so that to a very good approximation, $I_p(\tau) = 0$. By using Equation 3.13 and Equation 3.14 and by neglecting spontaneous decay rates, we obtain

$$\frac{\partial I_p}{\partial z} + \frac{1}{v_g}\frac{\partial I_p}{\partial t} = \left(1 - f_p\right)\frac{hc}{\lambda_p}\frac{\partial N_g}{\partial t} - f_p \alpha_{p0} I_p.\qquad(3.21)$$

If Equation 3.13 is integrated from $t = -\infty$ to $t = \tau$, the ground state population N_g can be expressed in terms of the integrated energy density as

$$N_g = N_{Cr} \exp\left(-\frac{\bar{E}_p}{E_{sa}} \right),\qquad(3.22)$$

where the initial ground-state population density is $N_g(-\infty) = N_{Cr} = (\alpha_{p0}/\sigma_a)$ and E_{sa} ($E_{sa} = hc/\sigma_a\lambda_p$) is the saturation fluence for absorption. If Equation 3.21 is integrated from $t = -\infty$ to $t = \tau$, we obtain the differential equation that describes the spatial evolution of the energy density:

$$\frac{\partial \overline{E}_p}{\partial z} = -\left(1 - f_p\right)\alpha_{p0}E_{sa}\left[1 - \exp\left(-\frac{\overline{E}_p}{E_{sa}}\right)\right] - f_p\alpha_{p0}\overline{E}_p. \tag{3.23}$$

Finally, by assuming that the transverse energy distribution of the beam is described by Φ_p, we find that the total pulse energy E_p obeys

$$\frac{\partial E_p}{\partial z} = -\left(1 - f_p\right)\alpha_{p0}E_{sa}\int_0^\infty dr2\pi r\left[1 - \exp\left(-\frac{E_p\Phi_p}{E_{sa}}\right)\right] - f_p\alpha_{p0}E_p. \tag{3.24}$$

In the analysis of saturation data, f_p and E_{sa} can be used as the fitting parameters to determine σ_a and σ_{esa}.

3.3.4.4 Experimental Results

In this section, we present the results of saturation measurements performed with a polycrystalline Cr²⁺:ZnSe sample to determine the absorption cross section σ_a. In the measurements, a 1.85-mm-thick Cr²⁺:ZnSe sample with a peak absorption coefficient of 12.2 cm⁻¹ at 1775 nm and a fluorescence lifetime of 4.3 μsec was used. A cw Tm-fiber laser at 1800 nm and a pulsed KTP optical parametric oscillator at 1570 nm were used to measure the variation of the sample transmission as a function of the beam waist location. The pump waist has a radius of 30 μm and 70 μm for the cw and pulsed cases, respectively. Figure 3.14 shows the experimentally measured and calculated variation of the transmission as a function of the pump spot position for both cw and pulsed cases. For the cw case, at 1800 nm, best-fit between experiment and theory was obtained for $f_p = 0$ and $\sigma_a = 6.05 \times 10^{-19}$ cm². In the pulsed case, at 1570 nm, best-fit values of f_p and σ_a were 0.06 and 2.15×10^{-19} cm², respectively. For both 1570 nm and 1800 nm, the best value of f_p is very small indicating that excited-state absorption is negligible, in agreement with earlier work [108,116–118]. The calculated absorption cross section values at 1570 nm and 1800 nm correspond to a peak absorption cross section of 5.6×10^{-19} and 6.1×10^{-19} cm² at 1775 nm (using Equation 3.5), respectively. The agreement between these values obtained by using different techniques is reasonably good. The averaged best-fit value of the cross section at 1775 nm comes to 5.85×10^{-19} cm². Several other groups also determined the absorption cross section for Cr²⁺:ZnSe. The results of these measurements along with the pump wavelengths used appear in Table 3.2. The best-fit average obtained in our studies is lower than the average ($\sim 10 \times 10^{-19}$ cm²) of the previously reported values shown in Table 3.2.

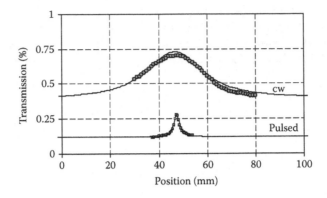

FIGURE 3.14

Measured and calculated variation of the transmission of a 1.85-mm thick Cr^{2+}:ZnSe sample as a function of the incident beam waist location. As pump sources, a 1570 nm KTP OPO producing 70 ns pulses (pulsed), and a Tm-Fiber laser at 1800 nm (cw) was used.

TABLE 3.2

Reported Peak Absorption Cross-Section Values (σ_o) and the Relative Strength of the Excited-State Absorption (f_p) for Cr^{2+}:ZnSe

σ_o ($\times 10^{-19}cm^2$)	f_p	Pump Wavelength	Reference
~11	0	Pulsed, 1534 nm	108
~7.5	—	Pulsed, 2017 nm	145
~10	< 0.04	Pulsed, 1598 nm	116
~9.5	—	cw, 2000 nm	146
~12	0	Pulsed, 1534 nm	117
~11.5	—	*	10
~8	—	*	15
~10	~0	—	Average
~5.5	0.06	Pulsed, 1570 nm	This work
~6	0	cw, 1800 nm	This work

Note: Pump wavelengths used in the measurements are also indicated. Those marked by * use the known chromium concentration value of the prepared samples to calculate σ_o.

3.4 Pulsed Operation of Cr^{2+}:ZnSe Lasers

3.4.1 Review of Earlier Work on Pulsed Cr^{2+}:ZnSe Lasers

Cr^{2+}:ZnSe lasers can be operated in pulsed mode by using a second pulsed pump laser whose wavelength overlaps with the absorption band. The peak powers obtainable from pulsed pump sources are typically high and lasing can thus be achieved at lower average pump powers than those needed for cw pumping. In this section, we review the work done in the development of pulsed Cr²⁺:ZnSe lasers. Continuous-wave and mode-locked regimes of operation will later be discussed in Section 3.5 and Section 3.6, respectively.

The first experimental demonstration of lasing action in Cr²⁺:ZnSe employed pulsed pumping [14,15]. In this work, Page et al. used a 10-Hz, 10-mJ, Co:MgF₂ laser operating at the wavelength of 1860 nm to excite the Cr²⁺:ZnSe gain medium. The pump pulsewidth was about 40 μsec. Plane-parallel Cr²⁺:ZnSe samples were placed at the center of a confocal resonator consisting of two curved mirrors (R = 20 cm). The output coupler had a transmission of 7.5%. The pump beam was focused to a spot of several millimeters and up to 600 μJ of output pulse energy was obtained with ~3.5 mJ of absorbed pump energy. The slope efficiency with respect to absorbed pump power was 22.6%. By using a diffraction grating in Littrow configuration, the laser output could be tuned between 2150 and 2800 nm. The same group also demonstrated the first diode-pumped, pulsed Cr²⁺:ZnSe laser [28]. In these experiments, an InGaAsP/InP diode laser array producing more than 70 W of peak power at 1.65 μm was used as the pump. The gain medium was excited from one side. Using a 10% output coupler, output energies as high as 18 μJ (0.34 W peak power) were obtained with about 150 μJ of absorbed pump energy. To date, pulsed operation of Cr²⁺:ZnSe lasers has been characterized in numerous other studies [28,42,106,109,119–122]. In the particular experiments reported by Zakel et al., a 10-mm-long cylindrical Cr²⁺:ZnSe sample (chromium concentration = 6×10^{18} ions/cm³) was pumped by a 30-W, 7-kHz, Tm:YALO laser at 1940 nm and average output powers as high as 18.5 W (2.64 mJ pulse energy) were obtained during gain-switched operation [122]. The slope efficiency exceeded 65%. An acousto-optic tunable filter was further used for rapid tuning of the laser output between 2.04 and 2.74 μm.

3.4.2 Dependence of the Power Performance on Chromium Concentration

In our studies, we investigated the gain-switched operation of a Cr²⁺:ZnSe laser pumped at 1.57 μm by a pulsed, eye-safe KTP optical parametric oscillator (OPO) [107]. This is a very attractive excitation scheme since the KTP OPO can be pumped by the 1064-nm Nd:YAG lasers which are widely available. One important issue that needs to be addressed in the case of 1570-nm pumping has to do with the poor overlap of the pump wavelength with the Cr²⁺:ZnSe absorption band (1500 nm to 2100 nm) whose center is at 1775 nm. If the crystal lengths are kept fixed, this requires the use of gain media with higher active ion concentrations to absorb the same amount of pump power. However, as explained in previous sections, high doping concentrations can increase the fluorescence quenching, passive losses at the lasing wavelength, and thermal loading. Hence, it is important to understand the effect of the active ion concentration on the lasing efficiency. In this work,

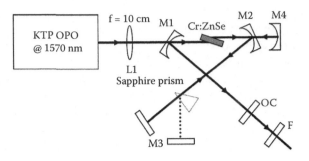

FIGURE 3.15
Schematic of the KTP OPO-pumped gain-switched Cr²⁺:ZnSe laser. See the text for a detailed description of the components. Mirror M4 was used only during the intracavity pumping experiments.

we investigated the dependence of the laser power performance on the active ion concentration in 1570-nm-pumped gain-switched Cr^{2+}:ZnSe lasers.

A schematic of the gain-switched Cr^{2+}:ZnSe laser is shown in Figure 3.15. In the experiments, we used a 1064-nm-pumped noncritically phase-matched KTP OPO capable of delivering 65-ns pulses at a pulse repetition rate of 1 kHz with up to 880 μJ of pulse energy. Cr^{2+}:ZnSe laser was configured as a standard x-cavity, consisting of two curved high reflectors (M1 and M2, R = 10 cm), a flat end high reflector (M3), and a flat output coupler (OC). The length of each resonator arm was around 15 cm. Several diffusion-doped Cr^{2+}:ZnSe samples were placed at Brewster incidence between the curved mirrors M1 and M2. The pump was focused to a beam waist of about 100 μm with a converging lens (L1, f = 10 cm). A filter (F) was used after the output coupler to block the residual pump beam. The cavity was not purged during the measurements and the relative humidity was around 40%.

We used three Cr^{2+}:ZnSe samples with thickness of ~3 mm and Cr^{2+} concentrations of 3.6×10^{18}, 14×10^{18}, and 25×10^{18} ions/cm³ to investigate the concentration dependence of the power performance. Table 3.3 lists the optical properties of the three samples. Three different output couplers with 5.5, 17.5, and 25% transmission were used to characterize the laser performance with each sample. Figure 3.16 shows the energy efficiency curves taken with the 25% output coupler. The highest output energy was obtained with Sample 2, which had a peak absorption coefficient of 16 cm⁻¹ at 1775 nm and a corresponding Cr^{2+} concentration of 14×10^{18} ions/cm³. In this case, the resonator produced 52 μJ of output energy with 340 μJ of absorbed pump energy. Note that the power performance degrades monotonically with increasing ion concentration. In particular, the slope efficiency of the resonator decreases from 25% to 12% when the ion concentration increases from 3.6×10^{18} to 25×10^{18} ions/cm³. This could be mainly attributed to an increase in passive losses at the lasing wavelength, as described in Subsection 3.2.5. Although the slope efficiency was higher with Sample 1, higher output energy was obtained with Sample 2 due the larger absorption of the latter.

TABLE 3.3

Properties of the Cr²⁺:ZnSe Samples Used in the KTP OPO-Pumped, Gain-Switched Lasing Experiments

Sample	Sample Length (mm)	α at 1775 nm (cm⁻¹)	Absorption at 1570 nm (%)	Average Cr²⁺ Concentration ($\times 10^{18}$ cm⁻³)	Lifetime (µsec)
1	3.03	4.1	51	3.6	5.0
2	3.04	16	75	14	4.4
3	3.06	29	99	25	1.5

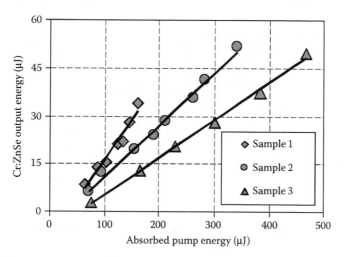

FIGURE 3.16

Measured variation of the output energy as a function of the absorbed pump energy for the three different Cr²⁺:ZnSe samples used in the experiment. Properties of the samples are listed in Table 3.3. A 2.9-µm mirror set and a 25% output coupler was used in the measurements.

Pulse forming dynamics during the gain-switched operation of the Cr²⁺:ZnSe laser depends on many factors such as the incident pump energy, cavity configuration, and the optical characteristics of the gain medium. In these experiments, we investigated the effect of the incident pump energy on the delay between the pump and the laser pulses, and the duration of the output pulse. Figure 3.17 shows the measured variation of the delay and the laser pulsewidth as a function of incident pump energy for Sample 1. As can be seen both the delay and the pulsewidth initially decrease with increasing pump energy as expected but then level off due to absorption saturation.

The passive loss analysis of the samples was also performed by using the power efficiency data obtained with three output couplers [123]. The slope efficiency η of the laser can be expressed in terms of the output coupler transmission T and the round trip passive loss L as

$$\eta = \eta_0 \frac{T}{T+L},$$

(3.25)

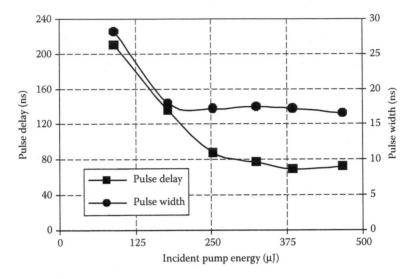

FIGURE 3.17
Measured variation of the delay and the output pulse width as a function of incident pump energy for the gain-switched Cr^{2+}:ZnSe laser. Sample 1 and the 2.9-μm mirror set (output coupler transmission = 25%) were used.

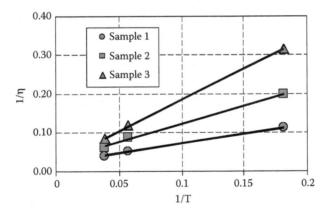

FIGURE 3.18
Measured variation of the inverse slope efficiency (1/η) as a function of the inverse output coupler transmission (1/T) for the three different Cr^{2+}:ZnSe samples used in the experiment.

where η_0 is the maximum slope efficiency that can be obtained at high output coupling. Factors including the finite quantum defect of the laser transition, mode matching between the pump and the laser beams, and luminescence quantum efficiency all affect η_0. Figure 3.18 shows the variation of the inverse slope efficiency (1/η) as a function of the inverse transmission (1/T) of the output couplers for the three samples. As can be seen from Equation 3.25, the intercept and the slope of the linear best fit can be used to determine the values of η_0 and L. Using this method, the single-pass loss was estimated to be 11, 15, and 31% for Samples 1, 2, and 3, respectively. Here, atmospheric and mirror losses were neglected. The estimated passive losses of the samples scale with chromium concentration, as expected from the spectroscopic results (See Subsection 3.2.5).

3.4.3 Intracavity-Pumped Cr²⁺:ZnSe Laser with Ultrabroad Tuning

In a second experiment, we used an intracavity pumping configuration and demonstrated ultrabroad tuning of the gain-switched Cr²⁺:ZnSe laser. A curved gold retro reflector (M4, R = 10 cm) was further placed after the curved high reflector (M2). In addition to enabling double-pass pumping of the gain medium [27,124], this configuration also provides feedback for the OPO setup, extending the effective resonator for 1570-nm oscillation from M4 up to the input high reflector of the optical parametric oscillator (not shown in Figure 3.15). To demonstrate tuning in different parts of the spectrum, four different mirror sets with central reflectivity wavelengths at 2, 2.25, 2.6, and 2.9 μm were used. All the high reflectors had transmission greater than 90% at 1570 nm. In the experiments, tuning was achieved by using a Brewster-cut sapphire prism. Most of the efficiency and the tuning data were obtained by using a Cr²⁺:ZnSe sample (Sample 4, referred to as S4 in the following graphs) which had a thickness of 1.94 mm and a Cr²⁺ concentration of 5.7×10^{18} ions/cm³. The measured fluorescence lifetime and the pump absorption were 5 μsec and 43%, respectively. Only in the tuning data at longer wavelengths (2600–3100 nm), a second sample (Sample 5, S5) with a thickness of 1.85 mm and concentration of 11×10^{18} ions/cm³ was used for improving the output efficiency. Figure 3.19 shows the energy efficiency

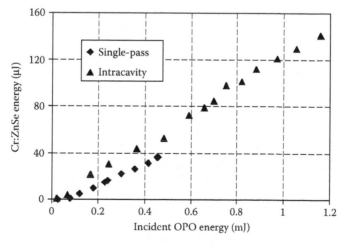

FIGURE 3.19

Energy efficiency curves taken with Sample 4 in the single-pass and intracavity pumping configurations. A 2.6-μm mirror set (output coupler transmission = 25%) was used.

curves for Cr²⁺:ZnSe Sample 4, taken with the 2.6 μm HR set. In the intracavity pumping configuration, output energies as high as 143 μJ was obtained with a 25% output coupler and 1.16 mJ of intracavity OPO pump energy. The pulse repetition rate was 1 kHz.

Tuning experiments were conducted by using four sets of resonator optics with center wavelengths of 2 (OC transmission = 7%), 2.25 (OC transmission = 25%), 2.6 (OC transmission = 25%), and 2.9 μm (OC transmission = 4%). The bracketed values indicate the output coupler transmission for each set of mirrors at the center of the reflectivity window. Measured variation of the output energy as a function of the wavelength is shown in Figure 3.20 for an input Nd:YAG pump energy of about 1.8 mJ. Here the input intracavity OPO pump energy did not remain constant due to the fact that different sets of the optics used in the tuning experiments had different overall transmission at 1570 nm. Due to the low damage threshold of the 2.25 μm HR mirrors, intracavity configuration was not applied with this set; hence lower output powers were obtained. The solid curve in Figure 3.20 shows the estimated single pass small signal reabsorption loss of the Cr²⁺:ZnSe gain medium (Sample 4). With intracavity pumping, the demonstrated fractional tuning range was 0.49 ($\Delta\lambda/\lambda_0 \cong (3100-1880)/2490 = 0.49$) comparable to that of Ti⁺³:Al₂O₃ ($\Delta\lambda/\lambda_0 \cong 0.57$) lasers (see Table 3.1). Also note from Figure 3.20 that use of Sample 5 with a 25% transmitting output coupler improves the laser efficiency in the long wavelength end of the tuning range.

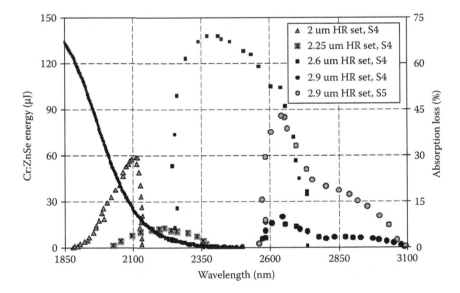

FIGURE 3.20
Tuning curve for the intracavity-pumped gain-switched Cr²⁺:ZnSe laser taken with four different mirror sets. Laser emission could be obtained in the 1880 to 3100 nm wavelength range. The solid line is the small-signal, single-pass loss of Sample 4 due to self absorption.

3.5 Continuous-wave Cr²⁺:ZnSe lasers

As noted in the Introduction, several favorable characteristics of the Cr²⁺:ZnSe medium enable efficient continuous-wave laser operation at room temperature. These include the four-level energy structure, absence of excited-state absorption, and high luminescence quantum efficiency near room temperature. On the other hand, the high thermal index coefficient (dn/dT = 70×10^{-6}/°K for Cr²⁺:ZnSe [125]) and temperature dependence of the fluorescence lifetime above room temperature cause unwanted thermal loading problems and could limit power scaling in room-temperature systems. In this section, we review the work done on the development of cw Cr²⁺:ZnSe lasers. We discuss the operation under various pumping configurations and the results obtained with regard to power efficiency and tuning performance. We also discuss the recent work done in our group on the characterization of a cw Cr²⁺:ZnSe laser pumped by a Tm-fiber laser at 1800 nm. Experimental data showing power efficiency, tuning, and the role of active ion concentration are presented.

The continuous-wave operation of a laser can be modeled by using the rate equation formalism. It can be shown that in the case of end pumping, the incident threshold pump power P_{th} required to achieve lasing is given by

$$P_{th} = \frac{\pi}{4} \frac{h\nu_P}{\sigma_{em}} \frac{\left(\omega_L^2 + \omega_P^2\right)}{\left(1 - \exp(-\alpha_P \ell)\right)} \left(L + L_{GSA} + T\right), \qquad (3.26)$$

where $h\nu_P$ is the energy of pump photons, σ_{em} is the stimulated emission cross section of the gain medium, ℓ is the length of the gain medium, α_p is the small-signal differential pump absorption coefficient, $(1 \text{ Exp}(-\alpha_p\ell))$ is the total absorption at the pump wavelength, L is the passive round-trip loss of the resonator, T is the transmission of the output coupler, and ω_L and ω_p are the average values of the laser and pump spot sizes in the gain medium. If present, L_{GSA} represents the round-trip ground state absorption loss in the gain medium. In addition, the slope efficiency η of the laser can be calculated by using the formula in Equation 3.25.

Room-temperature continuous-wave operation of a Cr²⁺:ZnSe laser was first reported by Wagner et al. [27]. A 1-W Tm:YALO laser operating at 1940 nm was used as the pump source. A three-mirror cavity was end-pumped with the Tm:YALO laser and different Cr²⁺:ZnSe samples with thicknesses in the 1–3-mm range were used. The pump was focused to a 60-μm waist. Best results were obtained with a 2.76-mm-long crystal that had a single-pass absorption of 35%. The curved gold high reflector of the laser cavity was used for double passing the pump beam, resulting in a total effective pump absorption of 58%. With a 7% transmitting output coupler, 250 mW of output was obtained by using 600 mW of absorbed pump power, giving a slope efficiency (with respect to the absorbed pump power) of 63%. By using a Brewster-cut ZnSe prism placed near the output coupler, the output wavelength of the laser could also be continuously tuned between 2138 and 2760 nm.

Following the first demonstration of cw lasing in Cr^{2+}:ZnSe lasers, several alternative pump sources were employed to obtain tunable cw output. These include Co^{2+}:MgF$_2$ [124,126], Tm:YALO [27], Tm:YLF [125], NaCl:OH⁻ [127], Er-doped fiber lasers [32,35,128–131], and others. In the particular experiments described by Sorokin and Sorokina, tunable diode-pumped cw operation was demonstrated by using InGaAsP-InP laser diodes operating at 1600 nm [132]. A three-mirror resonator was used and the gain crystal was end-pumped from two sides. The total available diode power was 1W. With 1% output coupler, lasing could be obtained with as low as 30 mW of absorbed pump power. Different output couplers with transmissions in the range 1–7% were used to investigate the dependence of the output power on pump power. By using the 7% output coupler and 460 mW of absorbed pump power, 70 mW of output could be obtained. The output wavelength of the laser could be varied by using an intracavity Lyot filter. In this case, the polarized output of the laser could be tuned over 350 nm from 2300 to 2650 nm.

In our experiments, we employed a thulium-doped fiber laser operating at 1800 nm to investigate the cw lasing characteristics of a Cr^{2+}:ZnSe laser. Pumping at 1800 nm has several advantages. First and foremost, because the pump wavelength is very close to the absorption resonance, a desired level of absorption can be obtained by incorporating the minimum possible amount of active ions into the ZnSe host. It is known that the amount of passive losses at the lasing wavelength as well as the rate of nonradiative decay increase with ion concentration. Pumping at 1800 nm and using crystals with low active ion concentration reduces passive laser losses, decreases the rate of nonradiative decay and also minimizes thermal loading effects that would arise due to the temperature dependence of the fluorescence lifetime. All these positive factors are expected to improve the power scaling capability of room-temperature Cr^{2+}:ZnSe lasers especially at higher pumping levels.

Figure 3.21 shows the experimental arrangement of the cw Tm-fiber-pumped Cr^{2+}:ZnSe laser. The pump source was a 5-W, commercial Tm-fiber laser at 1800 nm. The collimated beam diameter ($1/e^2$) and the M^2 were 4.5 mm and 1.03, respectively. In the first set of experiments, we used a single Cr^{2+}:ZnSe sample with a length 2.6 mm and a pump absorption coefficient of 10.7 cm^{-1} at 1800 nm. The measured fluorescence lifetime was 4.6 μsec. The round-trip loss was 4.1% at the lasing wavelength, giving a crystal figure of merit (FOM = $\alpha_{1800}/\alpha_{2500}$, α_{1800} = differential absorption coefficient at 1800 nm and α_{2500} = differential loss coefficient at 2500 nm) of 70. The sample was held in a copper holder maintained at 15°C. The cavity was a standard, astigmatically compensated, four-mirror x-cavity in which the gain crystal was placed between two curved high reflectors each with R = 10 cm (M1 and M2). The cavity was terminated with a flat high reflector (M3) on one end and a flat

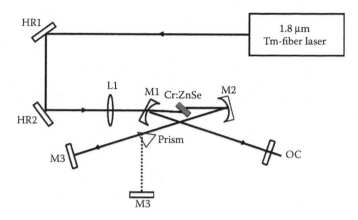

FIGURE 3.21
Schematic of the 1800-nm-pumped continuous wave Cr²⁺:ZnSe laser. The reflectivity band of the high reflectors was centered around 2.6-μm. See the text for a detailed description of the components.

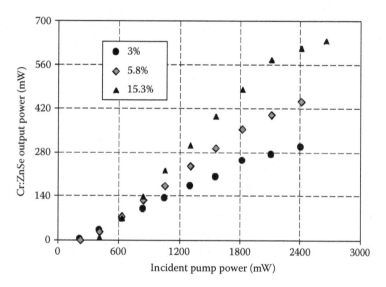

FIGURE 3.22
Continuous-wave power efficiency data for the cw Cr²⁺:ZnSe laser taken with 3, 5.8, and 15.3% transmitting output couplers.

output coupler (OC) on the other. In the experiments, three different output couplers with transmissions of 3, 5.8, and 15.3% at 2500 nm were used. The incident pump beam was focused to a waist of 36 μm inside the gain medium with an input lens (f = 10 cm). ABCD analysis of the cavity shows that the cavity beam waist at the center of the stability region was 45 μm at the wavelength of 2400 nm (total cavity length = 90 cm).

Figure 3.22 shows the power efficiency data for the Tm-fiber-pumped cw Cr²⁺:ZnSe laser. The best performance was obtained with the 15.3% transmitting output coupler. In this case, as high as 640 mW of output power was obtained with 2.5 W of pump. The corresponding slope efficiency with respect to incident power was 34%. The incident threshold power was measured to be 210, 280, and 410 mW for the 3, 5.8, and 15.3% output couplers, respectively. Findlay–Clay analysis was first used to determine the round-trip passive loss L of the resonator [133]. Here, it is assumed that the incident threshold pump power is proportional to (L + T). Best linear fit to the graph of P_{th} as a function of T gave 10.6% for L. Alternatively, Caird analysis can be used to determine L by using the fact that a plot of $1/\eta$ as a function of $1/T$ gives a straight line (see Equation 3.25). The slope efficiency was measured to be 16, 23, and 34% with respect to the incident pump power for the output couplers with respective transmissions of 3, 5.8, and 15.3%. L was determined to be 6.2% from the Caird analysis. Similar discrepancy between the Findlay–Clay and Caird analyses was also observed in previous laser studies with Cr²⁺:ZnSe [15,27,110,124], and attributed to the existence of residual ground state absorption [43,79,100,110,124] which affects the threshold power but not the slope efficiency [46,124]. By using Equation 3.26 and the lasing threshold data, the average value of the stimulated emission cross section was further determined to be 4.2×10^{-19} cm² at the laser wavelength (around 2500 nm). Using the peak emission cross value determined in Subsection 3.3.1 and the emission line shape function, the value of the emission cross section at 2500 nm is calculated to be 4.7×10^{-19} cm², in good agreement with the value determined from the laser threshold data.

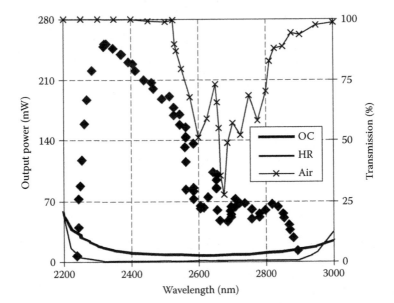

FIGURE 3.23
Tuning curve for the 1800-nm-pumped cw Cr²⁺:ZnSe laser. The variation of the transmission of the output coupler and the high reflectors is also shown. The laser could be tuned between 2240 and 2900 nm. The dips in the tuning curve correspond to the atmospheric absorption lines.

Tuning characteristics of the cw Cr^{2+}:ZnSe laser was further investigated by using a Brewster-cut MgF_2 prism placed in the high-reflector arm of the resonator. The output coupler had a transmission of 3%. At the pump power of 1.8 W, the output of the laser was reduced from 255 to 250 mW after the insertion of the prism. Figure 3.23 shows the tuning curve. Without purging the cavity, broad tunability could be obtained in the 2240–2900 nm wavelength range with a single set of optics. The black and gray solid lines in Figure 3.23 show the wavelength dependence of the high reflector (HR) and output coupler transmissions (OC), respectively. Tuning on both sides was limited with the reflectivity bandwidth of the high reflectors. In these measurements, the relative humidity was 67%. A rough estimate of the transmission of air at that humidity level is also shown [134]. Note that the observed dips near the wavelength of 2600 nm coincide with the atmospheric absorption peaks.

In the second set of experiments, we investigated the effect of active ion concentration on the cw power performance of the Cr^{2+}:ZnSe laser. Here, several Cr^{2+}:ZnSe samples with different Cr^{2+} concentration were used. Chromium was incorporated into 2-mm-thick polycrystalline ZnSe samples by diffusion doping. All of the samples were prepared at 1000°C. By varying the diffusion time, samples with average Cr^{2+} concentration between 0.8×10^{18} and 23.2×10^{18} ions/cm³ were prepared. The laser setup was similar to what was described above. The four-mirror x-cavity was end pumped by the Tm-fiber laser at 1800 nm and the output coupler with 3% transmission was used. Figure 3.24 shows the power efficiency data taken using four samples

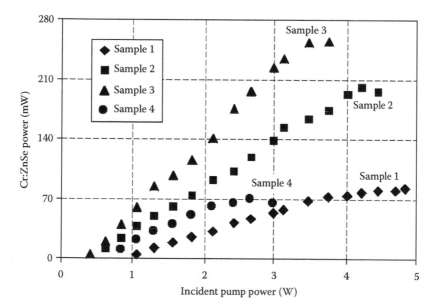

FIGURE 3.24

Power efficiency data for the cw Cr^{2+}:ZnSe laser taken with 4 samples with chromium concentrations of 0.8 (Sample 1), 2.2 (Sample 2), 4.0 (Sample 3) and 12.8 (Sample 4) × 10¹⁸ ions/cm³. The transmission of the output coupler was 3%.

with chromium concentrations of 0.8 (Sample 1), 2.2 (Sample 2), 4.0 (Sample 3) and 12.8 (Sample 4) $\times 10^{18}$ ions/cm^3. Note that the slope efficiency as well as the threshold pump power depends on the active ion concentration. At the fixed incident pump power of 2.1 W, maximum output power of 165 mW is obtained by using the sample with an average chromium concentration of 8.5×10^{18} ions/cm^3. Power performance was observed to degrade at lower and higher concentrations as can be seen from Figure 3.24. The optimum chromium concentration determined from laser power measurements comes close to that (6×10^{18} ions/cm^3) predicted from the fluorescence efficiency measurements described in Subsection 3.3.3.

3.6   Mode-Locked Cr^{2+}:ZnSe Lasers

As discussed in the previous sections, the Cr^{2+}:ZnSe gain medium has a very broad emission band extending from 2000 to 3000 nm, allowing for the possibility of ultrashort optical pulse generation in the picosecond and femtosecond time scales. The technique of mode locking is routinely employed to produce such short pulses. The principles of mode locking and its application to various solid-state laser systems are extensively discussed in other chapters of this handbook (see for example, Chapter 7 to Chapter 11). In this section, we focus on the recent mode-locking work performed with Cr^{2+}:ZnSe lasers. We discuss the first experimental demonstration of acousto-optic mode locking in Cr^{2+}:ZnSe lasers which enabled the generation of picosecond pulses. We then discuss more recent work in which passive mode-locking techniques have been used to generate picosecond and femtosecond pulses.

Active mode locking of Cr^{2+}:ZnSe lasers was first experimentally demonstrated by Carrig et al. [135]. In this experiment, acousto-optic mode locking was used to generate picosecond pulses. A schematic of the experimental setup is shown in Figure 3.25. The four-mirror, astigmatically compensated z-cavity contained a 2.3-mm-thick Cr^{2+}:ZnSe crystal (xtal in Figure 3.25) placed at Brewster incidence between two concave high reflectors (M1 and M2). The total absorption of the 2.3-mm-thick Cr^{2+}:ZnSe crystal varied with the location of the pump waist inside the gain medium but was typically around 65%. Spatial variation of the absorption over the crystal cross section results from the diffusion of the chromium ions during the doping process and was addressed in detail in Section 3.2. The crystal was held between two copper plates and no active cooling was employed. The crystal temperature was measured to be 26°C. The Cr^{2+}:ZnSe laser was end pumped by a cryogenic NaCl:OH$^-$ laser operating at 1.58 μm. A maximum pump power of 1175 mW was available in the experiments. By using a converging input lens (L1, f = 10 cm), the pump beam was passed through the input dichroic mirror M1 and focused to a waist of approximately 40 μm (1/e^2 radius) inside the gain medium. The flat output coupler (OC) of the resonator had a transmission of

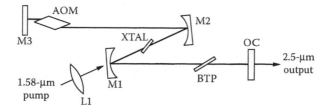

FIGURE 3.25
Schematic of the acousto-optically mode-locked Cr²⁺:ZnSe laser. See the text for a detailed description of the components.

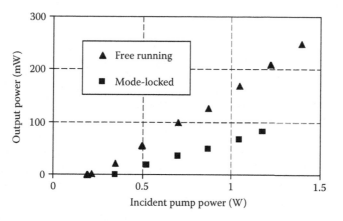

FIGURE 3.26
Power efficiency curves for the free-running and mode-locked Cr²⁺:ZnSe laser.

3%. The total cavity length was adjusted to give a pulse repetition rate of 81.2 MHz. The estimated cavity beam waist inside the Cr²⁺:ZnSe crystal was 35 μm. A single, 1.1-mm-thick, sapphire birefringent tuning plate (BTP) was also included in the cavity for bandwidth control. To obtain a train of mode-locked pulses, the cavity loss was modulated with an acousto-optic mode locker (AOM) consisting of a 1-cm-long Brewster-cut quartz slab driven at 40.62 MHz. The modulator was located near the end high reflector (M3) of the resonator. The peak transmission modulation was about 14% at the lasing wavelength.

Figure 3.26 shows the cw power efficiency curves for the free-running and mode-locked Cr²⁺:ZnSe laser. In the free-running mode without the tuning plate and the mode-locker, the incident threshold pump power and the slope efficiency were 190 mW and 24.2%, respectively. During mode-locked operation, the threshold pump power increased to 350 mW and the slope efficiency was reduced to 11.3%, mainly due to the insertion loss of the mode locker. In this case, the resonator produced as high as 82 mW of output power at the center wavelength of 2.47 μm. Matching the pulse repetition rate to the mode-locker modulation frequency readily gave a train of mode-locked picosecond pulses. The collinear intensity autocorrelation shown in Figure 3.27 was measured by

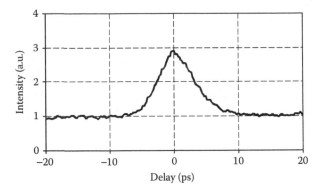

FIGURE 3.27
Collinear intensity autocorrelation of the output pulses obtained from the acousto-optically mode-locked Cr^{2+}:ZnSe laser. By assuming a Gaussian intensity profile, the duration (FWHM) of the pulses was determined to be 4.4 ps.

using two-photon absorption in a Ge detector. The full-width at half-maximum (FWHM) pulse duration τ_p was determined to be 4.4 ps by assuming a Gaussian pulse profile. The spectral width of the pulses was further measured by using a scanning Fabry–Perot interferometer with a finesse of 120 and a free-spectral range of 750 GHz. The spectral width (FWHM) came to 104 GHz, with a corresponding time-bandwidth product of 0.46, in good agreement with the theoretical limit of 0.44 for Gaussian pulses.

According to the amplitude modulation (AM) mode-locking theory of Siegman and Kuizenga [136], the duration (FWHM) of the output pulses is given by

$$\tau_p = \frac{\sqrt{2\ln 2}}{\pi}\left(g_0\right)^{1/4}\left(\frac{1}{\delta_1 f_m \Delta f_a}\right)^{1/2},$$

where g_0 is the saturated round-trip amplitude gain coefficient (0.075), δ_1 is the modulation depth parameter for the acousto-optic modulator (0.54), f_m is the modulator drive frequency (40.6 MHz), and Δf_a is the gain bandwidth of the laser (30 THz). The estimated values of these parameters for the setup discussed here are indicated in parentheses. With these values, the estimated duration came to 7.7 ps, in reasonable agreement with the experimentally measured value of 4.4 ps. Acousto-optically initiated mode locking of a Cr^{2+}:ZnSe laser was also reported in other studies [137]. Purely active and passive mode-locking regimes were experimentally demonstrated, resulting in the generation of pulses as short as 4 ps.

Continuous-wave passive mode locking of a Cr^{2+}:ZnSe laser was experimentally demonstrated by Pollock et al. by using a Semiconductor Saturable Absorbing Mirror (SESAM) [138]. In these experiments, a 10-mm-long cylindrical Cr^{2+}:ZnSe crystal was placed at one end of a folded cavity which was end-pumped by a Tm:YALO laser at 1940 nm. The pump beam was focused to a waist radius of approximately 75 μm inside the crystal. Passive mode locking was initiated by using a SESAM structure located at the other end of the resonator. An intracavity lens with a focal length of 2.25 cm was used to focus the beam onto the SESAM which consisted of a 50-layer sublattice of InAs/GaSb quantum wells grown on top of a dielectric Bragg stack. The small-signal round-trip absorption and saturation fluence of the SESAM were 12% and 40 μJ/cm², respectively. The output coupler also served as the folding mirror of the cavity and two output beams were produced. Up to 400 mW of total mode-locked output power could be generated. During mode-locked operation, 10.8-ps-long pulses (assuming a sech² profile) with a spectral bandwidth of 37 GHz were produced, giving a time-bandwidth product of 0.4. No dispersion compensation scheme was employed. The repetition rate was 100 MHz. By moving the SESAM to the other end of the cavity where the mode area is 8 times larger than that used for mode-locked operation and by shortening the cavity length, passively Q-switched operation could also be obtained, yielding 47–62-ns pulses at a repetition rate of 1 MHz. In more recent studies [139], Sorokina et al. reported on the generation of femtosecond pulses with a duration of about 100 fs from a passively mode-locked Cr²⁺:ZnSe by using a SESAM similar in design to that used in the earlier studies [138]. In this particular case, an intracavity sapphire plate was also included to compensate for dispersion and to reduce the pulsewidths to the femtosecond range.

3.7 Conclusions

In this chapter, we have provided a comprehensive review of the work on the development of Cr²⁺:ZnSe lasers. A detailed account of the preparation methods and spectroscopic characterization techniques was presented. Diffusion doping was described in detail and various methods for the determination of the diffusion coefficient were discussed. Pulsed, continuous-wave, and mode-locked operations of the Cr²⁺:ZnSe lasers were also reviewed. Overall, the studies performed in our group delineated the effect of active ion concentration on passive losses, fluorescence efficiency, fluorescence lifetime, and power performance. The results indicate that chromium concentrations of less than 10×10^{18} ions/cm³ are more suitable for obtaining efficient lasing especially in the cw regime. Related to this, we note that the best route to power scaling in cw Cr²⁺:ZnSe lasers is through use of pump lasers operating as close to the peak absorption wavelength at 1775 nm as possible. This makes it possible to obtain a desired level of pump absorption with the least possible chromium concentration and hence minimizes several deleterious thermal loading problems that can arise during high-power operation.

Acknowledgments

We thank Mehmet Somer, Adnan Kurt, Hamit Kalaycioğlu, Hüseyin Çankaya, Muharrem Güler, Natali Çizmeciyan, Nuray Dindar, and Doğan Efe for their assistance in the experiments. This project was supported in part by the Scientific and Technical Research Council of Turkey (TÜBITAK), project TBAG-2030 and the BAYG program, and the Network of Excellence in Micro-Optics (NEMO) funded by the European Union 6th Framework program. A. Sennaroglu also acknowledges the support of the Turkish Academy of Sciences in the framework of the Young Scientist Award Program AS/TUBA-GEBIP/2001-1-11.

References

1. S. Adachi and T. Taguchi, Optical properties of ZnSe, *Phys. Rev. B* 43, 9569–9577, 1991.
2. I.T. Sorokina, Cr^{2+}-doped II-VI materials for lasers and nonlinear optics, *Opt. Mater.* 26, 395–412, 2004.
3. E.M. Gavrushchuk, Polycrystalline zinc selenide for IR optical applications, *Inorg. Mater.* 39, 883–898, 2003.
4. H.A. Weakliem, Optical spectra of Ni^{2+}, Co^{2+}, and Cu^{2+} in tetrehedral sites in crystals, *J. Chem. Phys.* 36, 2117–2140, 1961.
5. J.M. Baranowski, J.W. Allen, and G.L. Pearson, Crystal-field spectra of $3d^n$ impurities in II-VI and III-V compound semiconductors, *Phys. Rev* 160, 627–632, 1967.
6. G.A. Slack, F.S. Ham, and R.M. Cherenko, Optical Absorption of Tetrahedral Fe^{2+} (cd^6) in Cubic ZnS, CdTe, and $MgAl_2O_4$, *Phys. Rev.* 152, 376–402, 1966.
7. G.A. Slack and B.M. Omera, Infrared luminescence of Fe^{2+} in ZnS, *Phys. Rev.* 163, 335–341, 1967.
8. G.A. Slack, S. Roberts, and F.S. Ham, Far-Infrared Optical Absorption of Fe^{2+} in ZnS, *Phys. Rev.* 155, 170–177, 1967.
9. R. Pappalardo and R.E. Dietz, Absorption spectra of transition ions in CdS crystals, *Phys. Rev.* 123, 1188–1203, 1961.
10. J.T. Vallin, G.A. Slack, S. Roberts, and A.E. Hughes, Infrared absorption in some II-VI compounds doped with Cr, *Phys. Rev. B* 2, 4313–4333, 1970.
11. J.T. Vallin and G.D. Watkins, EPR of Cr^{2+} in II-VI lattices, *Phys. Rev. B* 9, 2051–2072, 1974.
12. W. Low and M. Weger, Paramagnetic resonance and optical spectra of divalent iron in cubic fields. I. Theory, *Phys. Rev.* 118, 1119–1130, 1960.
13. W. Low and M. Weger, Paramagnetic resonance and optical spectra of divalent iron in cubic fields. II. Experimental results, *Phys. Rev.* 118, 1130–1136, 1960.
14. L.D. DeLoach, R.H. Page, G.D. Wilke, S.A. Payne, and W.F. Krupke, Transition metal-doped zinc chalcogenides spectroscopy and laser demonstration of a new class of gain media, *IEEE J. Quantum Electron.* 32, 885–895, 1996.

15. R.H. Page, K.I. Schaffers, L.D. DeLoach, G.D. Wilke, F.D. Patel, J.B. Tassano, Jr., S.A. Payne, W.F. Krupke, K.-T. Chen, and A. Burger, Cr²⁺ doped zinc chalcogenides as efficient, widely tunable mid-infrared lasers, *IEEE J. Quantum Electron.* 33, 609–619, 1997.

16. J. McKay and K.L. Schepler, Kilohertz, 2.6-μm Cr²⁺:CdSe laser, *Proc. Adv. Solid-State Lasers*, Vol. 26, M.M. Fejer, H. Injeyan, and U. Keller, Eds., OSA, Boston, MA, 1999, pp. 420–426.

17. J. McKay, K.L. Schepler, and G.C. Catella, Efficient grating-tuned mid-infrared Cr²⁺:CdSe laser, *Opt. Lett.* 24, 1575–1577, 1999.

18. J.B. McKay, W.B. Roh, and K.L. Schepler, Extended mid-IR tuning of a Cr²⁺:CdSe laser, *Proc. Adv. Solid State Photonics*, Vol. 68, M.E. Fermann and L.R. Marshall, Eds., OSA, Montreal, 2002, pp. 371–373.

19. A.G. Bluiett, U. Hommerich, R.T. Shah, S.B. Trivedi, S.W. Kutcher, and C.C. Wang, Observation of lasing from Cr²⁺:CdTe and compositional effects in Cr²⁺-doped II-VI semiconductors, *J. Electron. Mater.* 31, 806–810, 2002.

20. I.T. Sorokina, E. Sorokin, A.D. Lieto, M. Tonelli, B.N. Mavrin, and E.A. Vinogradov, A new broadly tunable room temperature continuous-wave Cr²⁺:ZnₓSe₁₋ₓ laser, *Proc. Advanced Solid State Photonics*, Vol. 98, C. Denman and I.T. Sorokina, Eds., OSA, Vienna, Austria, 2005, pp. 263–267.

21. U. Hommerich, X. Wu, V.R. Davis, S.B. Trivedi, K. Grasza, R.J. Chen, and S. Kutcher, Demonstration of room-temperature laser action at 2.5 μm from Cr²⁺:Cd₀.₈₅Mn₀.₁₅Te, *Opt. Lett.* 22, 1180–1182, 1997.

22. U. Hommerich, J.T. Seo, A. Bluiett, D. Temple, S.B. Trivedi, H. Zong, S.W. Kutcher, C.C. Wang, R.J. Chen, and B. Schumm, Mid-infrared laser development based on transition metal doped cadmium manganese telluride, *Luminescence*, 1143–1145, 2000.

23. S.B. Trivedi, S.W. Kutcher, C.C. Wang, G.V. Jagannathan, U. Hommerich, A. Bluiett, M. Turner, J.T. Seo, K.L. Schepler, B. Schumm, P.R. Boyd, and G. Green, Transition metal doped cadmium manganese telluride: a new material for tunable mid-infrared lasing, *J. Electron. Mater.* 30, 728–732, 2001.

24. J.T. Seo, U. Hommerich, H. Zong, S.B. Trivedi, S.W. Kutcher, C.C. Wang, and R.J. Chen, Mid-infrared lasing from a novel optical material: chromium-doped Cd₀.₅₅Mn₀.₄₅Te, *Phys. Stat. Sol.* 175, R3, 1999.

25. J.T. Seo, U. Hommerich, S.B. Trivedi, R.J. Chen, and S. Kutcher, Slope efficiency and tunability of Cr²⁺:Cd₀.₈₅Mn₀.₁₅Te mid-infrared laser, *Opt. Commun.*, 267–270, 1998.

26. P. Cerny, H. Sun, D. Burns, U.N. Roy, and A. Burger, Spectroscopic investigation and continuous wave laser demonstration utilizing single crystal Cr²⁺:CdZnTe, *Proc. Advanced Solid-State Photonics*, Vol. 98, C. Denman and I.T. Sorokina, Eds., OSA, Vienna, Austria, 2005, pp. 268–273.

27. G.J. Wagner, T.J. Carrig, R.H. Page, K.I. Schaffers, J. Ndap, X. Ma, and A. Burger, Continuous-wave broadly tunable Cr²⁺:ZnSe laser, *Opt. Lett.* 24, 19–21, 1999.

28. R.H. Page, J.A. Skidmore, K.I. Schaffers, R.J. Beach, S.A. Payne, and W.F. Krupke, Demonstrations of diode-pumped grating-tuned ZnSe:Cr⁺² lasers, *Proc. OSA TOPS, Advanced Solid-State Lasers*, Vol. 10, W.R. Rosenberg, Ed., Orlando, 1997, pp. 208–210.

29. T.J. Carrig, G.J. Wagner, A. Sennaroglu, J.Y. Jeong, and C.R. Pollock, Modelocked Cr²⁺:ZnSe laser, *Opt. Lett.* 25, 168–170, 2000.

30. E. Sorokin, S. Naumov, and I.T. Sorokina, Ultrabroadband infrared solid-state lasers, *IEEE J. Sel. Top. Quantum Electron.* 11, 690–712, 2005.

31. G.J. Wagner, B.G. Tiemann, W.J. Alford, and T.J. Carring, Single-Frequency Cr:ZnSe Laser, *Proc. OSA Advanced Solid-State Photonics*, Vol. 94, G.J. Quarles, Ed., OSA, New Mexico, 2004, pp. 371–375.

32. C. Fischer, E. Sorokin, I.T. Sorokina, and M.W. Sigrist, Photoacoustic monitoring of gases using a novel laser source tunable around 2.5 μm, *Opt. Lasers Eng.* 43, 573–582, 2005.

33. T.J. Carrig, A.K. Hankla, G.J. Wagner, C.B. Rawle, and I.T.M. Kinnie, Tunable infrared laser sources for DIAL, *Proc. Laser Radar Technology and Applications VII, Proceedings of SPIE*, Vol. 4723, G.W. Kamerman, Ed., SPIE, 2002, pp. 147–155.

34. E. Sorokin, I.T. Sorokina, C. Fischer, and M.W. Sigrist, Widely tunable Cr^{2+}:ZnSe laser source for trace-gas sensing, *Proc. Advanced Solid State Photonics*, Vol. 98, C. Denman and I.T. Sorokina, Eds., OSA, Vienna, Austria, 2005, pp. 826–830.

35. N. Picque, F. Gueye, G. Guelachvili, E. Sorokin, and I.T. Sorokina, Time-resolved Fourier transform intracavity spectroscopy with a Cr^{2+}:ZnSe laser, *Opt. Lett.* 30, 3410–3412, 2005.

36. V.A. Akimov, V.I. Kozlovskii, Y.V. Korostelin, A.I. Landman, Y.P. Podmar'kov, and M.P. Frolov, Spectral dynamics of intracavity absorption in a pulsed Cr^{2+}:ZnSe laser, *Quantum Electron.* 35, 425–428, 2005.

37. F.K. Tittel, D. Richter, and A. Fried, Mid-infrared laser applications in spectroscopy, in *Solid-State Mid-Infrared Laser Sources*, Vol. 89, Springer Topics in Applied Physics, I.T. Sorokina and K.L. Vodopyanov, Eds., Springer-Verlag, Berlin, 2003, pp. 445–516.

38. V. Girard, R. Farrenq, E. Sorokin, I. T. Sorokina, G. Guelachvili, and N. Picque, Acetylene weak bands at 2.5 μm from intracavity Cr^{2+}:ZnSe laser absorption observed with time-resolved Fourier transform spectroscopy, *Chem. Phys. Lett.* 419, 584–588, 2006.

39. V.A. Akimov, M.P. Frolov, Y.V. Korostelin, V.I. Kozlovsky, A.I. Landman, and Y.P. Podmar'kov, Cr^{2+}:ZnSe laser for application to intracavity laser spectroscopy, *Proceedings of Laser Optics 2003: Solid State Lasers and Nonlinear Frequency Conversion*, Vol. 5478, V.I. Ustugov, Ed., SPIE, 2003, pp. 285–290.

40. E. Sorokin and I.T. Sorokina, Mid-IR high-resolution intracavity Cr^{2+}:ZnSe laser based spectrometer," Proceedings of Advanced Solid State Photonics, C. Denman and I.T. Sorokina, Eds., OSA, Vienna, Austria, 2005, Vol. 98, pp. 848–852.

41. M. Raybaut, A. Godard, R. Haidar, M. Lefebvre, P. Kupecek, P. Lemasson, and E. Rosencher, Generation of mid-infrared radiation by self difference frequency mixing in chromium-doped zinc selenide, *Opt. Lett.* 31, 220–222, 2006.

42. A. Zakel, G.J. Wagner, W.J. Alford, and T.J. Carrig, High-power, rapidly tunable dual-band CdSe optical parametric oscillator, *Proc. Advanced Solid-State Photonics*, Vol. 98, C. Denman and I.T. Sorokina, Eds., OSA, Vienna, Austria, 2005, pp. 433–437.

43. I.T. Sorokina, Crystalline mid-infrared lasers, in *Solid-State Mid-Infrared Laser Sources*, Vol. 89, Springer Topics in Applied Physics, I.T. Sorokina and K.L. Vodopyanov, Eds., Springer-Verlag, Berlin, pp. 255–349.

44. A. Sennaroglu, Broadly tunable Cr^{4+}-doped solid-state lasers in the near infrared and visible, *Prog. Quantum Electron.* 26, 287–352, 2002.

45. A.A. Kaminskii, Modern Developments in the physics of crystalline laser materials, *Phys. Stat. Sol.* 200, 215–296, 2003.

46. S. Kück, Laser-related spectroscopy of ion-doped crystals for tunable solid-state lasers, *Appl. Phys. B* 72, 515–562, 2001.

47. P.F. Moulton, Tunable solid-state lasers, *Proc. IEEE* 80, 348–364, 1992.

48. J.C. Walling, O.G. Peterson, H.P. Jenssen, R.C. Morris, and E.W. O'Dell, Tunable alexandrite lasers, *IEEE J. Quantum Electron.* 16, 1302–1315, 1980.

49. J.C. Walling, D.F. Heller, H. Samelson, D.J. Harter, J.A. Pete, and R.C. Morris, Tunable alexandrite lasers: development and performance, *IEEE J. Quantum Electron.* QE-21, 1568–1581 (1985).

50. T.H. Maiman, Stimulated optical radiation in ruby, *Nature* 187, 493–494, 1960.

51. L.F. Johnson, R.E. Dietz, and H.J. Guggenheim, Optical maser oscillation from Ni^{2+} in MgF_2 involving simultaneous emission of phonons, *Phys. Rev. Lett.* 11, 318–320, 1963.

52. L.F. Johnson, H.J. Guggenhem, and R.A. Thomas, Phonon-terminated optical masers, *Phys. Rev.* 149, 179–185, 1966.

53. L.F. Johnson, R.E. Dietz, and H.J. Guggenheim, Spontaneous and stimulated emission from Co^{2+} ions in MgF_2 and ZnF_2, *Appl. Phys. Lett.* 5, 21–23, 1964.

54. L.F. Johnson and H.J. Guggenheim, Phonon-terminated coherent emission from V^{2+} ions in MgF_2, *J. Appl. Phys.* 38, 4837–4839, 1967.

55. D.E. McCumber, Theory of phonon-terminated optical masers, *Phys. Rev.* 134, A299–A306, 1963.

56. D.E. McCumber, Theory of vibrational structure in optical spectra of impurities in solids. I. Singlets, *J. Math. Phys.* 5, 221–230, 1963.

57. D.E. McCumber, Theory of vibrational structure in optical spectra of impurities in solids. II. Multiplets, *J. Math. Phys.* 5, 508–521, 1963.

58. D.E. McCumber, Einstein relations connecting broadband emission and absorption spectra, *Phys. Rev.* 136, A954–A957, 1964.

59. P.F. Moulton, A. Mooradian, and T.B. Reed, Efficient cw optically pumped Ni:MgF2 laser, *Opt. Lett.* 3, 1978.

60. J.C. Walling, H.P. Jenssen, R.C. Morris, E.W. O'Dell, and O.G. Peterson, Tunable-laser performance in BeAl2O:Cr3+, *Opt. Lett.* 4, 1979.

61. P. Zeller and P. Peuser, Efficient, multiwatt, continuous-wave laser operation on the $^4F_{3/2}$-$^4I_{9/2}$ transitions of Nd:YVO$_4$ and Nd:YAG, *Opt. Lett.* 25, 34–36, 2000.

62. B. Struve, G. Huber, V.V. Laptev, I.A. Scherbakov, and E.V. Zharikov, Tunable room-temperature cw laser action in Cr^{3+}:GdScGa-Garnet, *Appl. Phys. B* 30, 117–120, 1983.

63. J. Drube, B. Struve, and G. Huber, Tunable room-temperature cw laser action in Cr^{3+}:GdScAl-Garnet, *Opt. Commun.* 50, 45–48, 1984.

64. S.A. Payne, L.L. Chase, H.W. Newkirk, L.K. Smith, and W.F. Krupke, LiCaAlF$_6$:Cr^{3+}: a promising new solid-state laser material, *IEEE J. Quantum Electron.* 24, 2243–2252, 1988.

65. S.A. Payne, L.L. Chase, L.K. Smith, W.L. Kway, and H.W. Newkirk, Laser performance of LiSAIF$_6$:Cr^{3+}, *J. Appl. Phys.* 66, 1051–1056, 1989.

66. J.J. Adams, C. Bibeau, R.H. Page, D.M. Krol, L.H. Furu, and S.A. Payne, 4.0–4.5 μm lasing of Fe: ZnSe below 180 K, a new mid-infrared laser material, *Opt. Lett.* 24, 1720–1722, 1999.

67. J. Kernal, V.V. Fedorov, A. Gallian, S.B. Mirov, and V.V. Badikov, Room temperature 3.9–4.5 μm gain-switched lasing of Fe:ZnSe, *Proc. Advanced Solid State Photonics*, OSA, Nevada, 2006, p. MD6.

68. J. Kernal, V.V. Fedorov, A. Gallian, S.B. Mirov, and V.V. Badikov, 3.9–4.8 μm gain-switched lasing of Fe:ZnSe at room temperature, *Opt. Express* 13, 10608–10615, 2005.

69. A.A. Voronov, V.I. Kozlovskii, Y.V. Korostelin, A.I. Landman, Y.P. Podmar'kov, and M.P. Frolov, Laser parameters of a Fe:ZnSe crystal in the 85–255-K temperature range, *Quantum Electron.* 35, 809–812, 2005.

70. V.A. Akimov, A.A. Voronov, V.I. Kozlovsky, Y.V. Korostelin, A.I. Landman, Y.P. Podmar'kov, and M.P. Frolov, Efficient IR Fe:ZnSe laser continuously tunable in the spectral range from 3.77 to 4.40 μm, *Quantum Electron.* 34, 912–914, 2004.

71. J.J. Adams, C. Bibeau, R.H. Page, and S.A. Payne, Tunable laser action at 4.0 microns from Fe:ZnSe, *Proc. Advanced Solid State Lasers*, Vol. 26, M.M. Fejer, H. Injeyan, and U. Keller, Eds., OSA, Boston, 1999, pp. 435–440.

72. A.A. Voronov, V.I. Kozlovskii, Y.V. Korostelin, A.I. Landman, Y.P. Podmar'kov, V.G. Polushkin, and M.P. Frolov, Passive Fe^{2+}:ZnSe single-crystal Q switch for 3-μm lasers, *Quantum Electron.* 36, 1–2, 2006.

73. C.-H. Su, S. Feth, M.P. Volz, R. Matyi, M.A. George, K. Chattopadhyay, A. Burger, and S.L. Lehoczky, Vapor growth and characterization of Cr-doped ZnSe crystals, *J. Cryst. Growth* 207, 35–42, 1999.

74. A. Burger, K. Chattopadhyay, J.O. Ndap, X. Ma, S. H. Morgan, C.I. Rablau, C.H. Su, S. Feth, R.H. Page, K.I. Schaffers, and S.A. Payne, Preparation conditions of chromium doped ZnSe and their infrared luminescence properties, *Cryst. Growth* 225, 249–256, 2001.

75. S.B. Mirov, S. Wang, V.V. Fedorov, and R.P. Camata, Pulse laser deposition growth and spectroscopic properties of chromium doped ZnS crystalline thin films, *Proc. Advanced Solid State Photonics*, Vol. 5332, Optical Society of America, Santa Fe, NM, 2004, p. WB15.

76. B.L. Vanmil, A.J. Ptak, L. Bai, L. Wang, M. Chirila, N.C. Giles, T.H. Myers, and L. Wang, Heavy Cr doping of ZnSe by molecular beam epitaxy, *J. Electron. Mater.* 31, 770–775, 2002.

77. J.O. Ndap, O.O. Adetunji, K. Chattopadhyay, C.I. Rablau, S.U. Egarievwe, X. Ma, S. Morgan, and A. Burger, High-temperature solution growth of Cr^{2+}:CdSe for tunable mid-IR laser application, *J. Cryst. Growth* 211, 290–294, 2000.

78. J.O. Ndap, K. Chattopadhyay, O.O. Adetunji, D.E. Zelmon, and A. Burger, Thermal diffusion of Cr^{2+} in bulk ZnSe, *J. Cryst. Growth* 240, 176–184, 2002.

79. C.I. Rablau, J.O. Ndap, X. Ma, A. Burger, and N.C. Giles, Absorption and photoluminescence spectroscopy of diffusion-doped ZnSe:Cr^{2+}, *Electron. Mater.* 28, 678–682, 1999.

80. U.N. Roy, O.S. Babalola, J. Jones, Y. Cui, T. Mounts, A. Zavalin, S. Morgan, and A. Burger, Uniform Cr^{2+} doping of physical vapor transport grown CdS_xSe_{1-x} crystals, *J. Electron. Mater.* 34, 19–22, 2005.

81. S.B. Mirov, V.V. Fedorov, K. Graham, I.S. Moskalev, I.T. Sorokina, E. Sorokin, V. Gapontsev, D. Gapontsev, V.V. Badikov, and V. Panyutin, Diode and fiber pumped Cr^{+2}:ZnS mid-infrared external cavity and microchip lasers, *IEE Proc. Optoelectron.* 150, 340–345, 2003.

82. M.B. Johnson, S.B. Mirov, V. Fedorov, M.E. Zvanut, J.G. Harrison, V.V. Badikov, and G.S. Shevirdyaeva, Absorption and photoluminescence studies of $CdGa_2S_4$:Cr, *Opt. Commun.* 233, 403–410, 2004.

83. V.R. Davis, X. Wu, U. Hommerich, K. Grasza, S.B. Trivedi, and Z. Yu, Optical properties of Cr^{2+} ions in $Cd_{0.85}Mn_{0.15}Te$, *J. Luminescence* 72, 281–283, 1997.

84. U.N. Roy, O.S. Babalola, Y. Cui, M. Groza, T. Mounts, A. Zavalin, S. Morgan, and A. Burger, Vapor growth and chracterization of Cr-doped $CdS_{0.8}Se_{0.2}$ single crystals, *J. Cryst. Growth* 265, 453–458, 2004.

85. V.A. Akimov, M.P. Frolov, Y.V. Korostelin, V.I. Kozlovsky, A.I. Landman, Y.P. Podmar'kov, and A.A. Voronov, Vapor growth of II-VI single crystals doped by transition metals for mid-infrared lasers, *Phys. Status Solidi* (c) 3, 1213–1216, 2006.

86. K.L. Schepler, S. Kück, and L. Shiozawa, Cr^{2+} emission spectroscopy in CdSe, *Luminescence*, 116–117, 1997.

87. V. Kasiyan, Z. Dashevsky, R. Shneck, and E. Towe, Optical and transport properties of chromium-doped CdSe and $CdS_{0.67}Se_{0.33}$ crystals, *J. Cryst. Growth* 290, 50–55, 2006.

88. U.N. Roy, Y. Cui, C. Barnett, K.-T. Chen, A. Burger, and J.T. Goldstein, Growth of undoped and chromium-doped CdS_xSe_{1-x} crystals by the physical vapor transport method, *J. Electron. Mater.* 31, 791–794, 2002.

89. I.T. Sorokina, E. Sorokin, S. Mirov, V. Fedorov, V. Badikov, V. Panyutin, A.D. Lieto, and M. Tonelli, Tunable continuous-wave room temperature Cr^{2+}:ZnS laser, *Proc. Advanced Solid State Photonics*, Vol. 68, M.E. Fermann and L.R. Marshall, Eds., OSA, Montreal, 2002, pp. 358–363.

90. K. Graham, S.B. Mirov, V.V. Fedorov, M.E. Zvanut, A. Avanesov, V. Badikov, B. Ignat'ev, V. Panutin, and G. Shevirdyaeva, Spectroscopic characterization and laser performance of diffusion doped Cr^{2+}:ZnS, *Proc. Advanced Solid State Photonics*, Vol. 50, C. Marshall, Ed., OSA, San Jose, 2001, pp. 561–567.

91. Y.G. Sadofyev, V.F. Pevtsov, E.M. Dianov, P.A. Trubenko, and M.V. Korshkov, Molecular beam epitaxy growth and characterization of ZnTe:Cr^{2+} layers on GaAs(100), *J. Vac. Sci. Technol.* B 19, 1483–1487, 2001.

92. U. Demirbas, A. Sennaroglu, A. Kurt, and M. Somer, Preparation and spectroscopic investigation of diffusion-doped Fe^{2+}:ZnSe and Cr^{2+}:ZnSe, *Proc. Advanced Solid-State Photonics*, Vol. 98, C. Denman and I. T. Sorokina, Eds., The Optical Society of America, Vienna, Austria, 2005, pp. 63–68.

93. G.J. Wagner, T.J. Carrig, R.H. Jarman, R.H. Page, K.I. Schaffers, J.O. Ndap, X. Ma, and A. Burger, High-efficiency, broadly-tunable continuous-wave Cr^{2+}:ZnSe laser, *Proc. Advanced Solid State Photonics*, Vol. 26, M.M. Fejer, H. Injeyan, and U. Keller, Eds., OSA, Boston, 1999, pp. 427–434.

94. E. Sorokin and I.T. Sorokina, Mode-locked ceramic Cr^{2+}:ZnSe laser, *Proc. Advanced Solid-State Photonics*, Vol. 83, J.J. Zayhowski, Ed., OSA, San Antonio, 2003, pp. 227–230.

95. T.J. Carrig, Transition-metal-doped chalcogenide lasers, *J. Electron. Mater.* 31, 759–769, 2002.

96. V.I. Levchenko, V.N. Yakimovich, L.I. Postnova, V.I. Konstantinov, V.P. Mikhailov, and N.V. Kuleshov, Preparation and properties of bulk ZnSe:Cr single crystals, *Cryst.* Growth 198, 980–983, 1999.

97. U. Demirbas, A. Sennaroglu, and M. Somer, Synthesis and characterization of diffusion-doped Cr^{2+}:ZnSe and Fe^{2+}:ZnSe, *Opt. Mater.* 28, 231–240, 2006.

98. U. Hommerich, A.G. Bluiett, I.K. Jones, S.B. Trivedi, and R.T. Shah, Crystal growth and infrared spectroscopy of Cr:$Cd_{1-x}Zn_xTe$ and Cr:$Cd_{1-x}Mg_xTe$, *J. Cryst. Growth* 287, 243–247, 2006.

99. H.S. Carslaw and J.C. Jaeger, *Conduction of Heat in Solids*. Oxford: Oxford Science, 1959.

100. G. Goetz, H. Zimmermann, and H.-J. Schulz, Jahn-Teller interaction at $Cr^{2+}(d^4)$ centers in tetrahedrally coordinated II-VI lattices studied by optical spectroscopy, *Zeitschrift Für Physik* B 91, 429–436, 1993.

101. M. Godlewski and M. Kaminska, The chromium impurity photogeneration transitions in ZnS, ZnSe and ZnTe, *J. Phys. C: Solid State Phys.* 13, 6537–6545, 1980.

102. S. Bhaskar, P.S. Dobal, B.K. Rai, R.S. Katiyar, H.D. Bist, J.O. Ndap, and A. Burger, Photoluminescence study of deep levels in Cr-doped ZnSe, *Appl. Phys.* 85, 439–443, 1998.

103. M. Godlewski, Mechanisms of radiative and nonradiative recombination in ZnSe:Cr and ZnSe:Fe, *Low Temp. Phys.* 30, 891–896, 2004.

104. Y.F. Vaksman, V.V. Pavlov, Y.A. Nitsuk, Y.N. Purtov, A.S. Nasibov, and P.V. Shapkin, Optical absorption and chromium diffusion in ZnSe single crystals, *Semiconductors* 39, 377–380, 2005.

105. A. Gallian, V.V. Fedorov, J. Kernal, S.B. Mirov, and V.V. Badikov, Laser oscillation at 2.4 μm from Cr²⁺ in ZnSe optically pumped over Cr ionization transitions, *Proc. Advanced Solid State Photonics*, Vol. 98, C. Denman and I.T. Sorokina, Eds., OSA, Austria, 2005, pp. 241–245.

106. K. Graham, V.V. Fedorov, S.B. Mirov, M.E. Doroshenko, T.T. Basiev, Y.V. Orlovskii, V.V. Osiko, V.V. Badikov, and V.L. Panyutin, Pulsed mid-IR Cr²⁺:ZnS and Cr²⁺:ZnSe lasers pumped by Raman-shifted Q-switched neodymium lasers, *Quantum Electron.* 34, 8–14, 2004.

107. A. Sennaroglu, U. Demirbas, A. Kurt, and M. Somer, Concentration dependence of fluorescence and lasing efficiency in Cr²⁺:ZnSe lasers, *Opt. Mater.* (2006 in press).

108. V.E. Kisel, V.G. Shcherbitsky, N.V. Kuleshov, V.I. Konstantinov, V.I. Levchenko, E. Sorokin, and I.T. Sorokina, Spectral kinetic properties and lasing chracteristics of diode-pumped Cr²⁺:ZnSe single crystals, *Opt. Spectrosc.* 99, 663–667, 2005.

109. A.V. Podlipensky, V.G. Shcherbitsky, N.V. Kuleshov, V.P. Mikhailov, V.I. Levchenko, V.N. Yakimovich, L.I. Postnova, and V.I. Konstantinov, Pulsed laser operation of diffusion-doped Cr²⁺:ZnSe, *Opt. Commun.* 167, 129–132, 1999.

110. A.V. Podlipensky, V.G. Shcherbitsky, N.V. Kuleshov, V.I. Levchenko, V.N. Yakimovich, M. Mond, E. Heumann, G. Huber, H. Kretschmann, and S. Kück, Efficient laser operation and continuous-wave didode pumping of Cr²⁺:ZnSe single crystals, *Appl. Phys. B*, 253–255, 2001.

111. B. Henderson and G.F. Imbush, *Optical Spectroscopy of Inorganic Solids*, Oxford: Clarendon Press, 1989.

112. R.C. Powell, *Physics of Solid-State Laser Materials*, New York: Springer-Verlag, 1998.

113. V.E. Kisel, V.G. Shcherbitsky, N.V. Kuleshov, V.I. Konstantinov, L.I. Postnova, L.I. Levchenko, E. Sorokin, and I.T. Sorokina, Emission lifetime measurements and laser performance of Cr:ZnSe under diode pumping at 1770 nm, *Proc. Advanced Solid-State Photonics*, Vol. 94, G.J. Quarles, Ed., Optical Society of America, Santa Fe, New Mexico, 2004, pp. 367–370.

114. U. Hommerich, I.K. Jones, E.E. Nyein, and S.B. Trivedi, Comparison of the optical properties of diffusion-doped polycrystalline Cr:ZnSe and Cr:CdTe windows, *J. Cryst. Growth* 287, 450–453, 2006.

115. A. Sennaroglu, U. Demirbas, S. Ozharar, and F. Yaman, Accurate determination of saturation parameters for Cr⁴⁺-doped solid-state saturable absorbers, *J. Opt. Soc. Am. B* 23, 241, 2006.

116. A.V. Podlipensky, V.G. Shcherbitsky, N.V. Kuleshov, and V.P. Mikhailov, Cr²⁺:ZnSe and Co²⁺:ZnSe saturable-absorber Q switches for 1.54-μm Er:glass lasers, *Opt. Lett.* 24, 960–962, 1999.

117. V.G. Shcherbitsky, S. Girard, M. Fromager, R. Moncorge, N.V. Kuleshov, V.I. Levchenko, V.N. Yakimovich, and B. Ferrand, Accurate method of the measurement of absorption cross sections of solid-state saturable absorbers, *Appl. Phys. B* 74, 367–374, 2002.

118. R.D. Stultz, V. Leyva, and K. Spariosu, Short pulse, high-repetition rate, passively q-switched Er:yttrium-aluminum-garnet laser at 1.6 microns, *Appl. Phys. Lett.* 87, 241118, 2005.

119. M.P. Frolov, Y.V. Korostelin, V.I. Kozlovsky, A.I. Landman, and Y.P. Podmar'kov, Efficient laser operation from Cr²⁺:ZnSe crystals produced by seeded physical vapor transport method, *Proc. Laser Optics 2003: Solid State Lasers and Monlinear Frequency Conversion*, Vol. 5478, V.L. Ustugov, Ed., SPIE, 2003, pp. 55–59.

120. V.I. Kozlovsky, Y.V. Korostelin, A.I. Landman, Y.P. Podmar'kov, and M.P. Frolov, Efficient lasing of a Cr²⁺:ZnSe crystal grown from a vapour phase, *Quantum Electron*. 33, 408–410, 2003.

121. W.J. Alford, G.J. Wagner, A.C. Sullivan, J.A. Keene, and T.J. Carrig, High-power and Q-switched Cr:ZnSe lasers, *OSA TOPS, Advanced Solid-State Photonics* 83, 13–17, 2003.

122. A. Zakel, G.J. Wagner, A.C. Sullivan, J.F. Wenzel, W.J. Alford, and T.J. Carrig, High-brightness, rapidly tunable Cr:ZnSe lasers, *Proc. Advanced Solid State Photonics*, Vol. 98, C. Denman and I.T. Sorokina, Eds., OSA, Vienna, Austria, 2005, pp. 723–727.

123. J.A. Caird, L.G. DeShazer, and J. Nella, Characteristics of Room-Temperature 2.3-μm Laser Emission from Tm³⁺ in YAG and YAlO₃, *IEEE J. Quantum Electron*. QE-11, 874–881, 1975.

124. I.T. Sorokina, E. Sorokin, A.D. Lieto, M. Tonelli, R.H. Page, and K.I. Schaffers, Efficient broadly tunable continuous-wave Cr²⁺:ZnSe laser, *J. Opt. Soc. Am. B* 18, 926–930, 2001.

125. K.L. Schepler, R.D. Peterson, P.A. Berry, and J.B. McKay, Thermal effects in Cr²⁺:ZnSe thin disk lasers, *IEEE J. Sel. Top. Quantum Electron*. 11, 713–720, 2005.

126. A.D. Lieto and M.Tonelli, Development of a cw polycrystalline Cr²⁺:ZnSe laser, *Opt. Lasers Eng*. 39, 305–308, 2001.

127. A. Sennaroglu, A.O. Konca, and C.R. Pollock, Continuous-wave power performance of a 2.47 μm Cr²⁺: ZnSe laser: experiment and modeling, *IEEE J. Quantum Electron*. 36, 1199–1205, 2000.

128. M. Mond, D. Albrecht, H.M. Kretschmann, E. Heumann, G. Huber, S. Kück, V.I. Yakimovich, V.G. Shcherbitsky, V.E. Kisel, and N.V. Kuleshov, Erbium doped fiber amplifier pumped Cr²⁺:ZnSe laser, *Phys. Stat. Sol*. 188, 2001.

129. S.B. Mirov, V.V. Fedorov, K. Graham, I.S. Moskalev, V.V. Badikov, and V. Panyutin, Erbium fiber laser-pumped continuous-wave microchip Cr²⁺:ZnS and Cr²⁺:ZnSe lasers, *Opt. Lett*. 27, 909–911, 2002.

130. I.S. Moskalev, V.V. Fedorov, and S.B. Mirov, Multiwavelength mid-IR spatially-dispersive CW laser based on polycrystalline Cr²⁺:ZnSe, *Opt. Express* 12, 4986–4992, 2004.

131. R.D. Peterson and K.L. Schepler, 1.9 μm-fiber-pumped Cr:ZnSe laser, *Proc. Advanced Solid-State Photonics*, Vol. 98, C. Denman and I.T. Sorokina, Eds., The Optical Society of America, Vienna, Austria, 2005, pp. 236–240.

132. E. Sorokin and I.T. Sorokina, Tunable diode-pumped continuous-wave Cr⁺²:ZnSe laser, *Appl. Phys. Lett*. 80, 3289–3291, 2002.

133. D. Findlay and R.A. Clay, The measurement of internal losses in 4-level lasers, *Phys. Lett*. 20, 277–278, 1966.

134. L.L. Gordley, B.T. Marshall, and D.A. Chu, LINEPAK: algorithms for modeling spectral transmittance and radiance, *J. Quant. Spectrosc. Radiat. Transfer* 52, 563–580, 1994.

135. T.J. Carrig, G.J. Wagner, A. Sennaroglu, J.Y. Jeong, and C.R. Pollock, Mode-locked Cr²⁺:ZnSe laser, *Opt. Lett*. 25, 168–170, 1999.

136. D.J. Kuizenga and A.E. Siegman, FM and AM mode locking of the homogeneous laser-Part I: Theory, *IEEE J. Quantum Electron*. QE-6, 694–708, 1970.

137. I.T. Sorokina, E. Sorokin, A.D. Lieto, M. Tonelli, R.H. Page, and K.I. Schaffers, Active and passive mode-locking of the Cr^{+2}:ZnSe laser, *Proc. Advanced Solid-State Lasers*, Vol. 50, C. Marshall, Ed., OSA, Munich, 2001, pp. 157–161.
138. C.R. Pollock, N.A. Brilliant, D. Gwin, T.J. Carrig, W.J. Alford, J.B. Heroux, W.I. Wang, I. Vurgaftman, and J.R. Meyer, Mode locked and Q-switched Cr: ZnSe laser using a semiconductor saturable absorbing mirror (SESAM), *Proc. Advanced Solid State Photonics*, C. Denman and I.T. Sorokin, Eds., OSA, Vienna, Austria, 2005, Vol. 98, pp. 252–256.
139. I.T. Sorokina, E. Sorokin, T.J. Carrig, and K.I. Scaffers, A SESAM passively mode-locked Cr:ZnS laser, *Proc. Advanced Solid-State Photonics*, Optical Society of America, Incline Village, Nevada, 2006, TuA4.
140. P.F. Moulton, Spectroscopic and laser characteristics of Ti:Al$_2$O$_3$, *J. Opt. Soc. Am. B* 3, 125–133, 1986.
141. V.G. Baryshevskii, M.V. Korzhik, A.E. Kimaev, M.G. Livshits, V.B. Pavlenko, M.L. Meil'man, and B.I. Minkov, Tunable chromium forsterite laser in the near IR region, *J. Appl. Spectrosc. (USSR)* 53, 675–676, 1990.
142. S. Kuck, K. Petermann, U. Pohlmann, U. Schonhoff, and G. Huber, Tunable room-temperature laser action of a Cr-4$^+$-doped Y$_3$Sc$_x$Al$_{5-x}$O$_{12}$, *Appl. Phys. B* 58, 153–156, 1994.
143. D. Welford and P.F. Moulton, Room-temperature operation of a Co:MgF$_2$ laser, *Opt. Lett.* 13, 975–977, 1988.
144. I.T. Sorokina, E. Sorokin, S. Mirov, V. Fedorov, V. Badikov, V. Panyutin, and K.I. Schaffers, Broadly tunable compact continuous-wave Cr^{2+}:ZnS laser, *Opt. Lett.* 27, 1040–1042, 2002.
145. T.-Y. Tsai and M. Birnbaum, Q-switched 2-μm lasers by use of a Cr $^{2+}$:ZnSe saturable absorber, *Appl. Opt.* 40, 6633–6637, 2001.
146. F.Z. Qumar and T.A. King, Passive q-switching of Tm-silica fibre laser near 2 μm by a Cr^{2+}:ZnSe saturable absorber crystal, *Opt. Commun.* 248, 501–508, 2005.

4

All-Solid-State Ultraviolet Cerium Lasers

Shingo Ono, Zhenlin Liu, and Nobuhiko Sarukura

CONTENTS

4.1   Introduction

The ultraviolet (UV) tunable lasers have become one of the most important tools in many fields of science and technology. The most impressive applications for them include environmental sensing, engine combustion diagnostics, semiconductor processing, micromaching, optical communications, and medicinal and biological applications.

To illustrate the use of UV lasers in such diverse areas, we may consider an example. The behavior of trace constituents in the Earth's upper atmosphere, governed by chemical, dynamical, and radiative processes, is of particular importance for the overall balance of the stratosphere and mesosphere. In particular, ozone plays a dominant role by absorbing the short-wavelength UV radiation which might damage living organisms and by maintaining the radiative budget equilibrium. The measurement of the total ozone column content and vertical profile by a ground-based UV spectrometer network or by satellite-borne systems constitute the fundamental basis for global observations and trend analysis. Remote measurements of the trace constituents using an active technique such as lidar have been made possible by the rapid development of powerful tunable laser sources, which have opened a new field in atmospheric spectroscopy by providing sources which can be tuned to characteristic spectral features of atmospheric constituents [1].

Tunable UV laser sources are used for atmospheric differential absorption lidar (DIAL) measurements from airplanes to analyze the global distribution of O_3 radicals which is directly relevant to the "ozone hole" and global climate formation problems [2]. In the airborne UV DIAL system, two frequency-doubled Nd:YAG lasers are used to pump two high-conversion-efficiency, frequency-doubled, tunable dye lasers. They used the UV lasers in the wavelength region from 289 to 311 nm. All-solid-state cerium lasers cover this wavelength region.

4.1.1   Cerium-Doped Fluoride Crystals

To provide tunable or ultrashort UV laser radiation in a reliable and efficient way, the best choice at the moment would be to use directly-pumped solid-state UV-active media, based on the electrically-dipole-allowed interconfigurational 5d-4f transitions of rare-earth ions in wide band-gap fluoride crystals: $YLiF_4$ (YLF) [3], LaF_3 [4], $LuLiF_4$ (LLF) [5,6], $LiCaAlF_6$ (LiCAF) [7–9], and $LiSrAlF_6$ (LiSAF) [10,11]. In fact, this is the only option which also allows independent control of tunable radiation bandwidth or, when necessary, even provides multiwavelength UV output in one laser beam from the oscillator [5,6]. This option was proposed in 1977 from purely spectroscopic considerations [12] and then confirmed experimentally on Ce^{3+} ion in UV region by D. J. Ehrlich et al. [3,4] and Nd^{3+} ion in VUV region by R. W. Waynant [13].

As early as 1977, K. H. Yang and J. A. Deluca proposed a simple way of implementing a tunable laser capable of producing radiation directly in the UV and even vacuum ultraviolet (VUV) spectral ranges [12]. For this purpose, they proposed to use interconfiguration 5d-4f transitions of rare-earth ions in wide band-gap dielectric crystals. Because of the strong lattice interaction with 5d electrons, the fluorescence that results from 5d to 4f transitions of trivalent rare-earth ions in solid hosts is characterized by broad bandwidths and large Stokes shifts. Such fluorescence is particularly attractive for the development of tunable lasers. Powder samples of $Ce^{3+}:LaF_3$ and $Ce^{3+}:LuF_3$ were excited by the 253.7-nm radiation transmitted through a narrow-band interference filter inserted in front of a mercury lamp source. Broadband UV fluorescence were reported for $Ce^{3+}:LaF_3$ (276–312 nm) and $Ce^{3+}:LuF_3$ (288–322 nm). The fluorescence quantum yields account for the fact that not all of the atoms raised to the pump bands subsequently decay to the upper laser level. Some of these atoms can in fact decay from the pump bands straight back to the ground state or perhaps to other levels which are not useful. The pump quantum efficiency or fluorescence quantum yields $\eta_q(\lambda)$ is defined as the ratio of the number of atoms which decay to the upper laser level to the number of atoms which are raised to the pump band by a monochromatic pump at wavelength λ. The fluorescence quantum yields of $LaF_3:1\%Ce^{3+}$ and $LuF_3:0.1\%Ce^{3+}$ are 0.9 and 0.82, respectively. Estimates of the threshold power for lasing action suggested that a laser system tunable from 276 to 322 nm is feasible with noble-gas-halide lasers as pumping sources [12].

After that, $Ce^{3+}:Y_3Al_5O_{12}$ (YAG) was investigated as a model system for a 5d-4f solid-state tunable laser [14]. This system was chosen because YAG had been extensively studied as a laser host and good quality crystals were readily available. Despite providing apparently adequate conditions to achieve stimulated emission, it was not possible to detect laser action in $Ce^{3+}:YAG$. It was found that there was strong excited-state absorption (ESA) in this material at the wavelengths of its fluorescence. The ESA was sufficiently strong to completely quench any possible laser action. The crystal showed a net optical loss instead of optical gain at the wavelength of the fluorescence transition. This self-absorption may explain the failure of all attempts to obtain stimulated emission in this material.

A laser of the type which has a 5d-4f transition was originally implemented with $Ce^{3+}:YLiF_4$ (Ce:YLF) as a laser medium [3]. It should be noted that Ce^{3+} ions are the most promising activators for the UV spectral range. However, in spite of a large number of studied Ce-activated materials [15,16], the investigations performed before 1992 revealed only two laser-active media [3,4].

In 1979, D. J. Ehrlich et al. reported the first observation of stimulated emission from a 5d-4f transition in triply ionized rare earth-doped crystal Ce:YLF, optically pumped at 249 nm, and emitted at 325.5 nm [3]. Because the Ce^{3+} ion has only one electron in the 4f state, the impurity energy levels of Ce-doped crystals are particularly simple. The ground state is split into a $^2F_{5/2}$ and a $^2F_{7/2}$ levels by the spin–orbit interaction. The first excited state is

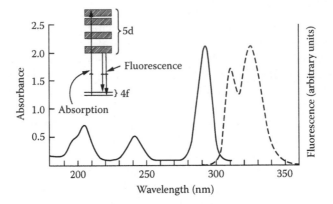

FIGURE 4.1
Absorption (solid line) and fluorescence (dotted line) spectra of Ce:YLF. (From D.J. Ehrlich et al., *Opt. Lett.*, **4**, 184–186, 1979.)

a 5d state, which interacts strongly with the host lattice because of the large spatial extent of the 5d wave function. Thus, the crystal-field interaction dominates over the spin–orbit interaction and the 5d state splits into four levels as a result of the S_4 site symmetry for the rare-earth ion in YLF. Figure 4.1 shows the absorption and fluorescence spectra for Ce:YLF.

The broad absorption bands that peak at 195, 205, 240, and 290 nm result from transitions from the 4f ground state to the crystal-field split 5d levels of the Ce^{3+} ion. The fluorescence spectrum has two peaks, the result of transitions from the lowest 5d level to the two spin–orbit ground states $4f(^2F_{5/2})$ and $4f(^2F_{7/2})$. The 40-ns radiative lifetime of the 5d level results from the electric-dipole-allowed character of the 5d → 4f transition. The potential tuning range of the Ce:YLF laser estimated from the half-power points of the fluorescence spectrum is from 305 to 335 nm. The maximum output energy observed was ~1 μJ in a pulse width of 35 ns for an absorbed pump energy of 300 μJ.

However, the operation of the Ce:YLF laser is hampered by several poor performance characteristics. These include an early onset of saturation and roll-off in the above-threshold gain and power output as well as a drop in the output for pulse repetition rates above 0.5 Hz. It has been shown that an excited state absorption of the UV pump light is responsible for a photoionization of the Ce^{3+} ions, which in turn leads to the formation of transient and stable color centers. The color centers have a deleterious effect on the lasing characteristics of Ce:YLF because they absorb at the cerium emission wavelengths. The growth and relaxation of these centers influence the gain saturation and pump rate limitation of the Ce:YLF laser [17,18]. This experiment is of historic significance, but it is of small practical use due to the existence of solarization.

In 1980, the same group mentioned above reported the operation of an optically pumped Ce^{3+}:LaF_3 laser [4]. The unpolarized near-UV absorption and emission spectra for a 0.05% Ce-doped LaF_3 crystal are shown in Figure 4.2.

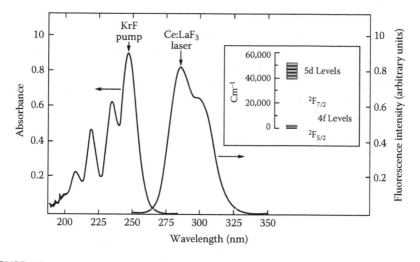

FIGURE 4.2

Ultraviolet absorption and spontaneous emission of a 700-μm-thick LaF$_3$ crystal with 0.05 at. % Ce^{3+} ion doping. (From D.J. Ehrlich et al., *Opt. Lett.*, **5**, 339–341, 1980. With permission.)

Because of the rapid internal relaxation to the lowest 5d state, the fluorescence spectrum was not noticeably different for ArF (193 nm), KrF (248 nm), or frequency-doubled Ar$^+$ ion laser (257 nm) pumping. Similarly, the fluorescence lifetime was identical for 248-nm or 193-nm excitation. For the 0.05%-doped crystal, the lifetime was 18 ± 2 ns. The approximate potential tuning range is from 275 to 315 nm. The output of a small commercial excimer laser, producing 40 mJ at 248 nm or 10 mJ at 193 nm in a 25-ns (FWHM) pulse, was used for optical pumping. The primary difficulties encountered, i.e., low output power and high threshold, can probably be ascribed to initial difficulties in crystal growth. However, because no one has been able to reproduce the results of this experiment, it is difficult to comment on the properties of Ce^{3+}:LaF$_3$ crystal as a laser medium.

Subsequent studies showed that the attempts to find laser-active media among Ce-activated materials failed because of absorption from the excited state of the 5d configuration of Ce^{3+} ions in Ce^{3+}:YAG [19,20] and Ce^{3+}:CaF$_2$ [21], formation of stable or transient color centers in Ce-activated samples [16,17,21], and other complex processes occurring in such media under the high-power UV pumping.

The spectrally broad vibronic emission bands in impurity-doped solids serve as the basis for wavelength-tunable laser operation in these materials. But because of the broad emission and absorption bands, these materials are susceptible to ESA which can significantly reduce the performance characteristics of the laser materials. The ESA is a two-step process, where the first photon absorbed promotes an electron from the 4f ground state to the lowest 5d state of the trivalent cerium ion. Within the lifetime of this excited 5d state, a second photon is absorbed which then photoionizes the ion by promoting that electron to the conduction band. The free electron subsequently traps out at an electron acceptor site forming a stable color center. These color centers are absorptive at the wavelengths for stimulated emission of the trivalent cerium ions, and hence they serve as a quenching mechanism for laser gain in this crystal. The color centers produced are photochromic in that they can be optically bleached [16]. Over a long period of time, difficulties in over-coming these problems, which are inherent in well-known materials used for producing UV light [3,4], made investigators believe that this scheme of UV lasers was of little promise.

However, recent investigations showed that, by an appropriate choice of activator–matrix complexes and active medium–pump source combinations, one can create efficient tunable lasers using d-f transitions in Ce^{3+} ion in the UV spectral range [5–9] and in Nd^{3+} ion in the VUV range [22]. Furthermore, such lasers proved to be stable under the pumping.

In 1992, M. A. Dubinskii et al. of Russia reported UV laser medium, Ce^{3+}:LuLiF$_4$ (Ce:LLF) crystal, which can be pumped by KrF excimer laser. This crystal has almost the same optical properties with Ce:YLF. But Ce:LLF has smaller solarization effect, and it is expectedly more effective in practical use [5,6]. In 1993, the same group reported Ce^{3+}:LiCaAlF$_6$ (Ce:LiCAF) crystal which can be pumped by the fourth harmonic of Nd:YAG laser. No solar-ization effect of this crystal was observed [7–9].

Ce^{3+}:LiSrAlF$_6$ (Ce:LiSAF) crystal was reported in 1994, and it also can be pumped by the fourth harmonic of Nd:YAG laser [10,11]. It has the similar laser properties with Ce:LiCAF crystal.

All these new Ce-doped crystals have a broad gain-width in the ultraviolet region, which is especially attractive for ultrashort-pulse generation and amplification. Figure 4.3 shows the tunable wavelength regions of the five known Ce-doped laser crystals.

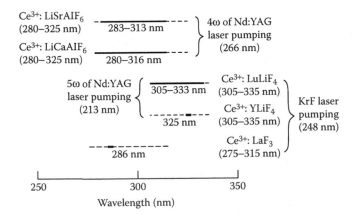

FIGURE 4.3

Various tunable lasers in ultraviolet region. Solid lines and dots indicate the confirmed tunable wavelength region, dotted lines show potential tunable wavelength region.

4.1.2 Basic Properties of Ce:LLF Laser Medium

Ce:LLF is a tunable solid-state laser material in ultraviolet region which was first reported in 1992 by Prof. Dubinskii in Russia [3,4]. The choice of this material for the experiments in UV is more reasonable because of its structural and chemical similarity to comprehensively studied and grown commercially YLF single crystal. The LLF crystals belong to the scheelite structural type. Similar to all rare-earth ions, Ce^{3+} ions take part in activation, substituting for Lu^{3+} ions in the position with the point group S_4.

In the early stages, $Ce^{3+}:LuLiF_4$ single crystals were grown from a carbon crucible using the Bridgman–Stockbarger method in a fluorinated atmosphere. Recently, high-quality, large Ce:LLF crystals (φ 18 mm × 10 mm in length) were successfully grown by the Czochralski (CZ) method.

Ce:LLF has a potential tuning region of around 305 to 340 nm, so it is especially attractive for use in spectroscopy of wide band-gap semiconductors for blue laser diodes, such as GaN [23]. The fluorescence spectrum of Ce:LLF crystal has two peaks at 311 nm and 328 nm, respectively (Figure 4.4). KrF excimer laser (248-nm), 5th harmonics of Nd:YAG laser, and frequency-doubled copper vapor laser (CVL) can be used as the pumping source for Ce:LLF crystal. The fluorescence lifetime was about 40 ns.

The small-signal gain and saturation fluence of Ce:LLF were evaluated using the second harmonic (325 nm) of a nanosecond DCM-dye laser as a probe. A KrF excimer laser was used to pump the Ce:LLF. The single-pass gain in the small-signal region (~1 mJ/cm²) was over 6-dB/cm (4.3 times)

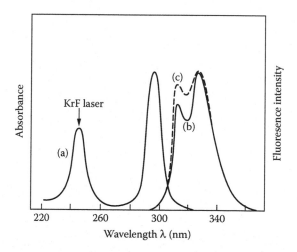

FIGURE 4.4

π-Polarized absorption (a) and normalized fluorescence (b: π-polarization, c: σ-polarization) spectra of Ce:LLF single crystal. (From M.A. Dubinskii, et al., in *18th International Quantum Electronics Conference*, OSA Technical Digest, Optical Society of America, Washington, D.C., 1992, paper FrL2, pp. 548–550; M.A. Dubinskii et al., *Laser Phys.*, **4**, 480–484, 1994. With permission.)

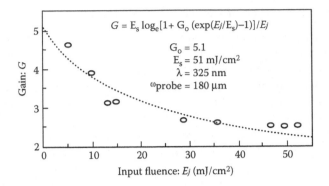

FIGURE 4.5

Ce:LLF gain dependence on input fluence of 10-ns, 325-nm probe pulses. The solid line indicates the gain saturation curve assuming the Frantz–Nodvik relation, which was fitted by the least-square method. The saturation fluence and the small signal gain were fitted to be 50 mJ/cm² and 5.1 (pumping fluence ~ 0.5 J/cm²). (From M.A. Dubinskii, et al., in *18th International Quantum Electronics Conference*, OSA Technical Digest, Optical Society of America, Washington, D.C., 1992, paper FrL2, pp. 548–550; M.A. Dubinskii et al., *Laser Phys.*, 4, 480–484, 1994. With permission.)

with 0.5-J/cm² pumping flux. The saturation fluence of Ce:LLF assuming the Frantz–Nodvik relation was estimated to be ~50 mJ/cm² at 325 nm (Figure 4.5), which is about two orders of magnitude higher than that of organic dyes (~1 mJ/cm²). The emission cross section estimated from this saturation fluence was ~10^{-17} cm² by the relation of $E_s = h\nu/\sigma$.

4.1.3 Basic Properties of Ce:LiCAF Laser Medium

In contrast to excimer laser pumped UV solid-state laser media, such as Ce:LLF and Ce:YLF, the Ce:LiCAF crystal, which was first reported by M. A. Dubinskii et al. in 1993 [7–9], is the first known tunable UV laser directly pumped by the fourth harmonic of a standard Nd:YAG laser. In that sense, this is the first truly all-solid-state tunable UV laser.

For the first time, the Ce:LiCAF were grown in a fluorinated atmosphere using the Bridgeman–Stockbarger technique from carbon crucibles. The crystals grown had the colquiriite structure and the space group P31c. The non-polarized absorption spectrum of a 2.3 mm thick Ce:LiCAF sample, containing about 0.1% of Ce^{3+} ions, is shown in Figure 4.6a. Due to the first 4f-5d-absorption peak with a half width of 3000 cm⁻¹ centered at about 37,000 cm⁻¹, Ce:LiCAF can be pumped by the fourth harmonics of various commercially available Nd-lasers (e. g., YAG, YAP, YLF, and GSGG). The Ce:LiCAF fluorescence spectrum (Figure 4.6b) displays the nearly two-humped shape characteristic of Ce^{3+} ions in most known hosts, due to the allowed 5d-4f-transitions terminating at the $^2F_{7/2}$ and $^2F_{5/2}$ components of the spin–orbit split

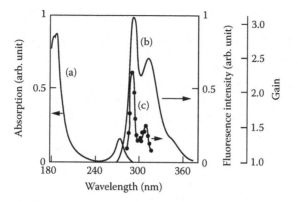

FIGURE 4.6
(a) Nonpolarized absorption spectrum of a Ce:LiCAF sample (0.1 at. %; 2.3 mm in length); (b) nonpolarized fluorescence spectrum of Ce:LiCAF (0.9 atomic percent); (c) single-pass small-signal gain dependence on the probe beam wavelength for a Ce:LiCAF sample (0.9 atomic percent; 2.3 mm in length). (From N. Sarukura et al., *IEEE J. Sel. Top. Quantum Electron.* **1**, 792–804, 1995. With permission.)

ground term. Ce:LiCAF has a potential tuning range from 280 to 320 nm. Ce:LiCAF has sufficiently higher effective gain cross section (6.0×10^{-18} cm^2) compared with Ti:sapphire [24], which is favorable for designing laser oscillators. Ce:LiCAF has also larger saturation fluence (115 mJ/cm^2) [11] than organic dyes, which is attractive for designing power amplifiers. The fluorescence lifetime was reported to be 30 ns, which is too short for constructing regenerative amplifiers. However, it is long enough for designing multipass amplifiers. The nonpolarized small-signal single-pass gain dependence on the probe beam wavelength for a 2.3-mm thick Ce:LiCAF sample with Ce^{3+} ion concentration of 0.9 atomic percent is shown in Figure 4.6c. The sample optical axis orientation with respect to the direction of observation was the same as for obtaining the fluorescence spectra (Figure 4.6b). The pump fluence of 0.3 J/cm^2 was used to obtain the above dependence. The probe fluence was less than 1 mJ/cm^2. From a comparison of Figure 4.6b and Figure 4.6c, it is evident that the small-signal gain curve shape is similar to the fluorescence spectrum observed. This means that induced absorption is small in the gain spectral region. The small-signal gain reaches a value of 2.5 in the vicinity of the main fluorescence peak [25].

Ce:LiSAF laser, which has similar property to CeLiCAF laser, was also demonstrated by C. D. Marshall in 1994 [11]. As Ce:LiSAF has well-matched absorption band and emission peak to Ce:LiCAF, it is can be directly pumped by the fourth harmonic of a standard Nd:YAG laser. However, Marshall determined that Ce:LiCAF shows a superior property in its lower propensity to form color centers. Both transient and permanent color centers have been observed in Ce:LiSAF lasers when pumped at 266 nm, though identical pumping conditions did not produce color centers in Ce:LiCAF.

4.2 Ultraviolet Tunable Pulse Generation from Ce^{3+}-Doped Fluoride Lasers

In most lasers, all of the energy released via stimulated emission by the excited medium is in the form of photons. Tunability of the emission in solid-state lasers is achieved when the stimulated emission of photons is intimately coupled to the emission of vibrational quanta (phonons) in a crystal lattice. In these "vibronic" lasers, the total energy of the lasing transition is fixed but can be partitioned between photons and phonons in a continuous fashion. The result is broad wavelength tunability of the laser output. In other words, the existence of tunable solid-state lasers is due to the subtle interplay between the Coulomb field of the lasing ion, the crystal field of the host lattice, and electron–phonon coupling permitting broadband absorption and emission. Therefore, the gain in vibronic lasers depends on transitions between coupled vibrational and electronic states; that is, a phonon is either emitted or absorbed with each electronic transition.

Rare-earth ions doped in appropriate host crystals exhibit vibronic lasing. The main difference between transition metal and rare-earth ions is that the former is crystal-field sensitive, and the latter is not. As distinct from transition metal ions, the broadband transition for rare-earth ions are quantum mechanically allowed and therefore have short lifetime and high cross sections. The Ce^{3+} ion laser, using a 5d-4f transition, has operated in the host crystals [19–22]. Such a system would be an alternative to the excimer laser as a UV source, with the added advantage of broad tunability.

Due to the vibronic nature of the Ce^{3+} ion laser, the emission of a photon is accompanied by the emission of phonons. These phonons contribute to thermalization of the ground-state vibrational levels. The laser wavelength depends on which vibrationally excited terminal level acts as the transition terminus; any energy not released by the laser photon will then be carried off by a vibrational phonon, leaving the Ce^{3+} ion at its ground state. The terminal laser level is a set of vibrational states well above the ground state. So the Ce^{3+} ion lasers belong to four-level lasers.

4.2.1 Tunable Ce:LLF Laser [29–33]

The schematic diagram of the tunable Ce:LLF laser pumped with 10-Hz frequency-doubled CVL is shown in Figure 4.7. Quasi-longitudinal pumping scheme was employed with a 2-mm long Ce:LLF crystal. Tunable operation is realized by rotating the single silica Brewster prism. The Ce:LLF laser tunability obtained at the pumping power level of 900 mW at 289 nm is shown in Figure 4.8. The tuning was achieved from 305 nm to 333.2 nm. Maximum output power of 150 mW at 311-nm wavelength was generated at room temperature. When the crystal was cooled to −3°C the output power increased approximately 50% over the full tuning range except around the 327-nm peak.

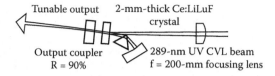

FIGURE 4.7

Experimental configuration for the prism-tuned Ce:LLF laser. (From A.J.S. McGonigle et al., *Electron. Lett.*, **35**, 1640–1641, 1999; A.J.S. McGonigle et al., *Appl. Opt.* **40**, 4326–4333, 2001. With permission.)

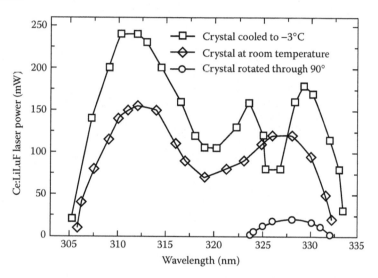

FIGURE 4.8

Prism-tuned Ce:LLF laser output power vs. wavelength at –3°C and 25°C. (From A.J.S. McGonigle et al., *Electron. Lett.*, **35**, 1640–1641, 1999; A.J.S. McGonigle et al., *Appl. Opt.* **40**, 4326–4333, 2001. With permission.)

The output power decreased and changed from π polarization to σ polarization at 327 nm. The decrease of output power is attributed to the ESA. At 327 nm, gain is highest for σ polarization, but the efficiency is lower.

4.2.2 Tunable Ce:LiCAF Laser [34]

The setup of the tunable short-cavity Ce:LiCAF laser is shown schematically in Figure 4.9. The cavity length was 25 mm. To obtain the tuning performance of this laser without consideration of its temporal characteristics, a high-Q laser with a low transmission flat output coupler (T = 20%) was designed. The laser consisted of a half-cut Brewster prism with a high reflection coating at one face used as the end mirror and a 10-mm-long, Brewster-cut at the end faces, 1% doped (in the melt) Ce:LiCAF crystal used as the gain medium without any special cooling. The 266-nm, horizontally polarized pumping pulses from a Q-switched Nd:YAG laser were focused longitudinally. In most cases, the laser operated at 2 Hz to reduce any possible thermal problems in

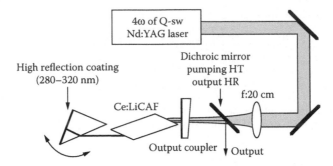

FIGURE 4.9
Experimental setup of the tunable, short-cavity Ce:LiCAF oscillator pumped by the fourth harmonics of a Q-switched Nd:YAG laser. (From Z. Liu et al., *Jpn. J. Appl. Phys.* **36**, L1384–L1386, 1997. With permission.)

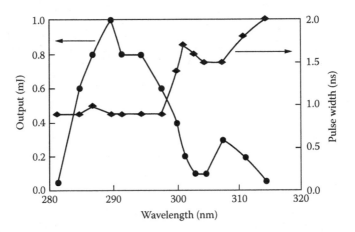

FIGURE 4.10
Tuning curve for the short-cavity Ce:LiCAF laser with 55% reflection output coupler.

the Ce:LiCAF crystal without mandatory cooling, and to obtain higher extraction efficiency. The c-axis of the Ce:LiCAF crystal was parallel to the direction of the pumping polarization. The tuning operation was realized by rotating the prism horizontally. A 55% reflection flat output coupler was used. A single-pulse output was generated with the pumping energy of 15 mJ. The demonstrated tuning range was 280 nm to 314 nm (Figure 4.10.).

4.3 Generation of Subnanosecond Pulses from Ce^{3+}-Doped Fluoride Lasers

Direct ultrashort pulse generation has not been obtained from ultraviolet solid-state lasers as it has been for near infrared tunable laser materials like

Ti:sapphire and Cr:LiSAF crystals. This is due to the difficulty of obtaining continuous-wave (CW) laser operation, which is required for Kerr lens mode-locking (KLM) schemes utilizing spatial or temporal Kerr type nonlinearity [37,38].

4.3.1 Short Pulse Generation with Simple Laser Scheme

A general technique for subnanosecond pulse generation from laser-pumped dye laser has been described [39]. The technique makes use of the resonator transients. These transients are in the form of damped relaxation oscillation or "spiking". These resonator transients are the consequences of the interaction between the excess population inversion and the photons in the cavity. Their durations can be controlled by proper choices of photon cavity decay time and pumping level. The decay lifetime (photon lifetime) t_c of a cavity mode is defined by means of the equation:

$$\frac{dE}{dt} = -\frac{E}{t_c}$$

(4.1)

where E is the energy stored in the mode. If the fractional (intensity) loss per pass is L and the length of the resonator is l_c, then the fractional loss per unit time is cL/nl_c, therefore

$$t_c = -\frac{nl_c}{cL}.$$

(4.2)

For the case of a resonator with mirrors' reflectivities R_1 and R_2,

$$t_c = \frac{nl_c}{c(1-\sqrt{R_1 R_2})}.$$

(4.3)

The quality factor of the resonator is defined universally as:

$$Q = \frac{\omega E}{P} = \frac{\omega E}{dE/dt}$$

(4.4)

where E is the stored energy, ω is the resonant frequency, and $P = dE/dt$ is the power dissipated. By comparing Equation 4.4 and Equation 4.1, we obtain

$$Q = \omega t_c.$$

(4.5)

We will consider a laser resonator with two mirrors pumped by a Q-switched laser. Normally, the pulse duration from the Q-switched laser is ~10 ns. During the pump pulse, inversion will build up in the laser-active medium; after threshold inversion is reached, a delayed laser pulse will develop. The proposed method of short pulse generation is based on the fact that this pulse may be shorter than the pump pulse, due to the transient characteristics of the laser oscillator.

The transient behavior of such a laser can be understood by solving numerical examples with a computer. The laser is adequately described by the well-known rate equations for a four-level system as follows:

$$\frac{dn}{dt} = W(t) - Bnq \qquad (4.6)$$

$$\frac{dq}{dt} = Bnq - \left(\frac{q}{t_c}\right) \qquad (4.7)$$

where n is the population inversion ($n_2 - n_1$) and nearly equal to the upper state population n_2, W(t) is the pumping rate, B is the Einstein B coefficient for stimulated emission, q is the total number of photons in the cavity, t_c is the resonator (photon) lifetime.

The pumping rate was assumed to have the form of a Gaussian pulse with full half-width T_1 and integrated photon number N. A large reduction of the over-all pulse duration demands a high value of T_1/t_c. The ratio T_1/t_c can be made large easily, because the laser resonator needs no switching elements and therefore need not be longer than the material itself. Further, it is possible to use a laser resonator of low mirror reflectivity, whose resonator lifetime is not markedly longer than the single-pass transit time l_c/c. To obtain short single pulses, it is necessary to use resonator lifetimes which are small compared to the pump duration in combination with controlling the level of pumping [40].

4.3.2 Subnanosecond Ce:LLF Laser Pumped by the Fifth Harmonic of Nd:YAG Laser [30]

The optical layout for a short-cavity Ce:LLF laser is shown schematically in Figure 4.11. The Ce:LLF single crystal with 0.2 atomic percentage doping level was cut to form a cylinder (5-mm diameter and 25-mm length) with a flat polished window on the side. The sample was oriented so that its optical axis was parallel to the side window and perpendicular to the cylinder longitudinal axis. No antireflection coatings were applied to the rod ends for this experiment. To increase the efficiency of side pumping, the effective pumping penetration depth was reduced by using the novel tilted-incidence-angle side pumping scheme instead of conventional normal-incidence side pumping. Using a 20-cm-focal-length cylindrical lens which was also tilted to be parallel to the side window of the laser crystal, the pumping pulse was focused down to a 1.2×0.15 cm^2 line-shaped area to provide the 140 mJ/cm^2

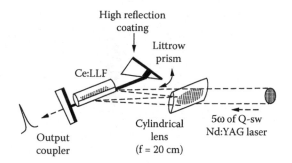

FIGURE 4.11

Short-cavity, tilted-incidence-angle (approximately 70°) side pumping tunable Ce:LLF laser optical layout. The pumping pulse was focused by a 20-cm focal length cylindrical lens tilted so that the cylinder axis is parallel to the side window of the laser crystal. (From N. Sarukura et al., *Appl. Opt.* **37**, 6446–6448, 1998. With permission.)

pumping fluence at nearly 70° incidence angle. Observed efficient penetration depth was estimated to be ~1 mm, meanwhile in conventional normal-incidence side pumping scheme, the pumping penetration depth was over 3 mm for the Ce:LLF crystal used here, which is too deep for obtaining a good output beam pattern. Obvious advantages of the above-mentioned (tilted-incidence-angle, side-pumping) scheme are very simple focusing geometry, reduced pumping fluence at the rod surface, reduced risk of damaging optics, and better matching of the excited rod volume and the laser cavity mode volumes (quite similar to the coaxial pumping scheme). This better matching resulted in a better output beam quality. The low-Q, short-cavity Ce:LLF laser consisted of a Littrow prism which was used as an end mirror, and a low reflection (20%) flat output coupler. The total length of the laser cavity was 6 cm. Typical spectrally and temporally resolved streak-camera images of the Ce:LLF laser output pulse are given in Figure 4.12. Using the 213-nm, 5-ns, 16-mJ pumping pulses, the σ-polarized, and satellite-free reproducible pulses were obtained with 880-ps and 77-μJ at 309 nm. It is worth mentioning here that pumping at 213 nm does not cause noticeable laser rod solarization or laser performance degradation during several hours of continuous operation at a 10 Hz repetition rate.

4.3.3 Short Pulse Generation from Ce:LiCAF Laser

4.3.3.1 *Short Pulse Generation from Ce:LiCAF Laser with Nanosecond Pumping*

Ce:LiCAF is a tunable UV laser medium which can be directly pumped by the fourth harmonic of a standard Nd:YAG laser. A 1% doped, 5-mm cubic Ce:LiCAF sample was used without any dielectric coatings on the polished surfaces. The optical layout of the subnanosecond Ce:LiCAF laser is shown in Figure 4.13. The 1.5-cm long laser cavity was formed by a flat high-reflection

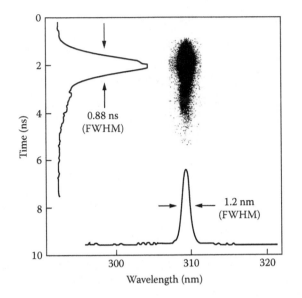

FIGURE 4.12
Temporally and spectrally resolved streak-camera image of UV short pulse from a low-Q, short-cavity Ce:LLF laser. (From N. Sarukura et al., *Appl. Opt.* **37**, 6446–6448, 1998. With permission.)

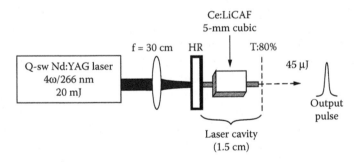

FIGURE 4.13
Optical layout of low-Q, short-cavity Ce:LiCAF oscillator with nanosecond pumping.

mirror and an 80% transmission flat output coupler. 20-mJ, 10-ns, 1-Hz, 266-nm, horizontally polarized pumping pulses (the fourth harmonic of a conventional 10-ns Q-switched Nd:YAG laser) were focused longitudinally from the high-reflection mirror side by a 30-cm focal-length lens with ~300 mJ/cm² pumping fluence inside the active medium. The c-axis of the Ce:LiCAF laser crystal sample was parallel to the direction of the pumping polarization. The absorbed energy was 5 mJ. The single output pulse has a energy of 45 μJ. The pulse duration was measured to be 600 ps using a streak camera.

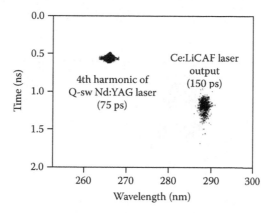

FIGURE 4.14

Streak camera image of the low-Q, short-cavity Ce:LiCAF laser pulse with picosecond pumping. (From Z. Liu et al., *J. Nonlinear Opt. Phys. Mater.*, 8, 41–54, 1999. With permission.)

4.3.3.2 Short Pulse Generation from Ce:LiCAF Laser with Picosecond Pumping [41]

For shorter pulse generation, a shorter pumping source was used: the fourth harmonic of a mode-locked Nd:YAG oscillator and regenerative amplifier system operated with the repetition rate of 10 Hz. A low-Q, short cavity Ce:LiCAF oscillator was formed by a flat high-reflection mirror and a 30% reflection flat output coupler. The cavity length was 1.5 cm. A 10-mm Brewster-cut, 1% doped (in the melt) Ce:LiCAF crystal was used. A typical spectrally and temporally resolved streak camera image of the output pulse of the Ce:LiCAF laser is shown in Figure 4.14. The Ce:LiCAF laser pulse width was measured to be 150 ps, whereas the pumping pulse width was 75 ps.

4.4 High-Power Ultraviolet Cerium Laser

4.4.1 High-Energy Pulse Generation from a Ce:LLF Oscillator [42]

A Ce:LLF laser resonator is established by using a flat high reflector and a flat output coupler. The length of the laser cavity was 6 cm. The layout is shown schematically in Figure 4.15. High-quality, large Ce:LLF crystals (φ 18 mm × 10 mm in length) were grown successfully by the Czochralski (CZ) method. The Ce:LLF sample used here was obtained by cutting the grown Ce:LLF crystal in the middle along its axis (a half-cut cylinder). The side window and two end surfaces were polished. No dielectric coatings were deposited on the polished end surfaces and the side window. The pumping

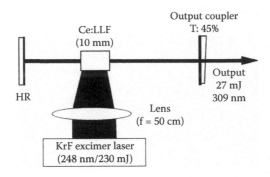

FIGURE 4.15
Experimental setup of the high-power Ce:LLF laser pumped by a randomly polarized KrF excimer laser operated at a repetition of 1 Hz. (From Z. Liu et al., *Jpn. J. Appl. Phys.*, **39**, L88–L89, 2000. With permission.)

pulses from a randomly polarized KrF excimer laser operated at 1 Hz were focused softly on the side window of the Ce:LLF crystal using a 50-cm-focal-length spherical lens under a normal-incidence side-pumping condition. Because the output pulse of the KrF excimer laser has a rectangular shape, it is not difficult to make a near-line-shape pumping area on the Ce:LLF crystal through a spherical lens. Almost all of the pumping pulse energy (maximum: 230 mJ) was absorbed by the Ce:LLF crystal. The coupler with 45% transmission was used. The maximum output pulse energy reached 27 mJ with the pumping pulse energy of 230 mJ, and the corresponding pumping fluence was approximately 0.6 J/cm^2. This is the highest output pulse energy ever achieved from this laser medium. The free-running Ce:LLF laser operated at the wavelength of 309 nm. The slope efficiency was approximately 17%.

4.4.2 High-Energy Pulse Generation from a Ce:LiCAF Oscillator [43,44]

The schematic diagram of a Ce:LiCAF laser resonator using the large-size Ce:LiCAF crystal is shown in Figure 4.16. The laser resonator is established by a flat high reflector and a flat output coupler with 30% reflection for 290 nm and 75% transmission for 266 nm separated by 4 cm. The large Ce:LiCAF crystal grown by the method described above (18 mm in diameter, clear aperture 15 mm, length 10 mm) is doped with 1.2 mol% Ce^{3+} ions. There is no coating on the parallel end faces of the crystal that are perpendicular to the optical axis of the resonator. To obtain a high-quality laser beam, the quasi-longitudinal pumping method was employed. For high-energy output without damage to the crystal and optics in the cavity, a large pump beam cross section is necessary. The three horizontally polarized pump beams were focused with a 40-cm-focal-length lens to produce a spot size of ϕ6 mm at the surface of the Ce:LiCAF crystal. To reduce the diffraction effects and disturbance to the beam uniformity, it is better to choose a ratio of crystal

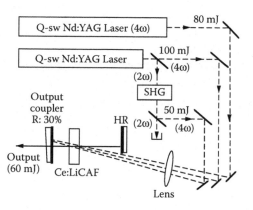

FIGURE 4.16

Experimental setup of the Ce:LiCAF laser oscillator pumped by the fourth harmonics of two Q-switched Nd:YAG lasers. The pulse energy as high as 60 mJ was achieved at 290 nm at 10 Hz repetition rate. (From Z. Liu et al., *Jpn. J. Appl. Phys.*, **39**, L446–L467, 2000. With permission.)

radius to beam radius as 2 mm or greater. Therefore, a much larger crystal diameter than the pump beam diameter is preferred. With the total pumping energy of 230 mJ, the output pulse energy as high as 60 mJ was achieved at 290 nm at 10 Hz repetition rate, which is the highest performance reported for a Ce:LiCAF oscillator until now, as far as we know. In this way, we demonstrated the generation of high-energy pulse at 290 nm very easily and efficiently.

4.4.3 High-Energy Pulse Generation from a Ce:LiCAF Laser Amplifier [45]

A high quality, large Ce:LiCAF crystal was successfully grown by the Czochralski (CZ) method. [46] A large-sized Brewster-plate gain module ($1 \times 2 \times 2$ cm^3) with a 3.2-cm^{-1} absorption coefficient at 266 nm and a 0.2-cm^{-1} absorption coefficient at 290 nm was cut from a 5-cm diameter Ce:LiCAF boule grown in CF$_4$ atmosphere by the Czochralski method. The experimental setup of a coaxially pumped large-aperture Ce:LiCAF double-pass power-amplifier module is illustrated in Figure 4.17. To generate seed pulses, a conventional Ce:LiCAF master oscillator with 8-mJ output energy and a preamplifier delivered horizontally polarized, 15-mJ, 3-ns probe pulses at 290 nm. Considering the 25-ns fluorescence lifetime and 3-ns probe-pulse duration, a bow-tie double-pass geometry was selected. It has 3-ns optical delay for the second amplification pass to achieve less gain relaxation and no significant temporal overlap of the probe pulse at the crystal. To pump the power-amplifier module, four beams of horizontally polarized, 266-nm, 8-ns pulses were prepared from three 10-Hz, Q-switched Nd:YAG lasers using Li$_2$B$_4$O$_7$ crystals. When the total fundamental energy from Nd:YAG lasers was 4.2 J, the total excitation energy at 266 nm was 0.38 J [47–49]. These beams delivered 182- and 60-mJ pump energy from one side and 75- and 65-mJ from the other. These

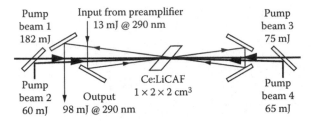

FIGURE 4.17

The experimental setup of a coaxially pumped large aperture Ce:LiCAF double-pass power-amplifier module. A conventional Ce:LiCAF master oscillator and preamplifier delivered 15-mJ, 3-ns probe pulses at 290 nm for the characterization of this module. A large Brewster plate $1 \times 2 \times 2$ cm^3 was used as the gain medium. A bow-tie double-pass geometry with 3-ns delay for the second amplification was selected to achieve less relaxation of gain and no significant temporal overlap of the probe pulse at the crystal. (From S. Ono et al., *Appl. Opt.*, **41**, 7556–7560, 2002. With permission.)

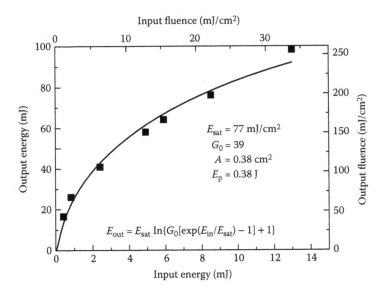

FIGURE 4.18

Output-energy dependence on input energy. The spot size and pump energy are 0.38 cm^2 and 0.38 J, respectively. Assuming the Frantz–Nodvik relation, saturation fluence and small signal gain coefficient are evaluated 77 mJ/cm^2 and 39 times, respectively. (From S. Ono et al., *Appl. Opt.*, **41**, 7556–7560, 2002. With permission.)

beams were projected into the gain medium with 8-mm diameter similar to the probe beam diameter of 7 mm. The crystallographic *c* axis of the gain medium is parallel to the probe beam polarization. Over 95% of the excitation energy was absorbed in the gain medium, and the pumping density was 0.7 J/cm^3. This pumping condition of less than 0.5 J/cm^2 from each side should provide an adequate safety margin for the optical damage threshold (~2 J/cm^2). From the output-energy dependence on different input energy as shown in Figure 4.18, the gain saturation was observed. The small signal gain and the saturation fluence were evaluated to be 39 times and 77 mJ/cm², assuming the Frantz–Nodvik relation. [50,51] This theoretical model for the slow decay of the gain medium assumption should still be applied, considering the 25-ns fluorescence lifetime, the 4-ns optical delay of the second pass, and the 8-ns pump pulse duration. The saturation fluence evaluated from the reported emission cross section was 115 mJ/cm². This deviation might be attributed to the oversimplified Frantz–Nodvik relation. The highest output energy reached 98 mJ for a 7.7-mJ transmitted input pulse with 13-mJ probe pulse from the preamplifier stage with an extraction efficiency of 25%. The pulse energy after the first amplification pass was 43 mJ. The amplified spontaneous emission was below the detection limit of .10 μJ when no probe pulse was input into the amplifier. This extraction efficiency was 83% of the theoretical limit for the present optical layout.

4.4.4 Ce:LiCAF Chirped-Pulse Amplification (CPA) System

The solid-state UV laser medium Ce:LiCAF was shown to have broad tunability, at least from 281 to 315 nm, corresponding to a bandwidth of a 3-fs pulse. Additionally, Ce:LiCAF has an attractively high saturation fluence of 115 mJ/cm² [11]. A 5-cm-diameter Ce:LiCAF crystal [46] was successfully grown for use in future terawatt Ce:LiCAF chirped-pulse amplification laser systems [52,53]. Moreover, owing to the recent development of new nonlinear crystals such as $CsLiB_6O_{10}$ (CLBO) and $Li_2B_4O_7$ (LB4), the conversion efficiency from the second harmonics of Nd:YAG lasers to the fourth harmonics is as high as 50% [47–49]. In terms of these energy-efficiency aspects, a Ce:LiCAF CPA laser system for use in high-peak-power systems is feasible, even when it is compared with a Ti:sapphire CPA laser system.

The setup of the Ce:LiCAF CPA system is illustrated in Figure 4.19. Femtosecond seed pulses at 290 nm with a 1-kHz repetition rate were provided by frequency tripling of the Ti:sapphire regenerative amplifier output. The duration of the seed pulse was 210 fs. The seed pulse has a spectrum bandwidth of 1.0 nm. Seed pulses without any dispersion compensation were guided to an eight-pass quartz-prism pulse stretcher with reflective optics, expanding the pulse duration to 2.6 ps and avoiding spatial chirping and aberration. Stretched seed pulses of 33-mJ energy were guided to a modified bow-tie-style four-pass Ce:LiCAF amplifier after being selected by a mechanical shutter [54]. A synchronously operated 10-Hz Q-switched Nd:YAG laser provided 100-mJ, 266-nm, 10-ns pump pulses. With the 1 cm × 1 cm × 1 cm Brewster-cut Ce:LiCAF crystal there was 82% absorption of the pump beam. The pump beam was softly focused with a 70-cm focal-length lens to a 4-mm diameter, with a pump fluence of 0.8 J/cm², which was well below the damage threshold of Ce:LiCAF crystal (.5 J/cm²). The diameter of the seed beam was approximately 1.5 mm for the first and second passes, and we expanded it to approximately 3 mm for the third and fourth passes to avoid deep gain quenching and possible damage to the optics in the amplifier. The

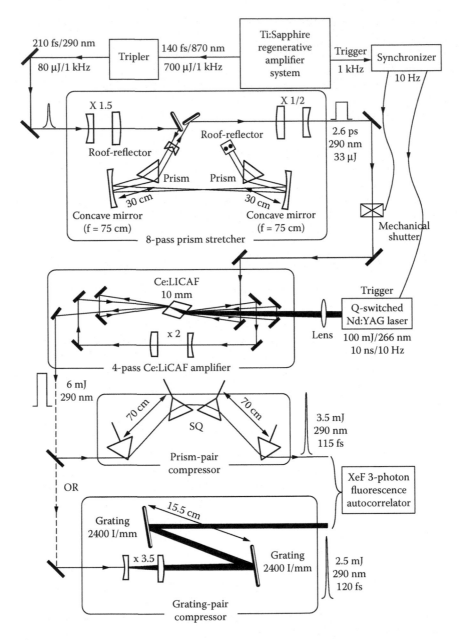

FIGURE 4.19
Experimental setup of the Ce:LiCAF CPA laser system. (From Z. Liu et al., *Opt. Lett.*, **26**, 301–303, 2001. With permission.)

maximum output energy was 6 mJ for this amplifier, with a net gain of 180 for the 33-mJ input pulse. The corresponding fluence was approximately 90 mJ/cm². which was near the saturation fluence of the Ce:LiCAF crystal. The amplified spontaneous emission was below the detection limit of 20 mJ when no seed pulse was input into the amplifier. The energy-extraction efficiency for the absorbed pump power reached 7.3%. The total gain factor was 370 for the transmission pulse when the output energy was 4.5 mJ. The gain factors for the four passes are estimated to be approximately 7, 3, 2, and 7. The gain-factor difference should be due to the spatial and temporal overlap quality of the pump and signal beams. The amplified pulse had a slightly broadened spectrum bandwidth of 1.6 nm [Figure 4.20(b)]. After amplification, the beam was guided to dispersion-compensating double-pass quartz Brewster prism pairs separated by 70 cm or to a 2400-line/mm-grating pair of prisms separated by 15.5 cm (Figure 4.19). The amplified pulse was compressed to 115 fs [Figure 4.20(a)] and 120-fs duration for each case, assuming a sech^2 pulse shape with time and bandwidth products of 0.66 and 0.69, respectively. These products are approximately twice the transform-limited

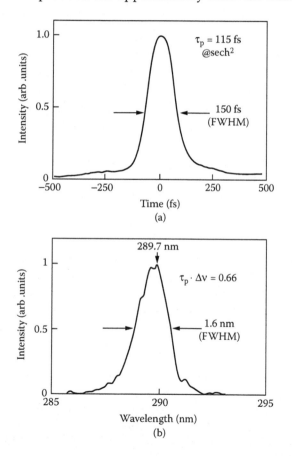

FIGURE 4.20
(a) Autocorrelation curve of the amplified pulse compressed by the prism compressor. The pulse duration was 115 fs. The width of the third-order autocorrelation trace is divided by a deconvolution factor of 1.29, assuming a sech^2 pulse shape. (b) Spectrum of the amplified pulse; the bandwidth was slightly broadened to 1.6 nm. (From Z. Liu et al., *Opt. Lett.*, **26**, 301–303, 2001. With permission.)

values. The throughput of the prism compressor was 60% and that of the grating compressor was 40%. To obtain a shorter pulse duration, we obviously need a more sophisticated compressor to compensate for the higher-order dispersion. The B integral [55] of this CPA system is estimated from the reported values of the nonlinear refractive index n_2 for quartz and $LiCaAlF_6$ to be less than 5, even when the compressor prism part is included [56,57]. Therefore a pulse-stretching ratio of 12 should be adequate for this amplification level. If the 5-cm-diameter Ce:LiCAF crystal could be utilized, with a pump fluence level similar to that reported above, output energy exceeding 0.5 J would be expected. Because the bandwidth of the gain medium is not fully utilized in the present system, a much shorter pulse duration of .10 fs may be achievable, with a shorter seed pulse and better dispersion [58].

4.5   Conclusions and Prospects

As mentioned at introduction, tunable lasers in the ultraviolet region attract special interests for applications relating to the remote sensing. The simple, compact, all-solid-state cerium laser can generate coherent radiation in this wavelength region. Due to the properties of longer-wavelength tuning region and the degradation-free, Ce:LLF and Ce:LiCAF lasers are especially attractive for use in spectroscopy of wide band-gap semiconductors for blue laser diodes, such as nitride-based laser diodes. Furthermore, the operating region of all-solid-state tunable UV lasers were expanded by using the sum-frequency mixing technique [59,60].

Ce:fluoride laser were also used to environmental sensing [61]. Narrow lasing linewidth (< 0.1 nm) operation was demonstrated in distributed-feedback tunable lasers and injection-seeded tunable laser [62,63]. Ce:LiCAF laser worked efficiently at 20 kHz repetition rate [64].

The Ce:LLF and Ce:LiCAF were proven to generate and amplify short pulses in UV region. Subnanosecond ultraviolet coherent pulses were generated directly from solid-state lasers from low-Q, short-cavity Ce:fluoride lasers pumped by the fifth and fourth harmonics of Nd:YAG lasers. The direct generation and efficient amplification of UV short pulses are also demonstrated from the simplest, all-solid-state, UV short-pulse, MOPA system composed of Ce:LiCAF crystals and conventional Q-switched Nd:YAG lasers [65]. In extension of the above short pulse work, Ce:LiCAF CPA was demonstrated. 30-GW peak power was delivered from this CPA system. Due to the available large Ce:LiCAF crystal, 3-ns pulse duration, 98-mJ output energy from the Ce:LiCAF amplifier were reported. A much higher output power can be expected by fully utilizing the crystal cross size while using a larger pumping source.

In the last 10 years, Ce:LLF and Ce:LiCAF have been developed due to the property of lower propensity to form color centers. However the quest for new UV laser materials goes on. Recently, several new Ce^{3+}-doped materials have been reported, including $Ce^{3+}:LiBaF_3$, $Ce^{3+}:LuPO_4$, $Ce^{3+}:KY_3F_{10}$ [29,66–68]. It is reasonable to expect cw solid-state ultraviolet lasers in future with the improvement of Ce:fluoride crystal qualities and high power ultraviolet cw pumping sources [69]. Further development of laser systems using these laser media will open up new possibilities of simple and compact tunable UV ultrashort-pulse laser light sources.

References

1. G.J. Megie, G. Ancellet, and J. Pelon, Lidar measurements of ozone vertical profiles, *Appl. Opt.*, **24**, 3454–3463, 1985.
2. E.V. Browell, Applications of lasers in remote sensing, in *Advanced Solid-State Lasers*, OSA Technical Digest, Memphis, paper MA1, 1995, pp. 2–4.
3. D.J. Ehrlich, P.F. Moulton, and R.M. Osgood, Jr., Ultraviolet solid-state Ce:YLF laser at 325 nm, *Opt. Lett.*, **4**, 184–186, 1979.
4. D.J. Ehrlich, P.F. Moulton, and R.M. Osgood, Jr., Optically pumped Ce:LaF₃ laser at 286 nm, *Opt. Lett.*, **5**, 339–341, 1980.
5. M.A. Dubinskii, R.Y. Abdulsabirov, S.L. Korableva, A.K. Naumov, and V.V. Semashko, New solid-state active medium for tunable ultraviolet lasers, in *18th International Quantum Electronics Conference*, OSA Technical Digest, Optical Society of America, Washington, D.C., 1992, paper FrL2, pp. 548–550.
6. M.A. Dubinskii, R.Y. Abdulsabirov, S.L. Korableva, A.K. Naumov, and V.V. Semashko, A new active medium for a tunable solid-state UV laser with an excimer pump, *Laser Phys.*, **4**, 480–484, 1994.
7. M.A. Dubinskii, V.V. Semashko, A.K. Naumov, R.Y. Abdulsabirov, and S.L. Korableva, Active medium for all-solid-state tunable UV laser, *OSA Proceedings on Advanced Solid-State Lasers*, Albert A. Pinto and Tso Yee Fan, Eds., Vol. 15, Optical Society of America, Washington, D.C., 1993, pp. 195–198.
8. M.A. Dubinskii, V.V. Semashko, A.K. Naumov, R.Y. Abdulsabirov, and S.L. Korableva, Spectroscopy of a new active medium of a solid-state UV laser with broadband single-pass gain, *Laser Phys.* **3**, 216–217, 1993.
9. M.A. Dubinskii, V.V. Semashko, A.K. Naumov, R.Y. Abdulsabirov, and S.L. Korableva, Ce^{3+}-doped colquiriite, a new concept of all-solid-state tunable ultraviolet laser, *J. Mod. Opt.* **40**, 1–5, 1993.
10. J.F. Pinto, G.H. Rosenblatt, L. Esterowitz, and G.J. Quarles, Tunable solid-state laser action in $Ce^{3+}:LiSrAlF_6$, *Electron. Lett.* **30**, 240–241, 1994.
11. C.D. Marshall, S.A. Payne, J.A. Speth, W.F. Krupke, G.J. Quarles, V. Castillo, and B.H.T. Chai, Ultraviolet laser emission properties of Ce^{3+}-doped $LiSrAlF_6$ and $LiCaAlF_6$, *J. Opt. Soc. Am. B* **11**, 2054–2065, 1994.
12. K.H. Yang and J.A. Deluca, UV fluorescence of cerium-doped lutetium and lanthanum trifluorides, potential tunable coherent sources from 2760 to 3220 A, *Appl. Phys. Lett.* **31**, 594–596, 1977.
13. R.W. Waynant, Vacuum ultraviolet laser emission from $Nd^{+3}:LaF_3$, *Appl. Phys. B* **28**, 205, 1982.

14. W.J. Miniscalco, J.M. Pellegrino, and W.M. Yen, Measurements of excited-state absorption in Ce^{3+}:YAG, *J. Appl. Phys.* **49**, 6109–6111, 1978.

15. R.R. Jacobs, W.F. Krupke, and M.J. Weber, Measurement of excited-state absorption loss for Ce^{3+} in $Y_3Al_5O_{12}$ and implications for 5d-4f rare-earth lasers, *Appl. Phys. Lett.* **33**, 410–412, 1978.

16. D.S. Hamilton, Trivalent cerium doped crystals as tunable system. Two bad apples, in *Tunable Solid State Lasers*, Springer-Verlag, Berlin, 1985, pp. 80–90.

17. K.-S. Lim and D.S. Hamilton, Optical gain and loss studies in Ce^{3+}:YLiF$_4$, *J. Opt. Soc. Am. B* **6**, 1401–1406, 1989.

18. K.S. Lim and D.S. Hamilton, UV-induced loss mechanisms in a Ce^{3+}:YLiF$_4$ laser, *J. Luminescence* **40, 41**, 319–320, 1988.

19. J.F. Owen, P.B. Dorain, and T. Kobayasi, Excited-state absorption in Eu^{2+}:CaF$_2$ and Ce^{3+}:YAG single crystals at 298 and 77 K, *J. Appl. Phys.* **52**, 1216–1223, 1981.

20. D.S. Hamilton, S.K. Gayen, G.J. Pogatshnik, and R.D. Ghen, Optical-absorption and photoionization measurements from the excited states of Ce^{3+}:$Y_3Al_5O_{12}$, *Phys. Rev. B* **39**, 8807–8815, 1989.

21. G.J. Pogatshnik and D.S. Hamilton, Excited-state photoionization of Ce^{3+} ions in Ce^{3+}:CaF$_2$, *Phys. Rev. B* **36**, 8251–8257, 1987.

22. M.A. Dubinskii, A.C. Cefalas, E. Sarantopoulou, S.M. Spyrou, C.A. Nicolaides, R.Y. Abdulsabirov, S.L. Korableva, and V.V. Semashko, Efficient LaF$_3$:Nd^{3+}-based vacuum-ultraviolet laser at 172 nm, *J. Opt. Soc. Am. B* **9**, 1148–1150, 1992.

23. S. Nakamura, M. Senoh, S. Nagahama, N. Iwasa, T. Yamada, T. Matsushita, Y. Sugimoto, and H. Kiyoku, Continuous-wave operation of InGaN multi-quantumwell-structure laser diodes at 233 K, *Appl. Phys. Lett.* **69**, 3034–3036, 1996.

24. P.F. Moulton, "Spectroscopic and laser characteristics of Ti:Al$_2$O$_3$," *J. Opt. Soc. Am. B* **3**, 125–133, 1986.

25. N. Sarukura, M.A. Dubinskii, Z. Liu, V.V. Semashko, A.K. Naumov, S.L. Korableva, R.Y. Abdulsabirov, K. Edamatsu, Y. Suzuki, T. Itoh, and Y. Segawa, Ce^{3+} activated fluoride crystals as prospective active media for widely tunable ultraviolet ultrafast lasers with direct 10-nsec pumping, *IEEE J. Sel. Top. Quantum Electron.* **1**, 792–804, 1995.

26. R.J. Lang, The spectrum of trebly ionized cerium, *Can. J. Res. A*, 14, 127–130, 1936.

27. P. Dorenbos, 5d-level energies of Ce^{3+} and crystalline environment. I. Fluoride compounds, *Phys. Rev. B*, **62**, 15640–15648, 2000.

28. B. Henderson and R.H. Bartram, *Crystal-Field Engineering of Solid-State Laser Materials*, Cambridge University Press, Cambridge, U.K., 2000.

29. D.W. Coutts and A.J.S. McGonigle, Cerium-doped fluoride lasers, *IEEE J. Sel. Top. Quantum Electron.* **40**, 1430–1440, 2004.

30. N. Sarukura, Z. Liu, S. Izumida, M.A. Dubinskii, R.Y. Abdulsabirov, and S.L. Korableva, All-solid-state tunable ultraviolet sub-nanosecond laser with direct pumping by the fifth harmonic of an Nd:YAG laser, *Appl. Opt.* **37**, 6446–6448, 1998.

31. K.S. Jonston, and D.W. Coutts, Influence of temperature dependent excited state absorption on a broadly tunable UV Ce:LLF laser, in *Tech. Dig. Conf. Lasers, Apps. Tech.* LAT 2002, p. 39.

32. A.J.S. McGonigle, S. Girard, D.W. Coutts, and R. Moncorge, 10-kHz continuously tunable Ce:LLF laser, *Electron. Lett.*, **35**, 1640–1641, 1999.

33. A.J.S. McGonigle, R. Moncorge, and D.W. Coutts, Temperature dependent polarization effects in Ce:LLF, *Appl. Opt.* **40**, 4326–4333, 2001.

34. Z. Liu, H. Ohtake, N. Sarukura, M.A. Dubinskii, R.Y. Abdulsabirov, S.L. Korableva, A.K. Naumov, and V.V. Semashko, Subnanosecond tunable ultraviolet pulse generation from a low-Q, short-cavity Ce:LiCAF laser, *Jpn. J. Appl. Phys.* **36**, L1384–L1386, 1997.

35. Z. Liu, N. Sarukura, M.A. Dubinskii, R.Y. Abdulsabirov, S.L. Korableva, A.K. Naumov, and V.V. Semashko, Tunable ultraviolet short-pulse generation from a Ce:LiCAF laser amplifier system and its sum-frequency mixing with an Nd:YAG laser, *Jpn. J. Appl. Phys.* **37**, L36–L38, 1998.

36. A.J.S. McGonigle, D.W. Coutts, and C.E. Webb, 530-mW 7-kHz cerium LiCAF laser pumped by the sum-frequency-mixed output of a copper-vapor laser, *Opt. Lett.*, **24**, 232–234, 1999.

37. D.E. Spence, P.N. Kean, and W. Sibbett, 60-fsec pulse generation from a self-mode-locked Ti:sapphire laser, *Opt. Lett.* **16**, 42–44, 1991.

38. N. Sarukura, Y. Ishida, and H. Nakano, Generation of 50-fsec pulses from a pulse-compressed, cw, passively mode-locked Ti:sapphire laser, *Opt. Lett.*, **16**, 153–155, 1991.

39. C. Lin and C.V. Shank, Subnanosecond tunable dye laser pulse generation by controlled resonator transients, *Appl. Phys. Lett.* **26**, 389–391, 1975.

40. D. Roess, Giant pulse shortening by resonator transients, *J. Appl. Phys.*, **37**, 2004–2006, 1966.

41. Z. Liu, N. Sarukura, M.A. Dubinskii, R.Y. Abdulsabirov, and S.L. Korableva, All-solid-state subnanosecond tunable ultraviolet laser sources based on Ce^{3+}-activated fluoride crystals, *J. Nonlinear Opt. Phys. Mater.*, **8**, 41–54, 1999.

42. Z. Liu, K. Shimamura, K. Nakano, N. Mujilatu, T. Fukuda, T. Kozeki, H. Ohtake, and N. Sarukura, Direct generation of 27-mJ, 309-nm pulses from a Ce^{3+}:LiLuF$_4$ oscillator using a large-size Ce^{3+}:LiLuF$_4$ crystal, *Jpn. J. Appl. Phys.*, **39**, L88–L89, 2000.

43. Z. Liu, S. Izumida, H. Ohtake, N. Sarukura, K. Shimamura, N. Mujilatu, S.L. Baldochi, and T. Fukuda, High-pulse-energy, all-solid-state, ultraviolet laser oscillator using large Czochralski-grown Ce:LiCAF crystal, *Jpn. J. Appl. Phys.*, **37**, L1318–L1319, 1998.

44. Z. Liu, K. Shimamura, K. Nakano, and T. Fukuda, High-pulse-energy ultraviolet Ce^{3+}:LiCaAlF$_6$ laser oscillator with newly designed pumping schemes, *Jpn. J. Appl. Phys.*, **39**, L446–L467, 2000.

45. S. Ono, Y. Suzuki, T. Kozeki, H. Murakami, H. Ohtake, N. Sarukura, H. Sato, S. Machida, K. Shimamura, and T. Fukuda, High-energy, all-solid-state, ultraviolet laser power-amplifier module design and its output-energy scaling principle, *Appl. Opt.*, **41**, 7556–7560, 2002.

46. K. Shimamura, N. Mujilatu, K. Nakano, S. L. Baldochi, Z. Liu, H. Ohtake, N. Sarukura, and T. Fukuda, Growth and characterization of Ce-doped LiCaAlF$_6$ single crystals, *J. Cryst. Growth* **197**, 896–900, 1999.

47. Y. Mori, S. Nakajima, A. Taguchi, A. Miyamoto, M. Inakaki, W. Zhou, T. Sasaki, and S. Nakai, Nonlinear optical properties of cesium lithium borate, *Jpn. J. Appl. Phys.*, **34**, L296–L298, 1995.

48. R. Komatsu, T. Sugawara, K. Sassa, N. Sarukura, Z. Liu, S. Izumida, S. Uda, T. Fukuda, and K. Yamanouchi, Growth and ultraviolet application of $Li_2B_4O_7$ crystals: generation of the fourth harmonic and fifth harmonics of $Nd:Y_3Al_5O_{12}$ lasers, *Appl. Phys. Lett.*, **70**, 3492–3494, 1997.

49. Y. Suzuki, S. Ono, H. Murakami, T. Kozeki, H. Ohtake, N. Sarukura, G. Masada, H. Shiraishi, and I. Sekine, 0.43-J, 10-Hz fourth-harmonic generation of Nd:YAG laser using large $Li_2B_4O_7$ crystals, *Jpn. J. Appl. Phys.*, **41**, L823–L824, 2002.

50. L.M. Frantz and J.S. Nodvick, Theory of pulse propagation in a laser amplifier, *J. Appl. Phys.*, **34**, 2346–2349, 1963.

51. A.E. Siegman, in *Lasers*, University Science, Mill Valley, California, 1987, p. 552.

52. D. Strickland and G. Mourou, Compression of amplified chirped optical pulses, *Opt. Commun.*, **56**, 219–221, 1985.

53. Z. Liu, T. Kozeki, Y. Suzuki, N. Sarukura, K. Shimamura, T. Fukuda, M. Hirano, and H. Hosono, Chirped pulse amplification for ultraviolet femtosecond pulses using Ce^{3+}:LiCaAlF$_6$ as a broad-band, solid-state gain medium, *Opt. Lett.*, **26**, 301–303, 2001.

54. W.H. Knox, Femtosecond optical pulse amplification, *IEEE J. Quantum Electron.*, **24** 388, 1988.

55. Y. Nabekawa, K. Kondo, N. Sarukura, K. Sajiki, and S. Watanabe, Terawatt KrF/Ti:sapphire hybrid laser system, *Opt. Lett.*, **18**, 1922–1924, 1993.

56. J.-C. Diels and W. Rudolph, in *Ultrashort Laser Pulse Phenomena*, Academic Press, San Diego, CA, 1996, p. 317.

57. G. Toci, M. Vannini, R. Salimbeni, M.A. Dubinskii, and E. Giorgetti, Proc. Int. Conf. Lasers 99, STS Press, McLean, 2000.

58. Z. Liu, S. Ono, T. Kozeki, Y. Suzuki, N. Sarukura, and H. Hosono, Generation of intense 25-fsec pulses at 290 nm by use of a hollow fiber filled with high-pressure argon gas, *Jpn. J. Appl. Phys.* **41**, L986–L988, 2003.

59. J.F. Pinto, L. Esterowitz, and T.J. Carrig, Extended wavelength conversion of a Ce^{3+}:LiCAF laser between 223and 243 nm by sum frequency mixing in β-barium borate, *Appl. Opt.*, **37**, 1060–1061, 1998.

60. Z. Liu, N. Sarukura, M.A. Dubinskii, V.V. Semashko, A.K. Naumov, S.L. Korableva, and R.Y. Abdulsabirov, Tunable ultraviolet short-pulse generation from a Ce:LiCAF laser amplifier system and its sum-frequency mixing with an Nd:YAG laser, *Jpn. J. Appl. Phys.*, **37**, L36–L38, 1998.

61. P. Rambaldi, M. Douard, J.-P. Wolf, New UV tunable solid-state lasers for lidar applications, *Appl. Phys. B* **61**, 117–120, 1995.

62. J.F. Pinto and L. Esterowitz, Distributed-feedback, tunable Ce^{3+}-doped colquiriite lasers, *Appl. Phys. Lett.*, **71**, 205–207, 1997.

63. M.A. Dubinskii, K.L. Schleper, R.Y. Abdulsabirov, and S.L. Korableva, All-solid-state injection-seeded tunable ultraviolet laser, *J. Mod. Opt.*, **45**, 1993–1998, 1998.

64. A.B. Petersen, All solid-state, 228-240 nm source based on Ce:LiCAF, Advanced Solid-State Lasers '97 (OSA), paper PD2, 1997.

65. N. Sarukura, Z. Liu, H. Ohtake, Y. Segawa, M.A. Dubinskii, R.Y. Abdulsabirov, S.L. Korableva, A.K. Naumov, and V.V. Semashko, Ultraviolet short pulses from an all-solid-state Ce:LiCAF master oscillator and power amplifier system, *Opt. Lett.*, **22**, 994–996, 1997.

66. M.A. Dubinskii, K.L. Schepler, V.V. Semashko, R.Y. Abdulsabirov, B.M. Galjautdinov, S.L. Korableva, and A.K. Naumov, Ce^{3+}:LiBaF$_3$ as new prospective active material for tunable UV laser with direct UV pumping, *OSA Trends in Optics and Photonics*, Vol. 10, Advanced Solid State Lasers, Optical Society of America, Washington, D.C., p. 30, 1997.

67. M. Laroche, S. Girard, R. Moncorge, M. Bettinelli, R.Y. Abdulsabirov, and V. Semashko, Beneficial effect of Lu^{3+} and Yb^{3+} ion in UV laser materials, *Opt. Mater.*, **22**, 147–154, 2003.

68. V.V. Semashko, M.A. Dubinskii, R.Y. Abdulsabirov, S.L. Korableva, and A.K. Naumov, Photodynamic processes in Ce-activated solid-state active media, in *Proc. OSA Trends in Optics and Photonics*, Advanced Solid State Lasers, 68, Washington, D.C., 2002, pp. 251–253.

69. D. Alderighi, G. Toci, M. Vannini, D. Parisi, and M. Tonelli, Experimental evaluation of the cw lasing threshold for a Ce:LiCaAlF$_6$ laser, *Opt. Express*, **13**, 7256–7264, 2005.

5

Eyesafe Rare Earth Solid-State Lasers

Robert C. Stoneman

CONTENTS

5.1 Introduction

Laser wavelengths greater than 1.4 μm provide enhanced eyesafety owing to the absorption in this region of the spectrum by liquid water in the eye. Operation at wavelengths within the eyesafe band is required for many applications for which the laser output beam is not contained. Several laser types operate within the eyesafe band:

- Bulk solid-state crystals and glasses doped with rare earth ions
- Erbium-doped fiber amplifier (EDFA)
- Crystals doped with transition metal ions

- Optical parametric oscillators
- Raman lasers

The scope of this chapter is restricted to bulk solid-state rare earth lasers operating within the eyesafe wavelength band. The primary focus is on laser transitions that can be pumped by laser diodes. The most significant are those doped with the trivalent holmium (Ho^{3+}), erbium (Er^{3+}), and thulium (Tm^{3+}) ions and operating within the wavelength range 1.5 to 2.1 μm. EDFAs are closely related to bulk erbium-doped lasers, and are discussed in depth in Reference 1 and Reference 2.

5.2   Eyesafe Wavelengths

Pulse energy, pulse duration, aperture size, and wavelength are the key factors that determine laser eyesafety. Of these factors, wavelength is typically the only one that can be modified without degrading performance. As a result, lasers with emission within the eyesafe wavelength band are commonly referred to as "eyesafe lasers," although lasers at any wavelength with restricted performance (for example, low output power) can be eyesafe. Wavelengths longer than approximately 1.4 μm are strongly absorbed by the cornea and the vitreous humor of the eye, which are composed primarily of liquid water. The laser intensity is thereby attenuated significantly before reaching the retina.

The absorption spectrum [3] of liquid water is shown in Figure 5.1. Combinations of the fundamental vibrations of the O–H bond produce the strong absorption bands. The fundamental vibrations are two stretch modes at 2.9 μm and a bending mode at 6.1 μm. The absorption band at 1.95 μm is a

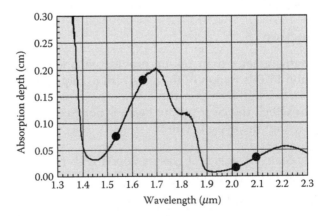

FIGURE 5.1
Absorption spectrum of liquid water.

TABLE 5.1

Absorption in Liquid Water of the Primary Eyesafe Rare Earth Laser Wavelengths

Rare Earth Laser	Wavelength (nm)	Absorption Depth (cm) in Liquid Water	OD in 2.5 cm of Liquid Water
Er:Glass	1535	0.075	15
Er:YAG	1645	0.181	6
Tm:YAG	2015	0.016	68
Ho:YAG	2097	0.035	31

combination of a stretch mode and a bending mode, and the absorption band at 1.45 μm is a combination of two stretch modes.

Table 5.1 lists several important eyesafe rare earth laser wavelengths and the corresponding absorption depth in liquid water. These wavelengths are also indicated on the absorption curve in Figure 5.1. The diameter of the human eye is approximately 2.5 cm. Table 5.1 shows the optical density (OD) in liquid water for this absorption length. The OD is very high for all wavelengths shown, and is greater than 5 for wavelengths greater than 1.38 μm for this absorption length.

5.3 Electronic Energy Structure

The lowest electronic states for the Ho^{3+}, Er^{3+}, and Tm^{3+} rare earth ions are shown in Figure 5.2. This diagram is adapted from the Dieke chart [4] of the trivalent rare earths. The electronic states are labeled with the $^{2S+1}L_J$ term notation. The arrows indicate the most significant laser transitions. The wavelengths shown are those for the yttrium aluminum garnet (YAG) crystal host. The 2940 nm Er^{3+} transition is nominally within the eyesafe band, but the absorption in liquid water at this wavelength is so strong (absorption depth is 1 μm) that it can damage the cornea and skin at low power. The other three laser transitions shown in Figure 5.2 are known as quasi-three-level lasers, because the lower laser level is within the ground electronic state.

The trivalent rare earths have a partially filled 4f shell. Table 5.2 shows the number of 4f electrons and the corresponding quantum numbers for the lowest electronic states of the Ho^{3+}, Er^{3+}, and Tm^{3+} ions. The electronic states with the highest S have the lowest energy, and for states with the same S those with the highest L have the lowest energy (Hund's rule, Reference 5, p. 251). For example, in Er^{3+} the 4f shell is missing three electrons. The electronic state with lowest energy has three unpaired electrons, for which S = 3/2 and L = 6. The J values range from L+S to L–S, and the states with the highest J have the lowest energy (for a shell more than half full, as is the case for the 4f shell in Ho^{3+}, Er^{3+}, and Tm^{3+}; see Reference 5, p. 268). Hund's rule is violated in Tm^{3+} owing to strong spin–orbit coupling that mixes the

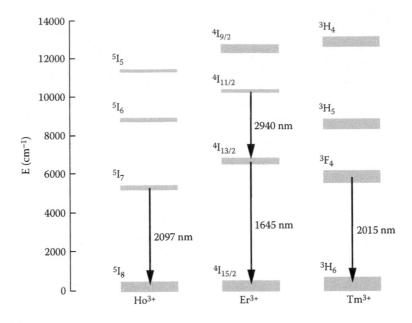

FIGURE 5.2
Lowest electronic energy states for the trivalent Ho, Er, and Tm rare earth ions.

TABLE 5.2

Quantum Numbers of the Lowest Electronic States of the Ho^{3+}, Er^{3+}, and Tm^{3+} Rare Earth Ions

Rare Earth Ion	Number of 4f electrons	Highest S	Highest L	J
Ho^{3+}	10	2	6 (I)	8, 7, 6, 5, 4
Er^{3+}	11	3/2	6 (I)	15/2, 13/2, 11/2, 9/2
Tm^{3+}	12	1	5 (H)	6, 5, 4

$^3H_4$ and $^3F_4$ states, resulting in a lower energy for the $^3F_4$ state relative to the $^3H_4$ and $^3H_5$ states [6]. The leading percentages of the $^3H_4$ and $^3F_4$ states are 60% and 63%, respectively [7].

5.4 Quasi-Three-Level Model

The electric field of the host crystal splits each electronic state into 2J+1 levels. The individual crystal-field levels are not shown in Figure 5.2, but the total extent of the splitting in YAG is shown to scale by the thickness of the gray bars [8]. The manifolds of crystal-field energy levels [8] for the two lowest electronic states of Er:YAG are shown in Figure 5.3. The crystal-field splittings are shown to scale, but the spacing between the $^4I_{15/2}$ and $^4I_{13/2}$ electronic

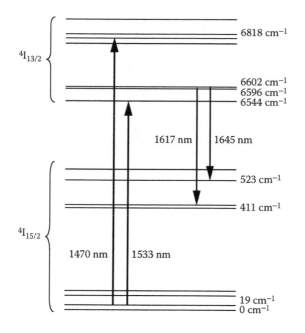

FIGURE 5.3
Crystal-field energy levels for the two lowest electronic states in Er:YAG.

states is compressed. The Er^{3+} crystal-field levels are doubly degenerate (Kramers degeneracy, which applies when the number of electrons is odd; Reference 5, p. 223), resulting in eight distinct levels for the $^4I_{15/2}$ state and seven distinct levels for the $^4I_{13/2}$ state.

The most direct method of pumping the 1645 nm Er:YAG laser is via the 1533 nm pump transition into the upper state manifold, as shown in Figure 5.3. An alternate pump transition at 1470 nm is also shown. In addition to the dominant 1645 nm laser transition, an alternate laser transition at 1617 nm is also shown. The energies of the relevant crystal-field levels for all four transitions are shown at the right of the figure. Figure 5.3 illustrates the "quasi" nature of the quasi-three-level laser transition. The lower laser level is a member of the manifold of crystal-field levels of the ground electronic state, but is not in general the lowest member of the manifold. Quasi-three-level lasers are modeled in references 9 to 16.

The key features that distinguish quasi-three-level laser operation from four-level laser operation can be illustrated by a simple model of the direct upper-state pumped Er:YAG laser. The model parameters are defined in Figure 5.4. The crystal-field levels shown are those of the 1645 nm Er:YAG laser pumped at 1533 nm, but the model can also be applied to the direct upper-state pumped 2097 nm Ho and 2015 nm Tm lasers in any crystalline or glass host material. The subscript 1 refers to the ground electronic state, and the subscript 2 refers to the first excited electronic state. N_1 (N_2) is the population density in the ground (first excited) electronic state. The fraction f of the electronic state

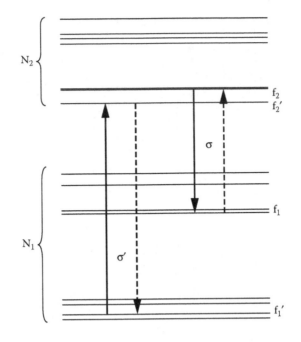

FIGURE 5.4
Parameters of the quasi-three-level laser model.

population in a particular crystal-field level is given by the Boltzmann distribution. Primed variables refer to the pump transition, and unprimed variables refer to the laser transition. For example, f_1' is the Boltzmann fractional population in the lower state of the pump transition. The cross sections for the pump and laser transitions are given by σ' and σ respectively. Solid arrows represent the "forward" directions of the transitions, and the dashed arrows represent the "reverse" directions. The reverse laser transition produces absorption from the ground state at the laser wavelength, an effect known as reabsorption loss. The reverse pump transition limits the extent to which the ground state can be depleted, thereby limiting the population density that can be stored in the upper laser state.

5.5 Reabsorption Loss

Reabsorption loss requires a minimum population density in the upper laser state to reach inversion on the laser transition. The laser inversion density is given by

$$\Delta N = f_2 N_2 - f_1 N_1. \tag{5.1}$$

Population in electronic states higher than the first excited state can be ignored in the upper-state pumped laser. Therefore

$$N = N_1 + N_2 \qquad (5.2)$$

where N is the Er^{3+} dopant concentration. It is convenient to work with the normalized upper-state population density β, defined by

$$\beta = N_2/N. \qquad (5.3)$$

Then, from Equation 5.2 and Equation 5.3,

$$N_1 = N (1 - \beta) \qquad (5.4)$$

$$N_2 = N \beta. \qquad (5.5)$$

Using Equation 5.1, Equation 5.4, and Equation 5.5, the normalized laser inversion density is

$$\Delta N/N = f_2 [\beta - f (1 - \beta)] \qquad (5.6)$$

where $f = f_1/f_2$ is the laser transition Boltzmann ratio.

Equation 5.6 shows that the laser inversion density does not become positive until the normalized upper-state population reaches a minimum value given by

$$\beta_{min} = f/(1 + f). \qquad (5.7)$$

The Boltzmann factor for the ith crystal-field level of the ground electronic state is given by

$$f_{1i} = \exp[-(E_{1i} - E_{11})/kT]/Z_1 \qquad (5.8)$$

where E_{1i} is the energy of the ith crystal-field level of the ground electronic state (e.g., E_{11} is the energy of the lowest crystal-field level of the ground electronic state), k is Boltzmann's constant, T is the laser rod temperature, and

$$Z_1 = \Sigma_j \exp(-E_{1j}/kT)$$

is the partition function for the ground electronic state.

Similarly, the Boltzmann factor for the ith crystal-field level of the first excited electronic state is given by

$$f_{2i} = \exp[-(E_{2i} - E_{21})/kT]/Z_2 \qquad (5.9)$$

where E_{2i} is the energy of the ith crystal-field level of the first excited electronic state, and

$$Z_2 = \Sigma_j \exp(-E_{2j}/kT)$$

is the partition function for the first excited electronic state.

TABLE 5.3

Crystal-Field Energy Levels and Boltzmann Factors (for T = 300 K) for the Lowest Two Electronic States of Er:YAG

I	E_{1i} (cm^{-1})	f_1	E_{2i} (cm^{-1})	f_2
1	0	0.264	**6544**	**0.276**
2	**19**	**0.241**	6596	0.215
3	57	0.201	**6602**	**0.209**
4	76	0.183	6779	0.089
5	411	0.037	6800	0.081
6	424	0.035	6818	0.074
7	**523**	**0.022**	6879	0.055
8	568	0.017	N/A	N/A

The crystal-field energy levels [8] and the corresponding Boltzmann factors (for T = 300 K) for the lowest two electronic states of Er:YAG are shown in Table 5.3. The bold cells are the relevant levels for the 1533 nm pump transition and the 1645 nm laser transition.

The laser transition Boltzmann ratio f is obtained by dividing Equation 5.8 by Equation 5.9 to give

$$f = (Z_2/Z_1) \exp[(1/\lambda_L - \Delta E_0)/kT] \tag{5.10}$$

where λ_L is the laser wavelength and $\Delta E_0 = E_{21} - E_{11}$ is the energy difference of the lowest crystal-field levels of the first excited electronic state and the ground electronic state.

Equation 5.10 has a simple dependence on wavelength and temperature. The ratio of partition functions Z_2/Z_1 is independent of wavelength; therefore f increases exponentially with decreasing wavelength. Figure 5.5 shows the wavelength dependence of f for Er:YAG at two temperatures in the region of the two primary laser wavelengths 1617 nm and 1645 nm.

Figure 5.6 shows the wavelength dependence of β_{min}, given by Equation 5.7, for Er:YAG at two temperatures. The minimum upper-state population required for inversion is substantially higher for the 1617 nm laser transition than for the 1645 nm transition (65% higher for 0°C, and 51% higher for 50°C). The wavelength dependence of reabsorption loss can have a significant impact on quasi-three-level laser performance.

Equation 5.10 shows that f has a significant exponential dependence on temperature provided $1/\lambda_L - \Delta E_0$ is not close to zero. In addition to the explicit exponential dependence, the variation of the partition function ratio with temperature is relatively insignificant, as shown in Figure 5.7 for Er:YAG.

Equation 5.10 shows that f increases as the temperature increases when $\lambda_L > 1/\Delta E_0$ (ignoring the slow variation of Z_2/Z_1 with temperature). For Er:YAG, $1/\Delta E_0 = 1528$ nm ($\Delta E_0 = 6544$ cm^{-1}, as seen in Table 5.3). Figure 5.8 shows the temperature dependence of f for the two primary laser wavelengths 1617 nm and 1645 nm, both of which are longer than $1/\Delta E_0$, hence the increase in f with temperature.

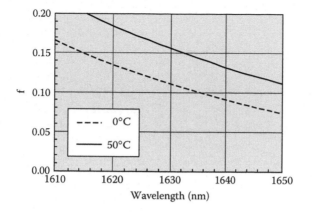

FIGURE 5.5
Wavelength dependence of f for Er:YAG at two temperatures.

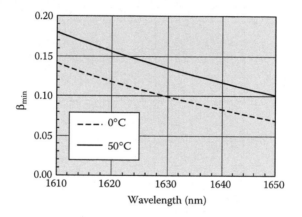

FIGURE 5.6
Wavelength dependence of β_{min} for Er:YAG at two temperatures.

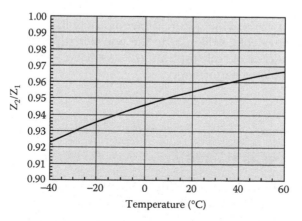

FIGURE 5.7
Temperature dependence of the partition function ratio Z_2/Z_1 in Er:YAG.

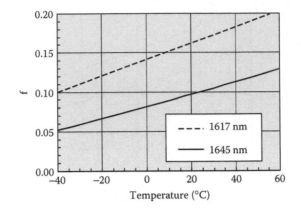

FIGURE 5.8
Temperature dependence of f for two Er:YAG laser wavelengths.

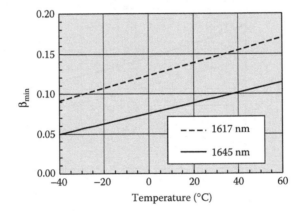

FIGURE 5.9
Temperature dependence of β_{min} for two Er:YAG laser wavelengths.

Figure 5.9 shows the temperature dependence of β_{min} for the two primary Er:YAG laser wavelengths. Similar to the wavelength dependence, the temperature dependence of reabsorption loss can have a significant impact on quasi-three-level laser performance.

5.6   Ground State Depletion

The analysis of ground state depletion [17–19] parallels that of reabsorption loss. The inversion density on the pump transition is given by

$$\Delta N' = f_2' \, N_2 - f_1' \, N_1. \tag{5.11}$$

Using Equation 5.4 and Equation 5.5, the normalized pump inversion density is

$$\Delta N'/N = f_2' \, [\beta - f' \, (1 - \beta)] \tag{5.12}$$

where $f' = f_1'/f_2'$ is the pump transition Boltzmann ratio. The reverse pump transition ensures that $\Delta N'$ can be no greater than zero. This places an upper limit on the normalized upper-state population, given by

$$\beta_{max} = f'/(1 + f'). \tag{5.13}$$

The pump transition Boltzmann ratio is derived in the same way as the laser transition Boltzmann ratio, and is given by

$$f' = (Z_2/Z_1) \, \exp[(1/\lambda_p - \Delta E_0)/kT] \tag{5.14}$$

where λ_p is the pump wavelength.

Figure 5.10 shows the wavelength dependence of f' for Er:YAG at two temperatures in the region of the two primary pump wavelengths 1470 nm and 1533 nm. As in Figure 5.5, the Boltzmann ratio increases exponentially with decreasing wavelength. As anticipated from the discussion preceding Figure 5.8, the 50°C curve lies below the 0°C curve in Figure 5.10 (the reverse of the relative positions of the curves in Figure 5.5) for wavelengths less than approximately 1528 nm ($1/\Delta E_0$). The actual crossing point of the curves is 1523 nm. The shift of the crossing point from the nominal value of 1528 nm is a result of the temperature dependence of Z_2/Z_1.

Figure 5.11 shows the wavelength dependence of β_{max} for Er:YAG at two temperatures. The maximum population density than can be stored in the upper state is substantially higher for the 1470 nm pump transition than for the 1533 nm pump transition.

Figure 5.12 shows the temperature dependence of f' for the two primary pump wavelengths 1470 nm and 1533 nm. As discussed above, the temperature

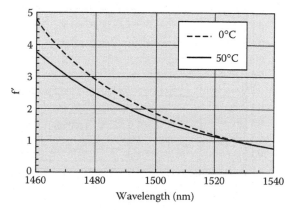

FIGURE 5.10
Wavelength dependence of f' for Er:YAG at two temperatures.

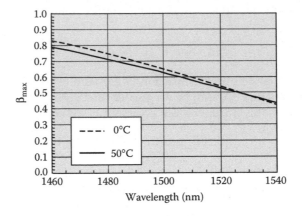

FIGURE 5.11
Wavelength dependence of β_{max} for Er:YAG at two temperatures.

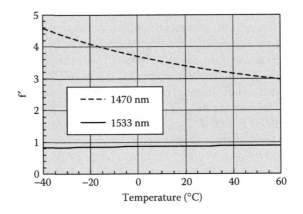

FIGURE 5.12
Temperature dependence of f' for two Er:YAG pump wavelengths.

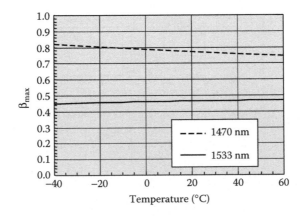

FIGURE 5.13
Temperature dependence of β_{max} for two Er:YAG pump wavelengths.

dependence for the 1470-nm pump wavelength has the opposite sign as the temperature dependence for the 1617-nm and 1645-nm laser wavelengths seen in Figure 5.8. The temperature dependence for the 1533-nm pump wavelength is nearly zero because this wavelength is close to $1/\Delta E_0$.

Figure 5.13 shows the temperature dependence of β_{max} for the two primary Er:YAG pump wavelengths.

5.7 Reciprocity of Emission and Absorption

The laser cross section σ shown in Figure 5.4 is not a directly measurable quantity. We can define effective cross sections by multiplying by the Boltzmann population factors. The effective absorption cross section for the laser transition is given by

$$\sigma_a = f_1 \, \sigma = \alpha/N \qquad (5.15)$$

where α is the absorption coefficient. Similarly, the effective emission cross section for the laser transition is given by

$$\sigma_e = f_2 \, \sigma. \qquad (5.16)$$

The absorption spectrum for Er:YAG with 1% Er concentration at room temperature is shown in Figure 5.14. The spectrum was measured with an FTIR spectrometer. The wavelength range covers the 1617-nm and 1645-nm laser lines.

The emission cross section spectrum is derived from the absorption spectrum by combining Equation 5.15 and Equation 5.16 to give the reciprocity relation between emission and absorption spectra for a quasi-three-level laser transition

$$\sigma_e = \alpha/f \, N. \qquad (5.17)$$

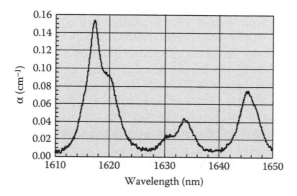

FIGURE 5.14
Absorption spectrum for 1% Er:YAG at room temperature.

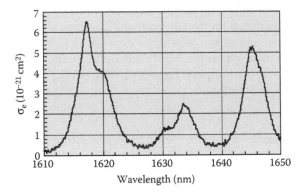

FIGURE 5.15
Emission cross section spectrum for Er:YAG at room temperature.

Reciprocity of emission and absorption spectra is discussed in references 20 to 25.

The emission cross section spectrum derived from the absorption spectrum of Figure 5.14 using Equation 5.17 is shown in Figure 5.15. The exponential dependence of the Boltzmann ratio f on wavelength results from reabsorption loss and is shown in Figure 5.5. This dependence accounts for the shift in the relative amplitudes of the 1617-nm and 1645-nm peaks between Figure 5.14 and Figure 5.15. The absorption coefficient of the 1617-nm line is twice that of the 1645-nm line, whereas the emission cross section of the 1617-nm line is only 25% greater than that of the 1645-nm line.

The laser gain cross section spectrum is given by

$$g = \sigma \, \Delta N / N. \tag{5.18}$$

Using Equation 5.6 and Equation 5.16, Equation 5.18 gives

$$g = \sigma_e \, [\beta - f \, (1 - \beta)]. \tag{5.19}$$

The gain cross section spectrum is shown in Figure 5.16 for two values of the normalized upper-state population density β. The explicit dependence of Equation 5.19 on f has a significant impact on the relative strengths of the 1617-nm and 1645-nm laser lines for different values of β. For lower values of β (for example, the $\beta = 0.25$ curve in Figure 5.16), the gain cross section is higher for the 1645 nm laser line than for the 1617 nm line. For higher values of β (for example, the $\beta = 0.50$ curve in Figure 5.16), the gain cross section is higher for the 1617-nm line than for the 1645-nm line. As a result, an Er:YAG laser with low output coupling transmission will oscillate on the 1645-nm line, but a laser with higher output coupling will shift to the 1617-nm line.

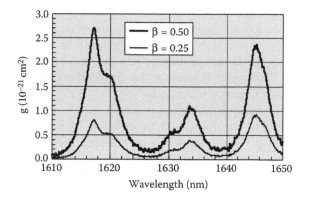

FIGURE 5.16
Gain cross section spectrum for Er:YAG at room temperature.

5.8 Thulium-Sensitized Holmium Laser

The relatively high levels of upper-state population density required for inversion of quasi-three-level laser transitions, as seen in Figure 5.6 and Figure 5.9, impose high pump intensity requirements on them. Prior to the past several years, pump intensities sufficient for direct upper-state pumping were not readily available. A technique that has been widely employed to reduce the pump intensity requirement is illustrated in Figure 5.17. In addition to the lasant ion, a second ion, known as the sensitizer, is introduced into the laser crystal. In this context, the lasant ion is known as the activator. In this scheme the pump light is absorbed by the sensitizer, which transfers its excitation to the activator. The sensitizer concentration can be relatively high, enabling efficient absorption of the pump, while the activator concentration is relatively low, minimizing reabsorption loss. In Figure 5.17 the sensitizer ion is Tm^{3+} and the activator ion is Ho^{3+}.

In the pumping scheme shown in Figure 5.17, the $^3H_4$ state of Tm^{3+} is pumped by a 792 nm AlGaAs laser diode. The $^3H_4 + {}^3H_6 \rightarrow {}^3F_4 + {}^3F_4$ cross-relaxation process converts the pump excitation into two ions in the first excited state of Tm^{3+}. The $^3F_4 \rightarrow {}^5I_7$ energy transfer process populates the upper state of the 2065 nm Ho^{3+} laser transition. The two upconversion processes are detrimental to the laser performance. The $^5I_7 + {}^5I_7 \rightarrow {}^5I_5 + {}^5I_8$ upconversion process removes a pair of Ho^{3+} ions from the upper laser state. The $^5I_7 + {}^3F_4 \rightarrow {}^5I_5 + {}^3H_6$ upconversion process removes a single Ho^{3+} ion from the upper laser state [26–32].

The pump and laser wavelengths shown in Figure 5.17 are those of the yttrium lithium fluoride (YLF) crystal host. Laser performance in the YLF host generally surpasses that in the YAG host owing to the larger upconversion coefficient in YAG relative to YLF. The cross-relaxation, energy transfer, and upconversion nonradiative processes exhibit energy mismatches that are bridged by crystal lattice phonons. The phonon spectrum is broader in YAG than in YLF, thereby increasing the upconversion coefficient in YAG owing to the smaller energy mismatch.

The rates for the nonradiative processes are faster for higher dopant concentrations, owing to the increased probability of interaction for shorter interaction distances. This imposes compromises on the choice of dopant concentrations for optimized performance of the sensitizer/activator laser scheme. Efficient population of the $^3F_4$ state by the cross-relaxation process requires a Tm^{3+} concentration of at least several percent, but the upconversion losses increase with increasing dopant concentrations. The upconversion losses are more significant for Q-switched than for CW laser operation, owing to the high upper-state population density of the energy stored prior to Q-switching. This effect is more pronounced when large pulse energies are required, owing to the long storage times required for operation at low pulse repetition frequency (PRF). The thermal equilibrium of the overlapped $^3F_4$ and $^5I_7$ state manifolds also plays a role in the optimization of the dopant concentrations. This places an effective upper limit on the ratio of the concentrations of Tm^{3+} to Ho^{3+} ions for Q-switched operation at low PRF, because the population stored in the Tm^{3+} ions is only available for Q-switched extraction after the time delay inherent in the transfer of population to the Ho^{3+} ion [33]. Typical optimum concentrations for the sensitizer/activator scheme are a few percent of Tm^{3+} and a few tenths of a percent of Ho^{3+}.

Early Ho:YAG lasers were pumped by xenon flashlamps and CW tungsten and mercury lamps [34,35]. The relatively low intensities of these pump sources necessitated cooling the laser rod with liquid nitrogen to overcome reabsorption loss. These lasers utilized the broadband absorption of Er^{3+} and Cr^{3+} sensitizers, in addition to the Tm^{3+} sensitizer shown in Figure 5.17, to match the broadband emission of the pumping lamps. Subsequent to these initial results, flashlamp-pumped Ho-doped lasers were demonstrated in a variety of crystalline hosts, including yttrium gallium garnet (YGG) [36], yttrium iron garnet (YIG) [36], yttrium aluminum oxide (YALO) [37,38], YLF [39–41], and barium yttrium fluoride (BaY_2F_8) [42] Flashlamp-pumped laser performance with improved efficiency was achieved by optimizing the concentrations of the Cr^{3+} and Tm^{3+} sensitizers and the Ho^{3+} activator in YAG [43–47]. Early applications of these lasers were in the field of remote sensing, including coherent laser radar and differential absorption lidar (DIAL) [48–50]. Sensitizer/activator Tm,Ho lasers have also been demonstrated in other garnet crystal hosts, including yttrium scandium gallium garnet (YSGG), gadolinium scandium aluminum garnet (GSAG), yttrium scandium aluminum garnet (YSAG) [51], and lutetium aluminum garnet (LuAG) [52].

The development of practical laser diodes pump sources led to the development of Tm,Ho sensitizer/activator lasers with significantly improved performance over that of lamp-pumped lasers. The primary benefit of laser diode pumping results from the close match between the narrowband laser diode output radiation and the narrow absorption lines of trivalent rare earth ions. The first demonstration of a diode-pumped rare earth laser with activator ion other than Nd^{3+} was a Ho:YAG laser sensitized with Er^{3+} and Tm^{3+} [53–55]. Shortly thereafter, the diode-pumped Er,Tm,Ho laser was demonstrated in the YLF crystal host [56,57]. These early results utilized cryogenic cooling of the laser rod to mitigate the temperature-dependent reabsorption loss.

Following these initial results, utilizing optimized concentrations of the sensitizer and activator ions, the diode-pumped Tm,Ho:YAG [58–61], Cr,Tm,Ho:YAG [62], and Tm,Ho:YLF [63,64] lasers were demonstrated at room temperature. These early results were limited to output powers of a few mW, but were soon followed by Tm,Ho:YAG lasers with output powers of tens of mW [65,66], followed by demonstrations of output powers greater than 10 W [67,68]. A mode-locked Tm,Ho:YAG laser has also been demonstrated [69].

Sensitizer/activator Tm,Ho lasers have also been operated by pumping with alternate lasers to simulate laser diode pumping. The Tm,Ho:YAG laser was pumped by a 785 nm alexandrite laser [70,71], Cr,Tm,Ho:YAG [62,72] and Cr,Tm,Ho:YSGG [72] lasers were pumped by a 647 nm krypton ion laser, the Tm,Ho:YAG laser was pumped by a 785 nm Cr:GSAG laser [73], and the Tm,Ho:LuAG laser was pumped by a 781 nm Ti:sapphire laser [74]. In addition to the garnet crystal hosts, the diode-pumped Tm,Ho laser has also been demonstrated in other oxide hosts such as YALO [75] and $GdVO_4$ (gadolinium vanadate) [76,77].

As described previously, the Tm,Ho:YLF laser benefits from weaker upconversion that the Tm,Ho:YAG laser [78]. Advances in diode-pumped Tm,Ho:YLF lasers were made by researchers at Coherent Technologies (now Lockheed Martin Coherent Technologies) [79], Lockheed Sanders (now BAE Systems) [80–83], and Schwartz Electro-Optics (now Q-Peak) [84–86]. Of particular note is the extensive research performed at NASA Langley Research Center (LaRC) on diode-pumped Tm,Ho:YLF lasers and amplifiers producing high pulse energies [87–94]. Pulse energies greater than 600 mJ have been demonstrated by the LaRC group in YLF [95–98], and more recently a pulse energy of 1 J has been demonstrated by the group in lutetium lithium fluoride (LuLF) [99,100]. Research has been performed on conductively cooled diode-pumped Tm,Ho:YLF and LuLF lasers for spaceborne applications [101–103]. As in the case of Tm,Ho:YAG lasers, alternate lasers have been used to simulate laser diode pumping [104]. In addition to YLF and LuLF [105–111] lasers, other fluoride crystal hosts have been utilized, including barium yttrium fluoride [112] and potassium yttrium fluoride (KYF_4) [113].

5.9 Thulium Laser

The upconversion processes shown in Figure 5.17 limit the storage of energy in the upper laser state. An alternate method for producing 2 μm laser output

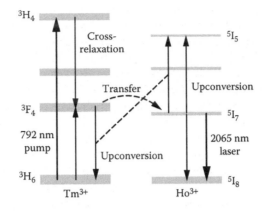

FIGURE 5.17
Energy level diagram of the diode-pumped Tm-sensitized 2.1 μm Ho laser.

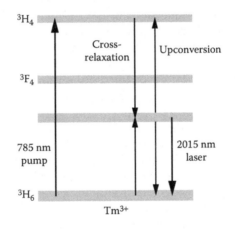

FIGURE 5.18
Energy level diagram of the diode-pumped 2.0 μm Tm laser.

is shown in Figure 5.18. The $^3H_6 \to {}^3H_4$ pump transition and the $^3H_4 + {}^3H_6 \to$ $^3F_4 + {}^3F_4$ cross-relaxation process are utilized as in Figure 5.17. In this scheme, however, the only dopant is Tm^{3+}, and the $^3F_4 \to {}^3H_6$ laser transition is utilized. The $^3F_4 + {}^3F_4 \to {}^3H_4 + {}^3H_6$ upconversion process is not shown in Figure 5.17 owing to the weakness of this process relative to the two upconversion processes involving the Ho^{3+} ion. This process is the inverse of the cross-relaxation process, and like the upconversion processes in Ho^{3+} it removes population from the upper laser state [114]. The pump and laser wavelengths shown in Figure 5.18 are those of the YAG crystal host.

As with Tm,Ho:YAG lasers, early Tm:YAG lasers were pumped by flashlamps [34]. Subsequent to this early work, efficiencies of flashlamp-pumped Tm:YAG lasers were improved by optimizing the concentrations of the Cr^{3+} sensitizer and the Tm^{3+} activator ions [115–119].

Shortly after the first demonstrations of diode-pumped Tm,Ho:YLF and Tm,Ho:YAG lasers, the diode pumping scheme of Figure 5.18 for the Tm:YAG laser was demonstrated at room temperature [120]. Subsequently the diode-pumped Tm:YAG was Q-switched [65], and applied to eyesafe remote sensing applications [121]. For these applications, the Tm:LuAG laser has been found to have more favorable atmospheric transmission than the Tm:YAG laser [122].

As in the case of Tm,Ho sensitizer/activator lasers, Tm lasers have also been operated by pumping with alternate lasers to simulate laser diode pumping. The Cr,Tm:YAG laser was pumped by a 647 nm krypton ion laser [123], the Tm:YAG and Tm:YSGG lasers were pumped by a 785 nm titanium sapphire laser [124,125], a mode-locked Tm:YAG laser was pumped by a titanium sapphire laser [126,127], a mode-locked Cr,Tm:YAG laser was pumped by a krypton ion laser [69], and the Tm:YAG laser was pumped by a 1064 nm Nd:YAG laser [128].

Diode-pumped Tm:YAG lasers have been operated with high efficiency, in particular in end-pumped geometries for which the pump and laser modes are well matched [129,130]. Of particular note is the end-pumping architecture developed by Lawrence Livermore National Laboratory (LLNL) [131–133]. In this architecture a lens duct is utilized to collect the pump light from a laser diode stack. The pump light is guided into the end and down the length of a Tm:YAG laser rod. This technique has yielded an output power of 115 W [134,135]. A group at DSO National Laboratories in Singapore has demonstrated 150 W from a diode-pumped Tm:YAG laser with a side pumping geometry [136–138].

Tm lasers have also been operated in a variety of oxide crystal hosts other than YAG. Early demonstrations of the Tm:YALO laser were flashlamp-pumped and cooled to liquid nitrogen temperatures [38,139]. The efficiency of the Tm:YALO laser was increased by pumping with a ti:sapphire laser [140,141] and laser diodes [142,145]. Output powers have been scaled to high levels with the utilization of fiber-coupled laser diode pumps [146,147], most recently to an output power of 50 W [148]. The Tm laser has been demonstrated in yttrium vanadate (YVO$_4$) [149–151] and GdVO$_4$ [152–154], in the tungstates KY(WO$_4$)$_2$ [155] and KGd(WO$_4$)$_2$, [156] and in yttria (Y$_2$O$_3$) [157–159] and scandia (Sc$_2$O$_3$) [159].

Tm lasers have also been operated in a variety of fluoride crystal hosts. Output powers have been scaled to high levels in YLF [160–162], with an output power exceeding 20 W [146]. The Tm laser has also been demonstrated in KYF$_4$ [163] and BaY$_2$F$_8$ [163–164].

5.10   Upper-State Pumped Holmium Laser

A third pumping scheme for producing 2 μm laser output is shown in Figure 5.19. This scheme is similar to that of Figure 5.17, except that the Tm^{3+} and

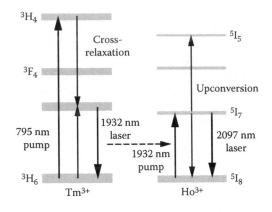

FIGURE 5.19
Energy level diagram of the upper-state pumped Ho laser.

Ho^{3+} ions are doped in separate crystals, thereby avoiding the $^5I_7 + {}^3F_4 \rightarrow$ $^5I_5 + {}^3H_6$ upconversion process that involves both Tm^{3+} and Ho^{3+} ions. This is the strongest of the three upconversion processes shown in Figure 5.17 and Figure 5.18, owing to the close match of the transition energies $^3F_4 - {}^3H_6$ and $^5I_5 - {}^5I_7$. In Figure 5.19 the upper state of the Ho laser is directly populated by pumping with the output of the Tm laser. This optical pumping process substitutes for the $^3F_4 \rightarrow {}^5I_7$ nonradiative energy transfer process of Figure 5.17, and is represented by a dashed line to indicate that the Tm^{3+} and Ho^{3+} ions are doped in separate laser media. The pump and laser wavelengths shown in Figure 5.19 are those of the Tm:YALO and Ho:YAG lasers.

In addition to the elimination of the Tm–Ho upconversion process, the direct upper-state pumping scheme also reduces the detrimental effects of the other two upconversion processes shown in Figure 5.17 and Figure 5.18 to a negligible level. The concentrations of Tm^{3+} and Ho^{3+} can be independently optimized because the two ions are doped into separate crystals. The advantages of the upper-state pumping scheme are particularly relevant when the Ho laser is Q-switched. Large pulse energies can be obtained while operating the Tm laser in the CW mode, owing to the long lifetime of the Ho^{3+} $^5I_7$ upper laser state (8 ms in YAG). The $^3F_4 + {}^3F_4 \rightarrow {}^3H_4 + {}^3H_6$ upconversion process does not significantly impair the performance of the Tm laser because CW operation does not entail energy storage in the upper laser state. The impact of the $^5I_7 + {}^5I_7 \rightarrow {}^5I_5 + {}^5I_8$ upconversion process [165,166] can be reduced to a negligible level by utilizing Ho^{3+} concentrations as low as a few tenths of a percent. In contrast, in the sensitizer/activator scheme the Ho^{3+} concentration must be kept at a level high enough to ensure efficient energy transfer from the Tm^{3+} ion.

The first demonstrations of direct upper-state pumping of a Ho laser by a Tm laser utilized the YAG crystal host for both lasers. Two pumping techniques were utilized in these initial demonstrations. In both cases the Ho:YAG laser was intracavity-pumped by the Tm laser. In the first case, the pump was a flashlamp-pumped Cr,Tm:YAG laser [167,168], and in the second case the pump was a Tm:YAG laser pumped by a Ti:sapphire laser [169,170].

The Ho laser has also been upper-state pumped by lasers other than Tm lasers. Direct upper-state pumping of the Ho:YAG laser by 1.9 μm laser diodes was demonstrated shortly after the initial demonstrations of Tm laser pumping [171–173]. By eliminating the intermediate Tm laser, direct pumping by laser diodes can significantly increase the efficiency of the Ho laser. Though high-power 1.9 μm laser diode devices have not yet proven to be sufficiently mature for this technique to be fully practicable, the potential for near-term improvement in efficiency is significant.

Most of the demonstrations of upper-state pumping of Ho lasers by Tm lasers have not utilized the intracavity pumping technique of the initial demonstrations, owing to the restrictions imposed by placing the Tm and Ho laser media in a common resonator. In particular, it is difficult to insert a Q-switch into the common resonator while operating the Tm laser in the CW mode. Continuous-wave Ho lasers, however, have exploited the intrinsic efficiency advantage of the intracavity pumping technique [174,175], with a CW output power of 7.2 W demonstrated from a Ho:YAG laser intracavity-pumped by a Tm:YAG laser with 53.4 W incident laser diode pump power [176]. The Ho:YAG laser has also been intracavity-pumped by a Tm:YLF laser [177].

Upper-state pumped Ho lasers with significantly higher output powers and pulse energies have been demonstrated by utilizing Tm:YALO and Tm:YLF pump lasers. These pump lasers are capable of operation with high output power at output wavelengths more closely matched to favorable absorption bands of Ho:YAG and Ho:YLF in comparison with the Tm:YAG pump laser. The stronger absorption in these pump bands eliminates the need for intracavity pumping.

The Ho:YAG laser was operated in the CW and Q-switched modes by pumping with a diode-pumped Tm:YALO laser [178]. The Ho:YAG laser was also pumped by a diode-pumped Tm:YLF laser [179,180], with an output power of 18.8 W from the Ho:YAG laser in the CW mode [181] and a pulse energy of 50 mJ from the Ho:YAG laser in the Q-switched mode at a PRF of 60 Hz [182].

The Ho:YLF laser has also been operated in the upper-state pumping mode. An alexandrite laser-pumped Tm:YALO pump laser was utilized to produce a pulse energy of 33 mJ from the Ho:YLF laser [183]. A diode-pumped Tm:YLF laser was utilized to produce a CW output power of 21 W and a Q-switched pulse energy of 37 mJ at a PRF of 100 Hz from a Ho:YLF laser [184].

More recently, upper-state pumped Ho lasers have been scaled to higher output power by pumping with diode-pumped Tm-doped fiber lasers. The Ho:YAG laser was upper-state pumped by a Tm fiber laser [185–187], yielding Ho:YAG output powers of 6.4 W in the CW mode [188,189] and 5.6 W in the Q-switched mode at a PRF of 20 kHz [190]. Single longitudinal mode operation of a CW Ho:YAG ring laser pumped by a diode-pumped Tm-doped fiber laser was demonstrated with an output power of 3.7 W [191,192]. Significant power scaling was demonstrated from a Ho:YLF laser pumped by a diode-pumped Tm-doped fiber laser, with an output power of 43 W in the CW mode and a pulse energy of 45 mJ in the Q-switched mode at a PRF of 200 Hz [193].

5.11 Ytterbium-Sensitized Erbium Laser

The sensitizer/activator scheme can also be applied to the 1.5 μm Er laser, as shown in Figure 5.20. In this case the Yb^{3+} ion is the sensitizer. The $^2F_{5/2}$ state is pumped by a 980 nm InGaAs laser diode. Energy transfer to the $^4I_{11/2}$ state [194] is followed by the $^4I_{11/2} \rightarrow {}^4I_{13/2}$ multiphonon decay process, which populates the upper state of the 1533 nm Er^{3+} laser transition. The $^4I_{13/2} + {}^4I_{13/2} \rightarrow {}^4I_{9/2} + {}^4I_{15/2}$ upconversion process removes a pair of Er^{3+} ions from the upper laser state [196–201]. Additional upconversion loss processes involving higher electronic energy states, some of which involve both the Yb^{3+} and Er^{3+} ions [202], are not shown. The pump and laser wavelengths in Figure 5.20 are those of a glass host.

The earliest Yb,Er:glass sensitizer/activator lasers were pumped by flashlamps [203–205]. Following these initial results, Yb,Er:glass lasers pumped by flashlamp-pumped 1 μm Nd lasers were demonstrated [206–208]. A review of the early flashlamp-pumped Yb,Er:glass laser developments is given in Reference 209. More recently, higher pulse energies have

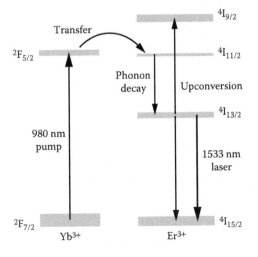

FIGURE 5.20
Energy level diagram of the diode-pumped Yb-sensitized Er laser.

been obtained from flashlamp-pumped Yb,Er:glass lasers by utilizing improved glass compositions [210,211]. A flashlamp-pumped Yb,Er:glass laser utilized a frustrated total internal reflection (FTIR) Q-switch to produce a pulse energy of 7 mJ [212]. A flashlamp-pumped Yb,Er:glass laser with an FTIR Q-switch has produced a pulse energy of 1 mJ in a single longitudinal mode for remote sensing applications [213]. A variety of flashlamp-pumped passively Q-switched Yb,Er:glass lasers have been demonstrated. Passive Q-switches doped with the saturable absorbers divalent uranium [214,215] and cobalt [214,216,217] have been utilized.

The CW diode-pumped Yb,Er:glass laser is modeled in Reference 218. Demonstrations of CW diode-end-pumped Yb,Er:glass lasers [219,220] were performed shortly after the development of the first diode-pumped Tm,Ho and Tm lasers. Single longitudinal mode demonstrations of CW diode-end-pumped microchip Yb,Er:glass lasers were conducted soon after [221–223], followed by broad tunability over the range 1528 nm to 1564 nm [224–226]. Over 300 mW CW has been demonstrated from a diode-end-pumped Yb,Er:glass laser, and 230 mW in a single longitudinal mode [227].

The pulsed diode-pumped Yb,Er:laser is modeled in Reference 228. Pulsed Yb,Er:glass lasers are typically operated at low PRF to avoid the strong thermal lens that arises from the low thermal conductivity of the glass host. Diode-pumped Yb,Er:glass lasers have been operated in the side-pumped configuration [229], producing a pulse energy of 1.2 J in the long-pulse mode [230] and a pulse energy of 14 mJ in the Q-switched mode [231]. In the end-pumped configuration, diode-pumped Yb,Er:glass lasers have also been operated in the long-pulse [232] and Q-switched modes [233,234]. As in the case of flashlamp-pumped Yb,Er:glass lasers, novel Q-switching techniques have been exploited for diode-pumped Yb,Er:glass lasers. The FTIR Q-switch has been utilized [235–237], producing a pulse energy of 26 mJ [238]. Passive Q-switches have also been utilized, including InGaAsP/InP semiconductor [239] and divalent uranium [240,241] and cobalt [242–245]. Q-switched diode-pumped Yb,Er:glass lasers have also been operated in a single longitudinal mode for coherent laser radar applications [246], producing a pulse energy of 11 mJ [247].

Yb,Er sensitizer/activator lasers have also been demonstrated in a variety of crystalline hosts. As in the case of Yb,Er:glass lasers, early Yb,Er crystal lasers were pumped by flashlamps [248]. Er-doped lasers, as with Tm,Ho and Tm lasers, have also been pumped with alternate lasers to simulate laser diode pumping. The nonsensitized Er:YAG and Er:YGG lasers were pumped by the 647 nm krypton laser [249,72,250]. A Ti:sapphire laser was utilized to pump Yb,Er lasers in the Y_2SiO_5 (YOS) [251], YAG [252], and $Ca_2Al_2SiO_7$ [253] crystal hosts.

Diode-pumped Yb,Er lasers have been operated in the YAG [254,255], YOS [255], $YCa_4O(BO_3)_3$ [256], and YVO_4 [257] crystal hosts. A diode-pumped Yb,Er:YAG laser has produced a pulse energy of 79 mJ in the long-pulse mode [258,259].

5.12 Upper-State Pumped Erbium Laser

The Er laser can also be directly upper-state pumped in a manner similar to the Ho:YAG laser shown in Figure 5.19. In Figure 5.20 the 1533 nm Er:glass laser of Figure 5.20 pumps the upper state of the 1645 nm Er:YAG laser. As in Figure 5.19, this optical pumping process is represented by a dashed line to indicate that the two Er^{3+} ions are doped in separate laser media.

As in Figure 5.20, the $^4I_{13/2} + {}^4I_{13/2} \rightarrow {}^4I_{9/2} + {}^4I_{15/2}$ upconversion process removes population from the upper laser state, and upconversion processes to higher energy states, some of which involve both the Yb^{3+} and Er^{3+} ions, are not shown. Also not shown is the $^4I_{13/2} + {}^4I_{13/2} \rightarrow {}^4I_{9/2} + {}^4I_{15/2}$ upconversion process in the Yb,Er:glass laser. Similar to the Ho:YAG upper-state pumping scheme of Figure 5.19, the impact of the $^4I_{13/2} + {}^4I_{13/2} \rightarrow {}^4I_{9/2} + {}^4I_{15/2}$ upconversion loss process can be reduced to a negligible level by utilizing concentrations of Er^{3+} as low as a few tenths of a percent. In contrast, in the Yb,Er:glass laser the concentrations of Yb^{3+} and Er^{3+} must be large enough to facilitate the $^2F_{5/2} \rightarrow {}^4I_{11/2}$ energy transfer process.

Early demonstrations of upper-state pumped nonsensitized Er-doped lasers utilized flashlamp-pumped pulsed Yb,Er:glass lasers as the pump source. The crystal hosts YAG [260], YOS [261], and YLF [262] were used for these demonstrations. In the case of Er:YLF pulse energies of 50 mJ in the Q-switched mode and 90 mJ in the long-pulse mode were obtained [262]. Upper-state pumped Er-doped lasers have also been operated in the intracavity pumping configuration. The Er:YAG laser was intracavity-pumped by a flashlamp-pumped Yb,Er:glass laser [263], and by a diode-pumped Yb,Er:glass laser [264,265].

More recent demonstrations of upper-state pumped Er:YAG lasers have utilized continuous-wave Yb,Er-doped fiber lasers as the pump source [266–269]. An output power of 60 W has been achieved in the CW mode [270], and a pulse

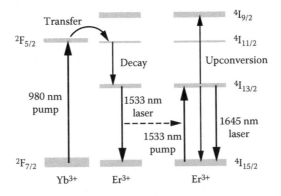

FIGURE 5.21
Energy level diagram of the upper-state pumped Er laser.

energy of 16 mJ in the Q-switched mode [271], for fiber-laser-pumped Er:YAG. Laser action in crystal hosts other than YAG has been demonstrated, including LuAG, YLF, YALO, and YVO$_4$ [272,273]. A single-frequency fiber-laser-pumped Q-switched Er:YAG laser has been utilized for coherent laser radar applications [274].

More recently, the Er:YAG laser has been directly upper-state pumped by laser diodes [275,276]. A pulse energy of 38 mJ has been achieved in the Q-switched mode [277], and 930 mJ in the long-pulse mode [278].

5.13 Summary

Lasers operating within the eyesafe wavelength band are becoming increasingly important for applications for which the laser output beam is not contained. Rare earth solid-state lasers doped with the trivalent Ho, Er, and Tm ions operate at wavelengths in the range 1.5 to 2.1 μm within the eyesafe band. Eyesafe rare earth lasers were first demonstrated shortly after the initial demonstrations of neodymium-doped 1 μm lasers in the mid-1960s. With the development of practical laser diode devices in the late 1980s, diode-pumped eyesafe rare earth lasers were demonstrated, producing increasingly higher output power levels as the performance of laser diode pump devices matured.

Several novel laser diode pumping schemes have been developed for the eyesafe rare earth lasers. The 2.1 μm Ho, 1.6 μm Er, and 2.0 μm Tm lasers are of the quasi-three-level type, in which the lower laser level resides within the ground electronic state manifold. These lasers exhibit reabsorption loss at the laser output wavelength. The sensitizer/activator pumping scheme was developed to overcome the high laser threshold imposed by reabsorption loss. Direct upper-state pumping of the 2.1 μm Ho laser and the 1.6 μm Er laser has been utilized to virtually eliminate the detrimental effects of upconversion loss in these lasers, thereby enabling energy storage at lifetimes approaching the intrinsic upper-state fluorescence lifetime. With the continued development of high-power laser diode pump devices, a range of opportunities exists for the further development of eyesafe rare earth solid-state lasers with improved performance.

References

1. Becker, P.C., Olsson, N.A., and Simpson, J.R., *Erbium-Doped Fiber Amplifiers*, Academic Press, San Diego, CA, 1999.
2. Baney, D.M., in *Fiber Optic Test and Measurement*, Ed., Derickson, D., Prentice Hall, Upper Saddle River, NJ, 1998.
3. Personal communication, Absorption spectrum measured by Greg Quarles at the Naval Research Laboratory, Washington, D.C.

4. Dieke, G.H. and Crosswhite, H.M., The spectra of the doubly and triply ionized rare earths, *Appl. Opt.* **2**, 675, 1963.
5. Landau, L.D. and Lifshitz, E.M., *Quantum Mechanics*, 3rd ed., Pergamon Press, New York, 1977.
6. Christensen, H.P., Spectroscopic analysis of $LiTmF_4$, *Phys. Rev. B* **19**, 6573, 1979.
7. NIST Atomic Spectra Database, http://physics.nist.gov.
8. Kaminskii, A.A., *Crystalline Lasers: Physical Processes and Operating Schemes*, CRC Press, Boca Raton, FL, 1996.
9. Fan, T.Y. and Byer, R.L., Modeling and CW operation of a quasi-three-level 946 nm Nd:YAG laser, *IEEE J. Quantum Electron.* **23**, 605, 1987.
10. Risk, W.P., Modeling of longitudinally pumped solid-state lasers exhibiting reabsorption losses, *J. Opt. Soc. Am. B* **5**, 1412, 1988.
11. Fan, T.Y., Optimizing the efficiency and stored energy in quasi-three-level lasers, *IEEE J. Quantum Electron.* **28**, 2692, 1992.
12. Peterson, P., Gavrielides, A., and Sharma, P.M., CW theory of a laser diode-pumped two-manifold solid state laser, *Opt. Commun.* **109**, 282, 1994.
13. Beach, R.J., Optimization of quasi-three level end-pumped Q-switched lasers, *IEEE J. Quantum Electron.* **31**, 1606, 1995.
14. Beach, R.J., CW theory of quasi-three level end-pumped laser oscillators, *Opt. Commun.* **123**, 385, 1996.
15. Rustad, G. and Stenersen, K., Modeling of laser-pumped Tm and Ho lasers accounting for upconversion and ground-state depletion, *IEEE J. Quantum Electron.* **32**, 1645, 1996.
16. Schellhorn, M. and Hirth, A., Modeling of intracavity-pumped quasi-three-level lasers, *IEEE J. Quantum Electron.* **38**, 1455, 2002.
17. Krupke, W.F. and Chase, L.L., Ground state depleted (GSD) solid state lasers: principles, characteristics, and scaling, *Proc. SPIE* **1040**, 68, 1989.
18. Beach, R., Solarz, R., Mitchell, S., Brewer, L., and Weinzapfel, S., Ground state depleted laser experiments, *Proc. SPIE* **1040**, 84, 1989.
19. Krupke, W.F. and Chase, L.L., Ground-state depleted solid-state lasers: principles, characteristics and scaling, *Opt. Quantum Electron.* **22**, S1, 1990.
20. McCumber, D.E., Theory of phonon-terminated optical masers, *Phys. Rev.* **134**, A299, 1964.
21. McCumber, D.E., Einstein relations connecting broadband emission and absorption spectra, *Phys. Rev.* **136**, A954, 1964.
22. Miniscalco, W.J. and Quimby, R.S., General procedure for the analysis of Er^{3+} cross sections, *Opt. Lett.* **16**, 258, 1991.
23. Chase, L.L., Payne, S.A., Smith, L.K., Kway, W.L., and Krupke, W.F., Emission cross sections and energy extraction for the mid-infrared transitions of Er, Tm, and Ho in oxide and fluoride crystals, *OSA Proc. Adv. Solid-State Lasers* **10**, 161, 1991.
24. Stoneman, R.C. and Esterowitz, L., Continuous-wave 1.50-μm thulium cascade laser, *Opt. Lett.* **16**, 232, 1991.
25. Payne, S.A., Chase, L.L., Smith, L.K., Kway, W.L., and Krupke, W.F., Infrared cross-section measurements for crystals doped with Er^{3+}, Tm^{3+}, and Ho^{3+}, *IEEE J. Quantum Electron.* **28**, 2619, 1992.
26. Kintz, G., Abella, I.D., and Esterowitz, L., Upconversion coefficient measurement in Tm^{3+}, Ho^{3+}:YAG at room temperature, *Proc. Int. Conf. Lasers*, 398, 1987.
27. French, V.A., Petrin, R.R., Powell, R.C., and Kokta, M., Energy-transfer processes in $Y_3Al_5O_{12}$:Tm, Ho, *Phys. Rev. B* **46**, 8018, 1992.

28. Hansson, G., Callenas, A., and Nelsson, C., Upconversion studies in laser diode pumped Tm,Ho:TLiF$_4$, *OSA Proc. Adv. Solid-State Lasers* **15**, 446, 1993.

29. Dinndorf, K.M., Cassanho, A., Yamaguchi, Y., and Jenssen, H.P., Relative upconversion rates in Tm-Ho doped crystals, *OSA Proc. Adv. Solid-State Lasers* **15**, 202, 1993.

30. Armagan, G., Walsh, B.M., Barnes, N.P., Modlin, E.A., and Buoncristiani, A.M., Determination of Tm-Ho rate coefficients from spectroscopic measurements, *OSA Proc. Adv. Solid-State Lasers* **20**, 141, 1994.

31. Dinnforf, K.M. and Jenssen, H.P., Lack of correlation between Tm,Ho upconversion measurements, *OSA TOPS Adv. Solid-State Lasers* **1**, 539, 1996.

32. Barnes, N.P., Filer, E.D., Morrison, C.A., and Lee, C.J., Ho:Tm lasers I: theoretical, *IEEE J. Quantum Electron.* **32**, 92, 1996.

33. Dinndorf, K.M. and Jenssen, H.P., Distribution of stored energy in the excited manifolds of Tm and Ho in 2 micron laser materials, *OSA Proc. Adv. Solid-State Lasers* **20**, 131, 1994.

34. Johnson, L.F., Geusic, J.E., and Van Uitert, L.G., Coherent oscillations from Tm^{3+}, Ho^{3+}, Yb^{3+} and Er^{3+} ions in yttrium aluminum garnet, *Appl. Phys. Lett.* **7**, 127, 1965.

35. Johnson, L.F., Geusic, J.E., and Van Uitert, L.G., Efficient high-power coherent emission from Ho^{3+} ions in yttrium aluminum garnet, assisted by energy transfer, *Appl. Phys. Lett.* **8**, 200, 1966.

36. Johnson, L.F., Dillon, J.F., Jr., and Remeika, J.P., Optical properties of Ho^{3+} ions in yttrium gallium garnet and yttrium iron garnet, *Phys. Rev. B* **1**, 1935, 1970.

37. Weber, M.J., Bass, M., Comperchio, E., and Riseberg, L.A., Ho^{3+} laser action in YAlO$_3$ at 2.119 μ, *IEEE J. Quantum Electron.* **QE-7**, 497, 1971.

38. Weber, M.J., Bass, M., Varitimos, T.E., and Bua, D.P., Laser action from Ho^{3+}, Er^{3+}, and Tm^{3+} in YAlO$_3$, *IEEE J. Quantum Electron.* **QE-9**, 1079, 1973.

39. Chicklis, E.P., Naiman, C.S., Folweiler, R.C., Gabbe, D.R., Jenssen, H.P., and Linz, A., High-efficiency room-temperature 2.06-μm laser using sensitized Ho^{3+} YLF, *Appl. Phys. Lett.* **19**, 119, 1971.

40. Chicklis, E.P., Naiman, C.S., Folweiler, R.C., and Doherty, J.C., Stimulated emission in multiply doped Ho^{3+} YLF and YAG: a comparison, *IEEE J. Quantum Electron.* **8**, 225, 1972.

41. Erbil, A. and Jenssen, H.P., Tunable Ho^{3+}:YLF laser at 2.06 μm, *Appl. Opt.* **19**, 1729, 1980.

42. Johnson, L.F. and Guggenheim, H.J., Electronic- and phonon-terminated laser emission from Ho^{3+} in BaY$_2$F$_8$, *IEEE J. Quantum Electron.* **QE-10**, 442, 1974.

43. Quarles, G.J., Marquardt, C.L., Esterowitz, L., High-efficiency 2.09-μm laser, *OSA Proc. Tunable Solid-State Lasers* **4**, 210, 1989.

44. Quarles, G.J., Rosenbaum, A., Marquardt, C.L., and Esterowitz, L., High-efficiency 2.09-μm flash lamp-pumped laser, *Appl. Phys. Lett.* **55**, 1062, 1989.

45. Barnes, N.P. and Cross, P., Performance of Ho:Tm:Cr:YAG and Ho:Tm:Er:YAG as a function of temperature, *OSA Proc. Tunable Solid-State Lasers* **5**, 215, 1989.

46. Bowman, S.R., Winings, M.J., Auyeung, R.C.Y., Tucker, J.E., Searles, S.K., and Feldman, B.J., Laser and spectral properties of Cr,Tm,Ho:YAG at 2.1 μm, *IEEE J. Quantum Electron.* **27**, 2142, 1991.

47. Hamlin, S.J., Myers, J.D., and Rexrode, T.R., High-efficiency, flashlamp-pumped CTH:YAG lasers operated above room temperature, *OSA Proc. Adv. Solid-State Lasers* **13**, 135, 1992.

48. Henderson, S.W., Hale, C.P., and Magee, J.R., Injection-seeded operation of a Q-switched Cr,Tm,Ho:YAG laser, *OSA Proc. Adv. Solid-State Lasers* **6**, 127, 1990.

49. Cha, S., Sugimoto, N., Chan, K., and Killinger, D.K., Tunable 2.1-μm Ho laser for DIAL remote sensing of atmospheric water vapor, *OSA Proc. Adv. Solid-State Lasers* **6**, 165, 1990.

50. Henderson, S.W., Hale, C.P., Magee, J.R., Kavaya, M.J., and Huffaker, A.V., Eye-safe coherent laser radar system at 2.1 μm using Tm,Ho:YAG lasers, *Opt. Lett.* **16**, 773, 1991.

51. Klimov, I.V., Shcherbakov, I.A., and Tsvetkov, V.B., Room temperature 2-μm laser action of Ho^{3+}-doped YSGG, GSAG, YSAG, and YAG crystals, *OSA Proc. Adv. Solid-State Lasers* **15**, 419, 1993.

52. Barnes, N.P. and Murray, K.E., Flashlamp pumped Ho:Tm:Cr:LuAG laser, *OSA Proc. Adv. Solid-State Lasers* **24**, 352, 1995.

53. Allen, R., Esterowitz, L., Goldberg, L., Weller, J.F., and Storm, M., Diode-pumped Ho:YAG laser, *Conference on Lasers and Electro-Optics*, paper FC1, OSA, 1986.

54. Allen, R., Esterowitz, L., Goldberg, L., Weller, J.F., and Storm, M., Diode-pumped 2-μm holmium laser, *Tunable Solid-State Lasers*, 144, OSA, 1986.

55. Allen, R., Esterowitz, L., Goldberg, L., Weller, J.F., and Storm, M., Diode-pumped 2-μm holmium laser, *Electron. Lett.* **22**, 947, 1986.

56. Hemmati, H., Diode laser-pumped Ho:YLF laser, *Conference on Lasers and Electro-Optics*, paper WI4, OSA, 1987.

57. Hemmati, H., Efficient holmium:yttrium lithium fluoride laser longitudinally pumped by a semiconductor laser array, *App. Phys. Lett.* **51**, 564, 1987.

58. Fan, T.Y., Huber, G., Byer, R.L., and Mitzscherlich, P., Continuous-wave diode-laser-pumped 2-μm Ho:YAG laser at room temperature, *Conference on Lasers and Electro-Optics*, paper FL1, OSA, 1987.

59. Kintz, G.J., Esterowitz, L., and Allen, R., CW diode-pumped Tm^{3+}, Ho^{3+}:YAG 2.1 μm room-temperature laser, *Electron. Lett.* **23**, 616, 1987.

60. Fan, T.Y., Huber, G., Byer, R.L., and Mitzscherlich, P., Continuous wave operation at 2.1 μm of a diode laser pumped, Tm-sensitized Ho:$Y_3Al_5O_{12}$ laser at 300 K, *Opt. Lett.* **12**, 678, 1987.

61. Fan, T.Y., Huber, G., Byer, R.L., and Mitzscherlich, P., Spectroscopy and diode laser-pumped operation of Tm,Ho:YAG, *IEEE J. Quantum Electron.* **24**, 924, 1988.

62. Huber, G., Duczynski, E.W., and Petermann, K., Laser pumping of Ho,Tm,Er-doped garnet lasers at room temperature, *Tunable Solid-State Lasers*, 18, OSA, 1987.

63. Kintz, G., Esterowitz, L., and Allen, R., Cascade laser emission at 2.31 and 2.08 μm from laser diode pumped Tm,Ho:$LiYF_4$ at room temperature, *Tunable Solid-State Lasers*, 20, OSA, 1987.

64. H. Hemmati, 2.07-μm cw diode-laser-pumped Tm,Ho:$YLiF_4$ room-temperature laser, *Opt. Lett.* **14**, 435, 1989.

65. Kane, T.J. and Wallace, R.W., Performance of a diode-pumped Tm:Ho:YAG laser at temperatures between −55 and +20°C, *Conference on Lasers and Electro-Optics*, paper FB3, OSA, 1988.

66. Kane, T.J. and Kubo, T.S., Diode-pumped single-frequency lasers and Q-switched laser using Tm:YAG and Tm,Ho:YAG, in *OSA Proc. on Adv. Solid-State Lasers* **6**, 136, 1990.

67. Bowman, S.R., Lynn, J.G., Searles, S.K., Feldman, B.J., McMahon, J., Whitney, W., and Epp, D., High-average-power operation of a Q-switched diode-pumped holmium laser, *Opt. Lett.* **18**, 1724, 1993.

68. Bowman, S.R., Lynn, J.G., Searles, S.K., Feldman, B.J., McMahon, J., Whitney, W., Marquardt, C., and Epp, D., Comparative study of diode-pumped two micron laser materials, *OSA Proc. on Advanced Solid-State Lasers* **15**, 415, 1993.
69. Heine, F., Heumann, E., Huber, G., and Schepler, K., CW mode-locking of Tm and Ho lasers, *OSA Proc. on Adv. Solid-State Lasers* **13**, 101, 1992.
70. Kintz, G.J., Allen, R., and Esterowitz, L., Two for one photon conversion observed in alexandrite pumped Tm^{3+},Ho^{3+}:YAG at room temperature, *Conference on Lasers and Electro-Optics*, postdeadline paper ThU4, OSA, 1987.
71. Petrin, R.R., Powell, R.C., Jani, M.G., Kokta, M., and Aggarwal, I.D., Comparison of laser-pumped Tm,Ho laser systems, *OSA Proc. on Adv. Solid-State Lasers* **13**, 139, 1992.
72. Huber, G., Duczynski, E.W., and Petermann, K., Laser pumping of Ho-, Tm-, Er-doped garnet lasers at room temperature, *IEEE J. Quantum Electron.* **24**, 920, 1988.
73. Kim, K.H., Choi, Y.S., Hess, R.V., Blair, C.H., Brockman, P., Barnes, N.P., Henderson, G.W., and Kokta, M.R., Experiments and theory for a Tm:Ho:YAG laser end pumped by a Cr:GSAG laser, *OSA Proc. on Adv. Solid-State Lasers* **6**, 155, 1990.
74. Filer, E.D., Barnes, N.P., Naranjo, F.L., and Kokta, M.R., Spectroscopy and lasing in $Ho:Tm:Lu_3Al_5O_{12}$, *OSA Proc. on Adv. Solid-State Lasers* **15**, 411, 1993.
75. Elder, I.F. and Payne, M.J.P., Lasing in diode-pumped thulium and thulium, holmium YAP, *OSA TOPS on Adv. Solid-State Lasers* **1**, 319, 1996. Yttrium aluminum perovskite (YAP) is an alternate name for YALO.
76. Sato, A. and Asai, K., Pulsed operation of a diode-side-pumped $Tm,Ho:GdVO_4$ laser at room temperature, *OSA TOPS on Adv. Solid-State Photonics* **83**, 298, 2003.
77. Urata, Y., Machida, H., Higuchi, M., Kodaira, K., and Wada, S., Diode-pumped $Tm,Ho:GdVO_4$ laser at room temperature, *OSA TOPS on Adv. Solid-State Photonics* **98**, 196, 2005.
78. McGuckin, B.T., Menzies, R.T., and Hemmati, H., Efficient energy extraction from a diode-pumped Q-switched $Tm,Ho:YLiF_4$ laser, *Appl. Phys. Lett.* **59**, 2926, 1991.
79. Hale, C.P., Henderson, S.W., and Suni, P.J.M., Single-longitudinal-mode and Q-switched diode-pumped Tm:Ho:YLF oscillators, in *OSA Proc. on Adv. Solid-State Lasers* **15**, 407, 1993.
80. Budni, P.A., Knights, M.G., Chicklis, E.P., and Jenssen, H.P., Performance of a diode-pumped high PRF Tm,Ho:YLF laser, *IEEE J. Quantum Electron.* **28**, 1029, 1992.
81. McCarthy, J.C., Budni, P.A., Labrie, G.W., and Chicklis, E.P., High efficiency, pulsed diode-pumped two micron laser, *OSA Proc. on Adv. Solid-State Lasers* **20**, 120, 1994.
82. Ketteridge, P., Budni, P., Lee, I., Schunemann, P., and Pollak, T., 8 micron ZGP OPO pumped at 2 microns, *OSA TOPS on Adv. Solid-State Lasers*, **1**, 168, 1996.
83. Pomeranz, L.A., Budni, P.A., Schunemann, P.G., Pollak, T.M., Ketteridge, P.A., Lee, I., and Chicklis, E.P., Efficient power scaling in the mid-IR with a $ZnGeP_2$ OPO, *OSA TOPS on Adv. Solid State Lasers* **10**, 259, 1997.
84. Finch, A. and Flint, J.H., Diode-pumped, 6-mJ, repetitively Q-switched Tm,Ho:YLF laser for clear air turbulence detection, *Conference on Lasers and Electro-Optics*, paper ThD3, OSA, 1995.
85. Flint, J.H. and Rines, D.M., High energy, diode-pumped Tm,Ho:YLF laser transmitter, *Conference on Lasers and Electro-Optics*, paper ThD4, OSA, 1995.

86. Finch, A., Flint, J.H., and Rines, D.M., 2.5-watt single-frequency CW Tm,Ho:YLF ring laser, *OSA TOPS on Adv. Solid-State Lasers* **1**, 312, 1996.

87. Jani, M., Barnes, N.P., Murray, K.E., and Lockard, G.E., Diode-pumped, long pulse length Ho:Tm:YLiF$_4$ laser at 10 Hz, *OSA Proc. on Adv. Solid-State Lasers* **24**, 362, 1995.

88. Rodriguez, W.J., Storm, M.E., and Barnes, N.P., Small signal gain of Ho:Tm:YLiF$_4$ at high pump fluences, *OSA Proc. on Adv. Solid-State Lasers* **24**, 392, 1995.

89. Lee, C.J., Han, G., and Barnes, N.P., Ho:Tm lasers II: experiments, *IEEE J. Quantum Electron.* **32**, 104, 1996.

90. Yu, J., Singh, U.N., Barnes, N.P., and Petros, M., 125-mJ diode-pumped injection-seeded Ho:Tm:YLF laser, *Opt. Lett.* **23**, 780, 1998.

91. Yu, J., Petros, M., Singh, U.N., and Barnes, N.P., An efficient end-pumped Ho:Tm:YLF disk amplifier, *OSA TOPS on Adv. Solid-State Lasers* **34**, 174 (2000).

92. Yu, J., Petros, M., Singh, U.N., Barnes, J.C., and Barnes, N.P., A high energy double pulsed Ho:Tm:YLF 2-μm laser, *OSA TOPS on Adv. Solid-State Lasers* **50**, 383, 2001.

93. Braud, A., Yu, J., Petros, M., and Barnes, N., Direct measurement of upper laser level population dynamics in a laser-diode side-pumped Ho:Tm:YLiF$_4$ 2 μm laser, *OSA TOPS on Adv. Solid-State Lasers* **68**, 240, 2002.

94. Chen, S., Yu, J., Petros, M., Singh, U.N., and Bai, Y., A double-pass diode-pumped Tm:Ho:YLF laser amplifier at 2.05 μm, *OSA TOPS on Adv. Solid-State Photonics* **83**, 309, 2003.

95. Williams-Byrd, J.A., Singh, U.N., Barnes, N.P., Lockard, G.E., Modlin, E.A., and Yu, J., Room-temperature, diode-pumped Ho:Tm:YLF laser amplifiers generating 700 mJ at 2-μm, *OSA TOPS on Adv. Solid-State Lasers* **10**, 199, 1997.

96. Singh, U.N., Yu, J., Petros, M., Barnes, N.P., Williams-Byrd, J.A., Lockard, G.E., and Modlin, E.A., Injection-seeded, room-temperature, diode-pumped Ho:Tm:YLF laser with output energy of 600 mJ at 10 Hz, *OSA TOPS on Adv. Solid State Lasers* **19**, 194, 1998.

97. Yu, J., Braud, A., Petros, M., and Singh, U.N., 600 mJ, double-pulsed Ho amplifier, *OSA TOPS on Adv. Solid-State Lasers* **68**, 236, 2002.

98. Yu, J., Braud, A., and Petros, M., 600-mJ, double-pulse 2-μm laser, *Opt. Lett.* **28**, 540, 2003.

99. Chen, S., Bai, Y., Yu, J., Trieu, B.C., Kavaya, M.J., Singh, U.N., and Petros, M., One-joule double-pulsed Ho:Tm:LuLF master-oscillator-power-amplifier (MOPA), *OSA TOPS on Adv. Solid-State Photonics* **98**, 740, 2005.

100. Yu, J., Trieu, B.C., Modlin, E.A., Singh, U.N., Kavaya, M.J., Chen, S., Bai, Y., Petzar, P., and Petros, M., One-joule-per-pulse 2-micron solid state laser, *13th Coherent Laser Radar Conference*, 36, 2005.

101. Sims, N., Jr., Cimolino, M.C., Barnes, N.P., and Asbury, B.G., 10 Hz PRF operation and temperature estimation of a conductively cooled, room temperature, diode-pumped Ho:Tm:YLF laser, *OSA Proc. on Adv. Solid-State Lasers* **24**, 396, 1995.

102. Petros, M., Yu, J., Trieu, B., Singh, U.N., Chen, S., Bai, Y., and Melak, T., High energy totally conductive cooled, diode-pumped, 2 μm laser, *OSA TOPS on Adv. Solid-State Photonics* **98**, 623, 2005.

103. Mizutani, K., Itabe, T., Ishii, S., Aoki, T., Asai, K., and Sato, A., Development of conductive cooled 2 micron lasers, *13th Coherent Laser Radar Conference*, 32, 2005.

104. Lee, C.J., Han, G., Bair, C.H., Barnes, N.P., Brockman, P., and Hess, R.V., Alexandrite laser pumped Ho:Tm:YLF laser performance, *OSA Proc. on Adv. Solid-State Lasers* **15**, 427, 1993.

105. Jani, M.G., Barnes, N.P., Murray, K.E., Hart, D.W., Quarles, G.J., and Castillo, V.K., Diode-pumped Ho:Tm:LuLiF₄ laser at room temperature, *IEEE J. Quantum Electron.* **33**, 112, 1997.

106. Sudesh, V., Asai, K., Shimamura, K., and Fukuda, T., Room-temperature Tm,Ho:LuLiF₄ laser with a novel quasi-end-pumping technique, *Opt. Lett.* **26**, 1675, 2001.

107. Sudesh, V., Shimamura, K., Fukuda, T., and Asai, K., Pulsed Tm,Ho:LuLiF₄ 2 μm laser using a novel quasi end-pump technique, *OSA TOPS on Adv. Solid-State Lasers* **50**, 571, 2001.

108. Walsh, B.M., Barnes, N.P., Yu, J., and Petros, M., A comparison of Tm:Ho:YLF and Tm:Ho:LuLF for 2.0 micrometer lasers; experiment and modeling, *OSA TOPS on Adv. Solid-State Lasers* **68**, 245 (2002).

109. Petros, M., Yu, J., Chen, S., Singh, U.N., Walsh, B. M., Bai, Y., and Barnes, N.P., Diode pumped 135 mJ Ho:Tm:LuLF oscillator, *OSA TOPS on Adv. Solid-State Photonics* **83**, 315, 2003.

110. Sudesh, V., Asai, K., Sato, A., Singh, U.N., Walsh, B.M., and Barnes, N.P., Continuous-wave diode-pumped Ho,Tm:LuLF laser at room temperature, *OSA TOPS on Adv. Solid-State Photonics* **94**, 339, 2004.

111. Chen, S., Yu, J., Petros, M., Singh, U.N., Bai, Y., Barnes, N.P., and Trieu, B.C., Diode-pumped Ho:Tm:LuLF laser oscillator and laser amplifier at 2 μm, *OSA TOPS on Adv. Solid-State Photonics* **94**, 344, 2004.

112. Dinndorf, K.M., Miller, H., Tabirian, A., Jenssen, H.P., and Cassanho, A., Two micron diode-pumped laser operation of Tm,Ho:BaY₂F₈, *OSA TOPS on Adv. Solid-State Lasers* **26**, 506, 1999.

113. Galzerano, G., Taccheo, S., Laporta, P., Sani, E., Toncelli, A., and Tonelli, M., Single-mode diode-pumped Tm-Ho:KYF₄ laser widely tunable around 2 μm, *OSA TOPS on Adv. Solid-State Photonics* **94**, 350, 2004.

114. Voron'ko, Y.K., Gessen, S.B., Es'kov, N.A., Kiryukhin, A.A., and Ryabochkina, P.A., Interaction of Tm³⁺ ions in calcium-niobium-gallium and yttrium-aluminum garnet laser crystals, *Quantum Electron.* **23**, 958, 1994.

115. Quarles, G.J., Rosenbaum, A., Marquardt, C.L., Esterowitz, L., Efficient room-temperature operation of a flash-lamp-pumped Cr,Tm:YAG laser at 2.01 μm, *Opt. Lett.* **15**, 42, 1990.

116. Quarles, G.J., Marquardt, C.L., and Esterowitz, L., Efficient room-temperature operation of a flashlamp-pumped Cr;Tm:YAG laser at 2.014 μm, *OSA Proc. on Adv. Solid-State Lasers* **6**, 150, 1990.

117. Pinto, J.F. and Esterowitz, L., Tunable, flashlamp-pumped operation of a Cr;Tm:YAG laser between 1.945 and 2.014 μm, *OSA Proc. on Adv. Solid-State Lasers* **6**, 134, 1990.

118. Quarles, G.J., Pinto, J.F., and Esterowitz, L., Broad tunability of flash-pumped, Tm-activated garnet lasers, *OSA Proc. on Adv.Solid-State Lasers* **10**, 167, 1991.

119. Bowman, S.R., Quarles, G.J., and Feldman, B.J., Upconversion losses in flashlamp-pumped Cr,Tm:YAG, *OSA Proc. on Adv. Solid-State Lasers* **13**, 169, 1992.

120. Kintz, G.J., Allen, R., and Esterowitz, L., Continuous-wave laser emission at 2.02 μm from diode-pumped Tm³⁺:YAG at room temperature, *Conference on Lasers and Electro-Optics*, paper FB2, OSA, 1988.

121. Suni, P.J.M. and Henderson, S.W., 1-mJ/pulse Tm:YAG laser pumped by a 3-W diode laser, *Opt. Lett.* **16**, 817, 1991.

122. Kmetec, J.D., Kubo, T.S., Kane, T.J., and Grund, C.J., Laser performance of diode-pumped thulium-doped $Y_3Al_5O_{12}$, $(Y, Lu)_3Al_5O_{12}$, and $Lu_3Al_5O_{12}$ crystals, *Opt. Lett.* **19**, 186, 1994.

123. Becker, T., Clausen, R., Huber, G., Duczynski, E.W., and Mitzscherlich, P., Spectroscopic and laser properties of Tm-doped YAG at 2 μm, *OSA Proc. on Tunable Solid-State Lasers* **5**, 150, 1989.

124. Stoneman, R.C. and Esterowitz, L., Room-temperature cw Tm^{3+}:YAG anf Tm^{3+}:YSGG lasers continuously tunable over the range 1.85-2.16 μm, *OSA Proc. on Tunable Solid-State Lasers* **5**, 157, 1989.

125. Stoneman, R.C. and Esterowitz, L., Efficient, broadly tunable, laser-pumped Tm:YAG and Tm:YSGG cw lasers, *Opt. Lett.* **15**, 486, 1990.

126. Pinto, J.F., Esterowitz, L., and Rosenblatt, G.H., Continuous-wave mode-locked 2-μm Tm:YAG laser, *Opt. Lett.* **17**, 731, 1992.

127. Pinto, J.F., Esterowitz, L., and Rosenblatt, G., CW mode-locked 2-μm laser, *OSA Proc. on Adv. Solid-State Lasers* **13**, 91, 1002.

128. Phua, P.B., Lai, K.S., Wu, R.F., Lim, Y.L., and Lau, E., Room-temperature operation of a multiwatt Tm:YAG laser pumped by a 1-μm Nd:YAG laser, *Opt. Lett.* **25**, 619, 2000.

129. Bollig, C., Clarkson, W.A., Hayward, R.A., and Hanna, D.C., Efficient high-power Tm:YAG laser at 2 μm, end-pumped by a diode bar, *Opt. Commun.* **154**, 35, 1998.

130. Berner, N., Diening, A., Heumann, E., Huber, G., Voss, A., Karszewski, M., and Giesen, A., Tm:YAG: A comparison between endpumped laser-rods and the "thin-disk"-setup, *OSA TOPS on Advanced Solid-State Lasers* **26**, 463, 1999.

131. Beach, R.J., Sutton, S.B., Honea, E.C., Skidmore, J.A., and Emanuel, M.A., High power 2-μm diode-pumped Tm:YAG laser, *Proc. SPIE* **2698**, 168, 1996.

132. Beach, R.J., Sutton, S.B., Honea, E.C., Skidmore, J.A., and Emanuel, M.A., High power 2-μm wing-pumped Tm^{3+}:YAG laser, *OSA TOPS on Adv. Solid-State Lasers* **1**, 213, 1996.

133. Beach, R.J., Sutton, S.B., Skidmore, J.A., and Emanuel, M.A., High-power 2-μm wing-pumped Tm:YAG laser, *Conference on Lasers and Electro-Optics*, paper CWN1, OSA, 1996.

134. Honea, E.C., Beach, R.J., Sutton, S.B., Speth, J.A., Mitchell, S.C., Skidmore, J.A., Emanuel, M.A., and Payne, S.A., 115-W Tm:YAG diode-pumped solid-state laser, *IEEE J. Quantum Electron.* **33**, 1592, 1997.

135. Honea, E.C., Beach, R.J., Sutton, S.B., Speth, J.A., Mitchell, S.C., Skidmore, J.A., Emanuel, M.A., and Payne, S.A., 115 W Tm:YAG CW diode-pumped solid-state laser, *OSA TOPS on Adv. Solid-State Lasers* **10**, 307, 1997.

136. Lai, K.S., Phua, P.B., Wu, R.F., Lim, Y.L., Lau, E., Toh, S.W., Toh, B.T., and Chang, A., 120-W continuous-wave diode-pumped Tm:YAG laser, *Opt. Lett.* **25**, 1591, 2000.

137. Lai, K.S., Xie, W.J., Phua, P.B., Wu, R.F., Lim, Y.L., Lau, E., and Chang, A. Greater than 100 W diode-pumped Tm:YAG laser and its thermal issues, *OSA TOPS on Adv. Solid-State Lasers* **50**, 603, 2001.

138. Lai, K.S., Xie, W.J., Wu, R.F., Lim, Y.L., Lau, E., Chia, L., and Phua, P.B., A 150 W 2-micron diode-pumped Tm:YAG laser, *OSA TOPS on Adv. Solid-State Lasers* **68**, 535, 2002.

139. Ivanov, A.O., Mochalov, I.V., Tkachuk, A.M., Fedorov, V.A., and Feofilov, P.P., Spectral characteristics of the thulium ion and cascade generation of stimulated radiation in a YalO$_3$:Tu^{3+}:Cr^{3+} crystal, *Sov. J. Quant. Electron.* **5**, 117, 1975.
140. Esterowitz, L. and Stoneman, R.C., Comparative studies of diode-pumped two-micron laser materials, in *Novel Laser Sources and Applications*, 38, SPIE Press, Bellingham, WA, 1993.
141. Stoneman, R.C. and Esterowitz, L., Efficient 1.94-μm Tm:YALO laser, *IEEE J. Sel. Top. Quantum Electron.* **1**, 78, 1995.
142. Elder, I.F. and Payne, M.J.P., Characterization of a CW diode-pumped Tm:YAlO$_3$ laser, *OSA Proc. on Adv. Solid-State Lasers* **24**, 358, 1995.
143. Elder, I.F. and Payne, M.J.P., Lasing in diode-pumped thulium and thulium, holmium YAP, *OSA TOPS on Adv. Solid-State Lasers* **1**, 319, 1996.
144. Elder, I.F. and Payne, M.J.P., Comparison of diode-pumped Tm:YAP with Tm:YAG, *OSA TOPS on Adv. Solid-State Lasers* **19**, 212, 1998.
145. Kalaycioglu, H., Sennaroglu, A., and Kurt, A., Comparative investigation of diode-pumped Tm^{3+}:YAlO$_3$ lasers: influence of doping concentrations, *OSA TOPS on Adv. Solid-State Photonics* **98**, 208, 2005.
146. Pomeranz, L.A., Budni, P.A., Lemons, M.L., Miller, C.A., Mosto, J.R., Pollak, T.M., and Chicklis, E.P., Power scaling performance of Tm:YLF and Tm:YALO lasers, *OSA TOPS on Adv. Solid-State Lasers* **26**, 458, 1999.
147. Pomeranz, L.A., Ketteridge, P.A., Budni, P.A., Ezzo, K.M., Rines, D.M., and Chicklis, E.P., Tm:YAlO$_3$ laser pumped ZGP mid-IR source, *OSA TOPS on Adv. Solid-State Photonics* **83**, 142, 2003.
148. Sullivan, A.C., Zakel, A., Wagner, G.J., Gwin, D., Tiemann, B., Stoneman, R.C., and Malm, A.I.R., High power Q-switched Tm:YALO lasers, *OSA TOPS on Adv. Solid-State Photonics* **94**, 329, 2004.
149. Saito, H., Chaddha, S., Chang, R.S.F., and Djeu, N., Efficient 1.94-μm Tm^{3+} laser in YVO$_4$ host, *Opt. Lett.* **17**, 189, 1992.
150. Hauglie-Hanssen, C. and Djeu, N., Further investigations of a 2-μm Tm:YVO$_4$ laser, *IEEE J. Quantum Electron.* **30**, 275, 1994.
151. Zayhowski, J.J., Harrison, J., Dill, C., III, and Ochoa, J., Tm:YVO$_4$ microchip laser, *Appl. Opt.* **34**, 435, 1995.
152. Zagumennyi, A.I., Zavartsev, Y.D., Mikhailov, V.A., Studenikin, P.A., Shcherba-kov, I.A., Ostroumov, V.G., Heumann, E., and Huber, G., New high efficient Tm:GdVO$_4$ diode pumped microchip laser, *OSA TOPS on Adv. Solid-State Lasers* **10**, 205, 1997.
153. Mikhailov, V.A., Zavartsev, Y.D., Zagumennyi, A.I., Ostroumov, V.G., Studeni-kin, P.A., Heumann, E., Huber, G., and Shcherbakov, I.A., Tm^{3+}:GdVO$_4$ — a new efficient medium for diode-pumped 2-μm lasers, *Quantum Electron.* **27**, 13, 1997.
154. Sorokin, E., Alpatiev, A.N., Sorokina, I.T., Zagumennyi, A.I., and Shcherbakov, I.A., Tunable efficient continuous-wave room-temperature Tm^{3+}:GdVO$_4$ laser, *OSA TOPS on Adv. Solid-State Lasers* **68**, 347, 2002.
155. Troshin, A.E., Kisel, V.E., Shcherbitsky, V.G., Kuleshov, N.V., Pavlyuk, A.A., Dunina, E.B., and Kornienko, A.A., Laser performance of Tm:KY(WO$_4$)$_2$ crystal, *OSA TOPS on Adv. Solid-State Photonics* **98**, 214, 2005.
156. Petrov, V., Griebner, Guell, F. Massons, J., Gavalda, J., Sole, R., Aguilo, M., and Diaz, F., Tunable cw lasing of Tm:KGd(WO$_4$)$_2$ near 2 μm, *OSA TOPS on Adv. Solid-State Photonics* **98**, 224, 2005.

157. Diening, A., Dicks, B.-M., Heumann, E., Meyn, J.-P., Petermann, K., and Huber, G., Continuous-wave lasing of Tm^{3+} doped T_2O_3 near 1.95 μm, *OSA TOPS on Adv. Solid State Lasers* **10**, 194, 1997.

158. Ermeneux, F.S., Sun, Y., Cone, R.L., Equall, R.W., Hutcheson, R.L., and Morcorge, R., Efficient CW 2 μm Tm^{3+}:Y_2O_3 laser, *OSA TOPS on Adv. Solid-State Lasers* **26**, 497, 1999.

159. Fornasiero, L., Berner, N., Dicks, B.-M., Mix, E., Peters, V., Petermann, K., and Huber, G., Broadly tunable laser emission from Tm:Y_2O_3 and Tm:Sc_2O_3 at 2 μm, *OSA TOPS on Adv. Solid-State Lasers* **26**, 450, 1999.

160. Ketteridge, P.A., Budni, P.A., Knights, M.G., and Chicklis, E.P., An all solid-state 7 watt CW, tunable Tm:YLF laser, *OSA TOPS on Adv. Solid State Lasers* **10**, 197, 1997.

161. Dergachev, A., Wall, K., and Moulton, P.F., A CW side-pumped Tm:YLF laser, *OSA TOPS on Adv. Solid-State Lasers* **68**, 343, 2002.

162. Mackenzie, J.I., So, S., Shepherd, D.P., and Clarkson, W.A., Comparison of laser performance for diode-pumped Tm:YLF of various doping concentrations, *OSA TOPS on Adv. Solid-State Photonics* **98**, 202, 2005.

163. Cornacchia, F., Parisi, D., Sani, E., Toncelli, A., and Tonelli, M., Comparative analysis of the 2 μm emission in Tm^{3+}:BaY_2F_8 and Tm^{3+}:KYF_4: spectroscopy and laser experiment, *OSA TOPS on Adv. Solid-State Photonics* **98**, 219, 2005.

164. Galzerano, G., Taccheo, S., Laporta, P., Cornacchia, F., Parisi, D., Sani, E., Toncelli, A., and Tonelli, M., Widely tunable 2 μm Tm:BaY_2F_8 vibronic laser, *OSA TOPS on Adv. Solid-State Photonics* **98**, 229, 2005.

165. Shaw, L.B., Chang, R.S.F., and Djeu, N., Measurement of up-conversion energy-transfer probabilities in Ho:$Y_3Al_5O_{12}$ and Tm:$Y_3Al_5O_{12}$, *Phys. Rev. B* **50**, 6609, 1994.

166. Barnes, N.P., Walsh, B.M., and Filer, E.D., Ho:Ho upconversion: applications to Ho lasers, *J. Opt. Soc. Am. B* **20**, 1212, 2003.

167. Bowman, S.R. and Feldman, B.J., Demonstration and analysis of a holmium quasi-two level laser, *Proc. SPIE* **1627**, 46, 1992.

168. Bowman, S.R. and Feldman, B.J., Demonstration and analysis of a holmium quasi-two-level laser, *Conference on Lasers and Electro-Optics*, paper CWG40, 1992.

169. Stoneman, R.C. and Esterowitz, L., Intracavity-pumped 2.1-μm Ho^{3+}:YAG laser, *OSA Proc. on Adv. Solid-State Lasers* **13**, 114, 1992.

170. Stoneman, R.C. and Esterowitz, L., Intracavity-pumped 2.09-μm Ho:YAG laser, *Opt. Lett.* **17**, 114, 1992.

171. Nabors, C.D., Fan, T.Y., Choi, H.K., Turner, G.W., and Eglash, S.J., Holmium laser pumped by 1.9-μm diode laser, *Conference of Lasers and Electro-Optics*, postdeadline paper CPD8, OSA, 1993.

172. Nabors, C.D., Ochoa, J., Fan, T.Y., Sanchez, A., Choi, H., Turner, G., 1.9-μm-diode-laser-pumped, 2.1-μm Ho:YAG laser, *Conference on Lasers and Electro-Optics*, paper CTuU1, OSA, 1994.

173. Nabors, C.D., Ochoa, J., Fan, T.Y., Sanchez, A., Choi, H.K., and Turner, G.W., Ho:YAG laser pumped by 1.9-μm diode lasers, *IEEE J. Quantum Electron.* **31**, 1603, 1995.

174. Bollig, C., Hayward, R.A., Kern, M., Clarkson, W.A., and Hanna, D.C., High-power operation of an intracavity-pumped Ho:YAG laser at 2.1 μm, *Conference on Lasers and Electro-Optics*, paper CTuE4, OSA, 1997.

175. Bollig, C., Hayward, R.A., Clarkson, W.A., and Hanna, D.C., 2-W Ho:YAG laser intracavity pumped by a diode-pumped Tm:YAG laser, *Opt. Lett.* **23**, 1757, 1998.

176. Hayward, R.A., Clarkson, W.A., and Hanna, D.C., High-power diode-pumped room-temperature Tm:YAG and intracavity-pumped Ho:YAG lasers, *OSA Proc. on Adv. Solid State Lasers* **34**, 90, 2000.

177. Schellhorn, M., Hirth, A., and Kieleck, C., Ho:YAG laser intracavity pumped by a diode-pumped Tm:YLF laser, *Opt. Lett.* **28**, 1933, 2003.

178. Budni, P.A., Pomeranz, L.A., Miller, C.A., Dygan, B.K., Lemons, M.L., and Chicklis, E.P., CW and Q-switched Ho:YAG pumped by Tm:YALO, *OSA TOPS on Adv. Solid State Lasers* **19**, 204, 1998.

179. Budni, P.A., Pomeranz, L.A., Lemons, M.L., Schunemann, P.G., Pollak, T.M., Mosto, J.R., and Chicklis, E.P., Mid-IR laser based on ZnGeP$_2$ and unsensitized Ho:YAG, *OSA Proc. on Adv. Solid-State Lasers* **26**, 454, 1999.

180. Budni, P.A., Pomeranz, L.A., Lemons, M.L., Miller, C.A., Mosto, J.R., and Chicklis, E.P., Efficient mid-infrared laser using 1.9-μm-pumped Ho:YAG and ZnGeP$_2$ optical parametric oscillators, *J. Opt. Soc. Am. B* **17**, 723, 2000.

181. Budni, P.A., Lemons, M.L., Mosto, J.R., and Chicklis, E.P., High-power/high-brightness diode-pumped 1.9-μm thulium and resonantly pumped 2.1-μm holmium lasers, *IEEE J. Sel. Topics Quantum Electron.* **6**, 629, 2000.

182. Budni, P.A., Ibach, C.R., Setzler, S.D., Gustafson, E.J., Castro, R.T., and Chicklis, E.P., 50-mJ, Q-switched, 2.09-μm holmium laser resonantly pumped by a diode-pumped 1.9-μm thulium laser, *Opt. Lett.* **28**, 1016, 2003.

183. Petros, M., Yu, J., Singh, U.N., and Barnes, N.P., High energy directly pumped Ho:YLF laser, *OSA TOPS on Adv. Solid-State Lasers* **34**, 178, 2000.

184. Dergachev, A. and Moulton, P.F., High-power, high-energy diode-pumped Tm:YLF-Ho:YLF-ZGP laser system, *OSA TOPS on Adv. Solid-State Photonics* **83**, 137, 2003.

185. Barnes, N.P., Clarkson, W.A., Hanna, D.C., Matera, V., and Walsh, B.M., Tm:glass fiber laser pumping Ho:YAG and Ho:LuAG, *Conference on Lasers and Electro-Optics*, paper CThV3, OSA, 2001.

186. Barnes, N.P. and Walsh, B.M., Tm:glass laser pumping a Ho:YAG laser, *Proc. SPIE* **4893**, 176, 2003.

187. Abdolvand, A., Shen, D.Y., Cooper, L.J., Williams, R.B., and Clarkson, W.A., Ultra-efficient Ho:YAG laser end-pumped by a cladding-pumped Tm-doped silica fiber laser, *OSA TOPS on Adv. Solid-State Photonics* **83**, 7, 2003.

188. Shen, D.Y., Abdolvand, A., Cooper, L.J., and Clarkson, W.A., Efficient Ho:YAG laser pumped by a cladding-pumped tunable Tm:silica-fiber laser, *Appl. Phys. B* **79**, 559, 2004.

189. Shen, D., Sahu, J., and Clarkson, W.A., Efficient holmium-doped solid-state lasers pumped by a Tm-doped silica fiber laser, *Proc. SPIE* **5620**, 46, 2004.

190. Lippert, E., Arisholm, G. Rustad, G., and Stenersen, K., Fiber laser pumped mid-IR source, *OSA TOPS on Adv. Solid-State Photonics* **83**, 292, 2003.

191. Shen, D.Y., Clarkson, W.A., Cooper, L.J., and Williams, R.B., Efficient single-axial-mode operation of a Ho:YAG ring laser pumped by a Tm-doped silica fiber laser, *Opt. Lett.* **29**, 2396, 2004.

192. Shen, D.Y., Clarkson, W.A., Cooper, L.J., and Williams, R.B., 3.7-Watt single-frequency cw Ho:YAG ring laser end-pumped by a cladding-pumped Tm-doped silica fiber laser, *OSA TOPS on Adv. Solid-State Photonics* **94**, 362, 2004.

193. Dergachev, A., Moulton, P.F., and Drake, T.E., High-power, high-energy Ho:YLF laser pumped with a Tm: fiber laser, *OSA TOPS on Adv. Solid-State Photonics* **98**, 608, 2005.

194. Jarman, R.H., Wallenberg, A.J., Bennett, K.W., and Anthon, D.W., Effects of cerium doping on energy transfer in Yb,Er lasers, *OSA Proc. on Adv. Solid-State Lasers* **20**, 160, 1994.

195. Zhekov, V.I., Lobachev, V.A., Murina, T.M., and Prokhorov, A.M., Cooperative phenomena in yttrium erbium aluminum garnet crystals, *Sov. J. Quantum Electron.* **14**, 128, 1984.

196. Zhekov, V.I., Murina, T.M., Prokhorov, A.M., Studenikin, M.I., Georgescu, S., Lupei, V., and Ursu, I., Cooperative process in $Y_3Al_5O_{12}$:Er^{3+} crystals, *Sov. J. Quantum Electron.* **16**, 274, 1986.

197. Georgescu, S., Lupei, V., Lupei, A., Zhekov, V.I., Murina, T.M., and Studenikin, M.I., Concentration effects on the up-conversion from the $^4I_{13/2}$ level of Er^{3+} in YAG, *Opt. Commun.* **81**, 186, 1991.

198. Spariosu, K., Birnbaum, M., and Viana, B., Er^{3+}:$Y_3Al_5O_{12}$ laser dynamics: effects of upconversion, *J. Opt. Soc. Am. B* **11**, 894, 1994.

199. Spariosu, K., Birnbaum, M., and Viana, B., Upconversion effects on Er:YAG laser dynamics, *Proc. SPIE* **2115**, 45, 1994.

200. Wang, J. and Simkin, D.J., Energy transfer within a regular distribution of donors and acceptors: application to the upconversion dynamics of Er^{3+}:$YAlO_3$ and Er^{3+}:YAG, *Phys. Rev.* **52**, 3309, 1995.

201. Huang, X., Cutinha, N., Alcazar de Velasco, A., Chandler, P.J., and Townsend, P.D., Upconversion in erbium doped YAG ion-implanted waveguides, *Nucl. Instr. Meth. Phys. Res. B* **142**, 50, 1998.

202. Lacovara, P., Energy transfer and upconversion in Yb:Er:YAG, *OSA Proc. on Adv. Solid-State Lasers* **13**, 296, 1992.

203. Snitzer, E. and Woodcock, R., Yb^{3+}-Er^{3+} glass laser, *Appl. Phys. Lett.* **6**, 45, 1965.

204. Murzin, A.G. and Fromzel, V.A., Maximum gains of laser-pumped glasses activated with Yb^{3+} and Er^{3+} ions, *Sov. J. Quantum Electron.* **11**, 304, 1981.

205. Murzin, A.G., Prilezhaev, D.S., and Fromzel, V.A., Some features of laser excitation of ytterbium-erbium glasses, *Sov. J. Quantum Electron.* **15**, 349, 1985.

206. Gapontsev, V.P., Zhabotinskii, M.E., Izyneev, A.A., Kravchenko, V.B., and Rudnitskii, Y.P., Effective $1.054 \rightarrow 1.54 \, \mu$ stimulated emission conversion, *JETP Lett.* **18**, 251, 1973.

207. Kalinin, V.N., Mak, A.A., Prilezhaev, D.S., and Fromzel, V.A., Lasing properties of Yb^{3+} and Er^{3+} glass laser pumped by a laser, *Sov. Phys. Tech. Phys.* **19**, 835, 1974.

208. Hanna, D.C., Kazer, A., and Shepherd, D.P., A 1.54 μm Er glass laser pumped by a 1.064 μm Nd:YAG laser, *Opt. Commun.* **63**, 417, 1987.

209. Gapontsev, V.P., Matitsin, S.M., Isineev, A.A., and Kravchenko, V.B., Erbium glass lasers and their applications, *Opt. Laser Technol.* **14**, 189, 1982.

210. Hamlin, S.J., Myers, J.D., and Myers, M.J., High repetition rate Q-switched erbium glass lasers, *Proc. SPIE* **1419**, 100, 1991.

211. Tilleman, M.M., Jackel, S., and Moshe, I., High-power free-running eye-safe laser based on a high-strength Cr:Yb:Er:glass rod, *OSA TOPS on Adv. Solid State Lasers* **19**, 162, 1998.

212. Asaba, K., Hosokawa, T., Hatsuda, Y., and Ota, J., Development of 1.54 μm near-infrared Q-switched laser, *Proc. SPIE* **1207**, 164, 1990.

213. McGrath, A.J., Munch, J., Smith, G., and Veitch, P., Injection-seeded, single-frequency, Q-switched erbium: glass laser for remote sensing, *Appl. Opt.* **37**, 5706, 1998.

214. Stultz, R.D., Camargo, M.B., Birnbaum, M., and Kokta, M., Divalent uranium and cobalt saturable absorber Q-switches at 1.5 μm, *OSA Proc. on Adv. Solid-State Lasers* **24**, 460, 1995.

215. Wu, R., Rhonehouse, D., Myers, M.J., Hamlin, S.J., and Myers, J.D., Spectral bleaching and 1535 nm Q-switching of uranium glass, *OSA Proc. on Adv. Solid-State Lasers* **24**, 440, 1995.

216. Birnbaum, M., Camargo, M.B., Lee, S., Unlu, F., and Stultz, R.D., Co^{2+}:ZnSe saturable absorber Q-switch for the 1.54 μm Er^{3+}:Yb^{3+}:glass laser, *OSA TOPS on Adv. Solid-State Lasers* **10**, 148, 1997.

217. Denker, B., Galagan, B., Godovikova, E., Meilman, M., Osiko, V., Sverchkov, S., and Kertesz, I., The efficient saturable absorber for 1.54 μm Er glass lasers, *OSA TOPS on Adv. Solid-State Lasers* **26**, 618, 1999.

218. Laporta, P., Longhi, S., Taccheo, S., and Svelto, O., Analysis and modeling of the erbium-ytterbium glass laser, *Opt. Commun.* **100**, 311, 1993.

219. Laporta, P., De Silvestri, S., Magni, V., and Svelto, O., Diode-pumped cw bulk Er:Yb:glass laser, *Opt. Lett.* **16**, 1952, 1991.

220. Laporta, P., Taccheo, S., and Svelto, O., High-power and high-efficiency diode-pumped Er:Yb:glass laser, *Electron. Lett.* **28**, 490, 1992.

221. Laporta, P., Longhi, S., Taccheo, S., and Svelto, O., Single-mode cw erbium-ytterbium glass laser at 1.5 μm, *Opt. Lett.* **18**, 31, 1993.

222. Laporta, P., Taccheo, S., Longhi, S., and Svelto, O., Diode-pumped microchip Er-Yb:glass laser, *Opt. Lett.* **18**, 1232, 1993.

223. Thony, P. and Molva, E., 1.55 μm-wavelength CW microchip lasers, *OSA TOPS on Adv. Solid-State Lasers* **1**, 296, 1996.

224. Taccheo, S., Laporta, P., and Svelto, C., Widely tunable single-frequency erbium-ytterbium phosphate glass laser, *Appl. Phys. Lett.* **68**, 2621, 1996.

225. Taccheo, S., Laporta, P., and Svelto, O., Linearly polarized, single-frequency, widely tunable Er:Yb bulk laser at around 1550 nm wavelength, *Appl. Phys. Lett.* **69**, 3128, 1996.

226. Taccheo, S., Laporta, P., Longhi, S., Svelto, O., and Svelto, C., Diode-pumped bulk erbium-ytterbium lasers, *Appl. Phys. B* **63**, 425, 1996.

227. Taccheo, S., Sorbello, G., Laporta, P., Karlsson, G., and Laurell, F., 230-mW diode-pumped single-frequency Er:Yb laser at 1.5 μm, *IEEE Photonics Technol. Lett.* **13**, 19, 2001.

228. Tanguy, E., Larat, C., and Pocholle, J.P., Modeling of the erbium-ytterbium laser, *Opt. Commun.* **153**, 172, 1998.

229. Labranche, B., Levesque, M., Morin, M., Taillon, Y., Snell, K., Galarneau, P., Hart, D., Dreze, C., and Mathieu, P., Side-pumped eyesafe laser, *OSA Proc. on Adv. Solid-State Lasers* **20**, 151, 1994.

230. Georgiou, E., Musset, O., and Boquillon, J.P., Diode-pumped bulk Er:Yb:glass 1.54 μm pulsed laser with 1.2J output and 25% optical efficiency, *OSA TOPS on Adv. Solid-State Lasers* **68**, 226, 2002.

231. Georgiou, E., Musset, O., and Boquillon, J.-P., Free-running and Q-switched operation of a transversely diode-pumped Er:Yb:glass 1.54-μm laser with high pulse energy and efficiency, *OSA TOPS on Adv. Solid State Lasers* **34**, 194, 2000.

232. Danger, T., Huber, G., Petermann, K., and Seeber, W., Dependence of the 1.6 μm laser performance on the composition of Yb,Er-doped fluoride phosphate glasses, *OSA TOPS on Adv. Solid State Lasers* **19**, 465, 1998.

233. Mond, M., Diening, A., Heumann, E., and Huber, G., Diode-endpumped 1.5 μm Er,Yb:glass laser, *OSA TOPS on Adv. Solid State Lasers* **34**, 212, 2000.

234. Tanguy, E., Pocholle, J.P., Feugnet, G., Larat, C., Schwarz, M., Brun, A., and Georges, P., Mechanically Q-switched codoped Er-Yb glass laser under Ti:sapphire and laser diode pumping, *Electron. Lett.* **31**, 458, 1995.

235. Denker, B.I., Korchagin, A.A., Osiko, V.V., Sverchkov, S.E., Allik, T.H., and Hutchinson, J.A., Diode-pumped and FTIR Q-switched laser performance of novel Yb-Er glass, *OSA Proc. on Adv. Solid-State Lasers* **20**, 148, 1994.

236. Labranche, B., Mailloux, A., Levesque, M., Taillon, Y., Morin, M., and Mathieu, P., Q-switched side-pumped eyesafe laser, *OSA Proc. on Adv. Solid-State Lasers* **24**, 379, 1995.

237. Wu, R., Myers, J.D., and Hamlin, S.J., Comparative results of diode pumped Er:glass lasers Q-switched with BBO Pockels cell and FTIR methods, *OSA TOPS on Adv. Solid State Lasers* **19**, 159, 1998.

238. Wu, R., Myers, J.D., Myers, M.J., and Wisnewski, T., 50 Hz diode pumped Er:glass eye-safe laser," *OSA TOPS on Adv. Solid-State Lasers* **26**, 336, 1999.

239. Fluck, R., Keller, U., Gini, E., and Melchior, H., Eyesafe pulsed microchip laser, *OSA TOPS on Adv. Solid State Lasers* **19**, 146, 1998.

240. Stultz, R.D., Sumida, D.S., and Bruesselbach, H., Diode-pumped, passively Q-switched, 10 Hz eyesafe Er:Yb:glass laser, *OSA TOPS on Adv. Solid-State Lasers* **1**, 448, 1996.

241. Stultz, R.D., Camargo, M.B., Lawler, M., Rockafellow, D., and Birnbaum, M., Diode-pumped Er:Yb:glass mini-transmitter, *OSA TOPS on Adv. Solid State Lasers* **19**, 155, 1998.

242. Thony, P., Ferrand, B., and Molva, E., 1.55 μm passive Q-switched microchip laser, *OSA TOPS on Adv. Solid State Lasers* **19**, 150, 1998.

243. Karlsson, G., Pasiskevicius, V., Laurell, F., Tellefsen, J.A., Denker, B., Galagan, B.I., Osiko, V.V., and Sverchkov, S., Diode-pumped Er-Yb:glass laser passively Q switched by use of $Co^{2+}:MgAl_2O_4$ as a saturable absorber, *Appl. Opt.* **39**, 6188, 2000.

244. Denker, B., Galagan, B., Osiko, V., Sverchkov, S., Karlsson, G., Laurell, F., and Tellefsen, J., Yb-Er laser glass for high average power diode-pumped 1.54 μm lasers, *OSA TOPS on Adv. Solid-State Lasers* **68**, 232, 2002.

245. Hamlin, S.J., Hays, A.D., Trussell, C.W., and King, V., Eyesafe erbium glass microlaser, *Proc. SPIE* **5332**, 97, 2004.

246. Yanagisawa, T., Asaka, K., and Hirano, Y., 1.5-μm coherent lidar using a single-longitudinal-mode diode-pumped Q-switched Er,Yb:glass laser, *Conference on Lasers and Electro-Optics*, paper CThJ4, OSA, 2000.

247. Yanagisawa, T., Asaka, K., Hamazu, K., and Hirano, Y., 11-mJ, 15-Hz single-frequency diode-pumped Q-switched Er,Yb:phosphate glass laser, *Opt. Lett.* **26**, 1262, 2001.

248. White, K.O. and Schleusener, S.A., Coincidence of Er:YAG laser emission with methane absorption at 1645.1 nm, *Appl. Phys. Lett.* **21**, 419, 1972.

249. Duczynski, E.W., Huber, G., Petermann, K., and Stange, H., Continuous wave 1600 nm laser action in Er-doped garnets, *OSA Proc. on Tunable Solid-State Lasers*, 197, 1987.

250. Stange, H., Petermann, K., Huber, G., and Duczynski, E.W., Continuous wave 1.6 μm laser action in Er doped garnets at room temperature, *Appl. Phys. B* **49**, 269, 1989.

251. Li, C., Moncorge, R., Souriau, J.C., Borel, C., and Wyon, C., Room temperature cw laser action of Y_2SiO_5:Yb^{3+}, Er^{3+} at 1.57 μm, *Opt. Commun.* **107**, 61, 1994.

252. Shepherd, D.P., Hanna, D.C., Large, A.C., Tropper, A.C., Warburton, T.J., Borel, C., Ferrand, B., Pelenc, D., Rameix, A., Thony, P., Auzel, F., and Meichenin, D., A low threshold, room temperature 1.64 μm Yb:Er:$Y_3Al_5O_{12}$ waveguide laser, *J. Appl. Phys.* **76**, 7651, 1994.

253. Simondi-Teisseire, B., Viana, B., Lejus, A.-M., Benitez, J.-M., Vivien, D., Borel, C., Templier, R., and Wyon, C., Room-temperature CW laser operation at ~1.55 μm (eye-safe range) of Yb:Er and Yb:Er:Ce:$Ca_2Al_2SiO_7$ crystals, *IEEE J. Quantum Electron.* **32**, 2004, 1996.

254. Kubo, T.S. and Kane, T.J., Diode-pumped lasers at five eye-safe wavelengths, *IEEE J. Quantum Electron.* **28**, 1033, 1992.

255. Schweizer, T., Jensen, T., Heumann, E., and Huber, G., Spectroscopic properties and diode pumped 1.6 μm laser performance in Yb-codoped Er:$Y_3Al_5O_{12}$ and Er:Y_2SiO_5, *Opt. Commun.* **118**, 557, 1995.

256. Wang, P., Dawes, J.M., Burns, P., Piper, J.A., Zhang, H., Zhu, L., and Meng, X., Spectral characterization and laser operations of Er^{3+}:Yb^{3+}:YCOB crystals at 1.5~1.6 μm, *OSA TOPS Adv. Solid State Lasers* **34**, 207, 2000.

257. Sokolska, I., Heumann, E., Kuck, S., and Lukasiewicz, T., Er^{3+}:YVO_4 and Er^{3+},Yb^{3+}:YVO_4 crystals as laser materials around 1.6 μm, *OSA TOPS Adv. Solid-State Lasers* **50**, 378, 2001.

258. Georgiou, E., Kiriakidi, F., Musset, O., and Boquillon, J.-P., 80mJ/1.64 μm pulsed Er:Yb:YAG diode-pumped laser, *Proc. SPIE* **5460**, 272, 2004.

259. Georgiou, E., Kiriakidi, F., Musset, O., and Boquillon, J.-P., 1.65-μm Er:Yb:YAG diode-pumped laser delivering 80-mJ pulse energy, *Opt. Eng.* **44**, 64202, 2005.

260. Spariosu, K. and Birnbaum, M., Room-temperature 1.644 micron Er:YAG lasers, *OSA Proc. Adv. Solid-State Lasers* **13**, 127, 1992.

261. Spariosu, K., Stultz, R.D., Camargo, M.B., Montgomery, S., Birnbaum, M., and Chai, B.H.T., Er:Y_2SiO_5 eye safe laser at 300 K, *OSA Proc. Adv. Solid-State Lasers* **20**, 156, 1994.

262. Setzler, S.D., Budni, P.A., and Chicklis, E.P., A high energy Q-switched erbium laser at 1.62 microns, *OSA TOPS Adv. Solid-State Lasers* **50**, 309 (2001).

263. Spariosu, K. and Birnbaum, M., Intracavity 1.549-μm pumped 1.634-μm Er:YAG lasers at 300 K, *IEEE J. Quantum Electron.* **30**, 1044, 1994.

264. Nikolov, S. and Wetenkamp, L., Single-frequency diode-pumped erbium lasers at 1.55 and 1.64 μm, *Electron. Lett.* **31**, 731, 1995.

265. Nikolov, S. and Wetenkamp, L., Diode pumped erbium lasers at 1.55 μm and 1.64 μm in single frequency operation, *Proc. SPIE* **2772**, 78, 1996.

266. Stoneman, R.C. and Henderson, S.W., High-power eyesafe laser transmitter for microDoppler coherent lidar, *Proc. IEEE* **907**, 342, 2001.

267. Young, Y.E., Setzler, S.D., Snell, K.J., Budni, P.A., Pollak, T.M., and Chicklis, E.P., Efficient 1645-nm Er:YAG laser, *Opt. Lett.* **29**, 1075, 2004.

268. Young, Y.E., Setzler, S.D., Pollak, T.M., and Chicklis, E.P., Optical parametric oscillator pumped at 1645 nm by a 9 W, fiber-laser-pumped, Q-switched Er:YAG laser, *OSA TOPS Adv. Solid-State Photonics* **94**, 387, 2004.

269. Jander, P., Sahu, J.K., and Clarkson, W.A., High-power Er:YAG laser at 1646 nm pumped by an Er,Yb fiber laser, *Proc. SPIE* **5620**, 297, 2004.

270. Shen, D.Y., Jander, P.J., Sahu, J.K., and Clarkson, W.A., High-power and ultra-efficient operation of 1645 nm Er:YAG laser pumped by a 100W tunable Er, Yb fiber laser, *OSA TOPS Adv. Solid-State Photonics* **98**, 618, 2005.

271. Malm, A.I.R., Hartman, R., and Stoneman, R.C., High-power eye safe YAG lasers for coherent laser radar, *OSA TOPS Adv. Solid-State Photonics* **94**, 356, 2004.
272. Setzler, S.D., Snell, K.J., Pollak, T.M., Budni, P.A., Young, Y.E., and Chicklis, E.P., 5-W repetitively Q-switched Er:LuAG laser resonantly pumped by an erbium fiber laser, *Opt. Lett.* **28**, 1787, 2003.
273. Setzler, S.D., Young, Y.E., Snell, K.J., Budni, P.A., Pollak, T.M., and Chicklis, E.P., High peak power erbium lasers resonantly pumped by fiber lasers, *Proc. SPIE* **5332**, 85, 2004.
274. Stoneman, R.C., Hartman, R., Malm, A.I.R., Vetorino, S., Gatt, P., and Henderson, S., Coherent laser radar systems using eyesafe Er:YAG laser transmitters, *13th Coherent Laser Radar Conference*, 54, 2005.
275. Setzler, S.D., Francis, M.W., and Chicklis, E.P., Direct diode pumped eyesafe erbium lasers, *Proc. SPIE* **5707**, 104, 2005.
276. Garbuzov, D., Kudryashov, I., and Dubinskii, M., Resonantly diode laser pumped 1.6-μm-erbium-doped yttrium aluminum garnet solid-state laser, *Appl. Phys. Lett.* **86**, 131115, 2005.
277. Setzler, S.D., Francis, M.P., Young, Y.E., Konves, J.R., and Chicklis, E.P., Resonantly pumped eye safe erbium lasers, *IEEE J. Sel. Top. Quantum Electron.* **11**, 645, 2005.
278. Garbuzov, D., Kudryashov, I., and Dubinskii, M., QCW operation (110 W, 0.9 J) of 1.6-μm Er:YAG laser resonantly pumped with InGaAsP/InP diode lasers, *Solid State and Diode Laser Technology Review*, paper P-13, 2005.

6

High-Power Neodymium Lasers
--

Susumu Konno

CONTENTS

6.1   Introduction

Neodymium (Nd)-doped solid state lasers play important roles in industrial, scientific, and medical fields. Nd (Nd^{3+}) is a rare earth trivalent ion. Most of the industrial laser systems are based on near-infrared emission of four-level Nd lasers. In this section, we review basic properties and recent progress of high-power Nd-doped solid-state lasers. In the opening sections, we study laser crystals. Next, we study the various shapes of laser crystals and review practical applications of Nd-doped lasers. Finally, we focus on wavelength conversion of Nd-doped lasers.

After 1990, diode-pumped solid-state lasers have made rapid progress. Owing to the development of pumping configurations and pump beam optics, laser diodes can excite specific pump band with specific pump beam mode. Pumping efficiency is several times higher and generated heat is much smaller than lamp-pumped systems. Therefore, power supplies and cooling systems are compact. These merits result in high efficiency, high brightness, and compact beam source. Moreover, development of laser diodes accelerates industrial application of the diode-pumped solid-state laser systems. Laser diode bars about 800 nm wavelength are used to pump Nd-doped laser crystal. Single 1-cm laser diode bars with as high as 80 W and two-dimensionally stacked laser diode bars with more that 100 W of output are commercially available. More than 10 kW output power was reported with diode-pumped solid-state laser systems.

Although lamp-pumped systems still play and will continue to play important roles in industrial fields, in this section we focus on diode-pumped systems.

6.2 Host Materials

With proper selection of host materials, we can optimize laser performance. Basic mechanical and spectroscopic properties of Nd-doped laser materials are summarized in Table 6.1 [1,2]. Let us briefly discuss Table 6.1. The thermal

TABLE 6.1

Mechanical and Optical Properties of Commonly Used Nd-Doped Hosts

Properties	Nd:YAG	Nd:YVO$_4$	Nd:YLF
Thermal conductivity (W/cmK)	0.14	0.051 (a axis) 0.0523 (c axis)	0.06
Thermal expansion ($\times 10^{-6}$ (K^{-1}))	7.5	4.43 (a axis) 11.37 (c axis)	13.0 (a axis) 8.0 (c axis)
Densities (g/cm^3)	4.56	4.24	3.99
dn/dT (K^{-1})	7.3×10^{-6}	8.5×10^{-6}(dn$_0$/dT)	2.9×10^{-6}(dn$_e$/dT)
Stimulated emission cross section (cm^2)	6.5×10^{-19}	15.6×10^{-19}	1.8×10^{-19}(p) 1.2×10^{-19}(p)
Fluorescence lifetime (ms)	230	100	480
Index of refraction	1.82	n$_0$ = 1.958 n$_e$ = 2.168	n$_0$ = 1.4481 n$_e$ = 1.4704

Source: From W. Koechner, *Solid-State Laser Engineering*, 5th revised and updated edition, Springer-Verlag, New York, 1999; A.A. Kaminskii, *Laser Crystals: Physics and Properties*, Springer-Verlag, New York, 1990; Xiaoyuan Peng et al., *Appl. Opt.*, 40, 1396, 2001.

conductivity and dn/dT are very important parameters to be considered in the design and operation of high-power solid-state lasers. From Table 6.1, Nd:YAG is found to be the most favorable because of its high thermal conductivity. Nd:YVO$_4$ is not suitable for high-power operation because of its low thermal conductivity and high dn/dT. Stimulated emission cross section of Nd:YVO$_4$ is about three times and ten times higher than Nd:YAG and Nd:YLF, respectively. Hence, it is suitable for generating short pulse in Q-switching operation, at high repetition frequency. Nd:YLF is suitable for high pulse energy generation because of its long fluorescence lifetime.

6.2.1 Basic Properties and Recent Progress of Nd:YAG Lasers

Figure 6.1 shows the energy level diagram of Nd:YAG, (Neodymium-doped yttrium aluminum garnet), which is the most commonly used solid-state laser material in industrial fields. The transition from the $^4F_{3/2}$ upper laser level to the $^4I_{9/2}$ lower laser level is the main simulated emission line at 1064 nm. The strongest pump band wavelength is about 810 nm. At room temperature, $^4F_{3/2}$ (upper laser level) and $^4I_{9/2}$ (lower laser level) are at the sub levels R$_2$ and Y$_3$, according to Boltzmann's law. Because the lower level is well above the ground state, threshold condition is easily obtained, with relatively low pump power. Host material YAG is chemically stable, has good thermal, mechanical properties, and presents isotropic crystal structure. These properties make Nd:YAG the most versatile solid-state laser material.

Since 1995, researchers have attempted to fabricate ceramic YAG with low scattering loss and large size [4]. Many advantages are reported for ceramic YAG. In comparison with single crystal, ceramic YAG can be fabricated with larger size at lower cost. In addition, higher doping concentration is possible (over 1% doping level is difficult for single crystal). An attractive feature of ceramic YAG is that one can fabricate a variety of composite structures, combining Nd:YAG ceramics with different doping concentrations. Intensive research is now taking place in this area [3–9].

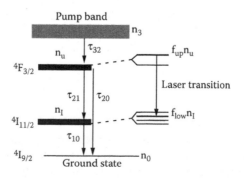

FIGURE 6.1
Energy level diagram of Nd:YAG.

6.2.2 Basic Properties and Recent Progress of Nd:YVO$_4$ Lasers[10]

Neodymium-doped yttrium vanadate (Nd:YVO$_4$) with its advantage of remarkably large emission cross section (three times as high as Nd:YAG) and strong pump-beam absorption results in a compact and efficient high-gain solid-state laser, generating short pulse, at high Q-switching repetition rate. Consequently, Nd:YVO$_4$ lasers are widely used in the microprocessing field.

Due to its natural birefringence, Nd:YVO$_4$ laser generates linearly polarized output (e-polarized) without any polarization optics. However, the fracture limit of Nd:YVO$_4$ is lower than that of Nd:YAG by a factor of four. This limits the maximum pump and output powers of Nd:YVO$_4$ crystal. Hence, Nd:YVO$_4$ lasers are preferred in low and middle power applications.

Recently, another vanadate crystal, Nd:GdVO$_4$ (Nd-doped gadolinium vanadate), has been reported. The thermal conductivity of Nd:GdVO$_4$ along its (110) direction is comparable with that of Nd:YAG. This feature is desirable for power scaling.

6.2.3 Nd:YLF

Neodymium-doped lithium yttrium fluoride (Nd:YLF) has a long fluorescence lifetime and is suitable for the generation of high-energy laser pulses. The crystal structure of Nd:YLF gives natural birefringence, thus providing linearly polarized output. The wavelength depends on polarizations (σ polarization: 1053 nm, π polarization: 1047 nm). The 1053-nm σ-polarized transition exhibits lower gain and positive thermal lens. The 1047-nm π-polarized higher gain transition exhibits negative thermal lens. In both wavelengths, the thermal lensing is much lower than in Nd:YAG.

6.3 Shape of Laser Crystals

In this section, we compare two shapes (rod and slab) of laser crystals for Nd-doped high-power solid-state laser systems. End-pumping and side-pumping schemes for each crystal shape are discussed.

6.3.1 Rod Crystals

A rod crystal is easy to use, fabrication cost is low, and cooling water management is simple (only two O-rings are required to seal the cooling water). Moreover, the circular cross section of rod crystals presents symmetric optical properties around the laser oscillation axis and provides a favorable circular cross-section beam. Rod crystals are most widely used in industrial laser systems because of these two features. In most cases, rod crystals are cooled by water flowing in glass tubes around them and, consequently, heat is extracted from the side surface of the crystals. Temperature distribution and refractive index distribution, which are perpendicular to the rod axis, modify the beam parameters of the laser. This so-called thermal lens and thermal birefringence limits scaling of output power with high-beam quality. Development of rod-shaped solid-state lasers is, in most cases, related to management of thermal lens and thermal birefringence.

6.3.1.1 End-Pumped Rod Lasers

Applying a laser diode as a pump source, one can control wavelength (absorption length), spot size, divergence, and incidence angle of the pump beam. End-pumping configuration can fully utilize these merits.

In an end-pumped system, the pump light enters through the "end" of the laser crystal. The pump-light axis and laser oscillation axis are parallel. End pumping results in higher mode matching overlap, higher absorption efficiency, and higher pump density. Optical-to-optical efficiency of single-mode end-pumped solid state lasers reaches more than 50%. However, absorbed pump light causes strong thermal distortion of laser crystals, and thermal fracture limits output power scaling. We will review typical end-pumped rod lasers in this section.

The resonator configuration shown in Figure 6.2 [13] is an example of a commercially available robust and efficient end-pumped laser system. Two 20-W diode bars are coupled to linear arrays of fiber optics. Coupling efficiency of fiber bundles, from the collimated diode bars to the output of the fiber bundle, is over 85%. The pump light is subsequently imaged into the faces of the Nd:YVO$_4$ active media through dichroic folding mirrors. The design shown in Figure 6.2 is a nearly confocal resonator and is highly insensitive to misalignment. The cavity length is about 14 cm. An output power of 13.8 W at 1064 nm with M^2 = 1.05 is obtained for a total pump power of 26 W from the fiber bundles. As a consequence of the thermal lensing, the laser mode size increases as a function of pump power, resulting in a nonlinear slope efficiency. The optical conversion efficiency at the maximum output power is 53%.

Q-switched, frequency converted end-pumped Nd:YVO$_4$ systems are in high demand for micromachining. Fiber-coupled diode bars avoid complexities and

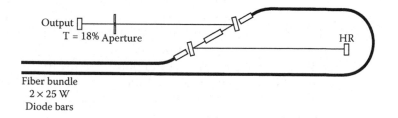

FIGURE 6.2
Schematic of an end-pumped solid-state laser system.

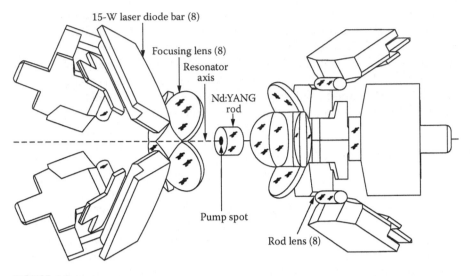

15-W laser diode bar (8)

Focusing lens (8)

Resonator
axis

Nd:YANG
rod

Pump spot

Rod lens (8)

FIGURE 6.3
Angularly multiplexed pump geometry used to focus the power of 8 15-W laser-diode bars
into each laser rod.

reduce maintenance costs. One can replace diode lasers without touching
the laser assembly.

Another example is a high-power end-pumped system producing 60-W
TEM_{00}-mode output power. Figure 6.3 shows the high average power angular-
ity multiplexed pump geometry used for end pumping [14]. Four 15-W laser
diode bars are placed around both ends of each rod. The two sets of diodes on
each rod are rotated 45° with respect to one another to produce a circular gain
distribution in the rod cross section. The divergence of each diode bar is from
40° to 10° in the plane perpendicular to the array by using a 2-mm-diameter
quartz rod lens. The diode light is then focused into the rod end by using a
14.2-mm focal length spherical lens. The pump light is incident upon the
Nd:YAG rod at an angle of 30°. The edge-cooled Nd:YAG rods have a diameter
of 6.35 mm, a length of 7.5 mm, and a doping level of 1.0 atomic percent.

The geometry described above enables efficient pumping. Both lenses have
antireflection coatings for the pump light, and the rod ends are antireflection
coated for the pump light and lasing wavelengths. A passive tuning scheme
is used to set the center wavelengths of the diodes to within 1 nm of the
optimum for absorption. The efficiency with which the pump light is transferred from the diode and absorbed in the rod is over 80%. Thus approximately 50 W of pump power is absorbed per rod end.

The pump light is concentrated in the central portion of the rod, as shown
in Figure 6.4. The result is high gain and a distribution that can be extracted
efficiently by the fundamental mode. The small signal gain in the central
area of each rod is approximately 0.5.

Multimode extraction tests using a single rod were performed with a 14.3-
cm-long resonator formed between 1.1-m concave mirrors (transmittance of

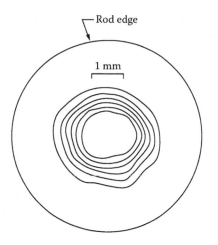

FIGURE 6.4
Fluorescence profile of a 6.35-mm-diameter end-pumped Nd:YAG rod. A total pump power of approximately 100 W is absorbed in the rod, 70 W of which is encircled within 2.4-mm mode diameter.

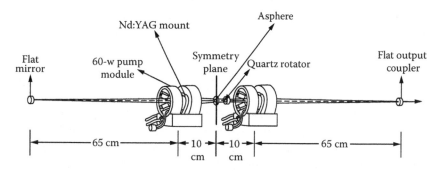

FIGURE 6.5
Symmetrical resonator used in TEM_{00} extraction experiments. Symmetry between the two laser rods allows straightforward correction of the thermal distortion and stress-induced birefringence.

the output coupler is 3%). An output power of 24 W is extracted from a rod pumped on a single end, whereas a rod pumped from both ends can produce 49.5 W. Multimode output power of 92 W is obtained with a total pump power of 235 W. The slope efficiency is 44% (based on the diode output power). The beam quality is 20–30 times diffraction limited. Figure 6.5 shows a symmetrical resonator designed for high-beam-quality extraction. The symmetry ensures that the mode is the same size and that rays pass through similar areas in both rods. The resonator is formed by flat-end mirrors separated from the rods by 65 cm. The rods are separated from one another by 20 cm. A 16-cm focal length lens located at the symmetry plane compensates for first-order thermal focusing. Neglecting aberrations, each rod has a thermal lens focal length of 25.7 cm and a TEM_{00}-mode diameter of 2.4 mm ($1/e^2$). A stable output power of 60 W is obtained with the flat–flat resonator and a total diode power of 235 W. The beam quality is 1.3 times diffraction limited.

6.3.1.2   Side-Pumped Rod Lasers

Side-pump configuration is applied for both diode-pumping and flash-lamp-pumping systems. The pump light strikes a laser crystal transversely, through the side surface of a laser crystal. Consequently, scaling of pump power and output power is much easier than end-pumped systems. However, because of low mode-matching overlap of the pump beam and oscillation laser beam, efficient operation of the TEM_{00} mode is difficult. Therefore, side-pump configurations are considered to be suitable for high-power multitransverse-mode systems. Side-pumping configurations do not include any pumping optics on the laser oscillation axis. Therefore, it is easy to align multicavities along an optical axis. The multicavity configuration is important in industrial high-power applications.

Theoretical and experimental investigations have proven that if a suitable arrangement of rods and mirrors are selected, the output power increases proportionally to the number of rods [15] without degrading beam quality. Figure 6.6 is a schematic drawing of the multirod configuration. The best beam quality, highest efficiency, and largest range of stable oscillation are obtained from a symmetric plane–plane resonator in which the two rods are separated by L/2 and the mirror–rod distance is L/4.

This type of simple and robust plane–plane resonator–amplifier multicavity configuration is widely used in industrial high-power applications. The most important application is welding. Five to ten identical lamp-pumped or laser-diode-pumped cavities (pump heads) are used. These employ Nd:YAG rods with diameter of 3 to 15 mm to produce several kW of CW output. The output power is coupled to several hundred micrometers core-diameter step index (SI) fibers to realize flexible beam delivery.

An amplifier configuration, to improve output beam mode stability against pump power, is reported [16]. This configuration stabilizes the thermal-lens-dependent variation in the output-beam parameters by extracting the laser beam from the collimating point of the periodic beam propagation. The configuration does not need any sophisticated technologies.

Figure 6.7 shows the setup of a preliminary experiment to confirm the laser performance of the proposed configuration. A resonator–amplifier system is

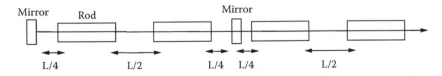

FIGURE 6.6
Amplification of the output power by several rods at a distance L/2. The beam quality remains constant because the rods form a type of lens waveguide.

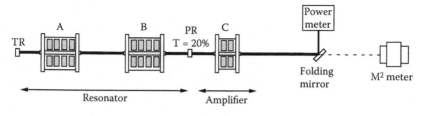

FIGURE 6.7
Schematic of the experimental setup.

composed of three pump cavities A, B, and C. The resonator consists of cavities A and B, along with a flat total reflector (TR) and a flat partial reflector (PR) (The reflectance of the out-coupling mirror was 80%). The pump cavities A and B are placed 400 mm apart from the TR and PR mirrors, respectively. The distance between rod ends of cavities A and B was set to 800 mm. Cavity A and cavity B contained a Nd:YAG rod (4 mm in diameter, 105 mm in length, 0.6% Nd doping) with 16 laser diodes (25 W rated diode output power), placed in fourfold symmetry around the rod axis. The pumped part length of the Nd:YAG rod in cavity A and B was 72 mm.

The amplifier pump cavity C was placed at a distance of 400 mm (optical path length) away from the PR mirror, along the optical axis of the output beam and contained a Nd:YAG rod (4 mm in diameter, 70 mm in length, 0.6% Nd doping) pumped by eight laser diodes (25 W rated diode output power), placed in fourfold symmetry. The pumped part length of the Nd:YAG rod in cavity C was 36 mm.

The total pump power of cavity A and B were maintained to be twice that of cavity C, so that pump intensity and thus the refractive index were maintained to be identical in the pumped region of all Nd:YAG rods. The output beam variation depending on the pump power was characterized by monitoring the waist beam diameter and the position of the beam waist. A two-pump-cavity resonator, consisting of cavity A and B, was used to characterize conventional system as a reference.

Figure 6.8 shows the pump-power dependence of the waist-beam position. Z_0 in vertical axis represents the distance between M^2 meter and the measured beam waist. The horizontal axis of Figure 6.8 and Figure 6.9 is the LD driving current of cavity B. From Figure 6.8, we can confirm that the measured output beam waist was maintained at a constant position, as was predicted from the calculation. Figure 6.9 compares the output waist-beam size of a conventional and a novel configuration. In the conventional system, the waist-beam diameter varied 23% (from 3.3 mm to 2.5 mm) as the pump power increased, and in the novel configuration, the beam diameter remained almost constant (beam size varied only 2% [3.4 to 3.6 mm]). Compared to the conventional configuration, the beam diameter variation of the novel configuration was suppressed to be less than 1/10. The experimental results in Figure 6.8 and Figure 6.9 confirmed the theoretical analysis.

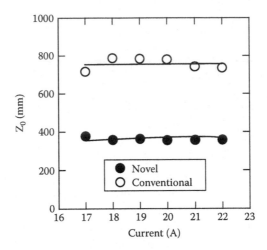

FIGURE 6.8
Position of the beam waist plotted against the pump power.

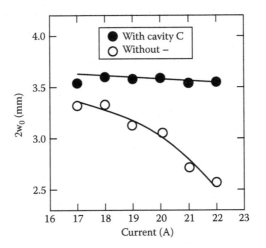

FIGURE 6.9
Variation of the waist-beam size against the pump power.

Let us now introduce a typical side-pumping configuration. A fiber-coupled diode-side-pumped system is described subsequently [17]. Figure 6.10 shows the experimental setup. The laser rod is placed inside a flow tube for cooling water, which is antireflection coated at 808 nm. The optical pump sources consist of fiber-coupled diode lasers with a nominal output power of 10 W each at 808 nm. The pump radiation is delivered through fibers to the Nd:YAG laser head. The optical quartz fibers have a core diameter of 800 μm, a total diameter of 1.5 mm, and 0.22 NA. Each pump module consists of 16 fibers. The fibers are mounted side by side with a spacing of 0.5 mm. The pump modules are arranged in a threefold symmetry around the laser rod, giving a total available pump power of approximately 370 W. The diode

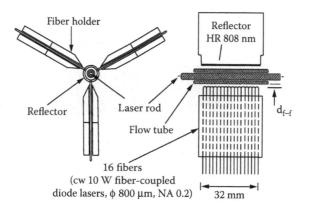

FIGURE 6.10
Laser head side-pumped by fiber-coupled diode lasers. The linear pump power density is 150W/cm.

laser radiation directly irradiates the laser rod without any additional focusing optics. The spacing d_{f-f} (Figure 6.10) between the fiber ends and the flow tube can be varied from 0.5 and 20 mm. For sufficient absorption of the diode-laser radiation, pump-light reflectors are mounted around the rod. Nearly 340 W of the total pump power is absorbed because of the double pass of the radiation in the laser rod. A maximum multimode output power of 160 W was obtained. In the TEM_{00} mode operation, 60 W was obtained at an optical-to-optical conversion efficiency of 25%.

Another example of side-pumped Nd:YAG rod laser aims at a simple and low-cost structure that avoids introducing complexity, such as precisely aligned and coated coupling optics of the laser diodes. Selection of the laser diode modules depends on the output power and wavelength [18].

For this purpose, a low-loss diffuse reflectivity pumping chamber is applied to pump laser rods directly. With this configuration, efficient pumping is realized with poor-wavelength-selected and broad spectral emission stacked arrays.

Figure 6.11 shows the schematic of the cross section of the pump module. The laser design combines the easy to scale stacked-array technique of the pump modules with the reliable and proven rod laser geometry. Different stack sizes can be adapted by the shape of the reflector.

Results are given from a double-rod system in which each cavity is pumped by a Cu microchannel-cooled stacked array consisting of 50 cw laser bars. All bars are driven electrically in series and at the same temperature. This leads to a 8-nm total line width (FWHM) of the stack diode. Differences in the output characteristics of the bars are averaged by the multiple passes of the pump light through the rod, resulting in a flat power deposition inside the Nd:YAG rod. With one cavity (size 10 × 5 × 5 cm), 350 W cw output power at 23% optical efficiency was achieved. Efficient, high-power operation of side pumped rod lasers are also reported [19–23].

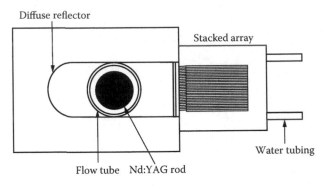

FIGURE 6.11
Experimental arrangement.

6.3.2 Slab Crystals

Compared with rod crystals, rectangular slab crystals have large cooling surfaces. Therefore, using slab-shaped laser media reduces thermally induced strain and aberration. For efficient cooling, two side surfaces (thin direction) of slab crystals are cooled uniformly. The other two side surfaces (thick direction) are thermally insulated. Hence, temperature gradient, thermal lens, and thermal birefringence are present only in one direction (thin direction). However, uniform cooling of rectangular slab crystals require complex cooling techniques. A zigzag path, bouncing on cooling surfaces (thin direction), efficiently eliminates the first-order thermal lens. Slab crystal with zigzag path has many polished surfaces. Consequently, the fabrication cost is relatively high.

6.3.2.1 End-Pumped Slab Lasers

To reduce the fabrication cost of the slab laser system, an end-pumped hybrid-resonator design was proposed [24]. The laser utilizes a low-cost slab crystal that has only two polished surfaces. Laser diode stacks and pump beam optics form a nearly homogeneous pumped volume with rectangular cross section in the center of the slab crystal.

Figure 6.12a shows the laser head. The slab is 3 mm × 8 mm × 16 mm. A diode laser stack consists of three 10 mm laser diode bars. Each diode bar emits 30 W pump light. The radiation emitted by each diode laser bar was individually collimated by microlenses. The three conditioned beams were parallel to one another in the plane defined by the fast axis and the propagation direction. To obtain a homogeneous line shaping, an Offner optic and a cylindrical lens (L2) were used. The Offner optic consisted of two almost concentric mirrors (M5 and M6) and had minimal imaging error.

The cylindrical lens was incorporated with its focal length apart from the exit facets of diode lasers to generate a virtual line source with dimensions

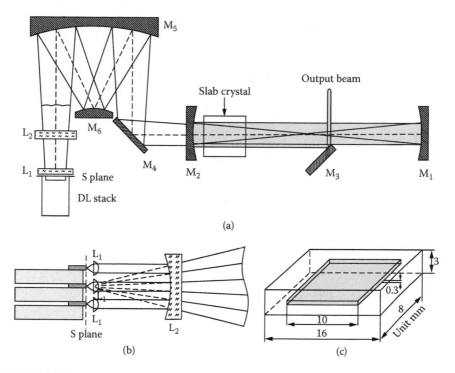

FIGURE 6.12

Schematic of the slab laser with a hybrid resonator: (a) complete laser head; (b) side view of the diode laser (DL) stack with the cylindrical lenses; (c) geometry of the pumped volume. M1, M2, resonator mirrors; M4, folding mirror for the pumping beam.

of 0.3 mm × 10 mm. Then the virtual line source was reimaged 1:1 near the entrance face (3 mm × 16 mm) of the slab crystal. The total transmission of the imaging optic including the microlens was 83%. Inside the pumped volume, the gain sheet had a nearly rectangular cross section of 0.3 mm × 10 mm as indicated by the shaded area within the slab in Figure 6.12c. The laser performance is shown in Figure 6.13. The beam power of 31 W at 1064 nm was generated at a beam propagation factor of $M^2 = 1.3$ (unstable direction) and 1.7 (stable direction).

A hybrid resonator was also employed for the Nd:YVO$_4$ system to produce high power at high beam quality. An Nd:YVO$_4$ laser with 100 W output power was reported with a partially end-pumped unstable resonator [25]. The resonator configuration and output characteristics are presented in Figure 6.14. The basic concept of the system is the same as described above. The Nd:YVO$_4$ slab laser crystal size was 1 mm × 10 mm × 12 mm. The system was electro-optically Q-switched at high repetition rate.

A maximum continuous-wave output power of 103 W was obtained at an incident power of 248 W with a beam quality faactor of $M^2 < 1.5$. The optical-to-optical conversion efficiency was 41.5%. In Q-switched mode, 1.66 mJ was obtained at 50 kHz repetition rate, with pulse length of 11.3 ns.

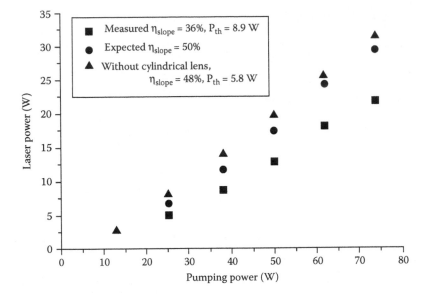

FIGURE 6.13
Dependence of laser output power on the diode pumping power.

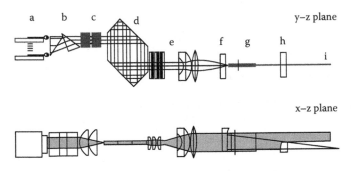

FIGURE 6.14A
Schematics of the laser; cross sections in y-z and x-z planes: (a) diode-stack; (b) pair of prisms; (c) two cylindrical lenses; (d) planar waveguide; (e) imaging group; (f) rear mirror, M1; (g) Nd:YVO4 slab; (h) output coupler M2; and (i) output laser beam.

6.3.2.2   Side-Pumped Slab Lasers

Let us now review a slab laser transversely pumped by fiber-coupled diode lasers [26].

Figure 6.15 shows a schematic of the laser head. The Nd:YAG slab is mounted in an aluminum frame and sealed at both ends. The O-rings are placed just as one would place O-rings on a rod laser crystal, with no care taken to locate them away from a bounce point because the slab is protected by a low-index coating. The top and bottom of the slab are insulated by placing gold-coated glass microscope slides in contact with the Nd:YAG slab. The glass slides are coated on the back with a thin RTV silicone layer for

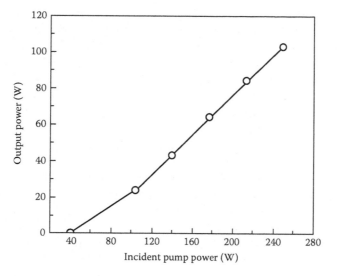

FIGURE 6.14B

Continuous-wave output power of the partially end-pumped Nd:YVO$_4$ slab laser vs. incident pump power into the slab crystal.

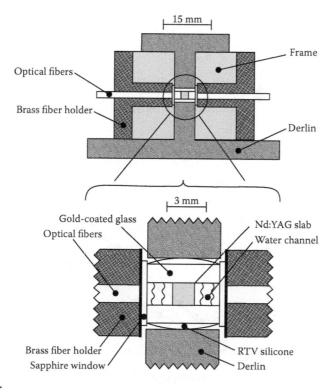

FIGURE 6.15

Schematic of the laser head. The thick black lines on the glass slides and the brass fiber holders represent gold coatings for confining the pump light.

stress relief. The last two sides of the frame contain the fiber pump modules. The Nd:YAG slab is water cooled with 2-mm-thick water channels flowing between the slab surfaces and the brass fiber holders. The water flows at a rate of 1 l/min, and the Reynolds number and flow geometry are selected to make the flow turbulent. The fibers are isolated from the water flow by a 0.5-mm antireflection-coated sapphire window glued onto the brass mount. The assembly of this laser head is simple and can be completed in a short time.

At a pump power of 235 W, the zigzag slab laser emitted 72 W of power in a square, multimode beam. The optical-to-optical slope efficiency was 36% with a threshold of 30 W. The laser was operated in a TEM_{00} mode configuration as shown in Figure 6.16. The asymmetric thermal lens is compensated by the astigmatism from an off-axis concave mirror. A 20-cm radius-of-curvature mirror was chosen to dominate the thermal lensing in the cavity, and the fold angle necessary to produce TEM_{00} mode operation at full power was 45°. A TEM_{00} mode power of 40 W was obtained at a pump power of 212 W. The output is polarized as a result of the Brewster slab faces, with a polarization ratio better than 100:1. Figure 6.17 shows the cw output power vs. the diode-laser input power for both multimode and TEM_{00} mode operation.

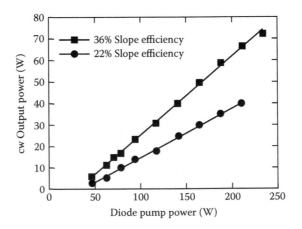

FIGURE 6.16
Cavity design for TEM_{00} mode operation.

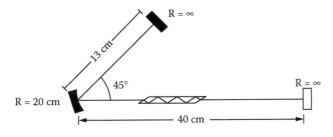

FIGURE 6.17
Input vs. output curve for multimode (squares) and TEM_{00} mode (circles) operation.

The pump beam absorption is highest at the surface where the pump beam enters. It is important to utilize the pump volume near the surface. We will review a side-pumped slab laser that utilizes a single high-angle-of-incidence reflection in a Nd:YVO$_4$ slab crystal to utilize the pump volume near the surface [27].

The experimental arrangement is shown in Figure 6.18. The laser medium was a 3%-doped Nd:YVO$_4$ parallelopiped bar with dimensions of 10.1 mm × 2.5 mm × 3.0 mm. Both a-cut ends were antireflection coated for the wavelength of 1.06 μm. Pumping was through the 10.1 mm × 3.0 mm face, which was polished flat and uncoated. The pump source was a 1-cm quasi-cw diode-laser bar (SDL Model 3230-TZ), which produced 200-ms square pump pulses with energies of as much as 12mJ at a repetition rate of 16 Hz. A 0.25-mm-diameter fiber lens was used to collimate the diode-laser output partially so as to control the spatial extent of the pumping. The laser cavity was formed between a concave high reflector (reflectivity $R_1 = 1$) and a plane output coupler ($R_2 = 0.475 - 0.985$) with a single bend that was due to the total internal reflection located at the center of the pump face. External angles, θ, ranged from 0° to 10°. To ensure that the curvature of the mode within the rod was small, the laser rod was located close to the flat output coupler. The radius of curvature of the high reflector and the total cavity length were chosen to produce the maximum output energy and best-quality laser mode.

For cavity parameters that gave a moderately small mode radius of 130 μm ($L_1 = 23$ mm, radius $r_1 = 100$ mm, $L_2 = 9$ mm), the best beam quality was obtained at an angle of $\theta = 4°$. At smaller angles, the output pulse energy was higher, but the beam showed a multimode structure caused by diffraction at the rod ends and the steep gradient in the gain across the beam cross section. At angles greater than 6°, the output energy was even larger, but the beam was stretched horizontally (i.e., in the plane of Figure 6.18) to a diameter at least twice that in the vertical direction. Some structure in the horizontal direction was also present. At an external angle of 4°, the output beam

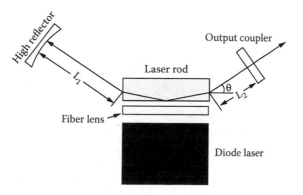

FIGURE 6.18

Experimental arrangement (top view). θ is the angle between the beam, external to the crystal and the pump face. The c axis of the Nd:YVO$_4$ laser rod was oriented perpendicular to the plane of the figure.

appeared to be TEM_{00} and was not sensitive to a small misalignment of the cavity mirror or the focusing of the pump light. Careful measurement of the beam waist at the focus of a diffraction-limited lens and the subsequent divergence gave values of M^2 of approximately 1.3 and 3.9 in the horizontal and vertical directions, respectively. The apparently poor beam quality in the vertical direction was unexpected from the measured far-field beam profile. Strong gain variations in the vertical direction as a result of nonuniform pumping in that direction are a likely source of this large divergence. Thermal lensing within the laser material appeared to have a negligible effect on the laser mode; no changes in the output beam were observed as the pump repetition rate was varied from 16 to 70 Hz.

The output pulse energy vs. the total uncorrected pump energy from the diode bar is shown in Figure 6.19 for output mirror reflectivities R_2 of 0.475, 0.815, and 0.985. The highest output was obtained with $R_2 = 0.475$, which indicates that the gain was high. As much as 2.3 mJ of energy was obtained for a pump energy of 12 mJ. The output pulses were polarized parallel to the crystal c axis and were approximately 200 μsec long with no observable oscillations. A maximum optical slope efficiency of 22% and an optical-to-optical conversion efficiency of 19% were obtained. If reflection losses at an uncoated fiber lens (n = 1.5) and the pump face (n = 1.96 for pump light polarized perpendicular to the c axis) are included, these values increase to 27% and 23%, respectively. High average power slab lasers are also reported [28,29].

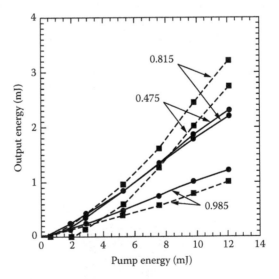

FIGURE 6.19
Output energy as a function of the diode pump energy. Results are shown for external angles θ of 4° (solid curves) and 10° (dotted curves) at output coupler reflectivities of $R_2 = 0.475, 0.815,$ and 0.985.

6.4 Wavelength Conversion Lasers

High-power Nd lasers are applied for various wavelength conversion systems. As wavelength conversion is a nonlinear process, conversion efficiency is a function of the fundamental beam intensity on the frequency conversion crystal. The Q-switched pulse laser beam is advantageous in achieving efficient frequency conversion.

The second-harmonic (wavelength is about 500 nm) of Nd-doped lasers can be used for laser display, microprocessing of metals, and large area processing, such as annealing of SiO_2 polycrystals. In the scientific field, second-harmonic lasers are used as a pump source for Ti:Sapphire lasers.

Frequency conversion ultraviolet (UV) laser systems are widely used for micromachining of various materials including copper, glasses, polymers, and ceramics. The high photon energy and high focusibility of UV lasers enables the extension of the processing capabilities of solid-state lasers.

Many organics absorb third- (wavelength is about 350 nm) and fourth-harmonic (wavelength is about 260 nm) of Nd lasers. The most important application of UV lasers is in the area of printed circuit board processing, where the work piece contains many organics. Laser drilling is a widely accepted method of creating microvias in high-density electronic interconnect and chip packaging devices. Commercially available laser sources used for microvia formation are of two types, CO_2 lasers and 355-nm UV solid-state lasers. High average power, high-throughput CO_2 lasers are advantageous for large-size drilling, and UV solid-state lasers are advantageous for high-precision, small-size drilling. Diameters down to 20–30 μm [30] are now achieved. In this area, UV solid-state lasers are required to have higher repetition Q-switching frequency and higher average power for high-speed processing.

Nd:vanadate (Nd:YVO_4) has become a desirable laser crystal in addition to Nd:YAG. Because glasses absorb fourth- and fifth-harmonic wavelength and not third harmonics, intensive research has been devoted to apply fourth and fifth harmonics for processing glass fibers and fiber grating. Next, we will focus on topics related to intracavity harmonic generation lasers.

6.4.1 Intracavity Second-Harmonic Generation

Figure 6.20 compares extracavity and intracavity schemes. SHG (second-harmonic generation) is obtained through intracavity and extracavity frequency conversion systems. Intracavity systems result in high conversion efficiency, but thermal distortion of frequency conversion crystals in the resonator causes unstable and inefficient laser performance. The problem limits scaling of average power with high beam quality.

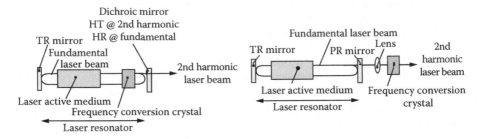

FIGURE 6.20
An intracavity frequency-conversion laser system and an extracavity frequency-conversion laser.

TABLE 6.2

Properties of Frequency Conversion Crystals

	KTP (KTiOPO$_4$)	LBO (LiB$_3$O$_5$)
Density (g/cm^3)	3.023	2.474
Specific heat (J/kg/K)	728	1060
Thermal conductivity (W/m/K)	2.0	3.5
	3.0	3.6
	3.3	
dn/dT	22.0	−1.9
(10^{-6}/K)	25.9	−13.0
	42.8	−8.3
Phase-matching acceptance temperature	$\Delta T = 17.5°$	$\Delta T = 6.2°$
Phase-matching acceptance angle	$\Delta\theta = 1.82°$	$\Delta\theta = 3.11°$
	$\Delta\phi = 0.42°$	$\Delta\phi = 0.18°$
d_{eff} (pm/V)	$\phi_{PM} = 23.3°$	$\theta_{PM} = 20.6°$
	$d_{15}\sin^2\phi_{PM} + d_{24}\cos^2\phi_{PM}$	$d_{31}\cos\theta_{PM} = 0.67 \times 0.936$
	$= 7.36$ (pm/V)	$= 0.627$ (pm/V)

The basic properties of SHG (second-harmonic generation) frequency conversion crystals are summarized in Table 6.2 [31]. Neglecting the complex phase-matching condition and employing planewave approximation, the conversion efficiency of second-harmonic generation is given by the equation

$$\frac{P_2}{P_1^2} = \frac{2\pi^2 d_{eff}^2 L^2}{\varepsilon_0 cn_1^2 n_2 \lambda_2^2 A} = 2.52 \times 10^{-9} \times \frac{L^2}{A}. \tag{6.1}$$

The conversion efficiency is proportional to the square of d_{eff}. From Table 6.2, it is seen that KTP(KTiOPO$_4$) is advantageous for low- and middle-power operation because of its high frequency conversion coefficient. The d_{eff} of KTP (Type II phase matching) is an order of magnitude larger than that of LBO (Type II phase matching) (see d_{eff} in Table 6.2). To increase average green output power, LBO(LiB$_3$O$_5$) is preferable because of its high damage threshold and good thermal and mechanical properties.

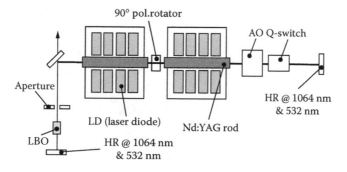

FIGURE 6.21
Schematic of the intracavity-doubled high-power green laser.

A typical intracavity-doubled high-power green laser that applies two side-pumped Nd:YAG rods, with quartz rotator, is presented in the following text.

Figure 6.21 shows a 138-W intracavity green pulse laser [32]. A quartz 90° polarization rotator is placed between two uniformly pumped Nd:YAG rods for polarization-dependent bifocusing compensation. The pump heads consist of four modules. Each module contains four 1-cm-long linear continuous-wave diode arrays (808-nm wavelength, 20-W output power). Each module is rotated 22.5° from the others around the optical axis to produce uniform pump–light distribution within the rod's cross section. An Nd:YAG rod (4 mm in diameter, 105 mm in length, with 0.6% Nd doping) is surrounded by a flow tube and a diffusive reflector. Figure 6.21 shows a schematic drawing of the L-shaped 0.7-m-long convex–convex cavity. The distance between the end mirrors (radius of curvature 1m) and the rod ends is 250 mm. The resonator is folded by a harmonic separator mirror (T > 98% at 532 nm, R > 99.5% at 1064 nm, where T is transmittance and R is reflectance) and a total reflector. The pumping head and acousto-optic Q-switches are placed in one arm of the resonator. The Q-switches are operated at 10-kHz repetition rate. A 15-mm-long Type-II phase-matched dual-wavelength antireflection-coated LBO(LiB$_3$O$_5$) crystal is placed in another arm of the resonator. The 532-nm green beam was extracted from the harmonic-separator mirror in one direction. In Figure 6.22, the laser performances at 1064-nm and 532-nm operation are compared. The 1064-nm operation is obtained by replacement of one of the total reflection end mirrors with a partial output coupler (transmittance, 15%) and removal of the LBO crystal from the resonator. At 800 W total diode-output power, 205 W of 1064 nm output power was obtained in cw operation, and 148 W in Q-switched operation. Then, the transmitting output coupler is replaced with a total reflector and the LBO crystal is placed in the resonator again. A maximum 532-nm green output power of 138 W was generated, with 800 W total diode output power and 1750 W electrical input power, corresponding to an optical-to-optical conversion efficiency of 17.3% and electrical-to-optical conversion efficiency of 7.9%. The ratio of green output to Q-switched IR output was approximately 90%. The beam

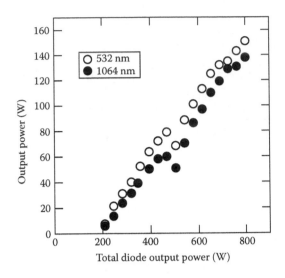

FIGURE 6.22
Power performance of the Nd:YAG laser at 532 and 1064 nm.

quality of the green output was $M^2 = 11$. The pulse width of the green beam was 70 ns at the maximum green output power. High-power, high-pulse-energy green lasers have also been reported [33–37].

6.4.2 Intracavity Third-Harmonic Generation

Compared to fourth-harmonic lasers, the third harmonic (350 nm in wavelength) of Nd-doped solid-state lasers can produce more output power with high conversion efficiency and longer lifetime of optical components. At the moment, the third harmonic of Nd lasers is commercially available as a beam source for micromachining. UV solid-state lasers are required for high average power for high-speed processing. High average power intracavity-tripled Nd:YAG and Nd:YLF lasers reported [38,39].

Figure 6.23 is a schematic of a Kr arc-lamp-pumped Nd:YAG laser. The linear cavity consists of a 145-mm Nd:YAG rod, an accousto-optic Q-switch, a Type-I phase-matched LBO crystal for second- harmonic generation, a Type-II LBO crystal for third-harmonic generation, and an intracavity dichroic mirror for harmonic separation. The cavity design is simple and compact and the overall length of the laser resonator is 74 cm. The repetition rate of the laser is continuously variable from 5 to 20 kHz. The THG output was 8 W at 5 kHz, 3 W at 20 kHz, and a maximum power of 8.8 W was achieved at 6 kHz. Nonlinear conversion efficiency was estimated at a moderate repetition of 10 kHz. We measured 20 W of SHG and 6.8 W of THG from the Nd:YAG oscillator. This corresponds to an effective SHG-to-THG conversion efficiency of 34%. Such high UV output and high efficiency were achieved by taking advantage of intracavity THG design. In addition to the high intracavity IR

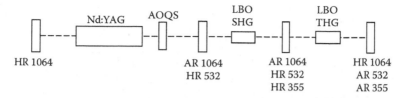

FIGURE 6.23
Schematic of an arc-lamp-pumped Nd:YAG laser for third harmonic generation.

power, the intracavity SHG is used in a multipass fashion. As a result, the interaction length of nonlinear optical crystal is effectively increased and high conversion efficiency is obtained.

This group further improved the ultraviolet output power by using Nd:YLF UV (351 nm) output power of 11.5 W and 23.2 W was obtained for TEM_{00} mode and multimode, respectively.

High average power fourth- and fifth-harmonic generation is mainly attempted with extracavity configuration, where the frequency conversion crystals are placed outside the resonator. We can thus separate the technical advances of the system, improvements of fundamental beam sources, and that of frequency conversion techniques. Intensive research has been reported in this area [40–44].

6.5 Summary

We have reported a variety of high average power Nd-doped lasers. Remarkable advances were obtained in pumping geometries, novel laser materials, and resonator configuration.

Recently, over 10 kW output power and several hundred watts of TEM_{00} mode output were reported with fiber laser systems. These systems will replace some application fields of solid-state lasers. However, it seems difficult to generate high pulse energy with fiber laser systems because of the small cross-sectional area of the core. In addition, because of complex pumping configuration, the cost is relatively higher than that for solid-state laser systems at the moment. In the next decade, these technical advances are expected to result in the development of industrial workhorses based on high-power solid- state laser systems.

References

1. W. Koechner, *Solid-State Laser Engineering*, 5th revised and updated edition, Springer-Verlag, New York, 1999.

2. A.A. Kaminskii, *Laser Crystals: Physics and Properties*, Springer-Verlag, New York, 1990.
3. X. Peng, A. Asundi, Y. Chen, and Z. Xiong, *Appl. Opt.*, 40, 1396, 2001.
4. A. Ikesue, T. Kinoshita, K. Kamata, and K. Yoshida, *J. Am. Ceram. Soc.*, 78, 1033, 1995.
5. J. Lu, M. Prabhu, J.Song, C. Li, J. Xu, K. Ueda, A. Kaminski, H. Yagi, and T. Yanagitani, *Appl. Phys.*, B 71, 469, 2000.
6. A. Ikesue, K. Yoshida, T. Yamamoto, and I. Yamaga, *J. Am. Ceram. Soc.*, 80, 1517, 1997.
7. A. Ikesue, K. Kamata, and K. Yoshida, *J. Am. Ceram. Soc.*, 79, 359, 1996.
8. L. Jianren, M. Prabhu, X. Jianqiu, K. Ueda, H. Yagi, and A.A. Kaminski, *Appl. Phys. Lett.*, 77, 3707, 2000.
9. L. Jianren, T. Murai, K. Takaichi, T. Uematsu, K. Misawa, M. Prabhu, J. Xu, K. Ueda, H. Yagi, T. Yanagitani, A.A. Kaminski, and A. Kudryashov, *Appl. Phys. Lett.*, 78, 3586, 2001.
10. Y. Sato and T. Taira, *Jpn. J. Appl. Phys.*, 41, 5999–6002, 2002.
11. T. Jensen, V.G. Ostroumov, J.P. Meyn, G. Huber, A.I. Zagumennyi, and I.A. Scherbakov, *Appl. Phys. B*, 58, 373, 1994.
12. H. Zhang, J. Liu, J. Wang, L. Zhu, Z. Shao, X. Meng, X. Hu, and M. Jain, *J. Opt. Soc. Am. B*, 19, 18, 2002.
13. N. Hodgson, K. Griswold, W. Jordan, S.L. Knapp, A.A. Peirce, C.C. Pohalski, E. Cheng, J. Cole, D.R. Dudley, A.B. Peterson, and W.L. Highan Jr., *Proc. SPIE*, 3611, 119, 1999.
14. S.C. Tidwell, J.F. Seamans, and M.S. Bowers, *Opt. Lett.*, 18, 116, 1993.
15. K.P. Driedger, R.M. Ifflander, and H. Weber, *IEEE J. Quantum Electron.* 24, 665, 1988.
16. S. Konno, T. Kojima, S. Fujikawa, and K. Yasui, *Appl. Opt.*, 41, 7569, 2002.
17. G. Dolla, M. Bode, S. Knoke, W. Shone, and A. Tunnermann, *Opt. Lett.*, 21, 210, 1996.
18. T. Brand and I. Shmidt, CLEO/Europe 1996, Technical Digest, CMA4, 1996.
19. T. Kojima and K. Yasui, *Appl. Opt.*, 36, 4981, 1997.
20. S. Fujikawa, T. Kojima, and K. Yasui, *IEEE J. Sel. Top. Quantum Electron.*, 3(1), 40, 1997.
21. S. Konno, S. Fujikawa, and K. Yasui, *Appl. Phys. Lett.*, 79, 2696, 2001.
22. S. Fujikawa, K. Furuta, and K. Yasui, *Opt. Lett.*, 26, 602, 2001.
23. A. Takada, Y. Akiyama, T. Takase, H. Yuasa, and A. Ono, OSA Advanced Solid-state lasers Topical Meeting, Technical Digest, MB18, 1999, pp. 69–71.
24. K. Du, N. Wu, J. Xu, J.Giesekus, P. Loosen, and R. Poprawe, *Opt. Lett.*, 23, 370, 1998.
25. K. Du, D. Li, H. Zhang, P. Shi, X. Wei, and R. Diart, *Opt. Lett.*, 28, 87, 2003.
26. R.J. Shine, Jr., A.L. Alfrey, and R. Byer, *Opt. Lett.*, 21, 869, 1996.
27. J.E. Bernard, and A.J. Allock, *Opt. Lett.*, 18, 968, 1993.
28. J. Machan, R. Moyer. D. Hoffmaster, J. Zamel, D. Burchman, R. Tinti, G. Holleman, L. Marabella, and H. Injeyan, OSA Advanced Solid-State Lasers Topical Meeting, Technical Digest, AWA2, 1998, pp. 263–265.
29. M. Sato, S. Naito, H. Machida, N. Iehisa, and N. Karube, OSA Advanced Solid State Lasers Topical Meeting, Technical Digest, MA2, 1999, pp. 3–5.
30. Y. Sun, C. Dunsky, H. Matsumoto, G. Simenson, *Proc SPIE*, 4915, 2002.
31. V.G. Dmitriev, G.G. Gurzadyan, and D.N. Nikogosyan, *Handbook of Nonlinear Optical Crystals*, 2nd ed., Springer-Verlag, New York, 1999.

32. S. Konno, T. Kojima, S. Fujikawa, and K. Yasui, *Opt. Lett.*, 25, 105, 2000.
33. B.J. Le Garrec, G.J. Raze, and M. Gilbert, *Opt. Lett.*, 21, 1990–1992, 1996.
34. J.J. Chang, E.P. Dragon, and I.L. Bass, 315 W pulsed green generation with a diode pumped Nd:YAG laser, Conference on Lasers and Electro Optics, Post Deadline Paper CPD2, 1998.
35. J.J. Chang, E.P. Dragon, C.A. Ebbers, I.L. Bass, and C.W. Cochran, in *Advanced Solid State Lasers*, C.R. Pollock and W.R. Bosenberg, Eds., Vol. 10 of OSA Trends in Optics and Photonics Series, Optical Society of America, Washington, D.C., 1997, p. 300.
36. E.C. Honea, C.A. Ebbers, R.J. Beach, J.A. Speth, J.A. Skidmore, M.A. Emanuel, and S.A. Payne, *Opt. Lett.*, 23, 1203, 1998.
37. S. Konno, Y. Inoue, T. Kojima, S. Fujikawa, and K. Yasui, *Appl. Opt.*, 40, 4341, 2001.
38. F. Zhou, M. Maikowski, and Q. Fu, Conference on Lasers and Electro-Optics, 1997 OSA, Technical Digest Series, Optical Society of America, Washington, D.C., CFF4, 1997, p. 484.
39. F. Zhou and M. Maikowski, Conference on Lasers and Electro-Optics, 1998 OSA, Technical Digest Series, Optical Society of America, Washington, D.C., CFL4, 1998, p. 542.
40. T. Kojima, S. Konno, S. Fujikawa, K. Yasui, K. Yoshizawa, Y. Mori, T. Sasaki, M. Tanaka, Y. Okada, 20-W ultraviolet-beam generation by fourth-harmonic generation of an all-solid-state laser, *Opt. Lett.*, 25, 58–60, 2000.
41. Y.K. Yap, M. Inagaki, S. Nakajima, Y. Mori, and T. Sasaki, *Opt. Lett*, 21, 1348–1350, 1996.
42. M. Oka, L.Y. Liu, W. Wiechmann, and S. Kubota, *IEEE J. Sel. Top. Quantum Electron.*, 1, 859, 1995.
43. U. Stamm, W. Zschocke, T. Schroder, N. Deutsch, and D. Basting, in *Advanced Solid State Lasers*, Vol. 10 of OSA Trends in Optics and Photonics Series, 1997, p. 7.
44. A. Finch, Y. Ohsako, J. Sakuma, K. Deki, M. Horiguchi, Y. Mori, T. Sasaki, K. Wall, J. Harrison, P.F. Moulton, and J. Manni, in *Advanced Solid State Lasers*, Vol.19 of OSA Trends in Optics and Photonics Series, 1998, p. 16.

7

Passively Mode-Locked Solid-State Lasers

Rüdiger Paschotta and Ursula Keller

CONTENTS

7.1   Introduction

The field of ultrashort pulse generation began in the mid-1960s, soon after the invention of the laser. Since then it has been developing rapidly, and this process is still continuing, with new regimes of operation being penetrated by ultrafast lasers and with new types of ultrafast lasers being developed. In Section 7.2, we present a historical review mainly of the developments of the last two decades. The main part of this paper contains an overview of a number of important technical aspects for ultrashort pulse generation with solid-state lasers. The emphasis of this chapter is to give an updated review of the progress in pulsed solid-state lasers (i.e., bulk but not fiber lasers) during the last 10 years. There has been a book chapter, "Ultrafast Solid-State Lasers," by the same authors, which appeared in the book *Ultrafast Lasers: Technology and Applications* in 2003 [1]; this has now been updated and somewhat expanded, e.g., concerning new performance records, Q-switched mode locking, and new saturable absorbers.

This chapter is organized as follows: Section 7.2 gives a historical review on mode-locked lasers. Section 7.3 contains a detailed discussion of the demands on gain media for ultrashort pulse generation and gives an overview on available media as well as the corresponding achievements. Technical issues of particular importance are covered in Section 7.4 to Section 7.7 and include the effects of dispersion and nonlinearities in laser cavities, and different mode-locking techniques. Several examples for ultrafast lasers are then discussed in Section 7.8. Finally, the article ends with a summary and outlook in Section 7.9.

7.2   History of Mode-Locked Lasers

Mode locking was first demonstrated in the mid-1960s using a HeNe laser [2], a ruby laser [3], and an Nd:glass laser [4]. Solid-state laser media were used for ultrashort pulse generation very early on. However, at that time solid-state lasers could not produce continuous-wave (cw) mode-locked output, i.e., a pulse train with essentially constant pulse repetition rate and pulse energy. Instead, only a burst of pulses were generated, where these bursts typically lasted for microseconds and were repeated with kilohertz frequencies. This resulted from the fact that the saturable absorber used for mode locking also drove those lasers into Q-switched operation; the resulting regime is called "Q-switched mode locking" (Figure 7.1). It leads to higher pulse energies and peak powers but is typically much less stable than cw mode locking and not suitable for many applications. This continued to be a problem for most passively mode-locked solid-state lasers until, in 1992, the first intracavity saturable absorber was designed correctly to prevent Q-switching instabilities in solid-state lasers [5].

For some time, the success of ultrafast dye lasers in the 1970s and 1980s diverted the research interest away from solid-state lasers. This was partly because Q-switching instabilities are not a problem for dye lasers but also because dye lasers soon allowed the generation of much shorter pulses. In 1974 the first sub-picosecond passively mode-locked dye lasers [6–8] and, in 1981, the first sub-100-fs colliding pulse mode-locked (CPM) dye lasers [9] were demonstrated. The CPM dye laser produced pulses as short as 27 fs with a typical average output power of about 20 mW [10], and for many years it became the workhorse of ultrafast laser spectroscopy in physics and chemistry, although working with the usually quite poisonous (partly carcinogenic) and short-lived dyes and their solvents was not convenient, and the output power remained quite limited. Shorter pulse durations, down to 6 fs, were achieved through additional amplification and external pulse compression but only at much lower repetition rates [11].

The development of diode lasers with higher average powers in the 1980s again stimulated a strong interest in solid-state lasers. Diode laser pumping provides dramatic improvements in efficiency, lifetime, size, and other important laser characteristics. For example, actively mode-locked diode-pumped Nd:YAG [12] and Nd:YLF [13–16] lasers generated 7–12 ps pulse durations for the first time. In comparison, flashlamp-pumped Nd:YAG and Nd:YLF lasers typically produced pulse durations of ≈100 ps and ≈30 ps, respectively [17,18]. Before 1992, however, all attempts to passively mode-lock diode-pumped solid-state lasers resulted in Q-switching instabilities that, at best, produced stable mode-locked pulses within longer Q-switched macropulses, as mentioned above.

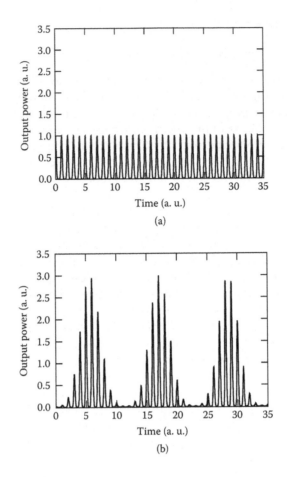

FIGURE 7.1
Schematic illustration of (a) continuous-wave (cw) mode locking and (b) Q-switched mode locking.

The breakthrough to ultrafast solid-state lasers was additionally sparked with the discovery of the Ti:sapphire laser medium [19], which was the first solid-state laser medium that was able to support a few femtosecond pulses. The existing passive mode-locking techniques, primarily developed for dye lasers, were inadequate because of the much longer upper state lifetime (in the μsec regime) and smaller gain cross section (in the 10^{-19} cm^2 regime) of Ti:sapphire compared to dyes (with nanosecond lifetimes and cross sections in the order of 10^{-16} cm^2). Therefore, passive pulse generation techniques had to be adapted to new solid-state laser materials. The strong interest in an all-solid-state ultrafast laser technology was the driving force, and formed the basis for many new inventions and discoveries.

Kerr lens mode locking (KLM) [20] of Ti:sapphire lasers was discovered in 1991 and produced the shortest laser pulses with less than 6 fs duration [21–23] directly from the laser cavity without any additional external cavity pulse compression. Sub-4-fs pulses have been demonstrated with external pulse compression [24] for the first time using cascaded hollow-fiber pulse compression. External pulse compression into few-optical-cycle regime [25] is either based on optical parametric amplification [26], compression of cavity-dumped pulses in a silica fiber [27], hollow-fiber pulse compression [28] or, more recently, occurs through filamentation [29]. Especially the latter two allow for pulse energies of more than 100 μJ with only few-optical-cycle, which fulfill a central task in the generation of attosecond extreme violet (XUV) pulses [30]. For such applications, using intense few-cycle pulses in the near-infrared driving extreme nonlinear processes, the electric field amplitude rather than the intensity envelope becomes the important factor. The electric field underneath the pulse envelope then needs to be located at a fixed position with respect to the peak of the envelope. This carrier-envelope-offset (CEO) [31] can be phase-locked in the laser oscillations with attosecond accuracy [32]. The CEO phase lock can be maintained through chirped-pulse amplification (CPA) Ti:sapphire system [33] followed by either hollow-fiber pulse compression [34] or filamentation [29], and also through chirped-pulse optical parametric amplification [35,36].

In 1992, semiconductor saturable absorber mirrors (SESAMs) [5,37] allowed for the first time self-starting and stable passive mode locking of diode-pumped solid-state lasers without Q-switching instabilities. This was soon demonstrated with a large number of different gain media, providing different output wavelengths, and pulse duration regimes [38–39]. Furthermore, we have developed the theoretical underpinnings of the performance of SESAMs in solid-state lasers [37,40,39], worked out design guidelines for application to practical laser systems either to prevent Q-switch-mode locking [41] or to optimize stable Q-switching [42], and taken this know-how to demonstrate unprecedented laser performance improvements in several key directions: shortest pulse widths (around 5 femtoseconds) [21,25], highest average and peak power from a passively mode-locked laser (nanojoules extended to microjoules) [43], and highest pulse repetition rate too (~1 Gigahertz extended to >160 Gigahertz) [44–46]. More recently, we have demonstrated passive mode locking of VECSELs (vertical external cavity surface emitting semiconductor lasers) [47], demonstrating more than 2 W average output power with close to transform-limited pulses which is something that semiconductor lasers had never before been able to do [48–49]. Full wafer-scale integration will make these sources extremely compact and cheap [50]. Today, SESAM mode-locked solid-state lasers have replaced KLM lasers for most applications. KLM is still required to obtain the shortest pulse duration in the few-optical-cycle regime and for extremely broad tunability. However, KLM has serious limitations because the mode-locking process is generally not self-starting and critical cavity alignment close to the cavity stability limit is required. Therefore, optimization of KLM lasers for short pulse durations usually leads to reduced stability, efficiency, and output power. These constraints become rather disturbing when high average output powers or very compact cavity designs are required. For these reasons, mode locking with intracavity SESAMs represents an attractive alternative to KLM.

Today, SESAMs are well established for ultrafast all-solid-state lasers. The main reason for their utility is that both the linear and nonlinear optical properties can be adjusted over wide ranges by suitable materials and design, allowing for more freedom in the specific laser cavity design. The main device parameters such as operation wavelength λ, modulation depth ΔR, saturation fluence F_{sat}, and absorber recovery time τ_A can be measured [51] and custom designed in a wide range for either stable cw mode locking [41,52], or pure Q-switching [42], or a combination of both [40]. The first SESAM device for solid-state lasers, the antiresonant Fabry–Perot saturable absorber (A-FPSA), consisted of a Fabry–Perot cavity filled with a saturable absorber with its thickness adjusted for antiresonance, i.e., the intensity in the cavity was substantially lower than the incident intensity [5]. Later, it was recognized that the top mirror element was not necessary, and that the absorber could be integrated in the lower Bragg mirror [53–54] or an appropriate spacer layer [37,54]. Earlier, a nonlinear Bragg reflector was introduced by Garmire's group but had too much loss modulation to be suitable for passively mode- locking solid-state lasers at the time [55]. Such a nonlinear Bragg reflector, which consists of alternating quarter-wave layers with a single quantum well absorber embedded, was later also sometimes called a saturable Bragg reflector (SBR) [56]. A close-to-resonant Fabry–Perot saturable absorber structure, referred to as the D-SAM (dispersive saturable absorber mirror) was used to optimize both negative group-delay dispersion (GDD) and the nonlinear modulation to construct more compact femtosecond sources [57].

One important parameter of a SESAM device is its saturation fluence, which has typical values in the range of several tens to hundreds of $\mu J/cm^2$. Lower saturation fluence is particularly relevant for fundamentally mode-locked solid-state lasers with an ultra-high pulse repetition rate (i.e., >>1 GHz) [44–46], and very high average output power [58,43]. Novel absorber materials with increased absorption cross sections are one alternative to reduce the saturation fluence. Quantum dots are promising candidates for this [50,59], and in the 1.3–1.5 μm wavelength regime GaInNAs absorbers can show decreased saturation fluence [60–62]. Two novel concepts for semiconductor saturable absorber structures result in decreased saturation fluence: the low-field-enhanced resonant-like SESAM device (LOFERS), which gives up to eleven times lower saturation fluence than the classical SESAM device, but at the price of reduced operation bandwidth and tightened growth tolerance, and the enhanced SESAM device (E-SESAM), which combines broadband operation, relaxed growth tolerances, and reduced saturation fluence [63].

The recovery times of SESAMs vary between a few nanoseconds for Q-switching applications and a few picoseconds for ultrafast lasers. For some time, it was believed that a pulse duration significantly below the absorber recovery time cannot be obtained because in this situation a time window with net gain appears behind the pulse. In dye lasers, this problem did not occur because not only the absorber but also the gain medium is strongly saturated by the circulating pulse. However, it was soon found that pulse durations far below the absorber recovery time are indeed possible also in solid-state lasers, where gain saturation by a single pulse is very weak. This was first understood for the regime of soliton mode locking [64–66] where the pulse formation is basically done by the soliton effects and the saturable absorber is only required to start and stabilize the mode locking. For example, a SESAM with a recovery time of 10 ps was sufficient to stabilize soliton pulses with only 300 fs duration in a Ti:sapphire laser [65]. In contrast to KLM, no critical cavity alignment is necessary, and mode locking is self-starting. Pulses as short as 13 fs have been produced with soliton mode-locked Ti:sapphire lasers [66]. It was also found that pulse durations well below the absorber recovery time are possible even without soliton pulse formation. In this regime, the pulse (but not any noise growing in the gain window behind the pulse) is constantly delayed by the absorber (which attenuates mostly the leading part and thus shifts the center of the pulse). This mechanism limits the time in which noise behind the pulse can grow [67].

For some time it was believed that mode locking with SESAMs would not allow for significant further increases of average output power. However, SESAM damage can be avoided even at very much high powers provided that suitable designs with low saturation fluence and not too high modulation depth are used, and Q-switching instabilities are safely suppressed [41,68]. The latter often appeared to be difficult when high-power laser heads with rather large laser mode areas in the gain medium are used. The Q-switching tendency could then be avoided only by rather strong focusing on the SESAM, which increases the risk of damage. However, high-power laser heads have been designed with significantly smaller mode areas, and this allowed the safe suppression of Q-switching instabilities without operating the SESAM under extreme conditions. We typically operate the device at three to five times the saturation fluence to reduce the residual losses. More suitable laser heads and cavity designs allowed us to demonstrate as much as 27 W of average power in 19-ps pulses [69], using three commercial Nd:YAG laser heads. Even significant further increases of power appear to be possible, e.g., just by using more laser heads.

In the sub-picosecond regime, it first appeared to be a far more challenging task to demonstrate multiwatt output powers. Gain media with the required broad amplification bandwidth often bring in other constraints: poorer thermal conductivity and lower laser cross sections, and sometimes a quasi-three-level laser transition. Although poor thermal properties are an obvious obstacle on the path towards higher powers, low laser cross sections can raise equally severe challenges by introducing a strong Q-switching tendency [41]. These problems have been solved for high powers in the sub-picosecond regime by using thin-disk laser heads (Subsection 7.8.2) [70] as developed in the 1990s at the University of Stuttgart. Only with optimized SESAMs as discussed earlier and the effect of spatial hole burning in the laser head [71] did it become possible to demonstrate a thin disk Yb:YAG laser that was passively mode-locked with a SESAM to generate 730-fs pulses with as much as 16 W of average power [72] and, more recently, 60 W [43] and 80 W [73] with similar pulse durations. For even shorter pulse durations, thin-disk laser heads can also be built with Yb^{3+}-doped tungstate crystals such as Yb:KYW, with which we obtained 22 W in 240-fs pulses so far [74]. We envisage that improved versions should soon allow for well over 30 W with pulse durations of 200 fs and even below. This new regime of power and pulse duration opens exciting perspectives for entirely new applications. This laser has been used to produce a red–green–blue (RGB) source with close to 10 W average output power in each color [73] and with external pulse-compression peak powers as high as 12 MW have been generated with 33-fs pulses and a pulse repetition rate of 34 MHz [75]. This could be focused to a peak intensity of 10^{14} W/cm^2, a regime where high field laser physics becomes possible, such as for example high-harmonic generation at 34 MHz to improve signal-to-noise ratio in XUV measurements.

Another active field of research is the generation of pulse trains with very high repetition rates. This also became possible using a systematic SESAM optimization with regard to Q-switching instabilities [52,76,63]. We have demonstrated quasi-monolithic miniature Nd:YVO$_4$ lasers which look more like previous Q-switched microchip lasers but generate stable mode-locked pulse trains with currently up to $\approx$160 GHz [44]. Such lasers provide the combination of high repetition rates with output powers far higher than those from mode-locked semiconductor or fiber lasers. This makes it possible also to drive nonlinear devices, e.g., parametric oscillators, with such lasers [77–79]. In this way, ultrabroadband tunable output into the S-, C- and L-band for telecommunication application is possible. Full C-band tuning in the multi-10-GHz regime can be obtained with diode-pumped Er:Yb:glass lasers [80–81] with up to 50 GHz [45] demonstrated so far. A wide area of new opportunities for applications, e.g., in telecommunications, arises and will be explored in the next few years.

Today, a large variety of reliable and compact all-solid-state ultrafast lasers is available with pulse durations ranging from picoseconds to well below 100 fs. A detailed table with all results using different solid-state lasers and different mode-locking techniques is provided elsewhere [39].

7.3   Gain Media for Ultrashort Pulse Generation

Gain media for ultrafast lasers have to meet a number of conditions. We first list those criteria which apply to continuous-wave (cw) lasers as well. Obviously the gain medium should have a laser transition in the desired wavelength range and a pump transition at a wavelength where a suitable pump source is available. Several factors are important to achieve good power efficiency: a small quantum defect, the absence of parasitic losses, and a high gain ($\sigma\tau$ product, where σ is the gain cross section and τ the upper state lifetime of the gain medium) are desirable. The latter allows for the use of an output coupler with relatively high transmission, which makes the laser less sensitive to intracavity losses. For high-power operation, we prefer media with good thermal conductivity, a weak (or even negative) temperature dependence of the refractive index (to reduce thermal lensing), and a weak tendency for thermally induced stress fracture.

For ultrafast lasers, in addition we require a broad emission bandwidth, as ultrashort pulses have a large bandwidth. More precisely, we need a large range of wavelengths in which a smoothly shaped gain spectrum is obtained for a fixed inversion level. The latter restrictions explain why the achievable mode-locked bandwidth is in some cases (e.g., some Yb^{3+}-doped media [82]) considerably smaller than the tuning range achieved with tunable cw lasers, particularly for quasi-three-level gain media. A less obvious requirement is that the laser cross sections should be high enough. While the requirement of a reasonably small pump threshold can be satisfied even with low laser cross sections if the fluorescence lifetime is large enough, it can be very difficult to overcome Q-switching instabilities (see section 7.3.5) in a passively mode-locked laser based on a gain material with low laser cross sections. Unfortunately, many broad-band gain media tend to have low laser cross sections, which can significantly limit their usefulness for passive mode locking, particularly at high pulse repetition rates and in cases where a poor pump beam quality or poor thermal properties necessitate a large mode area in the gain medium. Finally, a short pump absorption length is desirable because it permits the use of a small path length in the medium, which allows for operation with a small mode area in the gain medium and also limits the effects of dispersion and Kerr nonlinearity. The latter is particularly important for very short pulses.

Most gain media for ultrafast lasers belong to one of two groups. The first group has quite favorable properties for diode-pumped high-power cw operation, but can not be used for femtosecond pulse generation because of their relatively small amplification bandwidth. Typical examples are Nd^{3+}:YAG and Nd^{3+}:YVO$_4$. With high-power laser diodes, one or several conventional end-pumped or side-pumped laser rods and a SESAM (section 7.3.4.1) for mode locking, up to 27 W of average power in 19-ps pulses has been achieved with Nd^{3+}:YAG [69], or 20 W in 20-ps pulsed with Nd^{3+}:YVO$_4$ [83]. Significantly shorter pulse durations have been achieved at lower output powers, down to 1.5 ps with 20 mW [84], using the technique of additive pulse mode locking (APM, section 7.3.4.3). For all these Nd^{3+}-doped crystals, the relatively large laser cross sections usually make it relatively easy to achieve stable mode-locked operation without Q-switching instabilities, if the laser mode area in the gain medium is not made too large. See section 8.1 for typical cavity setups.

The second group of gain media is characterized by a much broader amplification bandwidth, typically allowing for pulse durations well below 0.5 ps, but also usually by significantly poorer thermal properties and lower laser cross sections. Ti^{3+}:sapphire [19] is a notable exception, combining nearly all desired properties for powerful ultrafast lasers, except that the short pump wavelength excludes the use of high-power diode pump lasers, and that the quantum defect is large. Using an argon-ion laser or a frequency-doubled solid-state laser as a pump source, Ti^{3+}:sapphire lasers have been demonstrated to generate pulses with durations below 6 fs and a few hundred milliwatts of average power [21,85]. For these pulse durations, KLM (section 7.3.4.2) is required, and self starting may be achieved with a SESAM in addition [21]. With SESAM alone, 13-fs pulses with 80 mW have been demonstrated [66]. If significantly longer pulse durations are acceptable, several watts of average power can be generated with a commercially available Ti^{3+}:sapphire laser, usually pumped with a frequency-doubled diode-pumped solid-state laser at ≈ 1 μm. Another option is to achieve rather high pulse energies and peak powers by using a very long laser cavity and limiting the peak intensities by the use of longer and chirped pulses in the cavity, which may be compressed externally. Such a laser has been demonstrated to produce 130-nJ pulses with <30 fs pulse duration and >5 mW peak power [86].

In recent years, Cr^{2+}:ZnSe [87] has been identified as another very interesting gain material which is in various ways similar to Ti^{3+}:sapphire, but emits at mid-infrared wavelengths around 2.2–2.8 μm. This very broad bandwidth should allow for pulse durations below 20 fs, although to date the shortest achieved pulse duration is much longer, ≈ 4 ps [88]. Apparently, the large Kerr nonlinearity of this medium is causing significant problems for short pulse generation.

Diode-pumped femtosecond lasers can be built with crystals like Cr^{3+}:LiSAF, Cr^{3+}:LiSGaF, or Cr^{3+}:LiSCAF which can be pumped at longer wavelengths than Ti^{3+}:sapphire. However, these media have much poorer thermal properties and thus can not compete with Ti^{3+}:sapphire in terms of output power; the achievable optical bandwidth is also lower. Cr^{3+}:LiSAF lasers have generated pulses as short as 12 fs [89], but only with 23 mW of output power, using KLM without self-starting ability. The highest achieved mode-locked power was 0.5 W in 100-fs pulses [90]. More recently, compact Cr^{3+}:LiSAF lasers with very low pump threshold have been developed, delivering e.g., 136-fs pulses with 20 mW average power for <100 mW optical pump power [91].

Cr^{4+}:forsterite emits around 1.3 μm and is suitable for pulse durations down to 14 fs with 80 mW [92], or for 800 mW in 78-fs pulses [93]. Normally, a Nd^{3+}-doped laser (which may be diode-pumped) is used for pumping of Cr^{4+}:forsterite. The same holds for Cr^{4+}:YAG, which emits around 1.4–1.5 μm and has allowed to generate pulses with 20 fs, 400 mW [94].

Other broad-band gain materials are phosphate or silicate glasses, doped with rare-earth ions such as Nd^{3+} or Yb^{3+}, for pulse durations down to ≈ 60 fs [95,96] and output powers of a few hundred milliwatts. The relatively poor thermal properties make high-power operation challenging. Up to 1.4 W of average power in 275-fs pulses [68], or 1 W in 175-fs pulses [97], have been obtained from Nd^{3+}:glass by using a specially adapted elliptical mode pumping geometry [98]. Here, a strongly elliptical pump beam and laser mode allow the use of a fairly thin gain medium which can be efficiently cooled from both flat sides. The resulting nearly one-dimensional heat flow reduces the thermal lensing compared to cylindrical rod geometries, if the aspect ratio is large enough. A totally different route toward high peak powers is to use a cavity-dumped laser; with such a device, based on Yb^{3+}:glass, 400-nJ pulses with more than 1 mW peak power have been generated [99].

Yb^{3+}:YAG has similar thermal properties as Nd^{3+}:YAG and at the same time a much larger amplification bandwidth. Another favorable property is the small quantum defect. However, challenges arise from the quasi-three-level nature of this medium and from the small laser cross sections, which favor Q-switching instabilities (see section 7.3.5). High pump intensities help in both respects. An end-pumped laser based on a Yb^{3+}:YAG rod has generated 340-fs pulses with 170 mW [100]. As much as 8.1 W in 2.2-ps pulses was obtained from an elliptical mode Yb^{3+}:YAG laser [101]. In 2000, the first Yb^{3+}:YAG thin disk laser [70] has been passively mode-locked, generating 700-fs pulses with 16.2 W average power [101]. The concept of the passively mode-locked thin disk layer has been demonstrated to be power scalable, which so far lead up to 80 W in 0.7-ps pulses [73] and up to 5-μm pulse energy [58]. An additive-pulse mode-locked Yb:YAG laser delivered 21 W in 0.58-ps pulses [102].

In recent years, a few Yb^{3+}-doped crystalline gain materials have been developed which combine a relatively broad amplification bandwidth (sufficient for pulse durations of a few hundred femtoseconds) with thermal properties which are better than those of other broad-band materials, although not as good as e.g., those of YAG or sapphire. Examples are Yb^{3+}:YCOB [103], Yb^{3+}:YGdCOB [104], Yb^{3+}:SFAP [105], Yb^{3+}:SYS [106], Yb^{3+}:BOYS [107,108], Yb^{3+}:KGW [109], Yb^{3+}:YVO$_4$ [110], Yb^{3+}:Y$_2$O$_3$ ceramics [111], and Yb^{3+}:CaF$_2$ [112]. With an end-pumped Yb^{3+}:KGW rod, 1.1 W of average power have been achieved in 176-fs pulses [109]. A Kerr lens mode-locked Yb^{3+}:KYW laser produced pulses as short as 107 fs [113], while a SESAM mode-locked Yb^{3+}:SYS laser reached 70 fs with 156 mW average power [106]. Note that some of these media exhibit rather low emission cross sections and therefore make stable passive mode locking difficult, while they might be very useful e.g., in regenerative amplifiers. Tungstate crystals (Yb^{3+}:KGW, Yb^{3+}:KYW) have been rather useful for passive mode locking since they have relatively high cross sections. Yb^{3+}:KYW has been applied in a thin disk laser, generating 22 W in 240-fs pulses [74]. With improved crystal quality, significant performance enhancements appear to be feasible. Another new class of materials with particular importance are the Yb^{3+}-doped sesquioxides [114] such as Y_2O_3, Sc_2O_3, and Lu_2O_3, which appear to be very suitable for high-power operation with short pulses.

Color center crystals can also be used for femtosecond pulse generation [115–117], but we do not discuss them here. Today, they are not frequently used any more because they need cryogenic conditions and other all-solid-state laser systems can cover and even exceed their performance.

7.4 Dispersion

7.4.1 Orders of Dispersion

When a pulse travels through a medium, it acquires a frequency-dependent phase shift. A phase shift that varies linearly with the frequency corresponds to a time delay, without any change of the temporal shape of the pulse. Higher-order phase shifts, however, tend to modify the pulse shape and are thus relevant for the formation of short pulses. The phase shift can be expanded in a Taylor series around the center angular frequency ω_0 of the pulse:

$$\varphi(\omega) = \varphi_0 + \frac{\partial\varphi}{\partial\omega}(\omega-\omega_0) + \frac{1}{2}\frac{\partial^2\varphi}{\partial\omega^2}(\omega-\omega_0)^2 + \frac{1}{6}\frac{\partial^3\varphi}{\partial\omega^3}(\omega-\omega_0)^3 + \ldots$$

Here, the derivatives are evaluated at ω_0. $T_g \equiv (\partial\varphi/\partial\omega)$ is the group delay, $D \equiv (\partial^2\varphi/\partial\omega^2)$ the group delay dispersion (GDD), $(\partial^3\varphi/\partial\omega^3)$ the third-order dispersion (TOD). The GDD describes a linear frequency dependence of the group delay and thus tends to separate the frequency components of a pulse; for positive GDD, e.g., the components with higher frequencies are delayed with respect to those with lower frequencies, which results in a positive "chirp" ("up-chirp") of the pulse. Higher orders of dispersion generate more complicated distortions.

Note that there is some confusion concerning the sign of group delay dispersion. The ultrafast community associates positive dispersion with the case $(\partial^2\varphi/\partial\omega^2) > 0$, but a convention with opposite sign is frequently used in fiber optics, where a wavelength rather than a frequency derivative occurs in the definition. To avoid confusion, one may talk of *normal* dispersion for $(\partial^2\varphi/\partial\omega^2) > 0$ and *anomalous* dispersion for $(\partial^2\varphi/\partial\omega^2) < 0$. This wording goes back to the observation that most transparent media exhibit $(\partial^2\varphi/\partial\omega^2) > 0$ in the range of visible wavelengths.

The broader the bandwidth of the pulse (i.e., the shorter the pulse duration), the more terms of this expansion are significant. GDD, which acts on an initially unchirped Gaussian pulse with FWHM (full width at half maximum) pulse duration τ_0, increases the pulse duration according to [18]

$$\tau = \tau_0 \cdot \sqrt{1 + \left(4(\ln 2) \cdot \frac{D}{\tau_0^2} \right)^2}. \qquad (7.1)$$

It is apparent that the effect of GDD becomes strong if $|D| > \tau_0^2$. Similarly, TOD becomes important if $(\partial^3 \varphi / \partial \omega^3) > \tau_0^3$. Note that dispersion within a laser cavity can have important effects even if it is not strong enough to significantly broaden a pulse during a single cavity round-trip.

7.4.2 Dispersion Compensation

If no dispersion compensation is used, the net GDD for one cavity round-trip is usually positive, mainly because of the dispersion in the gain medium. Other components like mirrors may also contribute to this. However, in lasers with >10 ps pulse duration the dispersion effects can often be ignored, as the total GDD in the laser cavity is typically at most a few thousand fs^2, much less than the pulse duration squared. For shorter pulse durations, the GDD has to be considered, and pulse durations well below 30 fs usually necessitate the compensation of TOD or even higher orders of dispersion depending on the thickness of the gain material. In most cases, the desired total GDD is not zero but negative, so that soliton formation (Section 7.6) can be exploited. Usually, one requires sources of negative GDD, and, in addition, appropriate higher-order dispersion for shorter pulses. The most important techniques for dispersion compensation are discussed in the following subsections.

7.4.2.1 *Dispersion from Wavelength-Dependent Refraction*

If the intracavity laser beam hits a surface of a transparent medium with nonnormal incidence, the wavelength dependence of the refractive index can cause wavelength-dependent refraction angles. In effect, different wavelength components will travel on slightly different paths, and this in general introduces an additional wavelength dependence to the round-trip phase, thus contributing to the overall dispersion. The most frequently used application of this effect is to insert a prism pair in the cavity [118], where the different wavelength components travel in different directions after the first prism and along parallel but separated paths after the second prism. The wavelength components can be recombined simply on the way back after reflection at a plane end mirror (of a standing-wave cavity) or by a second prism pair (in a ring cavity). Spatial separation of different wavelengths occurs only in a part of the cavity. The obtained negative GDD from the geometric effect is proportional to the prism separation, and an additional (usually positive) GDD contribution results from the propagation in the prism material. The latter contribution can be easily adjusted via the prism insertion, so that the total GDD can be varied in an appreciable range. Some higher-order dispersion is also generated, and the ratio of TOD and GDD can for a given prism material be varied only in a limited range by using different combinations of prism separation and insertion. Some prism materials with lower dispersion (e.g., fused quartz instead of SF10 glass) can help to reduce the amount TOD generated together with a given value of GDD, but these also necessitate a larger prism separation. The prism angles are usually not used as optimization parameters but rather are set to be near Brewster's angle in order to minimize reflection losses. The small losses and the versatility of the prism pair technique are the reasons why prism pairs are very widely used in ultrafast lasers. In a few millimeter thick Ti^{3+}:sapphire lasers, pulse durations around 10 fs can be reached with negative dispersion only from a fused quartz prism pair.

More compact geometries for dispersion compensation make use of a single prism only [119,120]. In this case, the wavelength components are spatially separated in the whole resonator, not only in a part of it. Even without any additional prisms, refraction at a Brewster interface of the gain medium can generate negative dispersion. In certain configuration, where the cavity is operated near a stability limit, the refraction effect can be strongly increased [121], so that significant negative GDD can be generated in a compact cavity. The amount of GDD may then also strongly depend on the thermal lens in the gain medium and on certain cavity dimensions.

7.4.2.2 Grating Pairs

Compared to prism pairs, pairs of diffraction gratings [122] can generate higher dispersion in a compact setup. However, because of the limited diffraction efficiency of gratings, the losses of a grating pair are typically higher than acceptable for use in a laser cavity, except in cases with a high gain (e.g., in fiber lasers). For this reason, grating pairs are normally only used for external pulse compression.

7.4.2.3 Gires–Tournois Interferometers (GTIs)

A compact device to generate negative GDD (even in large amounts) is the Gires–Tournois Interferometer (GTI) [123], which is a Fabry–Perot interferometer, operated in reflection. As the rear mirror is highly reflective, the GTI as a whole is highly reflective over the whole wavelength range, whereas the phase shift varies nonlinearly by 2π for each free spectral range, calculated as $\Delta v = c/2nd$ where n and d are the refractive index and the thickness of the spacer material, respectively. Within each free spectral range, the GDD oscillates between two extremes the magnitude of which is proportional to d^2 and also depends on the front mirror reflectivity. Ideally, the GTI is operated near a minimum of the GDD, and the usable bandwidth is some fraction (e.g., one tenth) of the free spectral range, which is proportional to d^{-1}. Tunable GDD can be achieved if the spacer material is a variable air gap which, however, must be carefully stabilized to avoid unwanted drifts. More stable but not tunable GDD can be generated with monolithic designs, based on thin films of dielectric media like TiO_2 and SiO_2, particularly for the use in femtosecond lasers. The main drawbacks of GTI are the fundamentally limited bandwidth (for a given amount of GDD) and the limited amount of control of higher-order dispersion.

7.4.2.4 Dispersive Mirrors

Dielectric Bragg mirrors with regular $\lambda/4$ stacks have negligible GDD when operated well within their reflection bandwidth, but they experience increasing dispersion at the edges of this range. Modified designs can be used to obtain well-controlled dispersion over a large wavelength range. One possibility already discussed is the use of a GTI structure (see Subsection 7.4.2.3). Another broad range of designs is based on the concept of the chirped mirror [124–125]: If the Bragg wavelength is appropriately varied within a Bragg mirror design, longer wavelengths can be reflected deeper in the structure, thus acquiring a larger phase change, which leads to negative dispersion. However, the straightforward implementation of this idea leads to strong oscillations of the GDD (as a function of frequency), which render such designs useless. These oscillations can be greatly reduced by numerical optimizations which introduce complicated (and not analytically explainable) deviations from the simple chirp law. A great difficulty is that the figure of merit to optimize is a complicated function of many layer thickness variables, which typically has a large number of local maxima and minima, and thus is quite difficult to optimize. Nevertheless, refined computing algorithms have lead to designs with respectable performance, which were realized with the precision growth of dielectric mirrors. Such mirrors can compensate the dispersion in Ti^{3+}:sapphire lasers for operation with pulses durations well below 10 fs [25,126].

The original chirped-mirror design was further refined by the double chirped mirror (DCM) concept which takes into account the impedance matching problem occurring at the air mirror interface and the grating structure in the mirror [127–128]. The impedance matching concept allowed much better insight into the design limitations and also allowed for the first time an analytical design with custom-tailored dispersion characteristics that required only minor numerical optimization [129–130]. These double-chirped mirrors resulted in new world-record pulse durations in the two-optical-cycle regime from KLM Ti:sapphire lasers [21–22,131]. However, the impedance matching to the air sets a limit. This impedance matching is based on a broadband antireflection (AR) coating that interferes with the rest of the multilayer mirror design and therefore has to be of very high quality with a very low residual power reflectivity of less than 10^{-4} [132]. However, this can only be achieved over a limited bandwidth and is impossible for more than ≈ 0.7 optical octaves in the near-infrared and visible spectrum [133].

The invention of the back-side coated (BASIC) mirrors [132] or later the tilted front-side mirrors [134] resolved this issue. In the back-side coated mirror the ideal DCM structure is matched to the low index material of the mirror, which ideally matches the mirror substrate material. This DCM structure is deposited on the back of the substrate, and the AR-coating is deposited on the front of the slightly wedged or curved substrate, so that the residual reflection is directed out of the beam and does not deteriorate the dispersion properties of the DCM structure on the other side of the substrate. Thus, the purpose of the AR-coating is only to reduce the insertion losses of the mirror at the air–substrate interface. For most applications it is sufficient to get this loss as low as 0.5%. Therefore, the bandwidth of such an AR-coating can be much broader. The trade-off is that the substrate has to be as thin as possible to minimize the overall material dispersion. In addition, the wedged mirror leads to an undesired angular dispersion of the beam.

Another possibility to overcome the AR-coating problem is given with the idea to use an ideal DCM under Brewster-angle incidence [135]. In this case, the low index layer is matched to air. However, under p-polarized incidence the index contrast and therefore the Fresnel reflectivity of a layer pair is reduced, and more layer pairs are necessary to achieve high reflectivity. This increases the penetration depth into the mirror, which has the advantage that these mirrors can produce more dispersion per reflection, but this means that scattering and other losses, and also fabrication tolerances, become even more severe. In addition, this concept is difficult to apply to curved mirrors. In addition, the spatial chirp of the reflected beam has to be removed by back reflection or an additional reflection from another Brewster-angle mirror.

Other methods to overcome the AR-coating problem are based on using different chirped mirrors with slightly shifted GTI oscillations that partially cancel each other. Normally, these chirped mirrors are very difficult to fabricate [132]. Many different growth runs normally result in strong shifts of those GTI oscillations so that a special selection of mirrors makes it ultimately possible to obtain the right dispersion compensation. Some tuning of the oscillation peaks can be obtained by the angle of incident [22]. A specially designed pair of DCMs has been used to cancel the spurious GTI oscillation [136] where an additional quarter wave layer between the AR-coating and the DCM structure was added in one of the DCMs. Also, this design has its drawbacks and limitations because it requires an extremely high precision in fabrication and restricts the range of angles of incidence.

After this overview it becomes clear that there is no perfect solution to the challenge of ultrabroadband dispersion compensation. At this point ultrabroadband chirped mirrors are the only way to compress pulses in the one- to two-optical-cycle regime [25]. For example, BASIC DCMs have been used for ultrabroadband dispersion compensation of a hollow-fiber supercontinuum [137].

7.4.2.5 *Dispersive SESAMs*

Negative dispersion can also be obtained from semiconductor saturable absorber mirrors (SESAMs) (Subsection 7.7.3.4.1) with specially modified designs. The simplest option is to use a GTI-like structure (Subsection 7.4.2.3) [57]. A double-chirped dispersive semiconductor mirror has also been demonstrated [138], and a saturable absorber could be integrated into such a device.

7.5 Kerr Nonlinearity

Due to the high intracavity intensities, the Kerr effect is relevant in most ultrafast lasers. The refractive index of, e.g., the gain medium, is modified according to

$$n(I) = n_0 + n_2 I \tag{7.2}$$

where I is the laser intensity and n_2, a material-dependent coefficient, which also weakly depends on the wavelengths. (The pump intensity is normally ignored because it is by far smaller than the peak laser intensity.) These nonlinear refractive index changes have basically two consequences. The first is a transverse index gradient, resulting from the higher intensities on the beam axis compared to the intensities in the wings of the transverse beam profile. This leads to a so-called Kerr lens with an intensity-dependent focusing effect (for positive, n_2) which can be exploited for a passive mode-locking mechanism as discussed in Subsection 7.7.3.4.2.

The second consequence of the Kerr effect is that the pulses experience self-phase modulation (SPM); the pulse center is delayed more (for positive n_2) than the temporal wings. For a freely propagating Gaussian transverse beam profile with radius w (defined so that at this radius we have $1/e^2$ times the peak intensity), the nonlinear coefficient γ_{SPM}, relating the on-axis phase change φ to the pulse power P according to $\varphi = \gamma_{SPM} P$, is given by

$$\gamma_{SPM} = \frac{2\pi}{\lambda} n_2 L \left(\frac{\pi}{2} w^2 \right)^{-1} = \frac{4 n_2 L}{\lambda w^2} \tag{7.3}$$

where L is the propagation length in the medium. Note that the peak intensity of a Gaussian beam is $I = P/(\pi w^2/2)$, and the on-axis phase change (and not an averaged phase change) is relevant for freely propagating beams. For guided beams, an averaged phase change has to be used that is two times smaller.

The most important consequence of SPM in the context of ultrafast lasers is the possibility of soliton formation (Section 7.6). Another important aspect is that SPM can increase or decrease the bandwidth of a pulse, depending on the original phase profile of the pulse. In general, positive SPM tends to increase the bandwidth of positively chirped (i.e., up-chirped) pulses, but negatively chirped pulses can be spectrally compressed. An originally unchirped pulse traveling through a nonlinear and nondispersive medium will experience an increase of bandwidth only to second order of the propagation distance, whereas a chirp grows in first order.

The simple picture of the Kerr nonlinearity as presented above is often sufficient for describing nonlinear effects in mode-locked laser cavities. However, the matter is actually more complicated. First of all, Equation 7.3 rests on an approximation for narrow-band light, the elimination of which results in more complicated equations with an additional term describing the phenomenon of self-steepening. This is sometimes relevant for ultrashort pulses. Besides, the nonlinear response of transparent media also has a noninstantaneous part, which results in the Raman effect. This can lead to the Raman self-frequency shift [139], as can be relevant in mode-locked fiber lasers, for example, but only rarely in bulk lasers.

7.6   Soliton Formation

Pulse propagation through a medium with both second-order dispersion (GDD) and a Kerr nonlinearity results in a strong interaction of both effects. A special case is that the intensity has a sech^2 temporal profile

$$P(t) = P_{\mathrm{p}} \, \text{sech}^2(t \, / \, \tau_{\mathrm{S}}) = \frac{P_{\mathrm{p}}}{\cosh^2(t \, / \, \tau_{\mathrm{S}})} \tag{7.4}$$

with the peak power P_p and the FWHM pulse duration $\tau_{\mathrm{FWHM}} \approx 1.76 \cdot \tau_{\mathrm{S}}$. If such a pulse is unchirped and fulfills the condition

$$\tau_{\mathrm{S}} = \frac{2|D|}{|\gamma_{\mathrm{SPM}}| E_{\mathrm{p}}} \tag{7.5}$$

where D and γ_{SPM} must have opposite signs and are calculated for the same propagation distance, and

$$E_{\mathrm{p}} \approx 1.13 \cdot P_{\mathrm{p}} \tau_{\mathrm{FWHM}} \tag{7.6}$$

is the pulse energy, then we have a so-called fundamental soliton. Such a pulse propagates in the medium with constant temporal and spectral shape, and only acquires an overall nonlinear phase shift. Higher-order solitons, where the peak power is higher by a factor that is the square of an integer number, do not preserve their temporal and spectral shape but evolve in such a way that the original shape is restored after a certain propagation distance, the so-called soliton period in the case of a second-order soliton.

Solitons are remarkably stable against various kinds of distortions. In particular, stable soliton-like pulses can be formed in a laser cavity although dispersion and Kerr nonlinearity occur in discrete amounts, and the pulse energy varies due to amplification in the gain medium and loss in other elements. As long as the soliton period amounts to many (at least about 5–10) cavity round-trips, the soliton simply "sees" the average GDD and Kerr nonlinearity, and this "average soliton" behaves in the same way as in a homogeneous medium. The soliton period in terms of the number of cavity round-trips is

$$N_S = \frac{\pi \tau_S^2}{2|D|} \approx \frac{\tau_{FWHM}^2}{2|D|} \tag{7.7}$$

where D is calculated for one cavity round-trip. N_S is typically quite large in lasers with pulse durations of > 100 fs, so that the average soliton is a good approximation. Once N_S becomes less than ten, the soliton is significantly disturbed by the changes of dispersion and nonlinearity during a round-trip, and this may lead to pulse breakup. In Ti^{3+}:sapphire lasers for pulse durations below 10 fs, the regime of small N_S is unavoidable and can be stabilized only by using a fairly strong saturable absorber. On the other hand, in cases with very large values of N_S (as sometimes encountered with very long pulses) it can be beneficial to decrease N_S by increasing both $|D|$ and γ_{SPM}, because stronger soliton shaping can stabilize the pulse shape and spectrum and make it less dependent on other influences.

Note that soliton effects can fix the pulse duration at a certain value even if other cavity elements (most frequently the laser gain with its limited bandwidth) tend to reduce the pulse bandwidth. The pulse will then acquire some positive chirp (assuming $n_2 > 0$ for the Kerr medium), and under these conditions SPM can generate the required extra bandwidth.

7.7 Mode Locking

7.7.1 General Remarks

Ultrashort light pulses are in most cases generated by mode-locked lasers. In this regime of operation, usually a single short pulse propagates in the laser cavity and generates an output pulse each time when it hits the output coupler mirror. The generated pulses are usually quite short compared to the round-trip time.

In the frequency domain, mode locking means operation of the laser on a number of axial cavity modes, where all these modes oscillate in phase (or at least with nearly equal phases). In this case, the mode amplitudes interfere constructively only at certain times, which occur with the period of the round-trip time of the cavity. At other times, the output power is negligibly small. The term *mode locking* resulted from the observation that a fixed phase relationship between the modes has to be maintained in some way to produce short pulses. The achievable pulse duration is then inversely proportional to the locked bandwidth, i.e., to the number of locked modes times their frequency spacing.

It is obvious in the frequency domain description that mode locking cannot be achieved if a significant amount of the laser power is contained in higher-order transverse modes of the cavity, because these usually have different resonance frequencies, so that the periodic recurrence of constructive addition of all mode amplitudes is not possible. Therefore, laser operation on the fundamental transverse cavity mode (TEM$_{00}$) usually is a prerequisite to stable mode locking.

In some cases, a mode-locked laser is operated with several equally spaced pulses circulating in the cavity. This mode of operation, called *harmonic mode locking* [18], can be used for the generation of pulse trains with higher repetition rates even if the cavity can not be made very short (as, e.g., in the case of fiber lasers). The main difficulty is that the timing between the pulses has to be maintained in some way, either by some kind of interaction between the pulses or with the aid of an externally applied timing information. A number of solutions have been found, but in this article we concentrate on fundamental mode locking, with a single pulse in the laser cavity, as it occurs in most ultrafast lasers.

So far we have assumed that the mode-locked laser operates in a steady state, where the pulse energy and duration may change during a round-trip (as an effect of gain and loss, dispersion, nonlinearity, etc.), but always return to the same values at a certain position in the cavity. This regime is called *continuous-wave (cw) mode locking* and is, indeed, very similar to ordinary cw (not mode-locked) operation, as the output power in each axial cavity mode is constant over time. Another important regime is Q-switched mode locking (Subsection 7.7.3.5), where the mode-locked pulses are contained in periodically recurring bunches that have the envelope of a Q-switched pulse. It is in some cases a challenge to suppress an unwanted tendency for Q-switched mode locking.

In addition to Q-switching tendencies, a number of other mechanisms can destabilize the mode-locking behavior of a laser. In particular, short pulses have a broader bandwidth than a competing cw signal and thus tend to experience less gain than the latter. For the pulses to be stable, some mechanism is required which increases the cavity loss for cw signals more than for pulses. Also, a mode-locking mechanism should ideally give a clear loss advantage for shorter pulses compared to any other mode of operation. We will discuss various stability criteria in the following sections.

Mechanisms for mode locking are grouped into active and passive schemes, and hybrid schemes that utilize a combination of both. Active mode locking (Subsection 7.7.2) is achieved with an active element (usually an acousto-optic modulator), generating a loss modulation which is precisely synchronized with the cavity round-trips. Passive schemes (Subsection 7.7.3) rely on a passive loss modulation in some type of saturable absorber. This passive loss modulation can occur on a much faster time scale: the shorter the pulse becomes, the faster the loss modulation can be — or, at least the reduction of the loss, although the recovery of most saturable absorbers takes a finite time even for very short pulses. Therefore, passive mode locking typically allows for the generation of much shorter pulses than achieved with active mode locking. Passive schemes are also usually simpler, as they do not rely on driving circuits and synchronization electronics, and allow more compact setups. The pulse timing is then usually not externally controlled. Synchronization of several lasers can be achieved with active mode-locking schemes, but it is also possible for passively mode-locked lasers using a feedback stabilization mechanism, which can act on the cavity length.

Mode-locked lasers are in most cases optically pumped with a cw source, e.g., one or several laser diodes. This requires a gain medium that can store the excitation energy over a time of more than one cavity round-trip (which typically takes a few nanoseconds). Typical solid-state gain materials fulfill this condition very well, even for rather low repetition rates, because the lifetime of the upper laser level is usually at least a few microseconds long or, in some cases, even more than a millisecond. Within a cavity round-trip time, the gain then undergoes only very minute changes. Synchronous pumping (with a mode-locked source), as is occasionally used for dye lasers, is therefore rarely applied to solid-state lasers and is not discussed in this article.

In the following sections we discuss some details of active and passive mode locking. The main emphasis is on passive mode locking, because such techniques are clearly dominating in ultrafast lasers.

7.7.2 Active Mode Locking

An actively mode-locked laser contains some kind of electrically controlled modulator, in most cases of acousto-optic and sometimes of electro-optic type. The former utilizes a standing acoustic wave with a frequency of 10s or 100s of MHz, generated with a piezoelectric transducer in an acousto-optic medium. In this way, one obtains a periodically modulated refractive index pattern at which the laser light can be refracted. The refracted beam is normally eliminated from the cavity, and the refraction loss is oscillating with twice the frequency of the acoustic wave, because the refractive index change depends only on the modulus of the pressure deviation.

A pulse circulating in the laser cavity is hardly affected by the modulator provided that it always arrives at those times where the modulator loss is at its minimum. Even then, the temporal wings of the pulse experience a somewhat higher loss than the pulse center. This mechanism tends to shorten the pulses. On the other hand, the limited gain bandwidth always tends to reduce the pulse bandwidth and thus to increase the pulse duration. Other effects like dispersion or self-phase modulation (Section 7.4) can have additional influences on the pulse. Usually, within 1000s of cavity round-trips a steady state is reached, where the mentioned competing effects, which are both rather weak, exactly cancel each other for every complete cavity round-trip.

The theory of Kuizenga and Siegman [17,140] describes the simplest situation with a modulator (in exact synchronism with the cavity) and gain filtering but no dispersion or self-phase modulation. The gain is assumed to be homogeneously broadened with Lorentzian shape. The simple result is the steady-state FWHM pulse duration

$$\tau \approx 0.45 \cdot \left(\frac{g}{M}\right)^{1/4} \left(f_m \cdot \Delta f_g\right)^{-1/2} \tag{7.8}$$

where g is the power gain at the center frequency, M is the modulation strength ($2M$ = peak-to-peak variation of power transmission), f_m is the modulation frequency, and Δf_g is the FWHM gain bandwidth. The peak gain g is slightly higher than the cavity loss l (for minimum modulator loss) according to

$$g = l + \frac{M}{4}\left(2\pi f_m \tau\right)^2, \tag{7.9}$$

so that the total cavity losses (including the modulator losses) are exactly balanced for the pulse train. Note that the losses would be higher than the gain for either a longer pulse (with higher modulator loss) or a shorter pulse (with stronger effect of gain filtering).

It is important to note that for non-Lorentzian gain spectra, the quantity Δf_g should be defined so that the corresponding Lorentzian reasonably fits to the gain spectrum within the pulse bandwidth, which may be only a small fraction of the gain bandwidth; only the "curvature" of the gain spectrum within the pulse bandwidth is important. This means that even weak additional filter effects caused, e.g., by intracavity reflections (etalon effects), can significantly reduce the effective bandwidth and thus lead to longer pulses. Such effects are usually eliminated by avoiding any optical interfaces that are perpendicular to the incident beam (even if they are antireflection coated). Also, note that the saturation of inhomogeneously broadened gain can distort the gain spectrum and thus affect the pulse duration.

With SPM in addition (but no dispersion), somewhat shorter pulses can be generated because SPM can generate additional bandwidth. However, with too much SPM the pulses become unstable. It has been shown with numerical simulations [141] that a reduction of the pulse duration by a factor in the order of two is possible. The resulting pulses are then chirped, so that some further compression with extracavity negative dispersion is possible.

The situation is quite different if positive SPM occurs together with negative GDD, because in this case soliton-like pulses (Section 7.6) can be formed [142]. If the soliton length corresponds to less than about 100 cavity round-trips, soliton shaping effects can be much stronger than the pulse shaping effect of the modulator. The pulse duration is then determined only by the soliton equation (Equation 7.6). The soliton experiences only a very small loss at the modulator. However, due to its bandwidth it also experiences less gain than a long background pulse, the so-called continuum, which can have a small bandwidth. The shorter the soliton pulse duration, the smaller the soliton gain compared to the continuum gain and the higher the modulator strength M must be to keep the soliton stable against growth of the continuum. Using results from Reference 64, it can be shown that the minimum pulse duration is proportional to

$$
\frac{g^{1/3}}{\Delta f_g^{2/3}} \left(\frac{N_s}{M} \right)^{1/6} \tag{7.10}
$$

where N_s is the soliton period in terms of cavity round-trips (Equation 7.7). This indeed shows that a smaller modulator strength is sufficient if N_s is kept small, but also that the cavity loss should be kept small (to keep g small), and the gain bandwidth is the most important factor.

With gain media like Nd^{3+}:YAG or Nd^{3+}:YLF, typical pulse durations of actively mode-locked lasers are in the 4–20 picosecond regime. A more detailed overview is given in Table 2 of Reference 143. Using Nd^{3+}:glass and the regime of soliton formation, pulse durations as short as 310 fs have been achieved [144]. For significantly shorter pulses, active mode locking is not effective because the time window of low modulator loss would be much longer than the pulse duration.

7.7.3 Passive Mode Locking

7.7.3.1 The Starting Mechanism

Passive mode locking relies on the use of some type of saturable absorber that favors the generation of a train of short pulses against other modes of operation, such as cw emission. Starting from a cw regime, the saturable absorber will favor any small noise spikes, so that those can grow faster than the cw background. Once these noise spikes contain a significant part of the circulating energy, they saturate the gain so that the cw background starts to decay. Later on, the most energetic noise spike, which experiences the least amount of saturable absorption, will eliminate all the others by saturating the gain to a level where these experience net loss in each roundtrip. In effect, we obtain a single circulating pulse. Due to the action of the saturable absorber, which favors the peak against the wings of the pulse, the duration of the pulse is then reduced further in each cavity round-trip, until broadening effects (e.g., dispersion) become strong enough to prohibit further pulse shortening. Note that other shortening effects, such as soliton shaping effects, can also become effective.

The described startup can be prevented if strong pulse broadening effects are present in an early phase. Particularly the presence of spurious intracavity reflections can be significant, because those tend to broaden (or split up) pulses even before they have acquired a significant bandwidth. A significantly stronger saturable absorber may then be required to get self-starting mode locking. These effects are usually difficult to quantify but can, in most bulk solid-state lasers, be suppressed by using suitable designs. For example, even antireflection coated surfaces in the cavity should be slightly tilted so that beams resulting from residual reflections are eliminated from the cavity.

Note that a saturable absorber with long recovery time (low saturation intensity) is most effective for fast self-starting mode locking, although a shorter recovery time may allow generation of shorter pulses. Other techniques to facilitate self-starting mode locking include the use of optical feedback from a moving mirror [145], which tends to increase the intracavity fluctuations in cw operation.

7.7.3.2 *Parameters of Fast and Slow Saturable Absorbers*

An important parameter of a saturable absorber is its recovery time. In the simplest case, we have a so-called fast saturable absorber, which can recover on a faster time scale than the pulse duration. In this case, the state of the absorber is largely determined by the instantaneous pulse intensity, and strong shaping of the leading, as well as of the trailing edge of the pulse, takes place. On the other hand, for short enough pulse durations we have the opposite situation of a slow absorber, with the absorber recovery occurring on a long time scale compared to the pulse duration. This regime is frequently used in ultrafast lasers, because the choice of fast saturable absorbers is very limited for pulse durations below 100 fs. We will discuss in Subsection 7.7.3.3 why it is possible that pulses with durations well below 100 fs can be generated even with absorbers which are much slower.

Concrete types of saturable absorbers are compared in Subsection 7.7.3.4. In the following we discuss some parameters which can be used to quantitatively characterize the action of saturable absorbers.

The intensity loss[*] q generated by a *fast saturable absorber* depends only on the instantaneous intensity I, the incident power divided by the mode area. Simple absorber models lead to a function

$$q(I) = \frac{q_0}{1 + I / I_{sat}} \tag{7.11}$$

that is a reasonable approximation in many cases. The response is then characterized by the parameter I_{sat}, called *saturation intensity*, and the unsaturated loss q_0. For pulses with a given peak intensity I_p, an average value of

---

[*] Note that in some publications q was defined as the amplitude (instead of intensity) loss coefficient, which is one half the value used here.

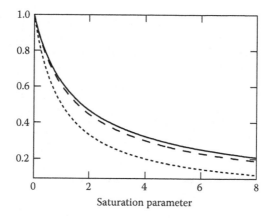

FIGURE 7.2

Solid curve: effective loss for a soliton pulse on a fast saturable absorber, as a function of the ratio of peak intensity to the saturation intensity; dashed curve: the same for a Gaussian pulse; dotted curve: loss for the peak intensity.

q can be calculated which represents the effective loss for the pulse. Figure 7.2 shows this quantity as a function of the normalized peak intensity for Gaussian and soliton (sech²) pulses, together with $q(I_p)$. We see that the pulse form has little influence on the average loss.

The behavior of a *slow saturable absorber* is described by the differential equation

$$\frac{dq}{dt} = -\frac{q - q_0}{\tau} - \frac{I}{F_{sat}} q \qquad (7.12)$$

with the recovery time τ, the unsaturated loss q_0, and the saturation fluence F_{sat}. (We assume $q \ll 1$.) If the recovery is so slow that we can ignore the first term, the value of q after a pulse with fluence F_{sat} is $q_0 \exp(-F_p/F_{sat})$ if the pulse hits an initially unsaturated absorber. The effective loss for the pulse is (independent of the pulse form)

$$q_p(F_p) = q_0 \left[1 - \exp(-F_p / F_{sat}) \right] \cdot F_{sat} / F_p = q_0 \left[1 - \exp(-S) \right] / S \qquad (7.13)$$

with the saturation parameter $S := F_p / F_{sat}$. For strong saturation ($S > 3$, as usual in mode-locked lasers), the absorbed pulse fluence is $\approx F_{sat} \Delta R$, and we have

$$q_p(S) \approx q_0 / S. \qquad (7.14)$$

Figure 7.3 shows a plot of this function, compared to the loss after the pulse. It is important to observe that the loss after the pulse gets very small for $S > 3$, but the average loss q_p for the pulse is still significant then.

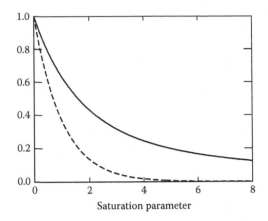

FIGURE 7.3
Solid curve: loss for a pulse on a slow saturable absorber, as a function of the saturation parameter (pulse energy divided by saturation energy); dotted curve: loss after the pulse.

The assumption of an exponential recovery of saturable absorbers is actually more a measure of convenience than a usually encountered element of reality. Semiconductor saturable absorbers (Subsection 7.7.3.4.1), for example, often exhibit biexponential or even more complicated recovery dynamics, which can also strongly depend on the excitation level, reflecting the complicated processes occurring in such devices.

Another important parameter for any saturable absorber is the damage threshold in terms of the applicable pulse fluence or intensity. Note that the absolute value of the damage threshold is actually less relevant than the ratio of the damage fluence and the saturation fluence, because the latter determines the typical operating parameters.

7.7.3.3   *Passive Mode Locking with Fast and Slow Saturable Absorbers*

Here we give some guidelines on what pulse durations can be expected from passively mode-locked lasers. First we consider cases with a fast saturable absorber and without significant influence of dispersion and self-phase modulation. We also assume that significant gain saturation does not occur during a pulse. For this situation, which can occur particularly in picosecond solid-state lasers mode-locked with SESAMs (Subsection 7.7.3.4.1), analytical results have been obtained [141–142]. These calculations are strictly valid only for weak absorber saturation, which is not the desired case, but numerical simulations showed that for a fully saturated absorber the obtained pulse duration can be estimated from

$$\tau \simeq \frac{0.9}{\Delta f_g} \sqrt{\frac{g}{\Delta R}} \tag{7.15}$$

where Δf_g is the FWHM gain bandwidth (assuming that a Lorentzian shape fits the gain spectrum well within the range of the pulse spectrum; see Subsection 7.7.2), g is the power gain coefficient (which equals the overall cavity losses), and ΔR is the modulation depth (maximum reflectivity change) of the absorber. Compared to the equation from the analytical results, we have increased the constant factor by about 10% because the analytical calculations are not accurate for a fully saturated gain and numerical simulations (which do not need to approximate the gain saturation with a linear function) result in typically ≈10% longer pulse durations. Note that significantly weaker or stronger absorber saturation results in longer pulses. If we introduce some self-phase modulation, the pulse duration can be somewhat reduced because this effect helps to increase the pulse bandwidth. However, the dynamics become unstable if too much self-phase modulation occurs. The same holds for phase changes which can arise in an absorber when it is saturated [39].

Note that an increase of ΔR also increases the required gain g, so that a value of ΔR larger than the linear cavity losses does not significantly reduce the pulse duration. The limit for the pulse duration is in the order of $1/\Delta f_g$.

With a slow saturable absorber, somewhat longer pulse durations are obtained. Without significant influence of dispersion and self-phase modulation, we can estimate the pulse duration with [67]

$$\tau \approx \frac{1.1}{\Delta f_g} \sqrt{\frac{g}{\Delta R}} \tag{7.16}$$

This equation, which is very similar to Equation 7.15, is an empirical fit to results from numerical simulations. It holds if the absorber is operated at roughly 3–5 times the saturation fluence. For significantly weaker or stronger absorber saturation, the pulses duration gets somewhat longer.

In contrast to the situation for fast absorbers, the influence of self-phase modulation always appears to make the pulses longer, apart from the instability occurring when the effect is too strong. Particularly in femtosecond lasers, it can be difficult to make the SPM effect weak enough; in this case, soliton mode locking [64,66] (Subsection 7.7.3.3) is a good solution.

As observed in many experiments, the pulse duration obtained with a slow saturable absorber can easily be more than an order of magnitude shorter than the absorber recovery time. This may seem quite surprising because a slow absorber can clean up only the leading part of the pulse, not the trailing part. Even more disturbing, there is usually a temporal window with net gain behind the pulse, so that one should expect any noise behind the pulse to exponentially grow in power and soon destabilize the pulse. However, the pulse maximum (but not the trailing part) is constantly delayed because the absorber attenuates only the leading part; in effect, the pulse tends to "eat up" the trailing part, and any noise there has only a limited time to grow before it merges with the pulse [67]. This mechanism alone can allow for pulse durations well below one tenths of the absorber recovery time. If the absorber is too slow, this can lead to instabilities, rather than simply to longer pulses.

In the femtosecond regime, many absorbers (including SESAMs) become very slow compared to the pulse duration. The question arises how stable mode-locked operation can be achieved with a pulse duration much shorter than the absorber recovery time. We have already seen that the temporal delay of the pulses, caused by the slow absorber, is a very important factor. However, there are other effects which can also significantly help to generate shorter pulses. In dye lasers, it was found that gain saturation during each pulse can very much shorten the time window with effective gain [147]: after the passage of the pulse, the absorber stays saturated for a while, but the gain is also exhausted and may need some longer time to recover. For this reason, dye lasers can be used to generate pulses with durations down to ≈ 27 fs [10], which is much less than the absorber recovery time (typically in the order of 0.1 ns to a few nanoseconds for dye absorbers).

In ion-doped solid-state gain media, this principle cannot be used because such gain media have relatively small laser cross sections, and the gain saturation fluence is much larger than the achievable pulse fluence. Therefore, gain saturation occurs only on a time scale in the order of the fluorescence decay time and is caused by the integral effect of many pulses, but gain saturation by a single pulse is very weak. Nevertheless it has been shown that even from solid-state media one can obtain pulses which are much shorter than the absorber recovery time. This is particularly the case if the mechanism of soliton mode locking [142,64–66] is employed. In the regime of negative intracavity dispersion, soliton effects (Section 7.6) can fix the pulse duration at a certain value. The saturable absorber is then required only to start the mode locking process and to stabilize the solitons. The latter means that two types of instability must be suppressed. First, a competing cw background (or long pulse pedestal) can have less bandwidth and thus experience more gain than the soliton, but this can be suppressed by the saturable absorber because this introduces a higher loss for the background. Second, pulse break-up can occur, particularly in cases with too strong or too weak soliton effects (Section 7.6) and/or with very strong saturation of the absorber. It turns out that soliton mode locking allows for somewhat shorter pulses than without dispersion and SPM, and it allows the use of an absorber with relatively small modulation depth (and accordingly small loss) because the limit for the pulse duration depends only weakly on the absorber parameters. Also note that operation with zero dispersion would often not be possible in the femtosecond domain (even if somewhat longer pulses would be acceptable) because the strong SPM effect would destabilize the pulses if it is not compensated with negative dispersion.

Note that apart from the shortest achievable pulse duration, self-starting mode locking is also a desirable goal. In this respect, a slow absorber is usually superior, because it has a lower saturation intensity which facilitates the mode-locking process in an early phase, as discussed in Subsection 7.7.3.1. Indeed, it is found that reliable self starting mode locking is usually easily achieved for lasers mode-locked with (usually slow) semiconductor absorbers (see the next section), though Kerr lens mode-locked lasers (Subsection 7.7.3.4.2) are usually not self starting, particularly when optimized for short pulses. Spatial hole burning can also help to start-up the mode locking process because the laser starts already with a larger bandwidth and the axial modes only need to get phase-locked [148–149].

7.7.3.4 Saturable Absorbers for Passive Mode Locking

In this section we discuss a number of different saturable absorbers (or artificial saturable absorbers) for passive mode locking, concentrating on the most important techniques for ultrashort pulse generation.

7.7.3.4.1 Semiconductor Absorbers

A semiconductor can absorb light if the photon energy is sufficient to excite carriers from the valence band to the conduction band. Under conditions of strong excitation, the absorption is saturated because possible initial states of the pump transition are depleted while the final states are partially occupied. Within typically 60–300 fs after the excitation, the carriers in each band thermalize, and this already leads to a partial recovery of the absorption. On a longer time scale — typically between a few ps and a few ns depending on defect engineering — they will be removed by recombination and trapping. Both processes can be used for mode locking of lasers. More details about the physics of semiconductor nonlinearities used for mode locking are given in a chapter of another book [150].

We typically integrate the semiconductor saturable absorber into a mirror structure, which results in a device whose reflectivity increases as the incident optical intensity increases. This general class of devices is called semiconductor saturable absorber mirrors (SESAMs) [37]. Semiconductor saturable absorbers are ideally suited for passive mode locking solid-state lasers because the large absorber cross section (in the range of 10^{-14} cm^2) and therefore small saturation fluence (in the range of $10\,\mu J/cm^2$) is ideally suited for suppressing Q-switching instabilities [41].

Most mode-locked lasers are operated in the wavelength regimes around 0.7–0.9 μm (e.g., Ti^{3+}:sapphire or Cr^{3+}:LiSAF) or 1.03–1.07 μm (e.g., Nd^{3+}:glass, Nd^{3+}:YAG, Yb^{3+}:YAG). For 0.8 μm, GaAs absorbers can be used together with Bragg mirrors made of AlAs and Ga$_x$Al$_{1-x}$As, where the gallium content x is kept low enough to avoid absorption. For wavelengths around 1 μm, the Bragg mirrors can be made of GaAs and AlAs, and the absorber is made of In$_x$Ga$_{1-x}$As, where the indium content x is adjusted to obtain a small enough band gap. Because In$_x$Ga$_{1-x}$As cannot be grown lattice-matched to GaAs, increasing indium content leads to increasing strain, which can lead to surface degradation by cracking and thus limits the absorber thickness. Nevertheless, such samples have been applied for mode locking of lasers operating even around 1.3 μm [151] and 1.5 μm [152]. In addition, InGaAsP [153] and AlGaAsSb [154] have been demonstrated as absorbers for the wavelength range from 1.3 μm to 1.5 μm. More recently, dilute nitrides have been successfully used as long-wavelength absorber material. This quaternary alloy has the advantage to be GaAs-based and to minimize the lattice mismatch by incorporating just a few percent of nitrogen to InGaAs while reducing the indium concentration [60]. GaInNAs SESAMs have less nonsaturable losses than InGaAs SESAMs and stable self-starting passive mode locking was demonstrated [155–156,62]. Other novel absorber materials with increased absorption cross sections can be used to reduce the saturation fluence. Quantum dots are promising candidates for this [50,59].

SESAMs can be produced either with molecular beam epitaxy (MBE) or with metal–organic chemical vapor deposition (MOCVD). The MOCVD process is faster and thus appears to be most suitable for mass production. It also leads to relatively small nonsaturable losses, so that such SESAMs are well suited for high-power operation. Similarly low losses can be achieved with MBE. However, MBE gives us the additional flexibility to grow semiconductors at lower temperatures, down to ≈200°C, whereas MOCVD is usually done at ≈600°C. Lower growth temperatures lead to microscopic structural defects which act as traps for excited carriers and thus reduce the recovery time, which can be beneficial for the use in ultrafast lasers. However, the nonsaturable losses of such samples also increase with decreasing growth temperature. This compromise between speed of recovery and quality of the surfaces can be improved by optimized annealing procedures or by beryllium doping [157,158].

The first SESAM device for solid-state lasers, the antiresonant Fabry–Perot saturable absorber (A-FPSA), consisted of a Fabry–Perot cavity filled with a saturable absorber, and with its thickness adjusted for antiresonance, i.e., the intensity in the cavity was substantially lower than the incident intensity [5]. The A-FPSA mirror is based on absorber layers sandwiched between the bottom AlAs/AlGaAs semiconductor and the top SiO_2/TiO_2 dielectric Bragg mirror. The top reflector of the A-FPSA provides an adjustable parameter that determines the intensity entering the semiconductor saturable absorber and therefore the saturation fluence of the saturable absorber device. Thus, this design allowed for a large variation of absorber parameters by simply changing absorber thickness and top reflectors [54,159]. This resulted in an even simpler SESAM design with a single quantum well absorber layer integrated into a Bragg mirror [53]. In the 10-femtosecond regime with Ti:sapphire lasers we have typically replaced the lower semiconductor Bragg mirror with a metal mirror to support the required large reflection bandwidth [160,161]. However, more recently, an ultrabroadband monolithically grown fluoride semiconductor saturable absorber mirror was demonstrated that covers nearly the entire gain spectrum of the Ti:sapphire laser. Using this SESAM inside a Ti:sapphire laser resulted in 9.5-fs pulses [162]. The reflection bandwidth was achieved with a $AlGaAs/CaF_2$ semiconductor Bragg mirror [163].

In a general sense, we then can reduce the design problem of a SESAM to the analysis of multilayered interference filters for a given desired nonlinear reflectivity response for both the amplitude and phase. More recently, novel design structures allowed to substantially lower the saturation fluence of SESAMs into the 1 μJ/cm^2 regime [63]. New terms "low-field-enhancement resonant-like SESAM device" (LOFERS) [164] and "enhanced SESAM device (E-SESAM)" [165] were introduced [63].

So far the SESAM is mostly used as an end mirror of a standing-wave cavity. Very compact cavity designs have been achieved for example in passively Q-switched microchip lasers [42,166–167] and passively mode-locked miniature lasers [168–169] where a short laser crystal defines a simple monolithic cavity. The SESAM attached directly to the laser crystal then formed one end-mirror of this laser cavity. As the laser cannot be pumped through the SESAM, the laser output needs to be separated from the collinear pump by a dichroic mirror. These examples suggest that there is need for a device that combines the nonlinear properties of the SESAM with an output coupler. This has been demonstrated before for a passively mode-locked fiber laser [170] and more recently for solid-state lasers [171].

Semiconductor doped dielectric films have been demonstrated for saturable absorber applications [172]. In this case, InAs-doped thin-film rf-sputtering technology was used which offers similar advantages as SESAMs, i.e., the integration of the absorber into a mirror structure. However, the saturation fluence of more than 10 mJ/cm^2 is in most cases much too high for stable solid-state laser mode locking. In comparison, epitaxially grown SESAMs typically have saturation fluences in the range of 1–100 μJ/cm^2 depending on the specific device structure.

The presence of two different time scales in the SESAM recovery, resulting from intraband thermalization and carrier trapping/recombination, can be rather useful for mode locking. The longer time constant results in a reduced saturation intensity for a part of the absorption, which facilitates self-starting mode locking. The faster time constant is more effective in shaping picosecond and even subpicosecond pulses. Therefore, SESAMs allow to easily obtain self-starting mode locking in most cases.

SESAMs are quite robust devices. The ratio of the damage fluence and the saturation fluence is typically in the order of 100, so that a long device lifetime can be achieved with the usual parameters of operation (pulse fluence ≈3–10 times the saturation fluence) [173]. For high-power operation, the thermal load can be substantial, but the relatively good heat conductivity e.g., of GaAs and the geometry of a thin disk allow for efficient cooling through the back side, which is usually soldered to a copper heat sink. The latter may be water-cooled in multiwatt lasers. Once the mode diameter is larger than the substrate thickness (typically 0.5 mm), further power scaling by increasing the mode area and keeping the pulse fluence constant will not substantially increase the temperature rise [68], assuming that the back side is sufficiently cooled. The temperature rise can be limited to below 100 K by the use of SESAM designs with low saturation fluence (100 μJ/cm^2 or less), because these allow operation with a larger mode area. SESAM damage can thus be avoided by using suitable designs and operation parameters, and further power scaling is not prohibited by such issues.

Perhaps surprisingly, SESAM heating is actually more of an issue in high repetition rate lasers than in high power lasers. For a given intracavity average power, a reduction of the cavity length (for increasing the repetition rate) demands reduction of the mode area (for sufficient saturation) and thus concentrates the fixed dissipated power on a smaller spot. Even though the temperature excursion scales only with the square root of the repetition rate (for fixed intracavity average power and saturation parameter), this can be a challenge for multiGHz lasers even with moderate output powers, and SESAMs with a low value of the saturation fluence are then needed.

7.7.3.4.2 Kerr Lens Mode Locking

Very fast effective saturable absorbers, suitable for the generation of pulses with durations below 10 fs, can be implemented using the Kerr effect. In typical gain media, the Kerr effect (i.e., the dependence of the refractive index on the light intensity) has a time constant in the order of only a few femtoseconds. The intensity gradients across the transverse mode profile, e.g., in the gain medium lead to a Kerr lens with intensity-dependent focusing power. This can be combined with an aperture to obtain an effective saturable absorber, which can be used for Kerr lens mode locking (KLM) [174–176] and which was the explanation for the earlier breakthrough result by Sibbett's group [20], also referred to as "magic mode locking" because the mechanism was not understood at that time. For example, a pinhole at a suitable location in the laser cavity leads to significant losses for cw operation but reduced losses for short pulses for which the beam radius at the location of the pinhole is reduced by the Kerr lens. This is also called hard-aperture KLM. Another possibility (soft-aperture KLM [177]) is to use a pump beam which has significantly smaller beam radius than the laser beam in the gain medium. In this situation, the effective gain is higher for short pulses if the Kerr lens reduces their beam radius in the gain medium, because then the pulses have a better spatial overlap with the pumped region.

In any case, KLM requires the use of a laser cavity which reacts relatively strongly to changes of the focusing power of the Kerr lens. This is usually only achieved by operating the laser cavity near one of stability limits, possibly with detrimental effects on the long-term stability.

Detailed modeling of the KLM action requires sophisticated three-dimensional simulation codes, including both the spatial dimensions and the time variable [178–180]. The basic reason for this is that the strength of the Kerr lens varies within the pulse duration, which leads to a complicated spatiotemporal behavior. However, a reasonable understanding can already be obtained simply by calculating the transverse cavity modes with a constant Kerr lens, using some averaged intensity value.

Self-starting mode locking with KLM can be achieved with specially optimized laser cavities [181–185]. Unfortunately, this optimization is in conflict with the goal of generating the shortest possible pulses. This is because the laser intensity changes by several orders of magnitude during the transition from cw to mode-locked operation, and the Kerr effect for very short pulses is too strong when the cavity is optimized so as to react sensitively to long pulses during the start-up phase. Basically, one would face a similar problem with any fast saturable absorber, but a slow saturable absorber is less critical in this respect because its action depends on pulse energy and not intensity.

Generation of very short pulses (below 6 fs duration) with a self-starting Ti^{3+}:sapphire laser has become possible by combining a SESAM with a cavity set up for KLM [186,21,162]. The SESAM is then responsible for fast self-starting, whereas the pulse shaping in the femtosecond domain is mainly done by the Kerr effect. Note, however, that this requires a SESAM with extremely broad reflection bandwidth, as can be achieved only with very special designs.

7.7.3.4.3 Additive Pulse Mode Locking

Already before the invention of Kerr lens mode locking (Subsection 7.7.3.4.2), the Kerr effect was used in a different technique called *additive pulse mode locking* (APM) [187–188]. A cavity containing a single-mode glass fiber is coupled to the main laser cavity. In this cavity, the Kerr effect introduces a larger nonlinear phase shift for the temporal pulse center, compared to the wings. When the pulses from both cavities meet again at the coupling mirror, they interfere in such a way that the pulse center is constructively enhanced in the main cavity, whereas its wings are reduced by destructive interference. A prerequisite for this to happen is, of course, that the cavity lengths are equal and matched with submicron precision. Because other methods (e.g., KLM or using saturable absorbers) do not need such a precise match, they are nowadays usually preferred to APM, although APM has been proven to be very effective particularly in picosecond lasers. For example, the shortest pulses (1.7 ps) from a Nd^{3+}:YAG laser have been achieved with APM [189].

A variant of APM which is also becoming obsolete is *resonant passive mode locking* (RPM) [190,191]. As with APM, a coupled cavity is used, but this contains a saturable absorber (i.e., an intensity nonlinearity) instead of a Kerr nonlinearity. With this methods, 2-ps pulses were generated with a Ti:sapphire laser [190] and 3.7-ps pulses in a Nd:YLF laser [191].

7.7.3.4.4 Nonlinear Mirror Mode Locking

$\chi^{(2)}$ Nonlinearities can also be used to construct effective saturable absorbers [192]. A nonlinear mirror based on this principle consists of a frequency doubling crystal and a dichroic mirror. For short pulses, a part of the incident laser light is converted to the second harmonic, for which the mirror is highly reflective, and converted back to the fundamental wave if an appropriate relative phase shift for fundamental and second-harmonic light is applied. On the other hand, unconverted fundamental light experiences a significant loss at the mirror. Thus, the device has a higher reflectivity at higher intensities. This has been used for mode locking, e.g., with up to 1.35 W of average output power in 7.9-ps pulses from a Nd^{3+}:YVO_4 laser [193]. The achievable pulse duration is often limited by group velocity mismatch between fundamental and second-harmonic light.

7.7.3.5 Q-Switching Instabilities

The usually desired mode of operation is cw mode locking, where a train of pulses with constant parameters (energy, duration, shape) is generated. However, the use of saturable absorbers may also lead to Q-switched mode locking (QML), where the pulse energy oscillates between extreme values (see Figure 7.1). The laser then emits bunches of mode-locked pulses, which may or may not have a stable Q-switching envelope.

The reason for the QML tendency is the following: Starting from the steady state of cw mode locking, any small increase of pulse energy will lead to stronger saturation of the absorber and thus to a positive net gain. This will lead to an exponential growth of the pulse energy until this growth is stopped by gain saturation. In a solid-state laser, usually exhibiting a large gain saturation fluence, this may take many cavity round-trips. The pulse energy will then drop even below the steady-state value. A damped oscillation around the steady state (and finally stable cw mode locking) is obtained only if gain saturation sets in fast enough.

The transition between the regimes of cw mode locking and QML has been investigated in detail [41,194] for slow saturable absorbers. If the absorber is fully saturated, the absorber always fully recovers between two cavity round-trips and soliton shaping effects do not occur. A simple condition for stable cw mode locking can be found [41]:

$$E_p^2 > E_{L,sat} E_{A,sat} \Delta R \qquad (7.17)$$

where E_p is the intracavity (not output) pulse energy and $E_{L,sat}$, $E_{A,sat}$ are the gain and absorber saturation energies, respectively. Using the saturation parameter $S := E_p/E_{A,sat}$, we can rewrite this to obtain

$$E_p > E_{L,sat} \frac{\Delta R}{S} \qquad (7.18)$$

This explains why passively mode-locked lasers often exhibit QML when weakly pumped and stable cw mode locking for higher pump powers. Normally, mode locking is fairly stable even for operation only slightly above the QML threshold, so that there is no need for operation far above this threshold. The QML threshold is high when $E_{L,sat}$ is large (laser medium with small laser cross sections and/or large mode area in the gain medium, enforced, for example, by poor pump beam quality or by crystal fracture); E_p can not be made large (limited power of pump source, high repetition rate, or large intracavity losses); or when a high value of ΔR is needed for some reason (e.g., a need exists for very short pulses). We thus find the following prescriptions avoid QML:

- Use a gain medium with small saturation fluence, and optimize the pumping arrangement for a small mode area in the gain medium.

- Minimize the cavity losses so that a high intracavity pulse energy can be achieved.

- Operate the saturable absorber in the regime of strong saturation, although this is limited by the tendency for pulse breakup or by absorber damage.

- Do not use a larger modulation depth ΔR than necessary.

- Use a cavity with low repetition rate.

Another interesting observation [41] is that in soliton mode-locked lasers the minimum intracavity pulse energy for stable cw mode locking is lower by typically a factor in the order of four. The reason for this is that a soliton acquires additional bandwidth if its energy rises for some reason. This reduces the effective gain, so that we have a negative feedback mechanism, which tends to stabilize the pulse energy. Thus, the use of soliton formation in a laser can help not only to generate shorter pulses but also to avoid QML.

Finally, we note that two-photon absorption (which may occur in a SESAM) can modify the saturation behavior in such a way that the QML tendency is reduced [195]. In this case, the two-photon absorption acts as an all-optical power limiter [196]; the reflectivity, plotted as a function of the pulse energy, is exhibiting a pronounced roll-over for high pulse energies. As the QML tendency is directly related to the slope of this saturation curve, it can be eliminated by operating near the maximum of the saturation curve, where the slope is near zero.

Two-photon absorption in SESAMs can be strong in the regime of pulse durations well below 1 ps, but it is usually too weak to be effective for pulse durations around 10 ps or longer, unless the SESAM is operated in the strongly saturated regime, close to its damage threshold. Surprisingly, it has been found [197] that SESAMs can exhibit a pronounced roll-over of the saturation curve even for pulse durations of several picoseconds; the reduction of the roll-over for longer pulses has been found to be much weaker than expected, at least for several SESAMs operated in the 1.5-μm regime. This means that apparently there is another physical effect, not identical to two-photon absorption but similarly affecting the saturation curves, although the nature of this effect has not yet been identified. The effect is nevertheless technologically very important, because it allows operation of passively mode-locked Er:Yb:glass lasers with extremely high pulse repetition rates, up to 80 GHz having been achieved so far (Subsection 7.8.4). This has been revealed partly by a careful characterization of SESAMs [63,197]

and partly by transfer function measurements on several passively mode-
locked lasers [76].

7.7.3.6   *Passive Mode Locking at High Repetition Rates*

The repetition rates of typical mode-locked solid-state lasers are in the order
of 30–300 MHz. Pulse trains with much higher repetition rates — higher than
1 GHz or even above 10 GHz — are required for some applications, e.g., in
telecommunications or for optical clocking of electronic circuits.

The are two basic approaches to realize such high repetition rates: either
one must use a very compact laser cavity, which has a short round-trip time
for the circulating pulse (fundamental mode locking), or one must arrange
for several pulses circulating in the cavity with equal temporal spacing. The
latter technique, called *harmonic mode locking*, requires some mechanism to
stabilize the pulse spacing. It is widely used in fiber lasers, because these
can usually not be made short enough for multi-GHz repetition rates with
fundamental mode locking, but it is rarely used in solid-state bulk lasers as
discussed in this chapter. A Cr^{4+}:YAG laser has been demonstrated with three
pulses in the cavity (i.e., harmonic mode locking), resulting in a repetition
rate of 2.7 GHz [198] which, to our knowledge, is the highest value achieved
with this technique, applied to diode-pumped solid-state lasers. A fundamentally mode-locked laser with 2.6 GHz repetition rate has also been demonstrated [199].

In the following, we concentrate on fundamentally (i.e., not harmonically) mode-locked bulk lasers, which have generated repetition rates of
up to nearly 160 GHz [44]. The difficulty is not the construction of a
sufficiently compact laser cavity; a cavity length of few millimeters (or even
well below 1 mm for 160 GHz) is easily obtained when a compact mode
locker such as a SESAM (Subsection 7.7.3.4.1) is used. The main challenge
is rather to overcome the tendency for Q-switched mode locking (Subsection 7.7.3.5) which becomes very strong at high repetition rates. As this
issue formerly constituted a challenge, even at repetition rates in the order
of 1 GHz, specially optimized laser cavities and SESAMs had to be developed for repetition rates > 10 GHz [63]. First, it was realized that Nd^{3+}:YVO$_4$
is a laser medium with a particularly high gain cross section; numbers
between about $100 \cdot 10^{-20}$ cm^2 and $300 \cdot 10^{-20}$ cm^2 have been quoted, with the
lower end of this range being more realistic. This leads to a small gain
saturation energy, which can be further minimized by using a laser cavity
with a rather small mode radius of 30 μm or less. The latter is facilitated
by good pump absorption of Nd^{3+}:YVO$_4$ (which relaxes the constraints on
the beam quality of the pump source) and the good thermal properties.
Other important design aspects are to minimize the intracavity losses, so
that operation with a small output coupler transmission and accordingly
high intracavity pulses energy is achieved, and of course to use a SESAM
with optimized parameters.

After the first milestone at 13 GHz [200], based on a cavity with a Brewster-cut Nd^{3+}:YVO_4 crystal, a SESAM and an air spacing, quasi-monolithic Nd^{3+}:YVO_4 lasers have been used for higher repetition rates of 29 GHz [168], later 59 GHz [201], 77 GHz [169] and finally 160 GHz [44]. More recently, diode-pumped lasers were demonstrated up to 40 GHz [202] and 80 GHz [79]. Such lasers are often built with a quasi-monolithic cavity design as discussed in Subsection 7.8.4. The repetition rate is limited not only by the Q-switching tendency but also because the finite pulse duration (typically around 3–6 ps) which leads to an increasing overlap of consecutive pulses. Note that the average output powers are at least a few tens of milliwatts in all cases, and sometimes a few hundreds of milliwatts. At a lower repetition rate of 10 GHz, a high power of 2.1 W was achieved with a more conventional laser design [44], although at the expense of a longer pulse duration of 14 ps.

Unfortunately, the choice of laser media for high pulse repetition rates at other wavelengths, e.g., 1.5 μm as required for telecom applications, is very limited: for this wavelength region, there is no gain medium with similarly favorable properties as Nd:YVO_4 for 1064 nm. Erbium-doped gain media have rather low laser cross sections, which initially seemed to greatly limit their use for passive mode locking at high repetition rates. Cr^{4+}:YAG is more favorable in terms of cross sections, but suffers from higher crystal losses and a quite variable crystal quality — particularly at high doping densities, as required for short crystals. Attempts to achieve repetition rates of 10 GHz or higher have so far not been successful; to our knowledge, the maximum is still at 2.6 GHz [198]. On the other hand, passively mode-locked Er:Yb:glass lasers turned out to operate stably even at very high pulse repetition rates of first 10 GHz [80], later 25 GHz [81], 40 GHz [203], 50 GHz [45] and more recently even at 77 GHz [219]. In this regime, the theoretical estimates had predicted that it would be hardly possible to avoid Q-switching instabilities. It took a few years to resolve this significant departure from theoretical expectations, as it turned out to be difficult even just to determine whether the discrepancy was due to effects in the erbium gain medium (with its comparatively complicated level structure) or in the SESAM. The explanation then turned out to be a modified saturation behavior of the SESAMs, as discussed in Subsection 7.7.3.5, although additional contributions from the gain medium can not strictly be ruled out. It appears, however, that these are not required for a quantitative explanation of the observations. Even the stabilizing effect of ordinary gain saturation, as considered in the original QML theory, is not important in these lasers, as the QML tendency is already eliminated by the modified SESAM saturation characteristics.

In recent years, surface-emitting external-cavity semiconductor lasers (VECSELs) have also been shown to be very suitable for pulse generation in the multi-GHz regime of repetition rates [220]. Compared to multi-GHz lasers based on more traditional solid-state gain media, they have a better (and already partly realized) potential for combining high repetition rates with high output powers. Also, they potentially offer important advantages for cheap mass production. We briefly discuss such lasers in Subsection 7.8.5. A more detailed review is given in Reference 220.

7.7.3.7   Summary: Requirements for Stable Passive Mode Locking

Here we briefly summarize a number of conditions which must be met to obtain stable passive mode locking:

- The laser must have a good spatial beam quality, i.e., it must operate on the fundamental transverse mode (see Section 7.1).

- The saturable absorber must be strong and fast enough so that the loss difference between the pulses and a cw background is sufficient to compensate for the smaller effective gain of the short pulses (Subsection 7.7.3.3). Also, it must be fast enough to suppress the growth of noise in the gain window behind the pulse (Subsection 7.7.3.3).

- The saturable absorber should be strong enough to guarantee for stable self starting. (A longer absorber recovery times helps in this respect, see Subsection 7.7.3.1.)

- The absorber must be significantly saturated, but should also not be too strongly saturated, because this would introduce a tendency for double pulses.

- Self-phase modulation must not be too strong, particularly in cases with zero or positive dispersion. In the soliton mode-locked regime, significantly stronger self-phase modulation can be tolerated and is even desirable.

- To avoid Q-switching instabilities, the conditions discussed in Subsection 7.7.3.5 must be fulfilled.

- In some cases where inhomogeneous gain saturation occurs (e.g., induced by spatial hole burning [71,148,149]), spectral stability may not be given: a spectral hole in the gain, generated by the current lasing spectrum, can lead to a situation where a shift of the gain spectrum to longer or shorter wavelengths increases the gain. Such a situation is unstable and must be avoided, possibly by the use of a suitable intracavity filter.

- Spurious intracavity reflections have to be eliminated with care, in so far as they do not leave the resonator.

7.8   Designs of Mode-Locked Lasers

In the previous sections we have discussed the most important physical effects which are relevant for the operation of mode-locked lasers. Here we give an overview on laser designs, with the emphasis on those designs which are currently used and/or are of promise to find widespread application in

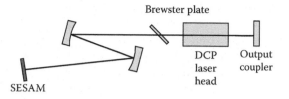

FIGURE 7.4

Setup of a passively mode-locked high-power Nd^{3+}:YAG laser, containing a DCP (direct coupled pump) laser head, two curved mirrors, a SESAM, and an output coupler mirror (OC) with 7% transmission.

the foreseeable future. We discuss a number of practical aspects for the design and operation of such lasers.

7.8.1   Picosecond Lasers

The setups of picosecond lasers typically do not differ very much from those of lasers for continuous-wave operation. Some mode locker is installed, which might be either an acousto-optic modulator (AOM) for active mode locking (Subsection 7.7.2) or, e.g., a SESAM (Subsection 7.7.3.4.1) for passive mode locking. Also, the cavity design needs to fulfill a few additional demands. As an example, we refer to Figure 7.4, which shows the setup of a high-power Nd^{3+}:YAG laser [69], passively mode-locked with a SESAM. The cavity design must provide appropriate beam radii both in the laser head (where the fundamental Gaussian mode should just fill the usable area) and on the SESAM, so as to use an appropriate degree of saturation. The latter depends on a number of factors: the output power, the output coupler transmission, the cavity length, and the saturation fluence of the SESAM. Obviously, the cavity length must be chosen to obtain the desired repetition rate. The equations given in Subsection 7.7.3.5 can be used to ensure that the chosen design will not suffer from Q-switching instabilities. The laser head is side-pumped in the mentioned example, but end-pumped laser heads can also be used, where the pump light is typically injected through a folding mirror which is highly reflective for the laser light.

The SESAM should typically be used as an end mirror, not as a folding mirror. Otherwise a tendency for the generation of multiple pulses, which would meet on the SESAM, might be induced.

Similar setups can be used for actively mode-locked lasers, where the SESAM is replaced by an AOM. The AOM should be placed close to an end mirror, for similar reasons as discussed above for the SESAM.

7.8.2   High-Power Thin-Disk Laser

By far, the highest average powers in the sub-picosecond domain can be obtained from thin disk Yb^{3+}:YAG lasers, passively mode-locked with a SESAM. The first result, with 16.2 W in 700-fs pulses [72], received a lot of attention for its unusually high output power. More importantly, this new approach introduced the first power-scalable technology for sub-picosecond lasers. For this reason, further big improvements became possible, first to 60 W average power [43] and later even to 80 W [73], in both cases with pulse durations near 700 fs. These lasers are operated in the soliton mode-locked regime (Subsection 7.7.3.3). Anomalous intracavity dispersion was originally obtained from a Gires–Tournois interferometer (Subsection 7.4.2.3), but for higher powers this turned out to be unsuitable due to strong thermal effects. The same applies to intracavity prism pairs; sufficient dispersion can only be achieved with high dispersion materials such as SF10 glass, which exhibit too high losses and thus thermal distortions. The method of choice is therefore the use of dispersive mirrors which, however, also have to be carefully optimized.

The power scalability of the passively mode-locked thin-disk laser rests on a combination of circumstances. First of all, the thin-disk laser head [70] itself is power scalable because of the nearly one-dimensional heat flow in the beam direction; thermal effects (such as thermal lensing) do not become more severe if the mode area is scaled up proportional to the power level. A possible problem is only the effect of stress, which has to be limited with refined techniques for mounting the crystal on the heat sink. The SESAM (Subsection 7.7.3.4.1) also has the geometry of a thin disk and thus does not limit the power; more power on an accordingly larger area does not significantly increase the temperature excursion or the optical intensities in the device. Finally, the tendency for Q-switching instabilities does not become stronger if, for instance, the pump power and the mode areas in the disk and the absorber are all doubled while leaving pump intensity and cavity length unchanged. Thus, the whole concept of the passively mode-locked thin-disk laser is power-scalable in the sense that the output power can be increased without making the following key challenges more severe: heating in the disk, heating or nonthermal damage of the SESAM, and Q-switching instabilities. Further power increases can introduce other challenges that are not problematic at lower powers, such as the difficulty to provide dispersion compensation with optical elements that can withstand the very high intracavity powers and additional sources of nonlinearities such as the air inside the cavity [58]. Simply increasing the mode areas everywhere also has its limits, because thermal effects on dispersive mirrors (having significantly higher absorption than ordinary highly reflecting mirrors) cannot be limited this way: larger beams do cause weaker thermal lenses, but become more sensitive to lensing. Also, laser cavities with very large mode areas are subject to various design restrictions. At the moment, one could build thin-disk laser heads with higher output power and still diffraction-limited mode quality, but lower-loss dispersive mirrors (or other dispersion-compensating elements) would be required to raise the average output power to levels well above 100 W. This may seem somewhat surprising, as the initial concerns were centered on diffraction-limited operation and SESAM damage, which then turned out not to be the limiting factors.

Apart from higher powers, one can go for still shorter pulse durations. For Yb:YAG, the available gain bandwidth will probably not allow significantly shorter pulses in a high-power thin-disk laser. (Low-power Yb:YAG lasers have generated pulses with durations down to 340 fs [100], but high-power lasers are subject to additional constraints.) Promising results have been achieved with ytterbium-doped tungstate crystals, particularly with Yb:KYW, which in a thin-disk laser generated 240-fs pulses with 22 W output power [74]. Although simulations show that Yb:KYW and Yb:KGW could in principle even be superior to Yb:YAG in terms of efficiency and output power, so far the difficulty to obtain Yb:KYW thin disks with sufficiently high quality has prevented the experimental demonstration of this superiority.

As the thin disk must be cooled from one side, a reflecting coating on this side is used, and the laser beam is always reflected at the disk. For mode-locked operation, the disk is preferably used as a folding mirror, rather than as an end mirror, because the two ends of the standing-wave cavity are then available for the output coupler mirror and the SESAM. In any case, a standing-wave pattern occurs in the gain medium, which leads to spatial hole burning. This has been shown to make stable mode locking possible only in a narrow range of pulse durations [71], except if the effect of spatial hole burning is reduced, e.g., by using a cavity with multiple reflections on the disk [204], or if the amplification bandwidth is strongly reduced by using a narrow-band gain medium or an additional etalon [204].

Passively mode-locked thin-disk lasers can serve as ideal pump sources for a variety of nonlinear conversion processes. Though laser design tends to become more difficult for higher powers, nonlinear conversion stages can greatly benefit from high peak powers. For example, 58% efficient single-pass frequency doubling has been achieved in critically phase-matched LBO at room temperature [72]; such performance can in the picosecond region usually be obtained only with noncritical phase matching, which requires a temperature-stabilized crystal oven. Also, high peak powers from thin disk Yb:YAG lasers allowed the realization of high gain fiber-feedback parametric oscillators [205–207], which have some significant advantages over low-gain devices and also allow efficient single-pass parametric generation in periodically poled $LiTaO_3$ [206]. A high power RGB laser source has been developed with 8 W in the red, 23 W in the green, and 10 W in the blue output beam, where all optical power originated from a thin disk Yb:YAG laser with no amplifier stages. Due to the extensive use of critical phase matching, only one of the nonlinear crystals, a parametric generator based on $LiTaO_3$, required a crystal oven. Even this last oven could be eliminated by using stoichiometric material, as recently demonstrated [208]. Note that such a system appears not to be possible with a picosecond laser system of similar power.

7.8.3 Typical Femtosecond Lasers

Most femtosecond lasers are based on an end-pumped laser setup, with a broadband laser medium such as Ti^{3+}:sapphire, Cr^{3+}:LiSAF, Nd:glass or

Yb³⁺:glass (see Section 7.3 and Table 2 in Reference 143 for an overview). In the case of Ti³⁺:sapphire, the pump source can be either an Ar ion laser or a frequency-doubled solid-state laser. In any case, one typically uses a few watts of pump power in a beam with good transverse beam quality, because the mode radius in the Ti³⁺:sapphire rod is usually rather small. Other gain media such as Cr³⁺:LiSAF, Nd:glass or Yb³⁺:glass are typically pumped with high-brightness diode lasers, delivering a few watts with beam quality M^2 factor in the order of 10 in one direction and <5 in the other direction, which allows for diffraction-limited laser output with typically a few hundred milliwatts.

The typically used laser cavities (see Figure 7.5 as an example) contain two curved mirrors in a distance of a few centimeters on both sides of the gain medium. The pump power is usually injected through one or both of these mirrors, which also focus the intracavity laser beam to an appropriate beam waist. One of the two "arms" of the cavity ends with the output coupler mirror, whereas the other one may be used for a SESAM as a passive mode locker. One arm typically contains a prism pair (Subsection 7.4.2.1) for dispersion compensation, which is necessary for femtosecond pulse generation. In most cases, femtosecond lasers operate in the regime of negative overall intracavity dispersion so that soliton-like pulses are formed.

Instead of a SESAM, or in addition to it, the Kerr lens in the gain medium can be used for mode locking (Subsection 7.7.3.4.2). In most cases, soft-aperture KLM is used. Here, the cavity design is made so that the Kerr lens reduces the mode area for high intensities and thus improves the overlap with the (strongly focused) pump beam. This is achieved by operating the laser cavity near one of the stability limits of the cavity, which are found by varying the distance between the above mentioned curved folding mirrors, or some other distance in the cavity.

For the shortest pulse durations around 5–6 fs [21–23], the strong action of KLM as an effective fast saturable absorber is definitely required. Also double-chirped mirrors (Subsection 7.4.2.4) are required for precise dispersion compensation over a very broad bandwidth. Typically, several dispersive mirrors are used in the laser cavity, and additional mirrors may be used for further external compression. A SESAM allows for self-starting mode locking [21,22].

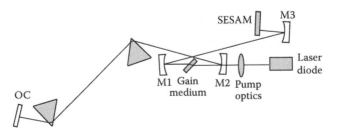

FIGURE 7.5
Typical setup of a femtosecond laser. The gain medium is pumped with a laser diode. A prism pair is used for dispersion compensation, and a SESAM as mode locker.

Higher pulse energies and peak powers have been generated by using laser setups with reduced repetition rates in the 10-MHz regime [58,209]. The long cavity length required for such repetition rates can be achieved by inserting a multipass cell. However, the limiting factor to the pulse energy is ultimately not the practically achievable cavity length but rather the non-linearity of the gain crystal and even of the air [58]. If self-phase modulation becomes too strong, this destabilizes the mode-locking process.

7.8.4 Lasers with High Repetition Rates

For ultrashort pulses with multi-GHz repetition rate, laser designs are required which are very different from those explained in the previous sections. In the following, we discuss some typical designs which have allowed us to realize extremely high pulse durations in different wavelength regimes.

In the 1-μm spectral region, Nd:YVO$_4$ has been found to be a particularly suitable gain medium for very high repetition rates, as already discussed in Section 7.3. For repetition rates of 40 GHz and above, a quasi-monolithic design [168] is useful, where the output coupler mirror is a dielectric coating directly fabricated on a curved and polished side of the laser crystal, whereas the SESAM is attached to the other side of the crystal (see Figure 7.6). The crystal may be antireflection-coated on the flat side or just uncoated. Note that with a reflecting coating on this side, there is a cavity effect. Depending on the exact size of the air gap between crystal and SESAM, the optical field can more or less penetrate the SESAM structure, and, accordingly, the effective modulation depth and saturation fluence of the SESAM are modified. In this case, for optimum performance, one has to manipulate the width of the air gap.

In the quasi-monolithic geometry, counter-propagating waves overlap in the crystal, leading to the phenomenon of spatial hole burning, which can have significant influence on the mode-locking behavior [71,148–149]. In particular, it allows to obtain shorter pulses, although often with some amount of chirp, because gain narrowing is effectively eliminated: even without a

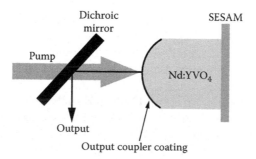

FIGURE 7.6
Quasi-monolithic setup as used for passively mode-locked Nd^{3+}:YVO$_4$ lasers with repetition rates above 20 GHz. SESAM: semiconductor saturable absorber mirror.

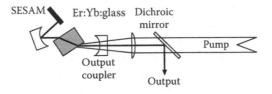

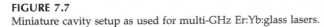

FIGURE 7.7
Miniature cavity setup as used for multi-GHz Er:Yb:glass lasers.

mode locker, such lasers tend to run with a significant emission bandwidth, and the mode locker more or less only has to lock the phases of the running cavity modes. Therefore, lasers of this design have achieved the shortest pulse duration of all Nd:YVO$_4$ lasers, despite the relatively small modulation depth of the SESAM. Short pulses are often desirable as such but also are important for reaching very high repetition rates, because otherwise there can be significant overlap of consecutive pulses.

For significantly higher output powers, a different kind of design is needed. The poor beam quality of the then-required pump laser enforces the use of a thinner gain medium together with some air spacing to obtain the required repetition rate. Such a laser has been developed for pumping multi-GHz parametric oscillators and generated 2.1 W in 14-ps pulses [44].

For multi-GHz Er:Yb:glass lasers, another design has been developed, where a thin plate of Er/Yb-doped glass is mounted within a very short laser cavity (Figure 7.7). For the highest repetition rate demonstrated so far, which is 77 GHz [219], the geometrical cavity length is ≈1.9 mm. Sufficient saturation of the SESAM is achieved by designing the laser cavity for rather small mode sizes, which implies the use of a folding mirror with a radius of curvature below 1 mm.

Another type of multi-GHz lasers, based on a semiconductor gain medium, is described in the following section.

7.8.5 Passively Mode-Locked Optically Pumped Semiconductor Lasers

Although we concentrate on lasers based on ion-doped crystals and glasses in this article, we note that semiconductor gain media are also very interesting to obtain high pulse repetition rates. The low gain saturation fluence of such a medium basically eliminates the problem of Q-switching instabilities in passively mode-locked lasers. The large amplification bandwidth allows for sub-picosecond pulse durations [220], and such devices can be developed for different laser wavelengths. With mode-locked edge-emitting semiconductor lasers, repetition rates of up to 1.5 THz [211] have been obtained, but typically at very low power levels and often not with transform-limited pulses. A general problem is that the achievable mode areas for single transverse mode propagation are rather limited, so that nonlinearities get too strong for high peak powers. Therefore, for a long time it seemed that semiconductor lasers could not play an important role in the domain of mode-locked lasers.

The optically pumped vertical-external-cavity surface-emitting semiconductor lasers (VECSEL) is a laser source that has sparked widespread interest in the past few years due to its capability of producing high average output powers in a diffraction-limited beam [212]. In this case the light emission is in a direction perpendicular to the chip surface, rather than along the surface (i.e., edge-emitting semiconductor laser). An external cavity (Figure 7.8) can be used to define the cavity modes, and it can host additional elements for various functions, in particular a SESAM for mode locking [213,220]. This concept allows for much larger mode areas and thus for higher powers in a diffraction-limited beam — particularly with optical pumping with a high power diode laser, which allows to obtain a smooth pumping distribution over large areas. Low power devices (well below 1 W average output power) can be cooled through the GaAs substrate. For higher powers, the thermal imped-ance has to be reduced, e.g., by thinning down the substrate after growth, by growth in reverse order with subsequent soldering to a heat sink and removal of the substrate by etching [48], or by attaching a transparent heat sink on the surface, where the heat sink may consist of diamond [214] or SiC [215].

The strong gain saturation in semiconductor materials limits the achievable pulse energies but also basically eliminates the Q-switching tendency and thus makes VECSELs suitable for very high average power at very high pulse repetition rates. In fact this combination, which is needed for some applications (e.g., for optical clocking or for pumping high repetition rate

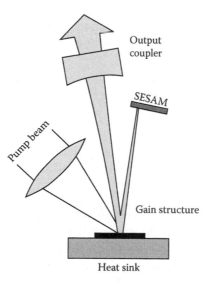

FIGURE 7.8
Cavity of a vertical external cavity surface-emitting semiconductor laser (VECSEL). The pump beam from a high-power diode laser comes from the side. The semiconductor gain structure serves as an amplifying folding mirror.

parametric oscillators) is rather difficult to achieve with traditional gain media such as ion-doped crystals or glasses, and VECSELs are very well placed here. On the other hand, VECSELs are not suitable for low repetition rates and high pulse energies, because those would lead to excessive gain saturation and consequently pulse breakup.

The first passively mode-locked VECSEL has been demonstrated in 2000 [47], at that time with a moderate performance; it generated strongly chirped 22-ps pulses with 22 mW of average power. Since then, the progress has been rapid. For example, a 200-mW device [216] has been shown to generate nearly transform-limited pulses with 3.2 ps duration, and recent achievements include 2.1 W in 4.7-ps pulses [49], a 10-GHz device with 0.5-ps pulses [210], a high power 10-GHz device with 1.4 W [217], and very high repetition rate VECSELs with up to 30 GHz [50] and 50 GHz [221]. The pulses in the early experiments were often strongly chirped, but in 2002 a study of the mode-locking dynamics in VECSELs revealed that a soliton-like pulse shaping mechanism in the positive dispersion regime can help to generate short pulses with low chirp [218].

One of the most application-relevant milestones that remain to be achieved in the field of passively mode-locked VECSELs is the integration of the semiconductor absorber into the gain structure, enabling the realization of ultracompact high-repetition-rate laser devices suitable for wafer-scale integration. We have recently succeeded in fabricating the key element in this concept, a quantum-dot-based saturable absorber with a very low saturation fluence, which for the first time allows stable mode locking of surface-emitting semiconductor lasers with the same mode areas on gain and absorber. Experimental results at high repetition rates of up to 30 GHz [50] and 50 GHz [221] are shown. This should lead to very compact ultrafast lasers which could be fabricated with a cheap wafer-scale mass production process.

7.9   Summary and Outlook

Although the field of ultrashort pulse generation had already been started in the 1960s, significant progress has been made only in the last two decades. Points of particular importance are:

- The transition from dye lasers to compact, powerful, efficient, and long-lived diode-pumped solid-state lasers.
- The development of semiconductor saturable absorber mirrors (SESAMs) that can be optimized for operation in very different parameter regimes concerning laser wavelength, pulse duration and power levels.
- The minimization of the pulse duration from Ti:sapphire lasers down to below 6 fs, using Kerr lens mode locking, sometimes assisted by a SESAM, and ultrabroadband dispersive mirrors.

- The breakthrough to unprecedented average and peak power levels from diode-pumped mode-locked lasers, based on improved laser heads and an improved understanding of absorber damage issues, Q-switching instabilities, and cavity designs, and most importantly the complicated interplay of such aspects.

- The demonstration of picosecond solid-state lasers with extremely high repetition rates of tens of GHz, based mainly on optimized cavity designs and an improved understanding of Q-switching instabilities.

For the next few years we expect the following developments in the field:

- New solid-state gain media will be utilized. Even after many years of solid-state lasers, new solid-state gain media with interesting properties are developed, which can lead to superior performance or even to entirely new achievements. For example, $Cr^{2+}:ZnSe$ should be suitable for the generation of pulses with 20 fs duration or less in a new spectral region around 2.7 μm, even though to our knowledge femtosecond pulses can not yet be produced, maybe as a consequence of excessive nonlinearity. New ytterbium-doped gain media can also have significant impact as, e.g., sesquioxides or improved tungstates for very high powers, or various other crystals for even shorter pulses from diode-pumped lasers.

- Passively mode-locked thin-disk lasers should reach even higher power levels beyond 100 W, and using novel gain media they should also reach significantly shorter pulse durations of perhaps below 200 fs. Amplifier devices for lower repetition rates will become important for material processing.

- As a relatively new type of lasers, optically pumped semiconductor lasers have a particular potential for improvements in terms of output power, pulse duration, tunability, as well as compactness and cheap production (with wafer-scale technologies) [220].

- Nonlinear frequency conversion stages (based on second-harmonic generation, sum frequency mixing, or parametric oscillation) will be pumped with high-power mode-locked lasers to generate short and powerful pulses at other wavelengths. This will be of interest, for example, in applications such as large-screen RGB display systems or for high-harmonic generation with unprecedented power levels. Nonlinear frequency conversion may also allow high repetition rate picosecond lasers to find new applications e.g., in telecommunications.

We thus firmly believe that the development of ultrafast laser sources has not come to its end but will continue to deliver new devices with superior properties for many applications. As the performance levels have relatively quickly reached regions which previously appeared to be unrealistic, there is now the opportunity to use the new sources for many new applications. It appears very realistic to expect that new research in this area will generate far reaching technical developments.

Acknowledgments

The authors are grateful to the entire all-solid-state laser group of Professor Keller at ETH, many students and post-docs having contributed in experiments and theory since 1993; the financial support through ETH, the Swiss Priority program in Optics, the Swiss National Science Foundation (SNF), the Swiss Innovation Promotion Agency (CTI/KTI), the Swiss National Center of Competence in Research (NCCR) in quantum physics, many industrial collaborators, and many other international academic collaborations.

References

1. R. Paschotta, U. Keller, Ultrafast solid-state lasers, in *Ultrafast Lasers, Technology and Applications* M.E. Fermann, A. Galvanauskas, G. Sucha, Eds., Marcel Dekker, New York, 2003, 1–60.

2. L.E. Hargrove, R.L. Fork, M.A. Pollack, Locking of HeNe laser modes induced by synchronous intracavity modulation, *Appl. Phys. Lett.*, Vol. 5, 4, 1964.

3. H.W. Mocker, R.J. Collins, Mode competition and self-locking effects in a Q-switched ruby laser, *Appl. Phys. Lett.*, Vol. 7, 270–273, 1965.

4. A.J. De Maria, D.A. Stetser, H. Heynau, Self mode-locking of lasers with saturable absorbers, *Appl. Phys. Lett.*, Vol. 8, 174–176, 1966.

5. U. Keller, D.A.B. Miller, G.D. Boyd, T.H. Chiu, J.F. Ferguson, M.T. Asom, Solid-state low-loss intracavity saturable absorber for Nd:YLF lasers: an antiresonant semiconductor Fabry-Perot saturable absorber, *Opt. Lett.*, Vol. 17, 505–507, 1992.

6. C.V. Shank, E.P. Ippen, Subpicosecond kilowatt pulses from a mode-locked cw dye laser, *Appl. Phys. Lett.*, Vol. 24, 373–375, 1974.

7. I.S. Ruddock, D.J. Bradley, Bandwidth-limited subpicosecond pulse generation in mode-locked cw dye lasers, *Appl. Phys. Lett.*, Vol. 29, 296–297, 1976.

8. J.C. Diels, E.W.V. Stryland, G. Benedict, Generation and measurement of pulses of 0.2 ps duration, *Opt. Commun.*, Vol. 25, 93, 1978.

9. R.L. Fork, B.I. Greene, C.V. Shank, Generation of optical pulses shorter than 0.1 ps by colliding pulse modelocking, *Appl. Phys. Lett.*, Vol. 38, 617–619, 1981.

10. J.A. Valdmanis, R.L. Fork, J.P. Gordon, Generation of optical pulses as short as 27 fsec directly from a laser balancing self-phase modulation, group-velocity dispersion, saturable absorption, and saturable gain, *Opt. Lett.*, Vol. 10, 131–133, 1985.

11. R.L. Fork, C.H.B. Cruz, P.C. Becker, C.V. Shank, Compression of optical pulses to six femtoseconds by using cubic phase compensation, *Opt. Lett.*, Vol. 12, 483–485, 1987.

12. G.T. Maker, A.I. Ferguson, Frequency-modulation mode locking of diode-pumped Nd:YAG laser, *Opt. Lett.*, Vol. 14, 788–790, 1989.
13. G.T. Maker, A.I. Ferguson, *Electron. Lett.*, Vol. 25, 1025, 1989.
14. U. Keller, K.D. Li, B.T. Khuri-Yakub, D.M. Bloom, K.J. Weingarten, D.C. Gerstenberger, High-frequency acousto-optic modelocker for picosecond pulse generation, *Opt. Lett.*, Vol. 15, 45–47, 1990.
15. K.J. Weingarten, D.C. Shannon, R.W. Wallace, U. Keller, Two gigahertz repetition rate, diode-pumped, mode locked, Nd:yttrium lithium fluoride (YLF) laser, *Opt. Lett.*, Vol. 15, 962–964, 1990.
16. T. Juhasz, S.T. Lai, M.A. Pessot, Efficient short-pulse generation from a diode-pumped Nd:YLF laser with a piezoelectrically induced diffraction modulator, *Opt. Lett.*, Vol. 15, 1458–1460, 1990.
17. D.J. Kuizenga, A.E. Siegman, FM and AM mode locking of the homogeneous laser — Part II: Experimental results in a Nd:YAG laser with internal FM modulation, *IEEE J. Quantum Electron.*, Vol. 6, 709–715, 1970.
18. A.E. Siegman, *Lasers*, University Science Books, Mill Valley, California, 1986.
19. P.F. Moulton, Spectroscopic and laser characteristics of Ti:Al$_2$O$_3$, *J. Opt. Soc. Am. B*, Vol. 3, 125–132, 1986.
20. D.E. Spence, P.N. Kean, W. Sibbett, 60-fsec pulse generation from a self-mode-locked Ti:sapphire laser, *Opt. Lett.*, Vol. 16, 42–44, 1991.
21. D.H. Sutter, G. Steinmeyer, L. Gallmann, N. Matuschek, F. Morier-Genoud, U. Keller, V. Scheuer, G. Angelow, T. Tschudi, Semiconductor saturable-absorber mirror-assisted Kerr-lens mode-locked Ti:sapphire laser producing pulses in the two-cycle regime, *Opt. Lett.*, Vol. 24, 631–633, 1999.
22. D.H. Sutter, L. Gallmann, N. Matuschek, F. Morier-Genoud, V. Scheuer, G. Angelow, T. Tschudi, G. Steinmeyer, U. Keller, Sub-6-fsec pulses from a SESAM-assisted Kerr-lens mode-locked Ti:sapphire laser: at the frontiers of ultrashort pulse generation, *Appl. Phys. B*, Vol. 70, S5–S12, 2000.
23. U. Morgner, F.X. Kärtner, S. H. Cho, Y. Chen, H.A. Haus, J.G. Fujimoto, E.P. Ippen, V. Scheuer, G. Angelow, T. Tschudi, Sub-two-cycle pulses from a Kerr-lens mode-locked Ti:sapphire laser: addenda, *Opt. Lett.*, Vol. 24, 920, 1999.
24. B. Schenkel, J. Biegert, U. Keller, C. Vozzi, M. Nisoli, G. Sansone, S. Stagira, S.D. Silvestri, O. Svelto, Generation of 3.8-fsec pulses from adaptive compression of a cascaded hollow fiber supercontinuum, *Opt. Lett.*, Vol. 28, 1987–1989, 2003.
25. G. Steinmeyer, D.H. Sutter, L. Gallmann, N. Matuschek, U. Keller, Frontiers in ultrashort pulse generation: pushing the limits in linear and nonlinear optics, *Science*, Vol. 286, 1507–1512, 1999.
26. A. Shirakawa, I. Sakane, M. Takasaka, T. Kobayashi, Sub-5-fsec visible pulse generation by pulse-front-matched noncollinear optical parametric amplification, *Appl. Phys. Lett.*, Vol. 74, 2268–2270, 1999.
27. A. Baltuska, Z. Wei, M.S. Pshenichnikov, D.A. Wiersma, R. Szipöcs, All-solid-state cavity dumped sub-5-fsec laser, *Appl. Phys. B*, Vol. 65, 175–188, 1997.
28. M. Nisoli, S. Stagira, S.D. Silvestri, O. Svelto, S. Sartania, Z. Cheng, M. Lenzner, C. Spielmann, F. Krausz, A novel high-energy pulse compression system: generation of multigigawatt sub-5-fsec pulses, *Appl. Phys. B*, Vol. 65, 189–196, 1997.
29. C.P. Hauri, W. Kornelis, F.W. Helbing, A. Heinrich, A. Couairon, A. Mysyrowicz, J. Biegert, U. Keller, Generation of intense, carrier-envelope phase-locked few-cycle laser pulses through filamentation, *Appl. Phys. B*, Vol. 79, 673–677, 2004.

30. M. Drescher, M. Hentschel, R. Kienberger, G. Tempea, C. Spielmann, G.A. Reider, P. B. Corkum, F. Krausz, X-ray pulses approaching the attosecond frontier, *Science*, Vol. 291, 1923–1927, 2001.

31. H.R. Telle, G. Steinmeyer, A.E. Dunlop, J. Stenger, D.H. Sutter, U. Keller, Carrier-envelope offset phase control: a novel concept for absolute optical frequency measurement and ultrashort pulse generation, *Appl. Phys. B*, Vol. 69, 327–332, 1999.

32. F.W. Helbing, G. Steinmeyer, U. Keller, Carrier-envelope offset phase-locking with attosecond timing jitter, *IEEE J. Sel. Top. Quantum Electron.*, Vol. 9, 1030–1040, 2003.

33. D. Strickland, G. Mourou, Compression of amplified chirped optical pulses, *Opt. Commun.*, Vol. 56, 219–221, 1985.

34. A. Baltuska, M. Uiberacker, E. Goulielmakis, R. Kienberger, V.S. Yakovlev, T. Udem, T.W. Hansch, F. Krausz, Phase-controlled amplification of few-cycle laser pulses, *J. Sel. Top. Quantum Electron.*, Vol. 9, 972–989, 2003.

35. C.P. Hauri, P. Schlup, G. Arisholm, J. Biegert, U. Keller, Phase-preserving chirped-pulse optical parametric amplification to 17.3 fsec directly from a Ti:sapphire oscillator, *Opt. Lett.*, Vol. 29, 1369–1371, 2004.

36. A. Baltuska, T. Fuji, T. Kobayashi, Controlling the carrier-envelope phase of ultrashort light pulses with optical parametric amplifiers, *Phys. Rev. Lett.*, Vol. 88, 133901, 2002.

37. U. Keller, K.J. Weingarten, F.X. Kärtner, D. Kopf, B. Braun, I.D. Jung, R. Fluck, C. Hönninger, N. Matuschek, J. Aus der Au, Semiconductor saturable absorber mirrors (SESAMs) for femtosecond to nanosecond pulse generation in solid-state lasers, *IEEE J. Sel. Top. Quantum Electron.*, Vol. 2, 435–453, 1996.

38. U. Keller, Recent developments in compact ultrafast lasers, *Nature*, Vol. 424, 831–838, 2003.

39. U. Keller, Ultrafast solid-state lasers, *Prog. Opt.*, Vol. 46, 1–115, 2004.

40. U. Keller, Semiconductor nonlinearities for solid-state laser mode locking and Q-switching, in *Nonlinear Optics in Semiconductors*, Vol. 59, E. Garmire, A. Kost, Eds., Academic Press, Boston, MA, 1999, pp. 211–286.

41. C. Hönninger, R. Paschotta, F. Morier-Genoud, M. Moser, U. Keller, Q-switching stability limits of continuous-wave passive mode locking, *J. Opt. Soc. Am. B*, Vol. 16, 46–56, 1999.

42. G.J. Spühler, R. Paschotta, R. Fluck, B. Braun, M. Moser, G. Zhang, E. Gini, U. Keller, Experimentally confirmed design guidelines for passively Q-switched microchip lasers using semiconductor saturable absorbers, *J. Opt. Soc. Am. B*, Vol. 16, 376–388, 1999.

43. E. Innerhofer, T. Südmeyer, F. Brunner, R. Häring, A. Aschwanden, R. Paschotta, U. Keller, C. Hönninger, M. Kumkar, 60 W average power in 810-fsec pulses from a thin-disk Yb:YAG laser, *Opt. Lett.*, Vol. 28, 367–369, 2003.

44. L. Krainer, R. Paschotta, S. Lecomte, M. Moser, K.J. Weingarten, U. Keller, Compact Nd:YVO$_4$ lasers with pulse repetition rates up to 160 GHz, *IEEE J. Quantum Electron.*, Vol. 38, 1331–1338, 2002.

45. S.C. Zeller, L. Krainer, G.J. Spühler, R. Paschotta, M. Golling, D. Ebling, K.J. Weingarten, U. Keller, Passively mode-locked 50-GHz Er:Yb:glass laser, *Electron. Lett.*, Vol. 40, 875–876, 2004.

46. R. Paschotta, L. Krainer, S. Lecomte, G.J. Spühler, S.C. Zeller, A. Aschwanden, D. Lorenser, H.J. Unold, K.J. Weingarten, U. Keller, Picosecond pulse sources with multi-GHz repetition rates and high output power, *N. J. Phys.*, Vol. 6, 174, 2004.

47. S. Hoogland, S. Dhanjal, A.C. Tropper, S.J. Roberts, R. Häring, R. Paschotta, U. Keller, Passively mode-locked diode-pumped surface-emitting semiconductor laser, *IEEE Photon. Technol. Lett.*, Vol. 12, 1135–1138, 2000.

48. R. Häring, R. Paschotta, A. Aschwanden, E. Gini, F. Morier-Genoud, U. Keller, High–power passively mode–locked semiconductor lasers, *IEEE J. Quantum Electron.*, Vol. 38, 1268–1275, 2002.

49. A. Aschwanden, D. Lorenser, H.J. Unold, R. Paschotta, E. Gini, U. Keller, 2.1-W picosecond passively mode-locked external-cavity semiconductor laser, *Opt. Lett.*, Vol. 30, 272–274, 2005.

50. D. Lorenser, H.J. Unold, D.J.H C. Maas, A. Aschwanden, R. Grange, R. Paschotta, D. Ebling, E. Gini, U. Keller, Towards wafer-scale integration of high repetition rate passively mode-locked surface-emitting semiconductor lasers, *Appl. Phys. B*, Vol. 79, 927–932, 2004.

51. M. Haiml, R. Grange, U. Keller, Optical characterization of semiconductor saturable absorbers, *Appl. Phys. B*, Vol. 79, 331–339, 2004.

52. R. Grange, M. Haiml, R. Paschotta, G. J. Spühler, L. Krainer, O. Ostinelli, M. Golling, U. Keller, New regime of inverse saturable absorption for self-stabilizing passively mode-locked lasers, *Appl. Phys. B*, Vol. 80, 151–158, 2005.

53. L.R. Brovelli, I.D. Jung, D. Kopf, M. Kamp, M. Moser, F.X. Kärtner, U. Keller, Self-starting soliton mode-locked Ti:sapphire laser using a thin semiconductor saturable absorber, *Electron. Lett.*, Vol. 31, 287–289, 1995.

54. I.D. Jung, L.R. Brovelli, M. Kamp, U. Keller, M. Moser, Scaling of the antiresonant Fabry-Perot saturable absorber design toward a thin saturable absorber, *Opt. Lett.*, Vol. 20, 1559–1561, 1995.

55. B.G. Kim, E. Garmire, S.G. Hummel, P.D. Dapkus, Nonlinear Bragg reflector based on saturable absorption, *Appl. Phys. Lett.*, Vol. 54, 1095–1097, 1989.

56. S. Tsuda, W.H. Knox, E.A. de Souza, W.Y. Jan, J.E. Cunningham, Low-loss intracavity AlAs/AlGaAs saturable Bragg reflector for femtosecond mode locking in solid-state lasers, *Opt. Lett.*, Vol. 20, 1406–1408, 1995.

57. D. Kopf, G. Zhang, R. Fluck, M. Moser, U. Keller, All-in-one dispersion-compensating saturable absorber mirror for compact femtosecond laser sources, *Opt. Lett.*, Vol. 21, 486–488, 1996.

58. S.V. Marchese, T. Südmeyer, M. Golling, R. Grange, U. Keller, Pulse energy scaling to 5 μJ from a femtosecond thin disk laser, *Opt. Lett.* to be published

59. S.W. Osborne, P. Blood, P.M. Smowton, Y.C. Xin, A. Stintz, D. Huffaker, L.F. Lester, Optical absorption cross section of quantum dots, *J. Phys.: Condens. Matter*, Vol. 16, S3749–S3756, 2004.

60. M. Kondow, T. Kitatani, S. Nakatsuka, M.C. Larson, K. Nakahara, Y. Yazawa, M. Okai, K. Uomi, GaInNAs: a novel material for long-wavelength semiconductor lasers, *IEEE J. Sel. Top. Quantum Electron.*, Vol. 3, 719–30, June 1997.

61. V. Liverini, S. Schön, R. Grange, M. Haiml, S.C. Zeller, U. Keller, A low-loss GaInNAs SESAM mode-locking a 1.3-μm solid-state laser, *Appl. Phys. Lett.*, Vol. 84, 4002–4004, 2004.

62. A. Rutz, R. Grange, V. Liverini, M. Haiml, S. Schön, U. Keller, 1.5 μm GaInNAs semiconductor saturable absorber for passively mode-locked solid-state lasers, *Electron. Lett.*, Vol. 41, 321–323, 2005.

63. G.J. Spühler, K.J. Weingarten, R. Grange, L. Krainer, M. Haiml, V. Liverini, M. Golling, S. Schön, U. Keller, Semiconductor saturable absorber mirror structures with low saturation fluence, *Appl. Phys. B.* 81, 27–32, 2005.

64. F.X. Kärtner, U. Keller, Stabilization of soliton-like pulses with a slow saturable absorber, *Opt. Lett.*, Vol. 20, 16–18, 1995.
65. I.D. Jung, F.X. Kärtner, L.R. Brovelli, M. Kamp, U. Keller, Experimental verification of soliton modelocking using only a slow saturable absorber, *Opt. Lett.*, Vol. 20, 1892–1894, 1995.
66. F.X. Kärtner, I.D. Jung, U. Keller, Soliton mode locking with saturable absorbers, *IEEE J. Sel. Top. Quantum Electron.*, Vol. 2, 540–556, 1996.
67. R. Paschotta, U. Keller, Passive mode locking with slow saturable absorbers, *Appl. Phys. B*, Vol. 73, 653–662, 2001.
68. R. Paschotta, J. Aus der Au, G.J. Spühler, F. Morier-Genoud, R. Hövel, M. Moser, S. Erhard, M. Karszewski, A. Giesen, U. Keller, Diode-pumped passively mode-locked lasers with high average power, *Appl. Phys. B*, Vol. 70, S25–S31, 2000.
69. G.J. Spühler, T. Südmeyer, R. Paschotta, M. Moser, K.J. Weingarten, U. Keller, Passively mode-locked high-power Nd:YAG lasers with multiple laser heads, *Appl. Phys. B*, Vol. 71, 19–25, 2000.
70. A. Giesen, H. Hügel, A. Voss, K. Wittig, U. Brauch, H. Opower, Scalable concept for diode-pumped high-power solid-state lasers, *Appl. Phys. B*, Vol. 58, 363–372, 1994.
71. R. Paschotta, J. Aus der Au, G.J. Spühler, S. Erhard, A. Giesen, U. Keller, Passive mode locking of thin disk lasers: effects of spatial hole burning, *Appl. Phys. B*, Vol. 72, 267–278, 2001.
72. J. Aus der Au, G.J. Spühler, T. Südmeyer, R. Paschotta, R. Hövel, M. Moser, S. Erhard, M. Karszewski, A. Giesen, U. Keller, 16.2 W average power from a diode-pumped femtosecond Yb:YAG thin disk laser, *Opt. Lett.*, Vol. 25, 859, 2000.
73. F. Brunner, E. Innerhofer, S. V. Marchese, T. Südmeyer, R. Paschotta, T. Usami, H. Ito, S. Kurimura, K. Kitamura, G. Arisholm, U. Keller, Powerful red-green-blue laser source pumped with a mode-locked thin disk laser, *Opt. Lett.*, Vol. 29, 1921–1923, 2004.
74. F. Brunner, T. Südmeyer, E. Innerhofer, R. Paschotta, F. Morier-Genoud, J. Gao, K. Contag, A. Giesen, V.E. Kisel, V.G. Shcherbitsky, N.V. Kuleshov, U. Keller, 240-fsec pulses with 22-W average power from a mode-locked thin-disk Yb:KY(WO₄)₂ laser, *Opt. Lett.*, Vol. 27, 1162–1164, 2002.
75. T. Südmeyer, F. Brunner, E. Innerhofer, R. Paschotta, K. Furusawa, J.C. Baggett, T.M. Monro, D.J. Richardson, U. Keller, Nonlinear femtosecond pulse compression at high average power levels using a large mode area holey fiber, *Opt. Lett.*, Vol. 28, 951–953, 2003.
76. A. Schlatter, S.C. Zeller, R. Grange, R. Paschotta, U. Keller, Pulse energy dynamics of passively mode-locked solid-state lasers above the Q-switching threshold, *J. Opt. Soc. Am. B*, Vol. 21, 1469–1478, 2004.
77. S. Lecomte, R. Paschotta, M. Golling, D. Ebling, U. Keller, Synchronously pumped optical parametric oscillators in the 1.5-μm spectral region with a repetition rate of 10 GHz, *J. Opt. Soc. Am. B*, Vol. 21, 844–850, 2004.
78. S. Lecomte, R. Paschotta, S. Pawlik, B. Schmidt, K. Furusawa, A. Malinowski, D.J. Richardson, U. Keller, Optical parametric oscillator with a pulse repetition rate of 39 GHz and 2.1-W signal average output power in the spectral region near 1.5-μm, *Opt. Lett.*, Vol. 30, 290–292, 2005.
79. S. Lecomte, R. Paschotta, S. Pawlik, B. Schmidt, K. Furusawa, A. Malinowski, D.J. Richardson, U. Keller, Synchronously pumped optical parametric oscillator with a repetition rate of 81.8 GHz, *IEEE Phot. Technol. Lett.*, Vol. 17, 483–485, 2005.

80. L. Krainer, R. Paschotta, G.J. Spühler, I. Klimov, C.Y. Teisset, K.J. Weingarten, U. Keller, Tunable picosecond pulse-generating laser with a repetition rate exceeding 10 GHz, *Electron. Lett.*, Vol. 38, 225–227, 2002.

81. G.J. Spühler, P.S. Golding, L. Krainer, I.J. Kilburn, P.A. Crosby, M. Brownell, K.J. Weingarten, R. Paschotta, M. Haiml, R. Grange, U. Keller, Novel multi-wavelength source with 25-GHz channel spacing tunable over the C-band, *Electron. Lett.*, Vol. 39, 778–80, 2003.

82. C. Hönninger, R. Paschotta, M. Graf, F. Morier-Genoud, G. Zhang, M. Moser, S. Biswal, J. Nees, A. Braun, G.A. Mourou, I. Johannsen, A. Giesen, W. Seeber, U. Keller, Ultrafast ytterbium-doped bulk lasers and laser amplifiers, *Appl. Phys. B*, Vol. 69, 3–17, 1999.

83. D. Burns, M. Hetterich, A.I. Ferguson, E. Bente, M.D. Dawson, J.I. Davies, S.W. Bland, High-average-power (>20 W) Nd:YVO$_4$ lasers mode locked by strain-compensated saturable Bragg reflectors, *J. Opt. Soc. Am. B*, Vol. 17, 919–926, 2000.

84. G.P.A. Malcolm, P.F. Curley, A.I. Ferguson, Additive pulse mode locking of a diode pumped Nd:YLF laser, *Opt. Lett.*, Vol. 15, 1303–1305, 1990.

85. U. Morgner, F.X. Kärtner, S.H. Cho, Y. Chen, H.A. Haus, J.G. Fujimoto, E.P. Ippen, V. Scheuer, G. Angelow, T. Tschudi, Sub-two-cycle pulses from a Kerr-lens mode-locked Ti:sapphire laser, *Opt. Lett.*, Vol. 24, 411–413, 1999.

86. A. Fernandez, T. Fuji, A. Poppe, A. Fürbach, F. Krausz, A. Apolonski, Chirped-pulse oscillators: a route to high-power femtosecond pulses without external amplification, *Opt. Lett.*, Vol. 29, 1366–1368, 2004.

87. R.H. Page, K.I. Schaffers, L.D. DeLoach, G.D. Wilke, F.D. Patel, J.B. Tassano, S.A. Payne, W.F. Krupke, K.-T. Chen, A. Burger, Cr^{2+}-doped zinc chalcogenides as efficient, widely tunable mid-infrared lasers, *IEEE J. Quantum Electron.*, Vol. 33, 609–619, 1997.

88. T.J. Carrig, G.J. Wagner, A. Sennaroglu, J.Y. Jeong, C.R. Pollock, Mode-locked Cr^{2+}:ZnSe laser, *Opt. Lett.*, Vol. 25, 168–170, 2000.

89. S. Uemura, K. Torizuka, Generation of 12-fsec pulses from a diode-pumped Kerr-lens mode-locked Cr:LiSAF laser, *Opt. Lett.*, Vol. 24, 780–782, 1999.

90. D. Kopf, K.J. Weingarten, G. Zhang, M. Moser, M.A. Emanuel, R.J. Beach, J.A. Skidmore, U. Keller, High-average-power diode-pumped femtosecond Cr:LiSAF lasers, *Appl. Phys. B*, Vol. 65, 235–243, 1997.

91. B. Agate, B. Stormont, A.J. Kemp, C.T.A. Brown, U. Keller, W. Sibbett, Simplified cavity designs for efficient and compact femtosecond Cr:LiSAF lasers, *Opt. Commun.*, Vol. 205, 207–213, 2002.

92. C. Chudoba, J.G. Fujimoto, E.P. Ippen, H.A. Haus, U. Morgner, F.X. Kärtner, V. Scheuer, G. Angelow, T. Tschudi, All-solid-state Cr:forsterite laser generating 14-fsec pulses at 1.3 μm, *Opt. Lett.*, 26, 292–294, 2001.

93. V. Petrov, V. Shcheslavskiy, T. Mirtchev, F. Noack, T. Itatani, T. Sugaya, T. Nakagawa, High-power self-starting femtosecond Cr:forsterite laser, *Electron. Lett.*, Vol. 34, 559–561, 1998.

94. D.J. Ripin, C. Chudoba, J.T. Gopinath, J.G. Fujimoto, E.P. Ippen, U. Morgner, F. X. Kärtner, V. Scheuer, G. Angelow, and T. Tschudi, Generation of 20-fs pulses by prismless Cr^{4+}:YAG laser, *Opt. Lett.* 27, 61–63, 2002.

95. J. Aus der Au, D. Kopf, F. Morier-Genoud, M. Moser, U. Keller, 60-fsec pulses from a diode-pumped Nd:glass laser, *Opt. Lett.*, Vol. 22, 307–309, 1997.

96. C. Hönninger, F. Morier-Genoud, M. Moser, U. Keller, L.R. Brovelli, C. Harder, Efficient and tunable diode-pumped femtosecond Yb:glass lasers, *Opt. Lett.*, Vol. 23, 126–128, 1998.

97. J. Aus der Au, F.H. Loesel, F. Morier-Genoud, M. Moser, U. Keller, Femtosecond diode-pumped Nd:glass laser with more than 1-W average output power, *Opt. Lett.*, Vol. 23, 271–273, 1998.

98. R. Paschotta, J. Aus der Au, U. Keller, Thermal effects in high power end-pumped lasers with elliptical mode geometry, *J. Sel. Top. Quantum Electron.*, Vol. 6, 636–642, 2000.

99. A. Killi, U. Morgner, M.J. Lederer, D. Kopf, Diode-pumped femtosecond laser oscillator with cavity dumping, *Opt. Lett.*, Vol. 29, 1288–1290, 2004.

100. C. Hönninger, G. Zhang, U. Keller, A. Giesen, Femtosecond Yb:YAG laser using semiconductor saturable absorbers, *Opt. Lett.*, Vol. 20, 2402–2404, 1995.

101. J. Aus der Au, S.F. Schaer, R. Paschotta, C. Hönninger, U. Keller, M. Moser, High-power diode-pumped passively mode-locked Yb:YAG lasers, *Opt. Lett.*, Vol. 24, 1281–1283, 1999.

102. M. Weitz, S. Reuter, R. Knappe, R. Wallenstein, B. Henrich, Passive mode-locked 21 W femtosecond Yb:YAG laser with 124 MHz repetition-rate, Conferences on Lasers and Electro-Optics 2004, paper CTuCC1.

103. G.J. Valentine, A.J. Kemp, D.J.L. Birkin, D. Burns, F. Balembois, P. Georges, H. Bernas, A. Aron, G. Aka, W. Sibbett, A. Brun, M.D. Dawson, E. Bente, Femtosecond Yb:YCOB laser pumped by narrow-stripe laser diode and passively mode locked using ion implanted saturable-absorber mirror, *Electron. Lett.*, Vol. 36, 1621–1623, 2000.

104. F. Druon, F. Balembois, P. Georges, A. Brun, A. Courjaud, C. Hönninger, F. Salin, A. Aron, F. Mougel, G. Aka, D. Vivien, Generation of 90-fsec pulses from a mode-locked diode-pumped Yb:Ca$_4$GdO(Bo$_3$)$_3$ laser, *Opt. Lett.*, Vol. 25, 423–425, 2000.

105. L.A.W. Gloster, P. Cormont, A.M. Cox, T.A. King, B.H.T. Chai, Diode-pumped Q-switched Yb:S-FAP laser, *Opt. Commun.*, Vol. 146, 177–180, 1997.

106. F. Druon, F. Balembois, P. Georges, Ultra-short-pulsed and highly-efficient diode-pumped Yb:SYS mode-locked oscillators, *Opt. Express*, Vol. 12, 5005–5012, 2004.

107. F. Druon, S. Chenais, P. Raybaut, F. Balembois, P. Georges, R. Gaume, G. Aka, B. Viana, S. Mohr, D. Kopf, Diode-pumped Yb:Sr3Y(BO$_3$)$_3$ femtosecond laser, *Opt. Lett.*, Vol. 27, 197–199, 2002.

108. F. Druon, S. Chénais, P. Raybaut, F. Balembois, P. Georges, R. Gaumé, G. Aka, B. Viana, D. Vivien, J.P. Chambaret, S. Mohr, D. Kopf, Largely tunable diode-pumped sub-100-fsec Yb:BOYS laser, *Appl. Phys. B*, Vol. 74, 201–203, 2002.

109. F. Brunner, G.J. Spühler, J. Aus der Au, L. Krainer, F. Morier-Genoud, R. Paschotta, N. Lichtenstein, S. Weiss, C. Harder, A.A. Lagatsky, A. Abdolvand, N.V. Kuleshov, U. Keller, Diode-pumped femtosecond Yb:KGd(WO$_4$)$_2$ laser with 1.1-W average power, *Opt. Lett.*, Vol. 25, 1119–1121, 2000.

110. V.E. Kisel, A.E. Troshin, N.A. Tolstik, V.G. Shcherbitsky, N.V. Kuleshov, V.N. Matrosov, T.A. Matrosova, M.I. Kupchenko, Spectroscopy and continuous-wave diode-pumped laser action of Yb3+:YVO4, *Opt. Lett.*, Vol. 29, 2491–2493, 2004.

111. A. Shirakawa, K. Takaichi, H. Yagi, J.-F. Bisson, J. Lu, M. Musha, K. Ueda, T. Yanagitani, T.S. Petrov, A.A. Kaminskii, Diode-pumped mode-locked Yb3+:Y2O3 ceramic laser, *Opt. Express*, Vol. 11, 2911–2916, 2003.

112. A. Lucca, M. Jacquemet, F. Druon, F. Balembois, P. Georges, P. Camy, J.L. Doualan, R. Moncorge, High-power tunable diode-pumped Yb3+:CaF2 laser, *Opt. Lett.*, Vol. 29, 1879–1881, 2004.

113. A.A. Lagatsky, C.T.A. Brown, W. Sibbett, Highly efficient and low threshold diode-pumped Kerr-lens mode-locked Yb:KYW laser, *Opt. Express*, Vol. 12, 3928–3933, 2004.

114. M. Larionov, J. Gao, S. Erhard, A. Giesen, K. Contag, V. Peters, E. Mix, L. Fornasiero, K. Petermann, G. Huber, J. Aus der Au, G.J. Spühler, R. Paschotta, U. Keller, A.A. Lagatsky, A. Abdolvand, N.V. Kuleshov, Thin disk laser operation and spectroscopic characterization of Yb-doped sesquioxides and potassium tungstates, OSA Tops Vol. 50, Advanced Solid State Lasers 2001, 625, 2001.

115. K.J. Blow, B.P. Nelson, Improved modelocking of an F-center laser with a nonlinear nonsoliton external cavity, *Opt. Lett.*, Vol. 13, 1026–1028, 1988.

116. P. Yakymyshyn, J.F. Pinto, C.R. Pollock, Additive pulse mode-locked NaCl:OH-laser, *Opt. Lett.*, Vol. 14, 621–623, 1989.

117. M.N. Islam, E.R. Sunderman, C.E. Soccolich, I. Bar-Joseph, N. Sauer, T.Y. Chang, B.I. Miller, Color center lasers passively mode locked by quantum wells, *IEEE J. Quantum Electron.*, Vol. 25, 2454–2463, 1989.

118. R.L. Fork, O.E. Martinez, J.P. Gordon, Negative dispersion using pairs of prisms, *Opt. Lett.*, Vol. 9, 150–152, 1984.

119. M. Ramaswamy-Paye, J.G. Fujimoto, Compact dispersion-compensating geometry for Kerr-lens mode-locked femtosecond lasers, *Opt. Lett.*, Vol. 19, 1756–1758, 1994.

120. D. Kopf, G.J. Spühler, K.J. Weingarten, U. Keller, Mode-locked laser cavities with a single prism for dispersion compensation, *Appl. Opt.*, Vol. 35, 912–915, 1996.

121. R. Paschotta, J. Aus der Au, U. Keller, Strongly enhanced negative dispersion from thermal lensing or other focussing elements in femtosecond laser cavities, *J. Opt. Soc. Am. B*, Vol. 17, 646–651, 1999.

122. E.B. Treacy, Optical pulse compression with diffraction gratings, *IEEE J. Quantum Electron.*, Vol. 5, 454–458, 1969.

123. F. Gires, P. Tournois, Interferometre utilisable pour la compression d'impulsions lumineuses modulees en frequence, *C. R. Acad. Sci. Paris*, Vol. 258, 6112–6115, 1964.

124. R. Szipöcs, K. Ferencz, C. Spielmann, F. Krausz, Chirped multilayer coatings for broadband dispersion control in femtosecond lasers, *Opt. Lett.*, Vol. 19, 201–203, 1994.

125. R. Szipöcs, A. Stingl, C. Spielmann, F. Krausz, in *Optics and Photonics News*, 1995, pp. 16–20.

126. A. Stingl, M. Lenzner, Ch. Spielmann, F. Krausz, R. Szipöcs, Sub-10-fsec mirror-dispersion-controlled Ti:sapphire laser, *Opt. Lett.*, Vol. 20, 602–604, 1995.

127. F.X. Kärtner, N. Matuschek, T. Schibli, U. Keller, H.A. Haus, C. Heine, R. Morf, V. Scheuer, M. Tilsch, T. Tschudi, Design and fabrication of double-chirped mirrors, *Opt. Lett.*, Vol. 22, 831–833, 1997.

128. N. Matuschek, F.X. Kärtner, U. Keller, Theory of double-chirped mirrors, *IEEE J. Sel. Top. Quantum Electron.*, Vol. 4, 197–208, 1998.

129. N. Matuschek, F.X. Kärtner, U. Keller, Exact coupled-mode theories for multilayer interference coatings with arbitrary strong index modulations, *IEEE J. Quantum Electron.*, Vol. 33, 295–302, 1997.

130. N. Matuschek, F.X. Kärtner, U. Keller, Analytical design of double-chirped mirrors with custom-tailored dispersion characteristics, *IEEE J. Quantum Electron.*, Vol. 35, 129–137, 1999.

131. R. Ell, U. Morgner, F.X. Kärtner, J.G. Fujimoto, E.P. Ippen, V. Scheuer, G. Angelow, T. Tschudi, M.J. Lederer, A. Boiko, B. Luther-Davis, Generation of 5-fsec

pulses and octave-spanning spectra directly from a Ti:sapphire laser, *Opt. Lett.*, Vol. 26, 373–375, 2001.

132. N. Matuschek, L. Gallmann, D.H. Sutter, G. Steinmeyer, U. Keller, Back-Side coated chirped mirror with ultra-smooth broadband dispersion characteristics, *Appl. Phys. B*, Vol. 71, 509–522, 2000.

133. J.A. Dobrowolski, A.V. Tikhonravov, M.K. Trubetskov, B.T. Sullivan, P.G. Verly, Optimal single-band normal-incidence antireflection coatings, *Appl. Opt.*, Vol. 35, 644–658, 1996.

134. G. Tempea, Tilted-front-interface chirped mirrors, *J. Opt. Soc. Am. B*, Vol. 18, 1747–1750, 2001.

135. G. Steinmeyer, Brewster-angled chirped mirrors for high-fidelity dispersion compensation and bandwidths exceeding one optical octave, *Opt. Express*, Vol. 11, 2385–2396, 2003.

136. F.X. Kärtner, U. Morgner, R. Ell, T. Schibli, J.G. Fujimoto, E.P. Ippen, V. Scheuer, G. Angelow, T. Tschudi, Ultrabroadband double-chirped mirror pairs for generation of octave spectra, *J. Opt. Soc. Am. B*, Vol. 18, 882–885, 2001.

137. G. Sansone, S. Stagira, M. Nisoli, S. DeSilvestri, C. Vozzi, B. Schenkel, J. Biegert, A. Gosteva, K. Starke, D. Ristau, G. Steinmeyer, U. Keller, Mirror dispersion control of a hollow fiber supercontinuum, *Appl. Phys. B*, Vol. 78, 551–555, 2004.

138. R. Paschotta, G.J. Spühler, D.H. Sutter, N. Matuschek, U. Keller, M. Moser, R. Hövel, V. Scheuer, G. Angelow, T. Tschudi, Double-chirped semiconductor mirror for dispersion compensation in femtosecond lasers, *Appl. Phys. Lett.*, Vol. 75, 2166–2168, 1999.

139. J.P. Gordon, Theory of the soliton self-frequency shift, *Opt. Lett.*, Vol. 11, 662–664, 1986.

140. D.J. Kuizenga, A.E. Siegman, FM und AM mode locking of the homogeneous laser — Part I: Theory, Part II: Experimental results, *IEEE J. Quantum Electron.*, Vol. 6, 694–715, 1970.

141. H.A. Haus, J.G. Fujimoto, E.P. Ippen, Structures for additive pulse mode locking, *J. Opt. Soc. Am. B*, Vol. 8, 2068–2076, 1991.

142. F.X. Kärtner, D. Kopf, U. Keller, Solitary pulse stabilization and shortening in actively mode-locked lasers, *J. Opt. Soc. Am. B*, Vol. 12, 486–496, 1995.

143. U. Keller, Ultrafast solid-state lasers, *Progress in Optics*, Vol. 46, 1–115, 2004.

144. D. Kopf, F. Kärtner, K.J. Weingarten, U. Keller, Pulse shortening in a Nd:glass laser by gain reshaping and soliton formation, *Opt. Lett.*, Vol. 19, 2146–2148, 1994.

145. P.W. Smith, Phase locking of laser modes by continuous cavity length variation, *Appl. Phys. Lett.*, Vol. 10, 51–53, 1967.

146. H.A. Haus, U. Keller, W.H. Knox, A theory of coupled cavity mode locking with resonant nonlinearity, *J. Opt. Soc. Am. B*, Vol. 8, 1252–1258, 1991.

147. G.H.C. New, Pulse evolution in mode-locked quasi-continuous lasers, *IEEE J. Quantum Electron.*, Vol. 10, 115–124, 1974.

148. B. Braun, K.J. Weingarten, F.X. Kärtner, U. Keller, Continuous-wave mode-locked solid-state lasers with enhanced spatial hole-burning, Part I: Experiments, *Appl. Phys. B*, Vol. 61, 429–437, 1995.

149. F.X. Kärtner, B. Braun, U. Keller, Continuous-wave-mode-locked solid-state lasers with enhanced spatial hole-burning, Part II: Theory, *Appl. Phys. B*, Vol. 61, 569–579, 1995.

150. U. Siegner, U. Keller, Nonlinear optical processes for ultrashort pulse generation, in *Handbook of Optics*, Vol. III, M. Bass, E.W. Stryland, D.R. Williams, W.L. Wolfe, Eds., McGraw-Hill, New York, 2000.

151. R. Fluck, G. Zhang, U. Keller, K.J. Weingarten, M. Moser, Diode-pumped passively mode-locked 1.3 μm Nd:YVO₄ and Nd:YLF lasers by use of semiconductor saturable absorbers, *Opt. Lett.*, Vol. 21, 1378–1380, 1996.

152. G.J. Spühler, L. Gallmann, R. Fluck, G. Zhang, L.R. Brovelli, C. Harder, P. Laporta, U. Keller, Passively mode-locked diode-pumped Erbium-Ytterbium glass laser using a semiconductor saturable absorber mirror, *Electron. Lett.*, Vol. 35, 567–568, 1999.

153. R. Fluck, R. Häring, R. Paschotta, E. Gini, H. Melchior, U. Keller, Eye safe pulsed microchip laser using semiconductor saturable absorber mirrors, *Appl. Phys. Lett.*, Vol. 72, 3273–3275, 1998.

154. R. Grange, O. Ostinelli, M. Haiml, L. Krainer, G.J. Spühler, S. Schön, M. Ebnöther, E. Gini, U. Keller, Antimonide semiconductor saturable absorber for 1.5-μm, *Electron. Lett.*, Vol. 40, 1414–1415, 2004.

155. V. Liverini, S. Schön, R. Grange, M. Haiml, S.C. Zeller, U. Keller, A low-loss GaInNAs SESAM mode-locking a 1.3-μm solid-state laser, *Appl. Phys. Lett.*, Vol. 84, 4002–4004, 2004.

156. H.D. Sun, G.J. Valentine, R. Macaluso, S. Calvez, D. Burns, M.D. Dawson, T. Jouhti, M. Pessa, Low-loss 1.3 μm GaInNAs saturable Bragg reflector for high-power picosecond neodymium lasers, *Opt. Lett.*, Vol. 27, 2124–2126, 2002.

157. M. Haiml, U. Siegner, F. Morier-Genoud, U. Keller, M. Luysberg, P. Specht, E.R. Weber, Femtosecond response times and high optical nonlinearity in Beryllium doped low-temperature grown GaAs, *Appl. Phys. Lett.*, Vol. 74, 1269–1271, 1999.

158. M. Haiml, U. Siegner, F. Morier-Genoud, U. Keller, M. Luysberg, R.C. Lutz, P. Specht, E.R. Weber, Optical nonlinearity in low-temperature grown GaAs: microscopic limitations and optimization strategies, *Appl. Phys. Lett.*, Vol. 74, 3134–3136, 1999.

159. L.R. Brovelli, U. Keller, T.H. Chiu, design and operation of antiresonant Fabry-Perot saturable semiconductor absorbers for mode-locked solid-state lasers, *J. Opt. Soc. Am. B*, Vol. 12, 311–322, 1995.

160. R. Fluck, I.D. Jung, G. Zhang, F.X. Kärtner, U. Keller, Broadband saturable absorber for 10 fsec pulse generation, *Opt. Lett.*, Vol. 21, 743–745, 1996.

161. I.D. Jung, F.X. Kärtner, N. Matuschek, D.H. Sutter, F. Morier-Genoud, Z. Shi, V. Scheuer, M. Tilsch, T. Tschudi, U. Keller, Semiconductor saturable absorber mirrors supporting sub-10 fsec pulses, *Appl. Phys. B: Special Issue on Ultrashort Pulse Generation*, Vol. 65, 137–150, 1997.

162. S. Schön, M. Haiml, L. Gallmann, U. Keller, Fluoride semiconductor saturable-absorber mirror for ultrashort pulse generation, *Opt. Lett.*, Vol. 27, 1845–1847, 2002.

163. S. Schön, M. Haiml, U. Keller, Ultrabroadband AlGaAs/CaF₂ semiconductor saturable absorber mirrors, *Appl. Phys. Lett.*, Vol. 77, 782, 2000.

164. K.J. Weingarten, G.J. Spühler, U. Keller, L. Krainer, US patent No. 6,538,298, (GigaTera AG, U.S.A., 2001).

165. K. J. Weingarten, G.J. Spühler, U. Keller, D.S. Thomas, US patent No. 6,826,219, (GigaTera AG, U.S.A., 2002).

166. B. Braun, F.X. Kärtner, U. Keller, J.-P. Meyn, G. Huber, Passively Q-switched 180 ps Nd:LSB microchip laser, *Opt. Lett.*, Vol. 21, 405–407, 1996.

167. R. Häring, R. Paschotta, R. Fluck, E. Gini, H. Melchior, U. Keller, Passively Q-switched Microchip Laser at 1.5 μm, *J. Opt. Soc. Am. B*, Vol. 18, 1805–1812, 2001.

168. L. Krainer, R. Paschotta, G.J. Spühler, M. Moser, U. Keller, 29 GHz mode-locked miniature Nd:YVO₄ laser, *Electron. Lett.*, Vol. 35, 1160–1161, 1999.

169. L. Krainer, R. Paschotta, M. Moser, U. Keller, 77 GHz soliton mode-locked Nd:YVO₄ laser, *Electron. Lett.*, Vol. 36, 1846–1848, 2000.
170. R.C. Sharp, D.E. Spock, N. Pan, J. Elliot, 190-fsec passively mode-locked thulium fiber laser with a low threshold, *Opt. Lett.*, Vol. 21, 881–883, 1996.
171. G.J. Spühler, S. Reffert, M. Haiml, M. Moser, U. Keller, Output-coupling semiconductor saturable absorber mirror, *Appl. Phys. Lett.*, Vol. 78, 2733–2735, 2001.
172. I.P. Bilinsky, J.G. Fujimoto, J.N. Walpole, L.J. Missaggia, InAs-doped silica films for saturable absorber applications, *Appl. Phys. Lett.*, Vol. 74, 2411–2413, 1999.
173. M. Haiml, A. Aschwanden, U. Keller, SESAM damage, in preparation for *Appl. Phys. B*.
174. U. Keller, G.W. 'tHooft, W.H. Knox, J.E. Cunningham, Femtosecond pulses from a continuously self-starting passively mode-locked Ti:sapphire laser, *Opt. Lett.*, Vol. 16, 1022–1024, 1991.
175. D.K. Negus, L. Spinelli, N. Goldblatt, G. Feugnet, Sub-100 femtosecond pulse generation by Kerr lens modelocking in Ti:Sapphire, in *Advanced Solid-State Lasers*, Vol. 10, G. Dubé, L. Chase, Eds., Optical Society of America, Washington, D.C., 1991, pp. 120–124.
176. F. Salin, J. Squier, M. Piché, Mode-locking of Ti:sapphire lasers and self-focusing: a Gaussian approximation, *Opt. Lett.*, Vol. 16, 1674–1676, 1991.
177. M. Piché, F. Salin, Self-mode locking of solid-state lasers without apertures, *Opt. Lett.*, Vol. 18, 1041–1043, 1993.
178. G. Cerullo, A. Dienes, V. Magni, Space-time coupling and collapse threshold for femtosecond pulses in dispersive nonlinear media, *Opt. Lett.*, Vol. 21, 65–67, 1996.
179. I.P. Christov, V.D. Stoev, M.M. Murnane, H.C. Kapteyn, Mode locking with a compensated space-time astigmatism, *Opt. Lett.*, Vol. 20, 2111–2113, 1995.
180. J.E. Rothenberg, Space-time focusing: breakdown of slowly varying envelope approximation in the self-focusing of femtosecond pulses, *Opt. Lett.*, Vol. 17, 1340–1342, 1992.
181. S. Chen, J. Wang, Self-starting issues of passive self-focusing mode locking, *Opt. Lett.*, Vol. 16, 1689–1691, 1991.
182. F. Krausz, T. Brabec, C. Spielmann, Self-starting passive mode locking, *Opt. Lett.*, Vol. 16, 235–237, 1991.
183. G. Cerullo, S. De Silvestri, V. Magni, Self-starting Kerr lens mode-locking of a Ti:sapphire laser, *Opt. Lett.*, Vol. 19, 1040–1042, 1994.
184. G. Cerullo, S. De Silvestri, V. Magni, L. Pallaro, Resonators for Kerr-lens mode-locked femtosecond Ti:sapphire lasers, *Opt. Lett.*, Vol. 19, 807–809, 1994.
185. M. Lai, Self-starting, self-mode-locked Ti:sapphire laser, *Opt. Lett.*, Vol. 19, 722–724, 1994.
186. I.D. Jung, F.X. Kärtner, N. Matuschek, D.H. Sutter, F. Morier-Genoud, G. Zhang, U. Keller, V. Scheuer, M. Tilsch, T. Tschudi, Self-starting 6.5 fsec pulses from a Ti:sapphire laser, *Opt. Lett.*, Vol. 22, 1009–1011, 1997.
187. P.N. Kean, X. Zhu, D.W. Crust, R.S. Grant, N. Landford, W. Sibbett, Enhanced mode locking of color center lasers, *Opt. Lett.*, Vol. 14, 39–41, 1989.
188. E.P. Ippen, H.A. Haus, L.Y. Liu, Additive pulse mode locking, *J. Opt. Soc. Am. B*, Vol. 6, 1736–1745, 1989.
189. J. Goodberlet, J. Jacobson, J.G. Fujimoto, P.A. Schulz, T.Y. Fan, Self-starting additive pulse mode-locked diode-pumped Nd:YAG laser, *Opt. Lett.*, Vol. 15, 504–506, 1990.

190. U. Keller, W.H. Knox, H. Roskos, Coupled-cavity resonant passive mode-locked (RPM) Ti:sapphire laser, *Opt. Lett.*, Vol. 15, 1377–1379, 1990.

191. U. Keller, T.H. Chiu, Resonant passive mode-locked Nd:YLF laser, *IEEE J. Quantum Electron.*, Vol. 28, 1710–1721, 1992.

192. K.A. Stankov, A mirror with an intensity-dependent reflection coefficient, *Appl. Phys. B*, Vol. 45, 191–195, 1988.

193. A. Agnesi, C. Pennacchio, G.C. Reali, V. Kubecek, High-power diode-pumped picosecond Nd:YVO$_4$ laser, *Opt. Lett.*, Vol. 22, 1645–1647, 1997.

194. F.X. Kärtner, L.R. Brovelli, D. Kopf, M. Kamp, I. Calasso, U. Keller, Control of solid-state laser dynamics by semiconductor devices, *Opt. Eng.*, Vol. 34, 2024–2036, 1995.

195. T.R. Schibli, E.R. Thoen, F.X. Kärtner, E.P. Ippen, Suppression of Q-switched mode locking and breakup into multiple pulses by inverse saturable absorption, *Appl. Phys. B*, Vol. 70, 41–49, 2000.

196. A.C. Walker, A.K. Kar, W. Ji, U. Keller, S. D. Smith, All-optical power limiting of CO$_2$ laser pulses using cascaded optical bistable elements, *Appl. Phys. Lett.*, Vol. 48, 683–685, 1986.

197. R. Grange, M. Haiml, R. Paschotta, G.J. Spühler, L. Krainer, M. Golling, O. Ostinelli, U. Keller, New regime of inverse saturable absorption for self-stabilizing passively mode-locked lasers, *Appl. Phys. B*, Vol. 80, 151–158, 2005.

198. B.C. Collings, K. Bergman, W.H. Knox, True fundamental solitons in a passively mode-locked short cavity Cr^{4+}: YAG laser, *Opt. Lett.*, Vol. 22, 1098–1100, 1997.

199. T. Tomaru, Two-element-cavity femtosecond Cr: YAG laser operating at a 2.6 GHz repetition rate, *Opt. Lett.*, 26, 1439–1441, 2001.

200. L. Krainer, R. Paschotta, J. Aus der Au, C. Hönninger, U. Keller, M. Moser, D. Kopf, K. J. Weingarten, Passively mode-locked Nd:YVO$_4$ laser with up to 13 GHz repetition rate, *Appl. Phys. B*, Vol. 69, 245–247, 1999.

201. L. Krainer, R. Paschotta, M. Moser, U. Keller, Passively mode-locked picosecond lasers with up to 59 GHz repetition rate, *Appl. Phys. Lett.*, Vol. 77, 2104–2105, 2000.

202. S. Lecomte, M. Kalisch, L. Krainer, G.J. Spühler, R. Paschotta, L. Krainer, M. Golling, D. Ebling, T. Ohgoh, T. Hayakawa, S. Pawlik, B. Schmidt, U. Keller, Diode-pumped passively mode-locked Nd:YVO$_4$ lasers with 40-GHz repetition rate, *IEEE J. Quantum Electron.*, Vol. 41, 45–52, 2005.

203. S.C. Zeller, L. Krainer, G.J. Spühler, K.J. Weingarten, R. Paschotta, U. Keller, Passively mode-locked 40-GHz Er:Yb:glass laser, *Appl. Phys. B*, Vol. 76, 1181–1182, 2003.

204. F. Brunner, R. Paschotta, J. Aus der Au, G.J. Spühler, F. Morier-Genoud, R. Hövel, M. Moser, S. Erhard, M. Karszewski, A. Giesen, U. Keller, Widely tunable pulse durations from a passively mode-locked thin disk Yb:YAG laser, *Opt. Lett.*, Vol. 26, 379–381, 2001.

205. T. Südmeyer, J. Aus der Au, R. Paschotta, U. Keller, P.G.R. Smith, G.W. Ross, D.C. Hanna, Femtosecond fiber-feedback OPO, *Opt. Lett.*, Vol. 26, 304–306, 2001.

206. T. Südmeyer, J. Aus der Au, R. Paschotta, U. Keller, P.G.R. Smith, G.W. Ross, D.C. Hanna, Novel ultrafast parametric systems: high repetition rate single-pass OPG and fiber-feedback OPO, *J. Phys. D: Appl. Phys.*, Vol. 34, 2433–2439, 2001.

207. T. Südmeyer, E. Innerhofer, F. Brunner, R. Paschotta, U. Keller, T. Usami, H. Ito, M. Nakamura, K. Kitamura, D.C. Hanna, High power femtosecond fiber-feedback OPO based on periodically poled stoichiometric LiTaO3, *Opt. Lett.*, Vol. 29, 1111–1113, 2004.

208. S.V. Marchese, E. Innerhofer, R. Paschotta, S. Kurimura, K. Kitamura, G. Arisholm, U. Keller, Room temperature femtosecond optical parametric generation in MgO-doped stoichiometric LiTaO3, *Appl. Phys. B.*, Vol. 23, 1049–1052, 2005.

209. S.H. Cho, B.E. Bouma, E.P. Ippen, J.G. Fujimoto, Low-repetition-rate high-peak-power Kerr-lens mode-locked Ti:Al$_2$O$_3$ laser with a multiple-pass cell, *Opt. Lett.*, Vol. 24, 417–419, 1999.

210. A. Garnache, S. Hoogland, A.C. Tropper, I. Sagnes, G. Saint-Girons, J.S. Roberts, <500-fsec soliton pulse in a passively mode-locked broadband surface-emitting laser with 100-mW average power, *Appl. Phys. Lett.*, Vol. 80, 3892–3894, 2002.

211. S. Arahira, Y. Matsui, Y. Ogawa, Mode-locking at very high repetition rates more than terahertz in passively mode-locked distributed-Bragg-reflector laser diodes, *IEEE J. Quantum Electron.*, Vol. 32, 1211–1224, 1996.

212. F. Kuznetsov, F. Hakimi, R. Sprague, A. Mooradian, High-POWER (>0.5-W CW) diode-pumped vertical-external-cavity surface-emitting semiconductor lasers with circular TEM$_{00}$ beams, *IEEE Photon. Technol. Lett.*, Vol. 9, 1063–1065, 1997.

213. A.C. Tropper, H.D. Foreman, A. Garnache, K.G. Wilcox, S.H. Hoogland, Vertical-external-cavity semiconductor lasers, *J. Phys. D: Appl. Phys.*, Vol. 37, R75–R85, 2004.

214. J.E. Hastie, J.-M. Hopkins, S. Calvez, C.W. Jeon, D. Burns, R. Abram, E. Riis, A.I. Ferguson, M.D. Dawson, 0.5-W single transverse-mode operation of an 850-nm diode-pumped surface-emitting semiconductor laser, *IEEE Photonics Technol. Lett.*, Vol. 15, 894–896, 2003.

215. W.J. Alford, T.D. Raymond, A.A. Allerman, High power and good beam quality at 980 nm from vertical external-cavity surface-emitting laser, *J. Opt. Soc. Am. B.*, 19, 663–666, 2002.

216. R. Häring, R. Paschotta, E. Gini, F. Morier-Genoud, H. Melchior, D. Martin, U. Keller, Picosecond surface-emitting semiconductor laser with >200 mW average power, *Electron. Lett.*, Vol. 37, 766–767, 2001.

217. A. Aschwanden, D. Lorenser, H.J. Unold, R. Paschotta, E. Gini, U. Keller, 10-GHz passively mode-locked surface emitting semiconductor laser with 1.4-W average output power, *Appl. Phys. Lett.*, Vol. 86, 131102, 2005.

218. R. Paschotta, R. Häring, U. Keller, A. Garnache, S. Hoogland, A.C. Tropper, Soliton-like pulse formation mechanism in passively mode-locked surface-emitting semiconductor lasers, *Appl. Phys. B*, Vol. 75, 445–451, 2002.

219. S.C. Zeller, R. Grange, T. Südmeyer, K.J. Weingarten, U. Keller, 77-GHz pulse train at 1.5 μm directly generated by a passively mode-locked high repetition rate. Er:yb:glass laser, submitted to *Electron. Lett.*

220. U. Keller, A.C. Tropper, Passively mode-locked surface-emitting semiconductor lasers, *Physics Reports*, Vol. 429, 67–120, 2006.

221. D. Lorenser, D.J.H.C. Maas, H.J. Unold, A.R. Bellancourt, B. Rudin, E. Gini, D. Ebeling, U. Keller, 50-GHz passively mode-locked surface-emitting semiconductor laser with 100-mW average output power, *IEEE J. Quantum Electron.* Vol. 42, 838–847, 2006.

8

Multipass-Cavity Femtosecond
Solid-State Lasers

Alphan Sennaroglu and James G. Fujimoto

CONTENTS

8.1 Introduction

There is a growing demand for compact, low-cost, high-energy femtosecond solid-state lasers in many branches of science and technology. Applications are diverse, including biomedical imaging [1], ultrafast spectroscopy [2,3], and high harmonic generation [4,5], to name a few. Let us consider a commercial state-of-the-art femtosecond laser that can be used in these applications. One of the most expensive components of such a system is the pump laser, the cost of which approximately scales with the output power. Hence, a major reduction in overall system cost can be achieved by reducing the pump power requirements. For example, use of novel resonator geometries with tight beam focusing can enable low-threshold laser operation and decrease the pump power needed to obtain mode-locked operation. This approach has been used by several groups and low-threshold femtosecond $Ti^{3+}:Al_2O_3$ lasers operating with less than 1 W of pump power have been demonstrated. For example, Read et al. [6] obtained sub-20-fs pulses with output powers of 35–100 mW by using pump powers of 400 mW–1 W. In other experiments, Kowalevicz et al. reported the generation of 14-fs pulses with an average power of 15 mW by using only 200 mW of pump power [7]. However, one drawback of this scheme is that the average output power of the femtosecond oscillator is also reduced. Because several applications in nonlinear optics require femtosecond pulses with high energy and high peak power, it becomes important to preserve or scale up the original pulse energies as the pump power is decreased. One possible approach is to use extended optical cavities to scale up the pulse energy. In this case, provided that the additional optics needed to extend the cavity length do not increase the losses by a considerable amount, the average output power of the short and long cavities remains nearly the same and extension of the cavity length scales up the pulse energy by lowering the pulse repetition rate. This technique provides an attractive alternative to direct generation of high-energy femtosecond pulses without resorting to more complex amplification schemes such as chirped-pulse amplification [8] or cavity damping [9,10]. In the experiments reported by Libertun et al., an extended cavity was used to generate 36-nJ pulses from a $Ti^{3+}:Al_2O_3$ laser by lowering the repetition rate to 15.5 MHz [11]. Simply extending the cavity length makes the resonator bulky, rendering it difficult to use the mode-locked laser in other applications or to integrate the device in measurement systems. As an example, when the repetition rate of a mode-locked laser is reduced from 100 MHz to 10 MHz, the length of the cavity increases from 1.5 to 15 m.

An elegant solution that maintains compactness in an extended resonator was first proposed by Cho et al. who used a Herriott-type multipass cavity (MPC) to increase the effective length of a mode-locked $Ti^{3+}:Al_2O_3$ laser, generating 16.5-fs pulses with a peak power of 0.7 MW at a pulse repetition rate of 15 MHz [12]. Further refinements in MPC design has recently led to the generation of pulses with as high as 0.5 μJ of output energy and peak powers of the order of 10 MW [13]. The simplest MPC configuration consists of two highly reflecting mirrors with a mechanism to inject and extract optical beams. An incident, off-axis beam undergoes multiple bounces before exit and hence, the desired extension of the resonator can be achieved with a compact arrangement of mirrors. Multipass optical cavities were first introduced by Herriott et al. in 1964 [14]. To date, MPCs have been widely used in many applications such as accurate optical loss measurements [15], stimulated Raman scattering [16,17], long-path absorption spectroscopy [18], high-speed path-length scanning [19], and construction of compact solid-state lasers [20,21]. More recently, MPC femtosecond lasers have become versatile sources of ultrashort optical pulses and have found applications in various fields such as micromachining of photonic devices in transparent media [22], and multiphoton ionization spectroscopy [23]. Micromachining applications and the interaction of intense light beams with transparent materials are discussed in greater detail in Chapter 9 of this handbook.

In this chapter, we give a detailed discussion of the design of multipass optical cavities and their application to pulse energy amplification in femtosecond solid-state lasers. In Section 8.2, we first summarize the general characteristics of mode-locked lasers. Multipass cavities are then introduced in Section 8.3 and the general design guidelines are outlined. First, we use paraxial ray tracing and discuss the stability condition for a general MPC. Conditions needed to obtain circular spot patterns on each mirror are derived. We then focus on special MPC configurations that leave invariant the original spotsize distribution of the short cavity. These are referred to as "q-preserving" MPCs. We give analytical design rules for the construction of q-preserving MPCs and provide illustrative examples. In particular, we prove that the angular advance of the bouncing beam after one round trip should equal π times a rational number in order for the MPC to be q-preserving. In the case of MPCs that use notches to inject and extract the beam, additional compensating optics is needed to make the overall MPC configuration q-preserving. This is further discussed in Subsection 8.3.4. The pulse repetition rate that can be obtained with a multipass cavity depends on the mirror separation and radii. We develop a simple model and show how the pulse repetition rates vary for different q-preserving configurations in the case of two-mirror MPCs consisting of flat-curved or curved-curved high reflectors. Finally, we discuss different experiments in which MPC configurations were utilized to build compact and/or high-energy mode-locked oscillators. Examples based on Ti:sapphire, Cr:LiSAF, and Nd:vanadate gain media are presented.

8.2　Mode-Locked Femtosecond Lasers

Most state-of-the-art femtosecond lasers are based on tunable solid-state gain media in which optical amplification over multi-THz bandwidths can be achieved by using insulating crystals or glasses doped with rare-earth or transition metal ions. Many examples of such gain media exist, including Ti^{3+}:sapphire [24], Cr^{4+}:forsterite [25,26], Cr^{4+}:YAG [27], ytterbium-doped hosts [28,29], and many others. Different ion–host combinations lead to laser operation in different spectral regions from the ultraviolet to infrared. The pulsewidths that can be obtained from mode-locked lasers depend on the bandwidth of the gain medium, cavity dispersion, and reflectivity band-width of the mirrors. With careful dispersion control, extremely short optical pulses below 6 fs have been generated [30–32].

Mode locking is the technique employed to obtain ultrashort optical pulses from solid-state lasers. Typical pulse durations are in the picosecond (10^{-12}) to femtosecond (10^{-15}) range. In a laser resonator, the standing-wave cavity supports the oscillation of many resonant longitudinal modes that are equally spaced in frequency and that overlap the amplification band of the gain medium. During mode-locked operation, these cavity modes are forced to oscillate with a constant phase, producing a periodic train of pulses. The repetition rate, f_{rep}, of the pulses, which is also equal to the frequency spacing between two adjacent longitudinal modes (this frequency spacing is also referred to as the free spectral range) is given by

$$f_{rep} = \frac{c}{2L}. \tag{8.1}$$

Here, c is the speed of light and L is the effective length of the optical resonator. The average output power P_{av} of the mode-locked laser and the energy W per pulse are further related through

$$P_{av} = W f_{ref}. \tag{8.2}$$

We note that during mode locking, the peak powers obtained from the resonator can be increased by orders of magnitude in relation to the average power. In particular, if the number of locked modes is N, then the pulse duration and peak power per pulse will be of the order of $1/N f_{FSR}$ and N P_{av}, respectively. To appreciate the effectiveness of this method in producing high-peak-power pulses, consider a modest example, where longitudinal modes of a 1.5-m-long $Ti^{3+}:Al_2O_3$ laser are locked over a 20-THz bandwidth. The free spectral range will be 100 MHz, corresponding to 200,000 locked modes and a mode-locked pulse duration of the order of 50 fs. Furthermore, assuming that an average power of 500 mW is generated from the resonator, mode locking will produce peak powers around 100 kW.

Active and passive mode locking techniques can be used to lock the phase of the oscillating modes. Active mode locking involves synchronized mod-ulation of the cavity loss or gain at the frequency equal to the free spectral range or one of its harmonics [33,34]. In the case of passive mode locking, pulse shaping is achieved by an intensity-dependent saturable loss mecha-nism [33,35]. Two variations of passive mode locking are widely used in state-of-the-art femtosecond solid-state lasers. In Kerr lens mode locking (KLM), Kerr nonlinearity-induced focusing inside the gain medium provides a saturable loss mechanism and aids in pulse shaping [33,36]. A second technique uses semiconductor saturable absorbers (SESAM) for initiating mode locking [37]. Chapter 7 of this handbook gives a comprehensive review of ultrafast lasers and SESAMs.

Dispersion compensation techniques are employed to obtain the shortest possible pulses from a femtosecond laser. If a large number of longitudinal modes are locked together, the pulse envelope $E_p(t)$ will be given in terms of the spectral amplitude distribution $E_p(\omega)$ as

$$E_p(t) = \int\limits_{-\infty}^{+\infty} d\omega \left[E_p(\omega) e^{i\Phi_p(\omega)} e^{i\omega t} \right]. \tag{8.3}$$

Here, ω is the angular frequency and $\Phi_p(\omega)$ is the phase. Ideally, it is desired to keep $\Phi_p(\omega)$ constant over the spectral extend of the pulse. In practice, however, as the locked bandwidth becomes large, material dispersion and self-phase modulation distort the phase and compensation techniques become necessary to minimize phase variations. Prism pairs [38] or chirped mirrors [39–42] are typically employed to introduce controlled amounts of second- and higher-order dispersion into the cavity. In the ideal case, where phase distortions are eliminated, it is possible to produce the shortest possible pulse for a given bandwidth, also known as the *transform-limited pulse*. In this case, the available spectral bandwidth Δv (FWHM) and the output pulsewidth τ_p (FWHM) satisfy

$$\Delta v \tau_p = K. \tag{8.4}$$

Here, K is a constant of the order of unity and depends on the specific pulse shape. For example, K = 0.315 for a hyperbolic secant-squared (sech²) pulse and 0.441 for a Gaussian pulse.

In the particular case of Kerr lens mode-locked lasers, negative dispersion is often introduced to balance self-phase modulation and form soliton-like pulses. Shortest pulses are generated when the net second-order dispersion of the resonator is negative. Under reasonable approximations, it can be shown that the pulse energy W and the net second-order dispersion D obey the scaling law [33]

$$\tau = \frac{4|D|}{W\delta}, \tag{8.5}$$

where 1.76τ is the full-width at half-maximum (FWHM) pulsewidth, and δ is the effective Kerr nonlinearity coefficient given by

$$\delta = \frac{2\pi}{\lambda} \frac{n_2 L}{A_{eff}}. \tag{8.6}$$

In Equation 8.6, λ is the vacuum wavelength, L is the length of the gain medium, n_2 is the nonlinear index of the gain medium, and A_{eff} is the effective mode cross-sectional area.

Equation 8.1 to Equation 8.6 provide the basic design rules for the construction of femtosecond KLM MPC lasers. We first note from Equation 8.1

and Equation 8.2 that as the cavity length is increased, the repetition rate decreases, and the corresponding output energy per pulse increases, provided that the average power of the laser remains nearly the same. In order to maintain the same mode-locked pulsewidth, the net cavity dispersion also needs to be scaled up proportional to the pulse energy, as can be seen from Equation 8.5. In practice, this is done by adding more negative dispersion into the cavity with a prism pair or additional chirped mirrors. If these conditions are satisfied, the peak power of the output pulses can be scaled up as the cavity length is increased.

8.3 General Characteristics of Multipass Cavities

8.3.1 What Is a Q-Preserving Multipass Cavity?

Let us consider as a specific example, a Kerr lens mode-locked femtosecond laser. In practice, the addition of an MPC to increase the output pulse energy involves two steps. First, a short laser resonator is constructed and optimized to obtain the best cw power performance. Focusing inside the gain medium is then adjusted to initiate KLM mode-locked operation. Here, the term *short* is relative and refers to the initial cavity being short in relation to the extended cavity containing the MPC. In the second step, the MPC is added to increase the effective cavity length. Consider, for example, a short cavity with a round-trip length of 2 m. In mode-locked operation, it will produce pulses at a repetition rate of 150 MHz. If the addition of the MPC increases the total round-trip length of the cavity to 20 m, the repetition rate will decrease to 15 MHz. Provided that the average output power remains nearly the same, this represents a pulse energy gain of ten. In such a case, because the length of the MPC laser is considerably larger than that of the short cavity, addition of an arbitrary MPC will, in general, change the spotsize distribution and hence the focusing inside the gain medium. In extreme cases, the new resonator may not even be stable or the KLM operating point of the cavity may be drastically modified. These problems can be mitigated if the MPC is designed in such a way as to leave invariant the initial spotsize distribution inside the gain medium. There are two important advantages of such a design. First, the optimum mode matching between the pump and the laser modes is maintained. As a result, similar power performance can be obtained with the extended cavity. Second, spotsize distribution needed to initiate KLM action remains nearly the same. Hence, stable mode-locked operation can be obtained by proportional scaling of the net cavity dispersion (see Equation 8.5) Figure 8.1 summarizes this important feature of the MPC laser design. To be specific, the short cavity was assumed to have an astigmatically compensated x configuration and consists of the mirrors M1, M2, M3, and M4. Cavity length extension is achieved with a two-mirror MPC consisting

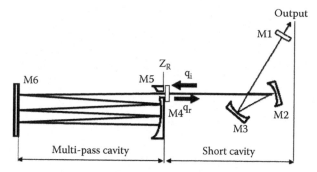

FIGURE 8.1

Schematic of a short laser cavity extended by using a multipass cavity (MPC). When the MPC is q-preserving, the q-parameter of the beam injected into the MPC (q_i) equals that of the returning beam (q_r) at the input reference plane z_R.

of mirrors M5 and M6. To extend the cavity, mirror M4 of the short cavity is removed. As a result, a single round trip of the laser also includes the transit of the beam through the MPC. Laser output is obtained through the mirror M_1. Here, z_R denotes the reference plane situated at the input of the MPC. Let q_i be the Gaussian beam parameter of the beam incident on the MPC. In general, the returning beam has a different beam parameter q_r. The ideal MPC design should be such that $q_i = q_r$. In our analysis of MPCs, we will refer to this class of designs as the q-preserving MPC configurations. In the case of a two-mirror multipass cavity, for example, there are infinitely many discrete mirror separations for which the MPC becomes q-preserving. These will be discussed in the next section.

8.3.2 Design Rules for Q-Preserving Multipass Cavities

In this section, we will analyze the propagation of a laser beam inside an MPC by using the matrix formalism of paraxial optics and obtain analytical design rules for the construction of q-preserving MPC's. A specific example of an MPC consisting of two mirrors (M1 and M2) is shown in Figure 8.2. As we discussed before, a mechanism is needed for injecting and extracting the beam. For the particular case shown in Figure 8.2, beam injection and extraction are done by using triangular notches cut on the mirrors. If the MPC mirrors are separated by a distance L_0, and the number of complete round-trips in the MPC is n, then the effective optical path length traversed after one transit becomes $2nL_0$. The MPC configuration needs to be stable so that the bouncing beam remains close to the optical axis after an arbitrary number of round trips. A stability condition similar to that of a standard two-mirror resonator guarantees the confinement of the bouncing beam near the optical axis of the MPC. If a single round trip is represented by a ray transfer matrix M_T of the form

$$M_T = \begin{bmatrix} A & B \\ C & D \end{bmatrix},$$

(8.7)

then, the elements A and D of M_T need to satisfy the inequality

$$\left| \frac{A+D}{2} \right| \le 1,$$

(8.8)

in order for the MPC to be stable. Note that because the bouncing beam returns to the same location after each round trip, the determinant of M_T is unity. If the stability condition given by Equation 8.8 is met, the two eigenvalues $\lambda_{1,2}$ of M_T have unit magnitude and can be expressed as $\lambda_1 = e^{i\theta}$ and $\lambda_2 = e^{-i\theta}$. In addition, the corresponding eigenvectors $\vec{V}_{1,2}$ are given by

$$\vec{V}_1 = \frac{1}{B(B-C)} \begin{bmatrix} B \\ -(A - e^{i\theta}) \end{bmatrix}$$

$$\vec{V}_2 = \frac{1}{B(B-C)} \begin{bmatrix} B \\ -(A - e^{-i\theta}) \end{bmatrix}$$

(8.9)

Here, $\cos\theta = (A + D)/2$. Standard diagonalization techniques can be used to calculate the nth power M_T^n of M_T, which gives the ray transformation matrix for n full round trips inside the MPC. M_T^n can be expressed in the form [43]

$$M_T^n = \begin{bmatrix} \dfrac{A-D}{2} \dfrac{\sin n\theta}{\sin\theta} + \cos n\theta & B\dfrac{\sin n\theta}{\sin\theta} \\ C\dfrac{\sin n\theta}{\sin\theta} & \dfrac{D-A}{2} \dfrac{\sin n\theta}{\sin\theta} + \cos n\theta \end{bmatrix}.$$

(8.10)

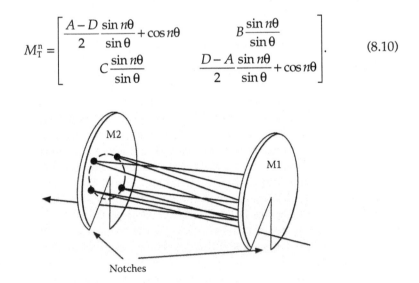

FIGURE 8.2
Sketch of a multipass cavity consisting of two reflecting mirrors (M1 and M2). In this particular case, notches are used for beam injection and extraction.

Let us next analyze the effect of the MPC on the propagation of rays to give a physical interpretation for the angle θ appearing in Equation 8.9 and Equation 8.10. We assume that the optical axis is along the z direction and the incident ray $\vec{r}_i$ is represented by the 2×1 column vector

$$\vec{r}_i = \begin{bmatrix} r_0 \\ r_0' \end{bmatrix}. \tag{8.11}$$

In Equation 8.11, r_0 and r_0' are the initial ray displacement from the optical axis and the initial angle of inclination, respectively. Neglecting astigmatic effects, the ray vector $\vec{r}_n$ after n complete round trips is given by

$$\vec{r}_n = M_T^n \vec{r}_i. \tag{8.12}$$

Let (x_0, x_0') and (y_0, y_0') be the initial offset and inclination of the incident ray in the x and y directions, respectively, at the input reference plane. By using Equation 8.10 and Equation 8.12, the transverse displacements x_n and y_n of the ray after the nth round trip become

$$x_n = x_0 \cos n\theta + \left(\frac{x_0(A-D) + 2Bx_0'}{2\sin\theta} \right) \sin n\theta$$

$$y_n = y_0 \cos n\theta + \left(\frac{y_0(A-D) + 2By_0'}{2\sin\theta} \right) \sin n\theta \tag{8.13}$$

Equation 8.13 shows that the spot pattern formed by the bouncing beam on the MPC mirrors will in general be elliptical. If the initial ray parameters are chosen such that

$$y_0 = 0$$

$$y_0' = \frac{x_0 \sin\theta}{B} \quad, \tag{8.14}$$

$$x_0' = \frac{x_0}{2B}(D-A)$$

then the bouncing rays form a circular spot pattern on each mirror and the coordinates (x_n, y_n) of the spots after n round trips are given by

$$x_n = x_0 \cos n\theta$$

$$y_n = x_0 \sin n\theta \tag{8.15}$$

This is schematically depicted in Figure 8.3, which shows three consecutive spots on one of the MPC mirrors. Note that the angle θ corresponds to the

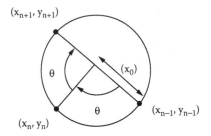

FIGURE 8.3
Sketch of the circular spot pattern on one of the MPC mirrors. When the initial offset and the angle of inclination of the beam are properly adjusted, a circular spot pattern can be obtained. θ is the angular advance of the spot pattern after one round trip and x_0 is the radius of the circular spot pattern.

angular advance of the spot pattern after one round-trip. The value of θ depends on the elements A and D of the ray transfer matrix M_T and hence on the specific configuration of the MPC.

We next consider the propagation of a Gaussian beam through an MPC and investigate the conditions that must be met in order for the MPC to be q-preserving. As is well known, information about the radius of curvature R and the spotsize ω of a Gaussian beam is contained in the complex beam parameter q, which is also referred to as the q-parameter; $1/q$ is given by [44]

$$\frac{1}{q} = \frac{1}{R} - i\frac{\lambda}{\bar{n}\pi\omega^2}.$$ (8.16)

In Equation 8.16, λ is the wavelength of light and $\bar{n}$ (not to be confused with the integer n) is the local refractive index of the medium. After passing through an optical system characterized by a ray transformation matrix $\overline{M}$ with elements $(\overline{A}, \overline{B}, \overline{C}, \overline{D})$ the q-parameter q' of the emerging Gaussian beam can be calculated from

$$q' = \frac{\overline{A}q + \overline{B}}{\overline{C}q + \overline{D}}.$$ (8.17)

Let us suppose that the injected beam leaves the MPC after n complete round trips. The matrix $\overline{M}$ representing n complete round trips will be given by M_T^n. If M_T^n is $\pm I$, where I is the 2×2 unity matrix, then we see from Equation 8.17 that $q' = q$ and the returning beam from the MPC will have the same q parameter as the incident beam. Looking at Equation 8.10, when

$$n\theta = m\pi,$$ (8.18)

where n and m are any two integers, $M_T^n = (-1)^m I$ and hence the MPC becomes q-preserving. The integer n denotes the number of round trips that the beam has undergone, and m designates the number of semicircular arcs that the spot pattern produced by the bouncing beam completes on one of the mirrors. As an example, if m = 3, then the bouncing beam traverses three semicircular arcs, thus covering a full angular sweep of 540° before exiting the MPC. Equation 8.18 is the key design rule for the construction of q-preserving MPCs.

We can provide an alternative interpretation of Equation 8.18 as follows: suppose the mirror radii and separation of a MPC are chosen so that the angular advance θ satisfies Equation 8.18. If the bouncing beam completes n round trips inside the MPC, then the MPC becomes q-preserving.

8.3.3   An Illustrative Design Example

We will outline the design of a specific two-mirror multipass cavity shown in Figure 8.4. Let the radii of curvature of the mirrors M_1 and M_2 be R_1 and R_2, respectively. The separation between the mirrors is L_0. To find the ray transfer matrix for one round trip, we need the ray matrices that describe a curved mirror and displacement in a uniform medium. The ray transfer matrix M_R of a curved mirror with radius R is given by

$$M_R = \begin{bmatrix} 1 & 0 \\ -\dfrac{2}{R} & 1 \end{bmatrix}.$$

(8.19)

For a displacement of d, the corresponding ray matrix M_d is

$$M_d = \begin{bmatrix} 1 & d \\ 0 & 1 \end{bmatrix}.$$

(8.20)

Starting immediately to the right of the mirror M_1 and proceeding to the right, the overall ray matrix M_T for one round trip can be calculated from the matrix product

$$M_T = M_{R_1} M_{L_0} M_{R_2} M_{L_0}.$$

(8.21)

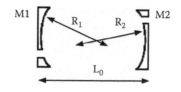

FIGURE 8.4
A two-mirror multipass cavity consisting of curved mirrors (M1 and M2) with radii R_1 and R_2. L_0 is the mirror separation.

It can be shown that the trace of M_T is independent of the starting position inside the resonator and $(A+D)/2$ is given by

$$\frac{A+D}{2} = 2\left(1-\frac{L_0}{R_1}\right)\left(1-\frac{L_0}{R_2}\right) - 1. \tag{8.22}$$

To consider a specific example, let M_1 be flat ($R_1 = \infty$) and R_2 be 2 m. If we choose θ as 24°, which is $2\pi/15$ (corresponding to m = 2 and n = 15), the MPC will be q-preserving if 15 full round trips are completed during the passage of the beam. By using Equation 8.22, we see that $(L_0/R_2) = 0.0432$ gives $\theta = 24°$. Because $R_2 = 2$ m, the mirror separation L_0 needs to be set at 0.0864 m for this q-preserving MPC configuration.

8.3.4 Compensating Optics for Non-q-Preserving MPCs

Long optical path lengths can still be obtained by using multipass cavities, which are not exactly q-preserving. In this case, it is of interest to know whether it is possible to use compensating optics to make the MPC q-pre-serving. In this section, we address this issue by looking at a specific MPC configuration in which notches carved on the mirrors are used to inject and extract the beam. We note once again that two conditions must be simulta-neously satisfied in order for the MPC to be q-preserving: θ must be adjusted to be $m\pi/n$ (m and n being integers) and n full round trips must be completed during the passage of the beam. When m is even, the bouncing spot pattern needs to complete a full circular sweep, or a multiple thereof, in order for the MPC to be q-preserving. However, it is not possible to extract such a beam with a notch, because it causes the incident beam to escape the MPC without undergoing any reflections. In this case, by adjusting the notch position on the second mirror, the bouncing beam can complete a maximum of n–1 round trips to maximize the effective optical path length. Therefore, one additional round trip must be introduced in order to make the MPC q-preserving.

In the case of a two-mirror MPC with notches, compensating optics can be added to make it q-preserving. To see how this is possible, let us consider the two-mirror MPC consisting of two curved reflectors (M_1 and M_2) as shown in Figure 8.5. The radii of M_1 and M_2 are R_1 and R_2, respectively. To analyze the propagation of the beam in the forward and backward directions, we introduce two reference planes in addition to the input reference plane z_R. Starting at the reference plane z_F and moving to the left, the round trip ray transfer matrix M_F can be expressed as

$$M_F = M_{L_0} M_{R_1} M_{L_0} M_{R_2}, \tag{8.23}$$

where the matrices for curved mirrors and displacements are located, as defined in Equation 8.19 and Equation 8.20. Similarly, starting at the reference

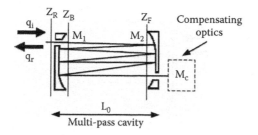

FIGURE 8.5

A two-mirror multipass cavity with notched mirrors. The mirrors M_1 and M_2 have radii R_1 and R_2. Because of a missing round-trip, a single transit is not q-preserving. After the mirror M_2, a compensating optical system represented by the ray transformation matrix M_c is added to make the multipass cavity q-preserving after a round trip. L_0 is the distance between mirrors M_1 and M_2.

plane z_B and moving to the right, the round trip ray transfer matrix M_B becomes

$$M_B = M_{L_0} M_{R_2} M_{L_0} M_{R_1}.$$ (8.24)

Because θ is adjusted to be $m\pi/n$, M_F and M_B both satisfy

$$M_F^n = M_B^n = (-1)^m.$$ (8.25)

Now, we analyze the case where the second notch is adjusted so that the beam completes only n–1 round trips inside the MPC before exit. We also assume that after the beam exits the MPC, it goes through a compensating optical system and returns to the MPC. Let M_c represent the transformation matrix of the compensating optics used to make the overall MPC q-preserving. Starting at the reference plane z_R, a complete round trip inside the MPC can now be expressed as

$$M_T = M_B^{n-1} M_{L_0} M_c M_F^{n-1} M_{L_0}$$
$$= M_{R_1}^{-1} M_{L_0}^{-1} M_{R_2}^{-1} M_c M_{R_2}^{-1} M_{L_0}^{-1} M_{R_1}^{-1}$$ (8.26)

A judicious choice of compensating optics can make the overall MPC q-preserving. In particular, if M_c is chosen as

$$M_c = M_{R_2} M_{L_0} M_{R_1} M_{R_1} M_{L_0} M_{R_2} = M_{R_2} M_{L_0} M_{R_1/2} M_{L_0} M_{R_2},$$ (8.27)

then M_T in Equation 8.26 equals the unity matrix and the q parameters of the incident (q_i) and returning beams (q_r) become identical at the input reference plane z_R. Practical realization of the compensating optics described by M_c is straightforward. The setup is shown in Figure 8.6. Upon exiting the second notch, the beam is first reflected by a curved mirror of radius R_2 and

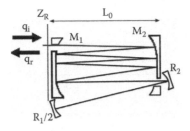

FIGURE 8.6
Compensating optics for a two-mirror multipass cavity with notched mirrors. After exiting the second mirror, the beam is first redirected by a curved mirror of radius R_2 and then retroreflected by a second curved mirror of radius $R_1/2$ situated at a distance of L_0 from the mirror M_2. L_0 is the distance between mirrors M_1 and M_2.

propagates a distance of L_0 up to the retroreflector. Note that the retroreflector has a radius of $R_1/2$, not R_1. The beam then traverses the same path back into the MPC. The total optical path length introduced by the MPC and the compensating optical system becomes $4nL_0$. The design discussed in this section gives a very practical q-preserving MPC example utilizing notches to inject and extract the beam. One important property of the design is that it is scalable. In other words, if all of the radii and mirror separations of the MPC are increased by a factor k, the overall path length increases by the same factor k, and, more important, the new MPC still remains q-preserving. For example, when R_1, R_2, and L_0 are doubled, the MPC remains q-preserving, the total optical path length is doubled, and the pulse repetition rate is halved (neglecting the length of the short cavity), without changing the original spotsize distribution of the short cavity.

If specific compensating optics is not available, non-q-preserving MPC configurations may sometimes be preferred due their simplicity. In such a case, it is of interest to know how much deviation occurs from the ideal q-preserving configuration. As a specific example, we consider an MPC consisting of a flat and a curved reflector with notches as shown in Figure 8.7. Again, mirrors M_1 and M_2 have radii R_1 and R_2. After n–1 round trips, the beam leaves the second mirror and is retroreflected by a flat mirror (M3 in Figure 8.7) that can also serve as an output coupler. The resulting configuration is not strictly q-preserving. As a result, if the lengths and radii of the MPC are scaled by some factor, the q-parameter of the returning Gaussian beam and hence the spotsize distribution inside the gain medium will in general be different. Reference [45] provides a thorough analysis of such non-q-preserving MPCs. It is shown that when $L_1 < L_0$ and the separation between the MPC mirrors is much less than the radius of curvature ($L_0/R_1 \ll 1$), the approximate effect of the non-q-preserving MPC is to shorten the arm length of the resonator by $|L_0 - L_1|$. It is well known that in a standard four-mirror laser resonator, the long arm determines the maximum spotsize of the beam at the center of the stability region. Hence, if this MPC is placed in either arm of a nearly symmetric four-mirror cavity or in the shorter arm of an

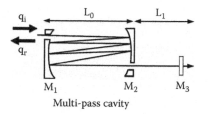

Multi-pass cavity

FIGURE 8.7
Schematic of a non-q-preserving multipass cavity. The exiting beam is retroreflected by a flat mirror M_3 placed at a distance of L_1 from the second mirror M_2. L_0 is the distance between mirrors M_1 and M_2.

asymmetric cavity, the maximum spotsize at the center of the stability region will remain approximately the same, provided that the short arm is initially longer than $|L_0 - L_1|$. In this case, the main effect of the deviation from ideal q-preserving operation is simply a shift in the stability region of the cavity. When the non-q-preserving cavity is added to the short cavity, the position of the focusing mirrors around the gain medium can then be readjusted in order to obtain lasing again.

8.3.5 Pulse Repetition Rates

As we discussed before, MPCs provide a versatile technique for increasing the effective optical path lengths with a compact arrangement of mirrors. It is important to note that the mirror separations cannot be arbitrary chosen due to two limitations. First, the multipass cavity arrangement must be stable as summarized by Equation 8.8. Second, mirror separations resulting in q-preserving conditions should be chosen. In general, only a discrete set of mirror separations satisfies both of these constraints. In this section, we derive analytical expressions for the mirror separations of two-mirror q-preserving MPCs and examine how the repetition rate of a femtosecond laser depends on the multipass cavity parameters. We also show that in the case of a two-mirror MPC consisting of a flat and a curved mirror, an optimum round trip number (n) exists, which minimizes the repetition rate for a given value of m (the integers n and m are as defined in Subsection 8.3.2).

Let us consider a two-mirror MPC. If the mirror separation is L_0, and the beam undergoes n round trips in a single transit, then the MPC introduces a total effective optical length of $4nL_0$. Neglecting the length of the short cavity, the repetition rate f_{rep} of the mode-locked laser containing the MPC becomes

$$f_{rep} = \frac{c}{4nL_0}. \tag{8.28}$$

Equation 8.28 is correct, whether or not the MPC is q-preserving. However, the q-preserving condition $\theta = m\pi/n$ is satisfied only at certain discrete values of L_0, as we discuss below for specific cases.

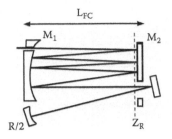

FIGURE 8.8
Two-mirror flat-curved multipass cavity with compensating optics. L_{FC} is the mirror separation.

We first consider a two-mirror q-preserving MPC consisting of a flat and a curved mirror as shown in Figure 8.8. Mirror M_1 has a radius of R and mirror M_2 is flat. Again, note that because notches were used in this particular design to inject and extract the beam, compensating optics was added after the flat mirror to make the overall MPC q-preserving. We will refer to this class of two-mirror MPCs as flat-curved (FC) MPCs and denote the q-preserving mirror separations and the corresponding repetition rates as L_{FC} and f_{FC}, respectively. Starting at the reference plane z_R, the ray transfer matrix M_T for one round trip becomes

$$
M_T = \begin{bmatrix} 1 - \dfrac{2L_{FC}}{R} & 2L_0\left(1 - \dfrac{L_{FC}}{R}\right) \\[2ex] -\dfrac{2}{R} & 1 - \dfrac{2L_{FC}}{R} \end{bmatrix}.
$$

(8.29)

Hence,

$$
\cos\theta = \frac{A+D}{2} = 1 - \frac{2L_{FC}}{R}.
$$

(8.30)

Because $\theta = m\pi/n$ for q-preserving configurations, we find, by using Equation 8.30, that L_{FC} is given by

$$
L_{FC} = \frac{R}{2}\left(1 - \cos\left(\frac{m\pi}{n}\right)\right).
$$

(8.31)

Here, R is the radius of the curved mirror. Note that for a fixed m, different q-preserving mirror separations are found by changing the value of n. For a fixed value of m, this gives infinitely many possible q-preserving mirror separations. Also, for large values of n, L_{FC} monotonically decreases and approaches zero. The corresponding repetition rate f_{rep} becomes

$$
f_{FC} = \frac{c}{2nR\left(1 - \cos\left(\dfrac{m\pi}{n}\right)\right)}.
$$

(8.32)

In Equation 8.32, c is the speed of light. Figure 8.9a and Figure 8.9b show how L_{FC} and f_{FC} vary as a function of n for m = 2 and R = 4 meters. Each solid square in Figure 8.9a and Figure 8.9b, corresponding to an integer value of n, represents a q-preserving configuration. Another interesting feature of flat-curved MPCs can be seen from Figure 8.9b. For a given value of m, there exists an optimum value n_{opt} of n, independent of R, which maximizes the effective optical length and minimizes the pulse repetition rate. This is of practical importance in the construction of flat-curved MPCs with very low repetition rates. For a fixed value of m, the optimum value n_{opt} can be found by setting to zero the derivative of f_{LC} with respect to n. This gives the following transcendental equation for n_{opt}:

$$1 = \cos\left(\frac{m\pi}{n_{opt}}\right) + \frac{m\pi}{n_{opt}}\sin\left(\frac{m\pi}{n_{opt}}\right). \tag{8.33}$$

In practice, n_{opt} will be the integer closest to the solution of Equation 8.33. For example, when m = 4, n_{opt} = 5 and the corresponding mirror separation

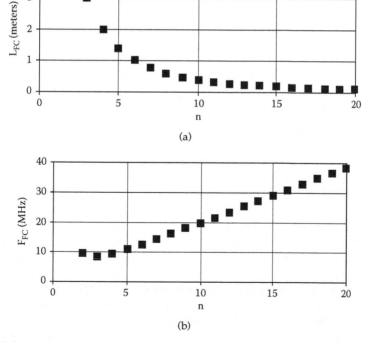

(a)

(b)

FIGURE 8.9
Calculated variation of the (a) mirror separation L_{FC} and (b) repetition rate f_{FC} as a function of n for a flat-curved, two-mirror multipass cavity. In the calculations, m = 2 and R = 4 meters. Each solid square, corresponding to an integer value of n, represents a q-preserving configuration.

and minimum repetition rates become 0.9R and c/18.09R, respectively. Variation of n_{opt} with m is tabulated in Reference 43.

In the case of curved-curved (CC) MPCs, the separation L_{CC} is obtained from

$$L_{CC} = R\left(1 - \cos\left(\frac{\theta}{2}\right)\right). \tag{8.34}$$

Different from the flat-curved case, two distinct sets of q-preserving solutions exist for the θ values. These can be expressed as $(m^{\pm}\pi/n)$, where $m^{+} = m$ and $m^{-} = m - 2n$. These correspond to angular advances of $(m\pi/n)$ and $(m\pi/n)$-2π, respectively. The resulting mirror separations $L_{CC}^{\mp}$ and repetition rates $f_{CC}^{\mp}$ of the q-preserving configurations become

$$L_{CC}^{\mp} = R\left(1 - \cos\left(\frac{m^{\pm}\pi}{2n}\right)\right)$$

$$= R\left(1 \mp \cos\left(\frac{m\pi}{2n}\right)\right) \tag{8.35}$$

and

$$f_{CC}^{\mp} = \frac{c}{4nR\left(1 \mp \cos\left(\frac{m\pi}{2n}\right)\right)}. \tag{8.36}$$

The resulting spot patterns on the MPC mirrors are different for the two different values of m. These are discussed in detail elsewhere [45].

Inspection of Equation 8.32 and Equation 8.36 shows that f_{FC}^{-} and f_{CC}^{-} have a similar dependence on n. In particular, the repetition rate attains a minimum value for both cases as n is varied. The only difference is that for the curved-curved MPC, the optimum value $(n_{opt})_{CC}$ is approximately the integer closest to $\frac{1}{2}$ $(n_{opt})_{FC}$. This choice of n gives the minimum possible repetition rate that can be obtained for these two configurations. The trends discussed above differ from the second set of solutions for the curved-curved MPC. As n is increased, f_{CC}^{+} monotonically decreases while L_{CC}^{+} asymptotically approaches 2R. Hence, it is preferable to use curved-curved MPCs to achieve the lowest possible repetition rates. Figure 8.10a and Figure 8.10b show the variation of the MPC separation and the corresponding repetition rate, respectively, for the two possible cases of the curved-curved MPCs. A practical constraint on the maximum allowed value of n comes from the loss per bounce introduced by the highly reflecting mirror. If the total number of round trips in one transit is n and if the reflective loss per bounce from each MPC mirror is l_R, then the addition of the MPC will introduce a total loss of

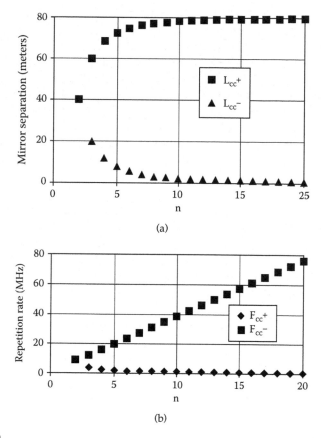

(a)

(b)

FIGURE 8.10

Calculated variation of the (a) mirror separation and (b) repetition rate as a function of n for the two possible cases of a curved-curved, two-mirror multipass cavity. In the calculations, $m = 2$ and $R = 4$ meters.

$4nl_R$ into the cavity. Hence, use of very high quality reflectors with ultralow reflective loss is critical in minimizing passive losses while achieving very long effective optical path lengths.

8.3.6 Practical Considerations

In this section, we discuss some of the important issues related to the construction of practical MPC femtosecond lasers. The most important point to remember is that the overall length of the MPC can be tens of meters, which makes the alignment of the laser extremely difficult by using the fluorescence of the gain medium alone. In practice, it is much easier to build the short laser cavity first. Once lasing is obtained, the power performance and mode matching of this cavity can be optimized further. Let us remember that, especially when a q-preserving MPC is added for cavity length extension, the laser spotsize distribution in the gain medium will remain invariant.

Suppose that the MPC will be placed in the arm containing the end high reflector of a standard four-mirror cavity built either in x or z configuration. To ease the alignment of the MPC, the high reflector is first replaced by a plane (not wedged) partial reflector with low output coupling. The leaking beam can then be used to align the MPC with the desired mirror separation and spot pattern. The partial reflector should be placed as close to the input reference plane of the MPC as possible. The end mirror of the MPC can then be aligned to provide feedback into the laser. At this stage, once the partial reflector of the short cavity is removed, the lasing of the extended cavity is readily obtained because the stability region should remain nearly the same as that of the short cavity.

The other important issue is about dispersion compensation in MPC femtosecond lasers. One possibility is to use prism pairs but they are undesirable in MPC configurations. The most important reason is that a large prism separation may be necessary to provide sufficient dispersion for high pulse energies, and this can make the setup rather bulky. In addition, if highly dispersive prisms are instead used to obtain the same amount of second-order dispersion with shorter prism separation, it may become difficult to control the higher order contributions to phase distortions. A more practical approach is to use double-chirped mirrors (DCMs) for dispersion compensation [41]. With careful design, controlled amounts of second- and higher-order dispersion can be introduced with a compact arrangement of MPCs.

8.4 Experimental Work on Multipass Cavity Femtosecond Lasers

In the previous sections, we discussed the general characteristics and the key design rules of multipass cavities. In practice, MPCs can be employed to improve the performance of femtosecond lasers in a number of ways. First, MPCs can be used to build very compact femtosecond resonators. This particular application of MPCs is first discussed in Susection 8.4.1. The technique was applied to a $Ti^{3+}:Al_2O_3$ oscillator and both q-preserving and non-q-preserving configurations were investigated. A 31-MHz femtosecond oscillator with a 30 cm × 45 cm footprint could be built and operated with pump powers as low as 1.5 W. Second, MPCs can be employed to increase the pulse energy of low-average-power femtosecond lasers. This is discussed in Subsection 8.4.2. In this case, experiments were performed with a Cr:LiSAF gain medium and 0.75-nJ pulses were obtained at 867 nm with 120 mW of pump power. Next, we discuss the application of MPCs in generating pulse energies above 100 nJ (Subsection 8.4.3) and in obtaining mode-locked operation from ultralong optical cavities (Subsection 8.4.4).

8.4.1 Compact Femtosecond Lasers Based on the MPC Concept

MPC configurations can be used to make very compact femtosecond lasers. This particular application of MPCs was experimentally demonstrated by using a $Ti^{3+}:Al_2O_3$ gain medium [45,46]. A schematic of the experimental setup is shown in Figure 8.11. The short cavity that extended up to the reference plane z_R, was a folded, astimatically compensated x cavity containing a short (2 mm) Brewster-cut $Ti^{3+}:Al_2O_3$ crystal with a pump absorption of about 75% at 532 nm. The radius of the concave high reflectors M1 and M2 was 3 cm. The resonator was end pumped by a diode-pumped frequency-doubled $Nd:YVO_4$ laser operating at 532 nm. The path length of the long arm extending from M2 to the output coupler (OC) was 35 cm, and the short arm (between M1 and z_R) was 21 cm. The estimated pump and laser beam radii inside the gain medium were 7 μm and 9 μm ($1/e^2$ radius), respectively. In order to maintain compactness, double-chirped mirrors (M1–M5) were used for dispersion compensation. The net round trip dispersion of the resonator was about -224 fs^2.

A flat-curved q-preserving MPC was used to extend the optical path length of the resonator. The curved mirror (M6) had a radius of 2 m. The beam entered the MPC through a notch on the curved mirror and exited through a second notch on the flat mirror (M7). The mirror separation was adjusted to be 23.4 cm. This corresponds to a 40° angular advance of the spot pattern after each round trip with m = 2 and n = 9 (see Equation 8.18). Because notches were used for beam injection and extraction, nine full round trips could not be completed and compensating optics was used to make the MPC q-preserving. The beam leaving the second flat mirror was reflected by a flat high reflector (M8) through the notch and then sent back by a curved retroreflector (M9) whose radius was 1 m. Although the overall path length of the resonator was 9.6 m, use of the MPC enabled the realization of a very compact design with a footprint of 30 cm × 45 cm. The corresponding pulse repetition rate was 31.25 MHz.

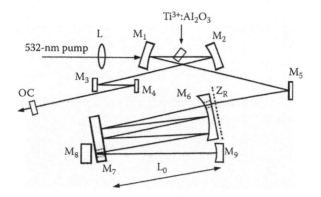

FIGURE 8.11
Schematic of the compact 31-MHz femtosecond $Ti^{3+}:Al_2O_3$ oscillator. A flat-curved multipass cavity was added for cavity extension. See the text for a description of the components.

Use of a tight-focusing resonator geometry resulted in low-threshold operation of the $Ti^{3+}:Al_2O_3$ laser. With an input pump power of 1.5 W and without the MPC, the resonator produced 178 mW of cw output power. The output coupler had a transmission of 11%. For this case, the threshold pump power and the slope efficiency were 510 mW and 20%, respectively. CW power measurements further showed that the optimum output coupling of the resonator was near 11% at a pump power of 1.5 W. The round-trip resonator losses were further estimated to be 6.8%. The multipass cavity was aligned by using another flat output coupler positioned before the first mirror (M6) of the MPC. The mirror tilts and separation were adjusted to obtain a nearly circular spot pattern on each of the MPC mirrors. The spot pattern on one of the MPC mirrors can be seen in Figure 8.12. Also seen in Figure 8.12 is the rectangular notch used for beam injection. The angular advance of the spot pattern after each round trip was 40°. With the compensating optics, a full round trip through the MPC gave a q-preserving configuration and lasing of the extended cavity could be readily achieved by removing the output coupler before the MPC. With the MPC, the cw output power decreased to 140 mW at the input pump power of 1.5 W. The round-trip insertion loss of the MPC was further estimated to be 3%.

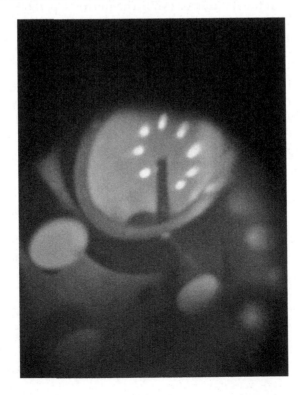

FIGURE 8.12
Spot pattern on one of the mirrors of the multipass cavity used in the 31-MHz femtosecond $Ti^{3+}:Al_2O_3$ oscillator.

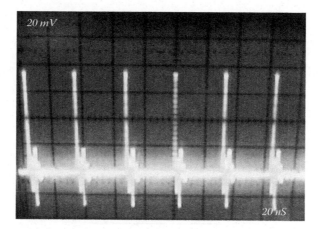

FIGURE 8.13
Oscilloscope trace of the pulse train at a repetition rate of 31.25 MHz.

Femtosecond pulse generation experiments were carried out with the 11% transmitting output coupler. Once the focusing of the cavity beam was optimized inside the gain medium, mode-locked operation could be readily obtained by moving one of the resonator end mirrors. Once initiated, stable, uninterrupted mode-locked operation could be sustained for many hours. Figure 8.13 shows the oscilloscope trace of the pulse train obtained. Characterization of the mode-locked pulses was done at a pump power of 1.5 W. The measured autocorrelation and spectrum of the pulses are shown in Figure 8.14a and Figure 8.14b, respectively. The spectral bandwidth (FWHM) was measured to be 40 nm. By assuming a hyperbolic secant pulse profile, the pulse duration was further determined to be 23 fs, with a corresponding time-bandwidth product of 0.45. With 1.5 W of pump power, 88 mW of mode-locked output power was obtained, corresponding to 2.8 nJ of pulse energy at a 31.25-MHz repetition rate. As the pump power was increased to 2.3 W, up to 4 nJ of output energy could be obtained.

A simpler non-q-preserving MPC design was also used to build a compact femtosecond $Ti^{3+}:Al_2O_3$ oscillator. The design was similar to what is shown in Figure 8.11 with $L_0 = 23.4$ cm, m = 2, and n = 9, except the missing round trip was not compensated and the flat 11% transmitting output coupler was placed at a distance of 6.5 cm from the exit mirror of the MPC. The net round-trip dispersion of this setup was -230 fs^2. We discussed non-q-preserving MPCs of this kind in Subsection 8.3.4. Note that here $L_1 = 6.5$ cm and $L_0 = 23.4$ cm. Because $L_1 < L_0$ and $L_0 << R$ (R = 200 cm), for both the short and the extended cavities, the beam waist at the center of the stability region remains nearly the same but the stability regions are shifted. This made the alignment of the extended cavity more difficult in comparison with the q-preserving cavity. Once the output coupler before the MPC was removed, laser action could not be readily obtained and the position of the focusing mirrors (M1 and M2) around the gain medium was adjusted to obtain lasing.

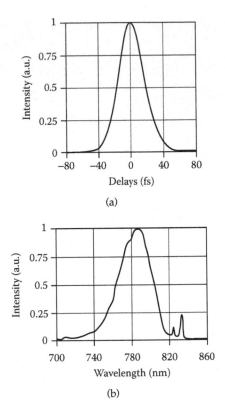

FIGURE 8.14

Measured (a) autocorrelation intensity and (b) spectrum of the femtosecond pulses generated with the 31-MHz multipass-cavity $Ti^{3+}:Al_2O_3$ oscillator. The input pump power is 1.5 W. The pulse duration (FWHM) and the spectral width are 23 fs and 40 nm, respectively.

The overall passive loss of the non-q-preserving configuration was slightly lower and hence somewhat better power performance was obtained. In particular, with 1.5 W of pump power, the mode-locked resonator produced 115 mW of average output power, corresponding to 3.65 nJ of pulse energy at a pulse repetition rate of 31.5 MHz. When the pump power was increased to 2.4 W, as high as 5.2 nJ of output pulse energy was obtained. At this pump power, 19-fs pulses with a time–bandwidth product of 0.39 were produced. The pulsewidth is shorter than that for the q-preserving case discussed above due to slightly higher available pulse energy as is predicted by the soliton mode-locking theory (see Equation 8.5).

8.4.2   Pulse Energy Scaling in Low-Average-Power Systems

In this section, we review another experiment in which an MPC was used to scale up the output energy of a mode-locked Cr:LiSAF laser with low average output power [47]. Cr:LiSAF is a broadly tunable solid-state gain medium that can be used to generate femtosecond laser pulses around

850 nm. One important advantage of this gain medium is that the absorption band around 670 nm allows direct diode pumping. Although broad-stripe diodes are available as pump sources, they are typically expensive and suffer from low mode quality, making it difficult to obtain good mode matching between the pump and the laser resonators. A cost-effective alternative is to use inexpensive, single-spatial-mode diodes to improve mode matching. However, the low output power of single-mode diodes has limited the directly generated output energy from the Cr:LiSAF oscillators to about 0.14 nJ [48].

In the experiments described by Prasankumar et al. [47], a flat-curved MPC was used to scale the output energy of a Cr:LiSAF laser. The short cavity contained a 5-mm-long, Brewster-cut 1.5% doped Cr:LiSAF crystal positioned between two curved mirrors with 10-cm radii. The short cavity, which had a repetition rate of 115 MHz was end pumped by 3 single-mode diodes producing a total of 120 mW. The MPC consisted of a flat and a curved mirror separated by 2 m. The radius of the curved mirror was 4 m, giving an angular advance of $\pi/2$ for the spot pattern after each round trip. So far, we mainly discussed the use of notches for beam injection and extraction. An alternative method is to use small pick-off mirrors to direct the incident and exit beams. In this particular case, small pick-off mirrors were used for beam injection and extraction. The overall repetition rate of the extended cavity was 8.6 MHz. Mode locking was initiated with a saturable Bragg mirror. Dispersion compensation was provided by DCMs or a prism pair. By using DCMs only, 39-fs pulses were produced with 20 nm bandwidth centered at 867 nm. The average output power was 6.5 mW. With a repetition rate of 8.6 MHz, this corresponds to pulse energy of 0.75 nJ and represents a fivefold enhancement compared to previous results.

8.4.3 High Pulse Energies with MPCs

Multipass cavities can also be used to obtain very high energy femtosecond pulses at megahertz repetition rates from standard lasers. In the experiments described by Kowalevicz et al. [49], a MPC was used to scale up the output energy of a Ti:Al$_2$O$_3$ laser pumped by a 10-W frequency-doubled Nd:vanadate laser. The short cavity was a standard x-folded configuration containing a 3-mm-long Ti:Al$_2$O$_3$ crystal between two curved high reflectors each with a radius of 10 cm. The pump absorption coefficient of the crystal was 2.76 cm^{-1} at 532 nm. The crystal was mounted on a copper block and maintained at 15°C. The short cavity was extended by using a flat-curved MPC. The radius of the curved mirror was 15 m and the mirror separation was set to 1 m, giving an angular advance of 30° per round trip. This corresponds to m = 2 and n = 12 (See Equation 8.18). Small pick-off mirrors were used for beam injection and extraction. The MPC introduced a total of 12 round trips per pass, giving a repetition rate of 5.85 MHz for the extended resonator. Dispersion compensation was achieved by using DCMs. In order to balance the chirp due to self phase modulation, the cavity was operated with a large amount of negative dispersion. The net dispersion of the cavity was estimated to be −1350 fs². Also, a 25% output coupler was used to decrease the intracavity intensity and to minimize instabilities due to excess Kerr nonlinearity. The absorption of the laser crystal was about 56%. In order to use the pump more efficiently, a curved retroreflector was used to double-pass the pump beam. During mode-locked operation, the resonator produced 877 mW of output power with 9.4W of input pump power. At a repetition rate of 5.85 MHz, this corresponds to an energy per pulse of 150 nJ. The output pulses had 43-fs duration (FWHM, assuming a hyperbolic secant squared pulse profile) and a spectral bandwidth of 16.5 nm (FWHM). The peak power of the pulses was 3.5 MW. This laser system has been used in the microfabrication of photonic components on transparent glass [22].

Direct scaling to even higher pulse energies becomes quite challenging due to the possibility of pulse instabilities caused by excessive nonlinearities in the gain medium. One alternative scheme is to build chirped-pulse oscillators with MPCs in order to avoid this problem. Fernandez et al. [50] used this method to build chirped-pulse multipass oscillator with a repetition rate of about 11 MHz. The resonator was operated with a small amount of positive dispersion. In combination with Kerr nonlinearities, positive dispersion led to the generation of strongly chirped picosecond pulses with a broad spectrum in the 760–820 nm range. The pulses extracted from the cavity were then compressed with an extracavity dispersive delay line consisting of a pair of LaK16 prisms. With about 10 W of input pump power, 220-nJ, 30-fs pulses could be produced at a pulse repetition rate of 11 MHz, corresponding to peak powers of 5 MW. More recently, the same group utilized a double MPC setup to generate 0.5 μJ, sub-40 fs pulses at a pulse repetition rate of 2 MHz, corresponding to a peak power in excess of 10 MW [13]. Similarly, researchers at the Max Planck Institute have obtained 0.5 μJ, 50-fs pulses at a repetition rate of 6 MHz. Tight focusing was employed to create peak intensities exceeding 10^{14} W/cm², and ionization of helium, which requires simultaneous absorption of at least 17 photons, was experimentally demonstrated [23].

8.4.4   MPC Lasers with Ultralow Repetition Rates

In this section, we review two studies in which MPCs were employed to build mode-locked solid-state lasers with ultralow repetition rates around 1 MHz. In the experiments described by Kolev et al. [51], a diode-pumped Nd:YVO₄ laser was constructed and passively mode-locked with semiconductor saturable absorber mirrors at 1064 nm. Cavity extension was achieved by incorporating a q-preserving MPC (referred to as a "zero-q-transformation" of ZqT system in Reference 51). Three different MPC designs were employed resulting in mode-locked operation with repetition rates of 4.1, 2.6, and 1.5 MHz. In the case of 1.5-MHz operation, there were up to 72 reflections from the MPC mirrors per round trip, and special mirrors with very high reflectivity (R > 99.997%) were used to minimize passive resonator losses. The duration of the output pulses was independent of the repetition rate and was measured to be 13 ps by assuming a sech^2 intensity profile. The average output power at 1.5 MHz repetition rate was 3.5 W, and the M^2 of the output beam was determined to be less than 1.1.

In another experiment, Papadopoulos et al. [52] demonstrated the operation of an ultralow-repetition-rate picosecond Nd^{3+}:YVO$_4$ MPC laser passively mode-locked with a SESAM. In this case, the MPC was folded with the help of two plane mirrors. By changing the alignment of the plane or concave mirrors, the number of passes and hence the repetition rate of the laser could be adjusted. In the experiments, a 5-mm-long 0.1%-doped Nd^{3+}:YVO$_4$ crystal was pumped by a 15-W diode laser at 808 nm. Passive mode locking was obtained with a SESAM having a modulation depth of about 6%. Stable mode-locked operation could be achieved with repetition rates as low as 1.2 MHz. The pulse duration and the average power were 16.3 ps and 470 mW, respectively, corresponding to a pulse energy of 392 nJ and peak power of 24 kW. The M^2 was further measured to be 1.1.

8.5 Conclusions

We have provided a detailed account of the design rules of multipass optical cavities and their application to pulse energy scaling in femtosecond solid-state lasers. Q-preserving multipass cavities were introduced, and their analytical design rules derived. We have proved that the angular advance of the bouncing beam after one round trip should equal π times a rational number in order for the MPC to be q-preserving. Illustrative design examples were given. In the case of MPCs with notched mirrors, the use of compensating optics to restore the q-preserving nature of the multipass cavity was discussed. Variation of the pulse repetition rates for different q-preserving configurations was also calculated for two-mirror MPCs consisting of flat-curved or curved-curved high reflectors. Finally, we discussed several experiments performed by different groups in which MPCs were utilized to build compact and/or high-energy mode-locked oscillators. MPC femtosecond lasers offer a simple alternative to pulse energy scaling and should find numerous applications in science and technology.

Acknowledgments

We are grateful to A. M. Kowalevicz, E.P. Ippen, and F.X. Kärtner for useful discussions. We also thank Adnan Kurt for the preparation of figures. This research was sponsored in part by the Air Force Office of Scientific Research Medical Free Electron Laser Program FA9550-040-1-0046 and FA9550-040-1-0011 and the National Science Foundation Programs ECS-01-19452, BES-01-19494 and ECS-0501478. A. Sennaroglu acknowledges the support of the Turkish Academy of Sciences in the framework of the Young Scientist Award Program AS/TUBA-GEBIP/2001-1-11 and the NSF-Tubitak travel grant (TBAG-U/110-104T247).

References

1. G.J. Tearney, M.E. Brezinski, B.E. Bouma, S.A. Boppart, C. Pitris, J.F. Southern, and J.G. Fujimoto, In vivo endoscopic optical biopsy with optical coherence tomography, *Science*, Vol. 276, 2037–2039, 1997.
2. M. Drescher, M. Hentschel, R. Kienberger, M. Uiberacker, V. Yakovlev, A. Scrinzi, T. Westerwalbesloh, U. Kleineberg, U. Heinzmann, and F. Krausz, Time-resolved atomic inner-shell spectroscopy, *Nature*, Vol. 419, 803–807, 2002.
3. A.H. Zewail, Femtochemistry: recent progress in studies of dynamics and control of reactions and their transition states, *J. Phys. Chem.*, Vol. 100, 12701, 1996.
4. C. Spielmann, N.H. Burnett, S. Santania, R. Koppitsch, M. Schnurer, C. Kan, and M. Lenzner, Generation of coherent x-rays in the water window using 5-femtosecond laser pulses, *Science*, Vol. 278, 661–664, 1997.
5. Z. Chang, A. Rundquist, H. Wang, M.M. Murnane, and H.C. Kapteyn, Generation of coherent soft x rays at 2.7 nm using high harmonics, *Phys. Rev. Lett.*, Vol. 79, 1997.
6. K. Read, F. Blonigen, N. Riccelli, M. Murnane, and H. Kapteyn, Low-threshold operation of an ultrashort-pulse mode-locked Ti:sapphire laser, *Opt. Lett.*, Vol. 21, 489–491, 1996.
7. A.M. Kowalevicz, T.R. Schibli, F.X. Kaertner, and J.G. Fujimoto, Ultralow-threshold Kerr-lens mode-locked $TiAl_2O_3$ laser, *Opt. Lett.*, Vol. 27, 2037–2039, 2002.
8. M.D. Perry and G. Mourou, Terawatt to petawatt subpicosecond lasers, *Science*, Vol. 264, 917–924, 1994.
9. A. Killi, U. Morgner, M.J. Lederer, and D. Kopf, Diode-pumped femtosecond laser oscillator with cavity dumping, *Opt. Lett.*, Vol. 29, 1288–1290, 2004.
10. M. Ramaswamy, M. Ulman, J. Paye, and J.G. Fujimoto, Cavity-dumped femtosecond Kerr-lens mode-locked $Ti:Al_2O_3$ laser, *Opt. Lett.*, Vol. 18, 1822–1824, 1993.
11. A.R. Libertun, R. Shelton, H.C. Kapteyn, and M.M. Murnane, A 36 nJ-15.5 MHz Extended-Cavity Ti:Sapphire Oscillator, presented at Conference on Lasers and Electro-Optics, Baltimore, MD, 1999.
12. S.H. Cho, B.E. Bouma, E.P. Ippen, and J.G. Fujimoto, Low-repetition-rate high-peak power Kerr-lens mode-locked $Ti:Al_2O_3$ laser with a multiple-pass cavity, *Opt. Lett.*, Vol. 24, 417–419, 1999.
13. S. Naumov, A. Fernandez, R. Graf, P. Dombi, F. Krausz, and A. Apolonski, Approaching the microjoule frontier with femtosecond laser oscillators, *N. J. Phys.*, Vol. 7, 216–226, 2005.
14. R. Herriott, H. Kogelnik, and R. Kompfner, Off-axis paths in spherical mirror interferometers, *Appl. Opt.*, Vol. 3, 523–526, 1964.

15. R. Herriott and H.J. Schulte, Folded optical delay lines, *Appl. Opt.*, Vol. 4, 883–889, 1965.
16. W.R. Trutna and R.L. Byer, Multiple-pass Raman gain cell, *Appl. Opt.*, Vol. 19, 301–312, 1980.
17. B. Perry, R.O. Brickman, A. Stein, E.B. Treacy, and P. Rabinowitz, Controllable pulse compression in a multi-pass-cell Raman laser, *Opt. Lett.*, Vol. 5, 288–290, 1980.
18. J.B. McManus, P.L. Kebabia, and M. S. Zahniser, Astigmatic mirror multipass absorption cells for long-path-length spectroscopy, *Appl. Opt.*, Vol. 34, 3336–3348, 1995.
19. P. Hsiung, X.D. Li, C. Chudoba, I. Hartl, T.H. Ko, and J.G. Fujimoto, High-Speed path-length scanning with a multiple-pass cavity delay line, *Appl. Opt.* — OT, Vol. 42, 640–648, 2003.
20. H.-Z. Cheng, P.-L. Huang, S.-L. Huang, and F.-J. Kao, Reentrant two-mirror ring resonator for generation of a single-frequency green laser, *Opt. Lett.*, Vol. 25, 542–544, 2000.
21. S.-L. Huang, Y.-H. Chen, P.-L. Huang, J.-Y. Yi, and H.-Z. Cheng, Multi-reentrant nonplanar ring laser cavity, *IEEE J. Quantum. Electron.*, Vol. 38, 1301–1308, 2002.
22. A.M. Kowalevicz, V. Sharma, E.P. Ippen, F.J.G., and K. Minoshima, Three-dimensional photonic devices fabricated in glass by use of a femtosecond laser oscillator, *Opt. Lett.*, Vol. 30, 1060–1062, 2005.
23. S. Dewald, T. Lang, C.D. Schröter, R. Moshammer, J. Ullrich, M. Siegel, and U. Morgner, Ionization of noble gases with pulses directly from a laser oscillator, *Opt. Lett.*, Vol. 31, 2072–2074, 2006..
24. P.F. Moulton, Spectroscopic and laser characteristics of Ti:Al$_2$O$_3$, *J. Opt. Soc. Am. B*, Vol. 3, 125–132, 1986.
25. V. Petricevic, S.K. Gayen, and R.R. Alfano, Laser action in chromium-doped forsterite, *Appl. Phys. Lett.*, Vol. 52, 1040–1042, 1988.
26. A. Sennaroglu, Broadly tunable Cr4+-doped solid-state lasers in the near infrared and visible, *Prog. Quantum Electron.*, Vol. 26, 287–352, 2002.
27. N.B. Angert, N.I. Borodin, V.M. Garmash, V.A. Zhitnyuk, V.G. Okhrimchuk, O.G. Siyuchenko, and A.V. Shestakov, Lasing due to impurity color centers in yttrium aluminum garnet crystals at wavelengths in the range 1.35–1.45 μm, *Sov. J. Quantum Electron.*, Vol. 18, 73–74, 1988.
28. W.F. Krupke, Ytterbium solid-state lasers: the first decade, *IEEE J. Sel. Top. Quant. Electron.*, Vol. 6, 1287–1296, 2000.
29. F. Brunner, E. Innerhofer, S.V. Marchese, T. Sudmeyer, R. Paschotta, T. Usami, H. Ito, S. Kurimura, K. Kitamura, G. Arisholm, and U. Keller, Powerful red-green-blue laser source pumped with a mode-locked thin disk laser, *Opt. Lett.*, Vol. 29, 1921–1923, 2004.
30. D.H. Sutter, G. Steinmeyer, L. Gallmann, N. Matuschek, F. Morier-Genoud, U. Keller, V. Scheuer, G. Angelow, and T. Tschudi, Semiconductor saturable-absorber mirror-assisted Kerr-lens mode-locked Ti:sapphire laser producing pulses in the two-cycle regime, *Opt. Lett.*, Vol. 24, 631–633, 1999.
31. O.D. Mucke, R. Ell, A. Winter, J. Kim, J.R. Birge, L. Matos, and F.X. Kärtner, Self-referenced 200 MHz octave-spanning Ti:sapphire laser with 50 attosecond carrier-envelope phase jitter, *Opt. Express*, Vol. 13, 5163–5169, 2005.
32. U. Morgner, F.X. Kärtner, S.H. Cho, H.A. Haus, J.G. Fujimoto, E.P. Ippen, V. Scheuer, G. Angelow, and T. Tschudi, Sub-two cycle pulses from a Kerr-Lens modelocked Ti:sapphire laser, *Opt. Lett.*, Vol. 24, 411–413, 1999.

33. H.A. Haus, Mode-locking of lasers, *IEEE J. Sel. Top. Quant. Electron.*, Vol. 6, 1173–1185, 2000.
34. A.E. Siegman and D.J. Kuizenga, Active mode-coupling phenomena in pulsed and continuous lasers, *Optoelectronics*, Vol. 6, 43–66, 1974.
35. E.P. Ippen, Principles of passive mode locking, *Appl. Phys. B*, Vol. 58, 159–170, 1994.
36. H.A. Haus, J.G. Fujimoto, and E.P. Ippen, Analytic Theory of additive pulse and Kerr lens mode locking, *IEEE J. Quant. Electron.*, Vol. 28, 2086–2096, 1992.
37. U. Keller, K.J. Weingarten, F.X. Kärtner, D. Kopf, B. Braun, I.D. Jung, R. Fluck, C. Hönninger, N. Matuschek, and J. Aus der Au, Semiconductor saturable absorber mirrors (SESAMs) for femtosecond to nanosecond pulse generation in solid-state lasers, *IEEE J. Sel. Top. Quantum Electron. (JSTQE)*, Vol. 2, 435–453, 1996.
38. R.L. Fork, O.E. Martinez, and J.P. Gordon, Negative dispersion using pairs of prisms, *Opt. Lett.*, Vol. 9, 150–152, 1984.
39. R. Szipöcs, K. Ferencz, C. Spielmann, and F. Krausz, Chirped multilayer coatings for broadband dispersion control in femtosecond lasers, *Opt. Lett.*, Vol. 19, 201–203, 1994.
40. R. Szipöcs and A. Kohazi-Kis, Theory and design of chirped dielectric laser mirrors, *Appl. Phys. B*, Vol. 65, 115–136, 1997.
41. F.X. Kärtner, N. Matuschek, T. Schibli, U. Keller, H.A. Haus, C. Heine, R. Morf, V. Scheuer, M. Tilsch, and T. Tschudi, Design and fabrication of double-chirped mirrors, *Opt. Lett.*, Vol. 22, 831–833, 1997.
42. N. Matuschek, F.X. Kärtner, and U. Keller, Theory of Double-Chirped Mirrors, *IEEE J. Sel. Top. Quantum Electron.*, Vol. 4, 197, 1998.
43. A. Sennaroglu and J.G. Fujimoto, Design criteria for Herriott-type multi-pass cavities for ultrashort pulse lasers, *Opt. Express*, Vol. 11, 1106–1113, 2003.
44. A.E. Siegman, *Lasers*, University Science Books, Mill Valley, California, 1986.
45. A. Sennaroglu, A.M. Kowalevicz, E.P. Ippen, and J.G. Fujimoto, Compact femtosecond lasers based on novel multi-pass cavities, *IEEE J. Quantum Electron.*, Vol. 40, 519–528, 2004.
46. A. Sennaroglu, A.M. Kowalevicz, F.X. Kärtner, and J.G. Fujimoto, High-performance, compact, prismless, low-threshold 30-MHz Ti:sapphire laser, *Opt. Lett.*, Vol. 28, 1674–1676, 2003.
47. R.P. Prasankumar, Y. Hirakawa, A.M. Kowalevicz, F.X. Kärtner, J.G. Fujimoto, and W. Knox, An extended cavity femtosecond Cr:LiSAF laser pumped by low cost diode lasers, *Opt. Express*, Vol. 11, 1265–1269, 2003.
48. B. Agate, B. Stormont, A.J. Kemp, C.T.A. Brown, U. Keller, and W. Sibbett, Simplified cavity designs for efficient and compact femtosecond Cr:LiSAF lasers, *Opt. Commun.*, Vol. 205, 207–213, 2002.
49. A.M. Kowalevicz, A.T. Zare, F.X. Kärtner, J.G. Fujimoto, S. Dewald, U. Morgner, V. Scheuer, and G. Angelow, Generation of 150-nJ pulses from a multiple-pass cavity Kerr-lens modelocked Ti:Al$_2$O$_3$ oscillator, *Opt. Lett.*, Vol. 28, 1597–1599, 2003.
50. A. Fernandez, T. Fuji, A. Poppe, A. Furbach, F. Krausz, and A. Apolonski, Chirped-pulse oscillators: a route to high-power femtosecond pulses without external amplification, *Opt. Lett.*, Vol. 29, 1366–1368, 2004.
51. V.Z. Kolev, M.J. Lederer, B. Luther-Davies, and A.V. Rode, Passive mode locking of a Nd:YVO$_4$ laser with an extra-long optical resonator, *Opt. Lett.*, Vol. 28, 1275–1277, 2003.
52. D.N. Papadopoulos, S. Forget, M. Delaigue, F. Druon, F. Balembois, and P. Georges, Passively mode-locked diode-pumped Nd:YVO$_4$ oscillator operating at an ultralow repetition rate, *Opt. Lett.*, Vol. 28, 1838–1840, 2003.

9

Cavity-Dumped Femtosecond Laser Oscillator and Application to Waveguide Writing

Alexander Killi, Max Lederer, Daniel Kopf, Uwe Morgner, Roberto Osellame, and Giulio Cerullo

CONTENTS

9.1   The Laser Light Source

9.1.1   Introduction

Femtosecond laser systems have proven their potential in research laboratories in a variety of applications that were previously unthinkable, in fields
as diverse as material processing, photonic device production, microscopy,
and biomedicine. The application of these lasers outside research laboratory
environments has been, up to now, very limited because of their complexity
and difficulty of operation, and because of the high prices of commercially
available laser systems.

In terms of both pulse energy and repetition rate, cavity-dumped laser
systems are in between oscillators and amplifier systems, and thus are ideal
laser sources for many applications such as microstructuring, laser surgery,
tissue manipulation, multiphoton microscopy, and laser spectroscopy. In the
past the main focus of research was on cavity-dumped TEM_{00}-pumped
Ti:sapphire laser systems [1–5]. However, because these lasers are pumped
in the green spectral region, where no laser diodes are available, their application is limited to a small community of users due to the high cost of the
green pump lasers. Recently, we became aware of the potential of diode-
pumped laser oscillators with cavity-dumping in terms of performance and
reliability. In two publications we have demonstrated the generation of femtosecond pulses with energies of 300 nJ and repetition rates higher than 1
MHz from a laser system based on Yb:glass [6,7]. However, this material is
limited in terms of power scaling because of its low absorption and emission
cross sections, which makes it necessary to use relatively long laser rods with
tight focusing [8]. Therefore, self-phase modulation is the limiting parameter
and cannot be further reduced easily [7,9]. Additionally, due to the poor
thermal conductivity of Yb:glass and the need of tight pump focusing, the
maximum pump power is limited to a few watts.

One possible way to overcome these disadvantages is to use a laser material with larger cross sections having a shorter pump absorption length at a
given doping concentration. This would allow one to increase the spot size
of the laser mode and also to shorten the crystal. $Yb:KY(WO_4)_2$ (Yb:KYW)
meets the desired advantages [10]: First, its absorption and emission cross
sections are larger than those of Yb:glass by an order of magnitude. Second,
the improved thermal conductivity leads to reduced thermal effects in comparison with Yb:glass. Therefore, it is possible to use low-brightness, high-
power multi-emitter diode bars for pumping, which reduces costs and
increases the reliability compared with high-brightness diode-pumping
options [11]. Even though the nonlinear refractive index is higher than in
Yb:glass, all the properties mentioned above make this crystal an almost
ideal candidate for high-peak-power laser systems.

The first part of this chapter is organized as follows: After a brief description
of the laser setup, we introduce three dynamic regimes which are dependent

on different dumping frequencies. The pulse-to-pulse stability, the transient spectra, and autocorrelations are discussed with respect to the theoretical model. A numerical evaluation of the laser dynamics is carried out and compared to the experimental results. The major pulse shaping mechanism is shown to be caused by solitary pulse propagation, but the spectral properties of the laser pulses are strongly influenced by the generation of Kelly sidebands.

9.1.2 Mode-Locked Yb:XXX Laser Oscillators

During the last decade, the research on Yb-doped laser materials has been very successful [12], which was triggered, on one hand, by the availability of high-brightness diodes emitting in the absorption band. On the other hand, the lasing transition of Yb^{3+} from $^2F_{7/2}$ to $^2F_{5/2}$ is rather simple and broad-band due to Stark degeneration without significant upconversion or cross relaxation channels; furthermore, the absence of concentration quenching allows for high doping levels [13]. The small quantum defect leads to self-absorption in the lower wavelength edge of the emission band, and Yb-doped laser materials can be described by a quasi-three-level model with the resulting requirements for the pump intensity.

Due to the large bandwidth, passive mode-locking was quite successful. The most advanced approaches are based on semiconductor saturable absorber mirrors (SESAMs) for achieving mode-locking, and on prisms or phase correcting mirrors to provide anomalous dispersion for solitary pulse shaping. Recent results have been: 90 fs pulses with 40 mW average output power and 0.4 nJ energy from a Yb:GdCOB laser oscillator, 68 fs pulses with 300 mW average output power, and 3 nJ energy from a Yb:SrYBO laser oscillator [14,15]; 340 fs pulses with 110 mW average output power and 1 nJ energy from a Yb:YAG laser oscillator, 60 fs pulses with 65 mW average output power and 0.6 nJ energy from a Yb:Glass laser oscillator [8,16,17]; 71 fs pulses with 120 mW average output power and 1.2 nJ energy from a Yb:KYW laser oscillator [18,19] and, finally, 810 fs pulses with 60 Watts of output power with 1.75 μJ from a mode-locked thin-disk laser oscillator [20,21]. High-power cw demonstrations with innovative Yb-doped materials are given in [22,23].

9.1.3 Numerical Model

The laser dynamics of the passively mode-locked oscillator with cavity-dumping is well described by three coupled equations: The first equation is Haus' master equation of mode locking [24] and describes the time evolution of the slowly varying field amplitude $A(t,T)$ on two time scales: the local time t in a moving frame around the pulse, and a global time T in a scale of many resonator round-trip times T_R including many linear and nonlinear pulse shaping effects

$$T_R \frac{\partial}{\partial T} A(t,T) = \left[g(T) - l + D_{g,f} \frac{\partial^2}{\partial t^2} + i \frac{\beta_2}{2} \frac{\partial^2}{\partial t^2} - q(t,T,A) - i\gamma \mid A(t,T) \mid^2 \right] A(t,T) \quad (9.1)$$

with the group delay dispersion (GDD) β_2, the so-called gain dispersion $D_{g,f} = g / \Omega_g^2 + 1 / \Omega_f^2$ which includes the gain bandwidth Ω_g and the bandwidth Ω_f of further cavity elements (both HWHM); g is the roundtrip amplitude gain, l and q the linear and nonlinear amplitude losses. The field amplitude is normalized in a way that $\mid A(t,T) \mid^2$ describes the power envelope. Then, the SPM coefficient (self-phase modulation) is given by:

$$\gamma = \frac{2\pi}{\lambda_0 A_L} n_2 l_L, \quad (9.2)$$

with the center wavelength λ_0, the effective mode area $A_L = \pi w^2_0$, the mode radius w_0, the nonlinear Kerr coefficient n_2, and the geometrical length l_L of the laser medium.

The second equation is the rate equation for the laser gain. Assuming only small changes of the gain $g(T)$ during one laser round-trip ($\tau_L \gg T_R$), which is well fulfilled in practically all cases using Yb-doped materials with their long upper-state lifetime τ_L, the rate equation is

$$T_R \frac{\partial}{\partial T} g(T) = -T_R \frac{g(T) - g_0}{\tau_L} - g(T) \frac{E(T)}{E_L} \quad (9.3)$$

with the small signal amplitude gain g_0, the pulse energy

$$E(T) = \int \mid A(t,T) \mid^2 dt, \quad (9.4)$$

and the saturation energy

$$E_L = \frac{hc}{M_{1as} \lambda_0 \sigma_L} A_L \quad (9.5)$$

with the number of passes M_{1as} through the gain medium per roundtrip and the emission cross section σ_L.

The third and last equation describes the dynamics of the saturable absorber mirror (SESAM):

$$\frac{\partial}{\partial t} q(t,T) = -\frac{q(t,T) - q_0}{\tau_q} - q(t,T) \frac{\mid A(t,T) \mid^2}{E_q}, \quad (9.6)$$

with the small signal amplitude losses q_0, the absorption life time τ_q, and the saturation energy

$$E_q = F_q A_q \qquad (9.7)$$

with the saturation flux density F_q, and the effective mode area A_q of the laser beam on the absorber mirror.

9.1.3.1 Numerical Solution

The master equation can be efficiently solved employing a split-step Fourier method [25], which applies the nonlinear effects like SPM and SESAM in the time domain, and the linear effects in the Fourier domain.

The gain $g(T)$ is slowly varying on time scales of the round-trip time and can be discretized as

$$\frac{\partial g(T)}{\partial T} := \frac{g(T_j) - g(T_{j-1})}{T_R}, \qquad (9.8)$$

leading to the discrete and recursive version of Equation 9.3 with $T_L = \tau_L / T_R$

$$g(T_j) = \frac{E_L(T_L g(T_{j-1}) + g_0)}{E_L T_L + E_L + E T_L}. \qquad (9.9)$$

The same can be done with the absorber equation (Equation 9.6) and a suitable time discretization Δt:

$$\frac{\partial q(t_k)}{\partial t} := \frac{q(t_k) - q(t_{k-1})}{\Delta t} = -\frac{q(t_k) - q_0}{\tau_q} - \frac{q(t_k) |A(t_k)|^2}{E_q}. \qquad (9.10)$$

With $t_q := \tau_q / \Delta t$ we obtain

$$q(t_k) = \frac{E_q(t_q(t_{k-1}) + q_0)}{E_q t_q + E_q + |A(t_k)|^2 \Delta t t_q}. \qquad (9.11)$$

9.1.3.2 Soliton Mode-Locking

In steady state, the effects on the pulse energy and the effects on the pulse shape cancel each other. The gain equals the loss, and the pulse broadening due to gain dispersion equals the shortening from the saturable absorber. Assuming that, in this case, their contribution to the pulse shaping can be neglected, Equation 9.1 reduces to the nonlinear Schrödinger equation

$$T_R \frac{\partial}{\partial T} A(t,T) = \left[i \frac{\beta_2}{2} \frac{\partial^2}{\partial t^2} - i\gamma |A(t,T)|^2 \right] A(t,T). \qquad (9.12)$$

With $\gamma > 0$ it can be solved in the case of anomalous dispersion ($\beta_2 < 0$) with a fundamental soliton

$$A(t,T) = A_0 \operatorname{sech}\left(\frac{t}{\tau_{\text{sech}}}\right) \exp\left(-i\psi \frac{T}{T_R}\right) \tag{9.13}$$

with the pulse peak amplitude and the nonlinear phase shift per roundtrip given by

$$A_0 = \sqrt{\frac{E}{2\tau_{\text{sech}}}}, \quad \text{and} \quad \psi = \frac{|\beta_2|}{2\tau_{\text{sech}}^2} = \frac{\gamma A_0^2}{2}. \tag{9.14}$$

The soliton fulfills the area theorem, saying that the product of amplitude/energy and duration is constant

$$A_0 \tau_{\text{sech}} = \sqrt{\frac{|\beta_2|}{\gamma}} \Leftrightarrow E\tau_{\text{sech}} = 2\frac{|\beta_2|}{\gamma}. \tag{9.15}$$

Later in the experimental part, we will show that the assumptions leading to the solitary solutions of the master equation are well fulfilled in our case.

9.1.3.3 Kelly Sidebands

The periodic disturbance of the soliton in a laser resonator due to the discrete ordering of the pulse-shaping elements leads to a continuous shedding of power from the pulse peak to the background, leading to a resonant enhancement of phase-matched spectral components. Because the phase of the background continuum is determined only by the linear cavity phase $\phi(\omega)$, the phase matching condition is given by [26–28]

$$\phi(\omega_{\pm n}) = \psi - n2\pi, \quad n \in \mathbf{N}, \tag{9.16}$$

and peaks at positions $\omega_{\pm n}$ can be observed in the optical spectra. Later, we will show that besides the sidebands, according to the laser roundtrip time, an additional set of Kelly sidebands will appear in the case of cavity dumping, where due to the periodic dumping an additional disturbance of the soliton on a second time scale is imposed.

9.1.4 Cavity-Dumping Dynamics of Picosecond Lasers

Because SPM and GDD terms can be neglected, the generation of picosecond pulses from a cavity-dumped laser oscillator can be modeled simply by solving the rate equations.

Figure 9.1a shows the calculated relaxation transient between two dumping events, together with some experimental data for verification. The dumping period of 20 μs is large enough to allow the laser to completely recover

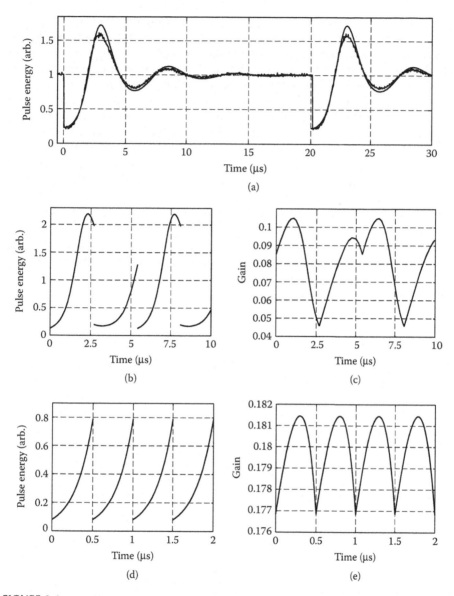

FIGURE 9.1

(a) Pulse energy transient between two dumping events in the picosecond oscillator (noisy line: experimental data; straight line: calculation); (b) and (c): calculated energy and gain transient for a dumping frequency in the resonant regime; (d) and (e): calculated energy and gain transient for a dumping frequency in the transient regime.

from the dumping to the steady state. This is a very stable mode of operation that we call the *relaxed regime*, where the dumping frequency is much smaller than the energy relaxation frequency, $f_{dump} \ll f_{energy}$. Things become more complicated in the so-called *resonant regime* with $f_{dump} \approx 2 f_{energy}$. Figure 9.1b and Figure 9.1c reveal the pulse energy as well as the gain transient. The pulse energy strongly changes with f_{dump}, depending on whether the dumping occurs in a minimum or a maximum of the energy relaxation oscillation. We observe a subperiodic behavior, where a pulse with high pulse energy is followed by a small one and vice versa. At even higher dumping frequencies, a third mode of operation can be identified, the *transient regime* with $f_{dump} \gg f_{energy}$ shown in Figure 9.1d and Figure 9.1e. The dumping events are imposed with a periodicity too small for the internal time constants, giving no time for the laser to show any pronounced gain dynamics at all. Therefore, we observe stable pulse trains with significant but decreasing energy enhancement. This is also an inherently stable mode of operation. The pulse energies of subsequent pulses as a function of the dumping frequency are subsumed in Figure 9.2a. The scale is normalized to the intracavity pulse energy without dumping; the three regimes are well distinguishable.

In conclusion, picosecond dumping is a stable scheme, in addition to the one resonance where the dumping frequency interferes with the energy relaxation oscillation. Things become much more complicated in the case of the femtosecond laser with solitary pulse shaping.

9.1.5 Cavity-Dumping Experiments With Picosecond Lasers

The picosecond Nd:YVO$_4$ laser with cavity dumping was an end-pumped system, which was passively mode-locked with a SESAM. It did not contain any components for dispersion control. The schematic setup of the device is outlined in Figure 9.3. The pump light was focused into the Nd:YVO$_4$ gain medium, which had a high reflection coating for the laser wavelength at 1064 nm on one side. Typically, 14 W of the pump power was absorbed inside the crystal. The other end of the 3.2 mm long crystal was cut at Brewster's

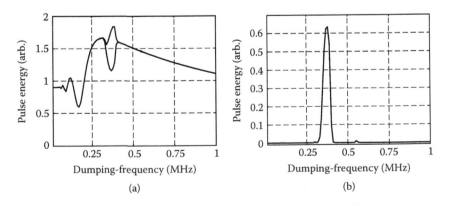

FIGURE 9.2
(a) Steady-state output pulse energy as a function of the dumping frequency of the picosecond oscillator (calculated); (b) difference in pulse energy between subsequent pulses as a function of the dumping frequency. The subharmonic behavior in the resonant regime is clearly visible in both graphs.

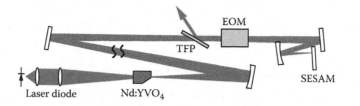

FIGURE 9.3
Schematic setup of the cavity-dumped picosecond Nd:YVO$_4$ laser system.

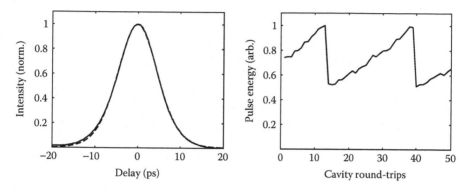

FIGURE 9.4
Left: typical autocorrelation of the output pulses at a dumping rate of 1.02 MHz. Right: normalized intracavity pulse energy in dependence on the number of round trips (the residual ripples in the curve are due to aliasing).

angle. The actual laser cavity of 27 MHz repetition rate had to be folded multiple times to allow for a compact setup. The dumping was performed by a combination of a Pockel's-cell (PC) and a thin-film polarizer (TFP). The PC was in a double-BBO-crystal configuration with a $\lambda/4$-voltage of 2.5 kV. For technical reasons the high voltage switches could only be driven up to 1.1 kV at dumping rates above 500 kHz, hence the dumping ratio was limited.

The system was analyzed in the stable "transient" regime at dumping frequencies of 0.71, 0.83, and 1.02 MHz, and the pulse energy remained constant around 1.8 μJ. The energy relaxation oscillation as characteristic intrinsic frequency of the system was well below the dumping frequencies ($f_{\text{energy}} \approx 180$ kHz); therefore, no destabilization was observed. At the repetition rate of 1 MHz, the dumping efficiency was as high as 48%, limited by the electronics as mentioned before. The intracavity pulse energy was determined as 3.8 μJ, corresponding to 102 W of average power. This suggests potentially higher output pulses if voltages closer to the $\lambda/4$-voltage of the PC could have been applied.

In Figure 9.4 some experimental data, which was taken at the repetition rate of 1 MHz, is shown. The autocorrelation trace of the dumped pulses on the left was fitted with the autocorrelation of a sech^2, giving a typical pulse with of 7.1 ps. On the right the measured intracavity pulse energy transient is displayed in dependence on the number of round trips.

9.1.6   Cavity-Dumping Dynamics of Femtosecond Lasers

Our investigations regarding the dynamics in the solitary parameter region have been performed with a mode-locked directly diode-pumped Yb:glass oscillator [9]. Besides the dumping frequency, two main inherent time constants are involved: the energy relaxation period $1/f_{energy}$ and the soliton relaxation period

$$n_R = \frac{2\pi}{\psi} = \frac{4\pi\tau_{\mathrm{sech}}^2}{|\beta_2|} \tag{9.17}$$

in numbers of round trips. Cavity dumping imposes an additional periodic perturbance of the solitary pulse. As all three time constants can be in the same order of magnitude, such a system displays an extraordinary dynamic behavior in certain parameter ranges, and a detailed knowledge about the pulsing dynamics is necessary for the identification and anticipation of stable regimes.

From an experimental point of view, because cavity dumping represents a periodic perturbance of the solitary pulse, the intracavity pulse shaping transient from one dumping event to the next is accessible for a measurement — at least, in the stable regimes. Here, we will compare the transient measurements to corresponding solutions of the master equation of mode locking. The excellent agreement between the measurement and the theory allows extrapolating the pulse parameters beyond the limits of the current experiment.

In order to characterize our laser in terms of stability, the system was analyzed in the same operational regimes as for the picosecond case, the "relaxed," "resonant," and "transient" regime for low, medium, and high dumping rates, respectively. Typical pulse energy and peak power evolutions between two dumping events are shown in Figure 9.5. For low repetition rates (< 20 kHz, Figure 9.5a and Figure 9.5b), the time between the dumping is long compared to the period of the energy relaxation oscillations, which allows the laser to relax to its steady state. In terms of pulse duration and chirp, the dumped pulses correspond to the unperturbed laser without cavity-dumping. It is possible to extend the relaxed regime to rather high frequencies (up to 20 kHz) by operating the system in terms of power and dispersion very close to the multiple pulse breakup. The overshoot after the dumping is limited considerably due to the optimized ratio between spectral filtering and absorber action, as was analyzed in detail in [29]. In this regime, the observed pulse to pulse reproducibility is very high. The rms-noise of the energy fluctuations was measured to be better than 0.1%, which is the resolution limit of the used oscilloscope. In the resonant regime with a dumping rate close to the energy relaxation oscillation frequency (around

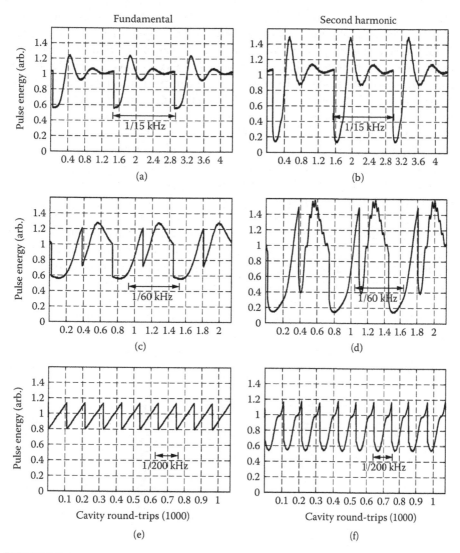

FIGURE 9.5
Intracavity energy and peak power (detected via second-harmonic energy) as monitored by a photodiode. The arrows indicate the dumping period.

40 kHz), a double-energy state for subsequent pulses appears and subharmonic behavior is observed. Intracavity pulse energy and peak power evolution between the dumping events in this regime are shown in Figure 9.5c and Figure 9.5d. This regime extends more or less from 20 to 80 kHz. The secondary oscillations visible on the crest of the peak power transient waveform are an effect of the third relevant time constant, namely the soliton relaxation period, which becomes important in the third, the transient regime: The "transient" regime is present at dumping frequencies higher than 80 kHz; the laser has no time to relax before the next dumping occurs.

This is depicted in Figure 9.5e and Figure 9.5f. Despite the saw-tooth shape of the fundamental, the structured second-harmonic trace, which reveals oscillations in the pulse duration, indicates the soliton relaxation period. As will be shown in the following, these oscillations can have a major effect on the stability of the system. To illustrate the complex interplay between the three relevant time constants in the transient regime, the energy relaxation period, the soliton relaxation period, and the dumping period, a stability diagram is measured and given in Figure 9.6. As a measure for the stability, the rms-noise of the second harmonic of dumped pulses is plotted as a function of dumping ratio and dumping frequency. Bright regions are operational parameters which lead to high stability, whereas dark regions represent poor stability. At certain dumping frequencies with high stability, the pulse to pulse rms-noise is better than 0.8%. It is obvious that a decreasing dumping ratio diminishes the perturbance and increases the stability, whereas at higher dumping ratios the darker regions of instability become broader, up to the point that only a few narrow stable areas remain. The instability can manifest itself as a subharmonic, multienergy state in the second-harmonic signal, irregular energy fluctuations, or both.

The three graphs in the lower part of Figure 9.6 show the peak power transients at different dumping frequencies in three neighboring stable areas of the transient regime as indicated in the stability plot. It is clearly visible that distinct regimes can be classified by the number of periods existing in

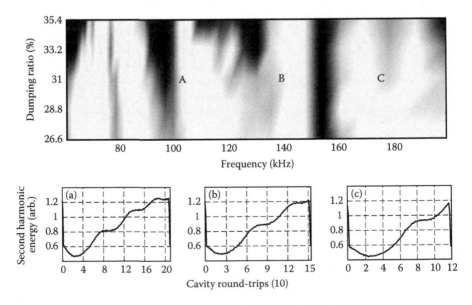

FIGURE 9.6
Upper part: measured stability regions of the system in the transient regime. Bright areas: low rms noise = high dynamical stability; dark areas: poor dynamical stability. Lower part: second-harmonic transient signal as monitored by a fast photo diode. The different regimes A (TR4), B (TR3), and C (TR2) are separated by unstable regions.

the transient. Because the second-harmonic transients unveil the soliton pulse width dynamics, an integer number of soliton phase periods fitting in the transient is necessary for the stability of the system. For clarification we denote the different regions of the transient regime TRn (where n is the number of minima/saddle points in the second-harmonic transient). At high repetition rates the system is in the regime TR2 (Figure 9.6c) and by lowering the dumping frequency, the laser becomes unstable in the area of 155 kHz and switches to TR3 (Figure 9.6b), where the laser is stable again. The same happens around 120 kHz, where TR4 is reached as shown in Figure 9.6a. The energy relaxation oscillation is responsible for a strong destabilization observed below 60 kHz, where we enter the "intermediate" regime with distinct subharmonic switching of the pulse energy (Compare with Figure 9.5c and Figure 9.5d). The system is dynamically unstable (period-doubling) until the "relaxed" regime below 20 kHz is entered. In contrast to the relaxed case, the characteristics of the dumped pulses in the transient regime are different from the unperturbed case for obvious reasons. The spectral and temporal properties of the dumped pulses are compared in Figure 9.7. Although the temporal pulse shape and width remain similar, the optical power spectrum in the transient regime is strongly modulated. From the shape of the autocorrelation we conclude that this structure is not due to multiple pulse instability. To clarify the details of the pulse shaping in the transient regime and to obtain further experimental insight into the dynamics

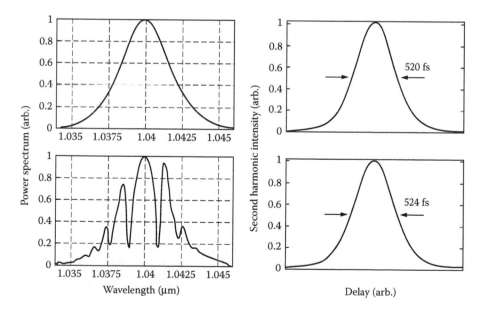

FIGURE 9.7
Measured spectra (left) and autocorrelation traces (right). The upper two plots are taken in the "relaxed" regime, they correspond to the results without cavity dumping, whereas shown below is the "transient" regime with a dumping rate of 180 kHz. The autocorrelation width of 520 fs (524 fs) corresponds to a pulse width of 337 fs (340 fs).

of the laser system we employed a more sophisticated method to measure transient spectra and autocorrelation traces. To achieve this, the intracavity pulses of the laser system were monitored by a background free autocorrelator and a scanning Fabry–Perot spectrometer. By detecting the output signal of both the autocorrelator and the spectrometer with fast photodiodes, subsequent laser pulses from the oscillator could be resolved. Figure 9.8 shows a schematic of the setup. A synchronization signal taken from the cavity dumper driver is used as the sampling clock for the oscilloscope (LeCroy Waverunner LT584). By adjusting the time delay with respect to the dumping event it was possible to record spectrum and autocorrelation trace for each roundtrip. For obvious reasons, this method can only be employed in stable regimes where the pulse evolution is reproducible in each dumping cycle.

In Figure 9.9 the transients of the autocorrelation and power spectra between two dumping events are shown for a dumping frequency of 180 kHz. Note the well-shaped autocorrelation trace despite the strongly structured spectra. Furthermore, an oscillation in the spectral and temporal width of the pulse can be observed. By comparing with Figure 9.6c, we can identify the laser to be operating in the TR2 regime. The periodicity of this oscillation will be discussed in the following section. The most prominent property in

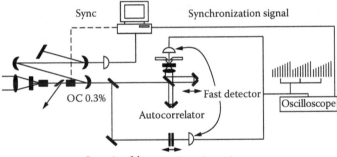

FIGURE 9.8
Experimental setup of the directly diode pumped Yb:glass laser with cavity dumping, employing a synchronous sampling scheme for autocorrelation and power spectrum of the pulse sequence during the transient.

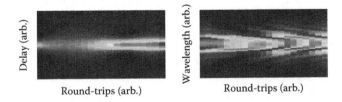

FIGURE 9.9
Measured autocorrelation and power spectrum of the pulse sequence between two dumping events for a dumping frequency of 180 kHz. Intensity is mapped to a grayscale from dark to bright.

the measured transient optical spectrum is the appearance of spectral side-bands moving toward the center of the spectrum. To understand this interesting behavior which was, to our knowledge, not yet reported in the literature, we employed numerical simulations of the pulse shaping.

The parameters used for the simulations are given in Table 9.1. The result for a dumping frequency of 180 kHz is depicted in Figure 9.10. A good agreement of the spectral and pulse evolution between the simulation and the experiment in Figure 9.9 can be observed. To stress this even further, the simulated (red) and the measured (blue) spectra directly before the dumping are compared in Figure 9.11. The agreement is excellent, and we can therefore conclude that our laser system is indeed very well described by the above equations. Figure 9.12 shows the simulated power spectrum and intensity of the pulse over a larger range on a logarithmic scale along with the corresponding phase. The remote sidebands in the power spectrum are the Kelly sidebands. The dips in the chirped temporal continuum background and the sidebands close to the peak of the spectrum (see also Figure 9.11) are a result of the cavity dumping. The simulations reveal the second set of Kelly sidebands due to the periodic perturbance of the system on a timescale much larger than the cavity round-trip time.

As was pointed out in the previous section, the laser system imposes two dominant perturbation periodicities on the pulse. One is the cavity round-trip time wherein we expect the "standard" Kelly sidebands. They appear at a spectral position where the nonlinear phase shift per round-trip modulo 2π is equal to the linear cavity phase due to dispersion. At these spectral positions,

TABLE 9.1

Parameter Values Used in the Simulations

Parameter	Value	Parameter	Value
g_0	0.21	E_L	3.8 mJ
L	0.027	Eq	138 nJ
q_0	0.0075	τ_L	1.3 ms
τ_q	1.5 ps	T_R	45 ns
β_2	−7200 fs²	Ω_g	2π 2 THz
γ	0.113 MW⁻¹	Ω_f	2π 10 THz

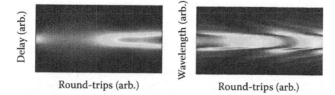

FIGURE 9.10
Simulated autocorrelation and power spectrum of the pulse sequence between two dumping events at a dumping frequency of 180 kHz. Intensity is mapped to a grayscale from dark to bright.

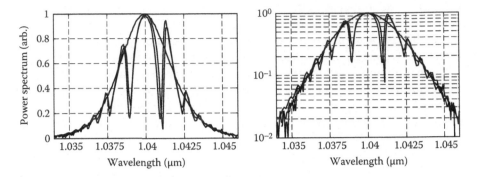

FIGURE 9.11
Linear and logarithmic plots of the spectra from simulation and measurement immediately before the dumping event, at 180 kHz repetition rate. The unmodulated curve is the measurement of the unperturbed laser.

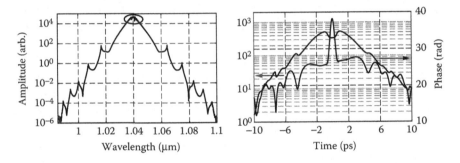

FIGURE 9.12
Simulated pulse spectrum (left) and pulse envelope (right and phase) immediately before the dumping event at 180 kHz repetition rate. The ellipse marks the dominant spectral structures, which are shown in more detail in Figure 9.11.

constructive interference of the pulse with the dispersive background occurs and phase matched continuum is generated at certain spectral positions. This is illustrated in the left part of Figure 9.13: Exactly at the points of intersection between the phase parabola due to the strong negative round-trip dispersion and the nonlinear phase modulo 2π, a sideband is present in the (calculated) optical spectrum. The standard Kelly sidebands have not been observed in the experiment because of the limited dynamics of the spectrometer used.

On the second timescale, namely the dumping period much longer than the cavity round-trip time, a second set of sidebands becomes apparent. As the pulse accrues a lot more dispersion between two dumping events, the phase parabola of the cavity dispersion is much steeper, and the sidebands appear much more pronounced around the center of the spectrum (see Figure 9.14). The right part of Figure 9.13 compares measured and calculated optical power spectra before dumping with the cavity phase. The sidebands appear exactly at the intersections between the phase parabola and the nonlinear phase modulo 2π, as predicted by theory.

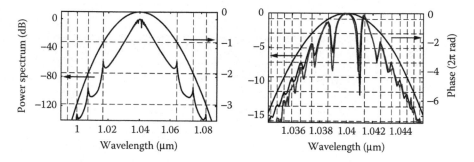

FIGURE 9.13

Phase matching between linear and nonlinear phase. The matching points become apparent in the optical spectrum as small peaks (Kelly sidebands). Left: sidebands due to the cavity roundtrip periodicity. Right: sidebands due to the dumping period (simulation and measurement). The cavity phase parabola is given by the solid curve, the nonlinear phase modulo 2π by the dotted horizontal lines. The vertical dotted lines indicate phase matching between the nonlinear and the linear phase, coinciding with the spectral peaks.

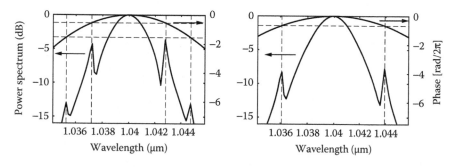

FIGURE 9.14

Spectral modulation for different dumping frequencies. Left: 540 kHz; right: 880 kHz. The cavity phase is given by the solid curve, the nonlinear phase by the dotted horizontal lines. The dumping ratio is 40%. A simple control of the Kelly sidebands can be accomplished by changing the dumping frequency. If the dumping frequency is increased, a reduced cavity GDD per dumping cycle results, and therefore a shifting of the Kelly sidebands away from the center of the spectrum. The simulation reveals that the sidebands become less pronounced with higher dumping frequencies due to the less efficient coupling to the continuum radiation.

The operation at higher dumping frequencies is also preferable in terms of stability. With a dumping frequency of more than 500 kHz, which is well beyond the energy relaxation oscillation (on the order of some 10 kHz) and the soliton relaxation periodicity (on the order of some 100 kHz), the system is in the stable TR1-regime. To anticipate the laser performance with high dumping rates the stability for varying dumping rates in the range of 80 kHz up to 2 MHz and dumping ratios in the range of 0.2 up to 0.4 has been analyzed with the numerical simulation. The results are shown in Figure 9.15. Again, the TRn regimes can readily be identified. We find good stability at frequencies of up to 1 MHz. The unstable region (shown in black) at very

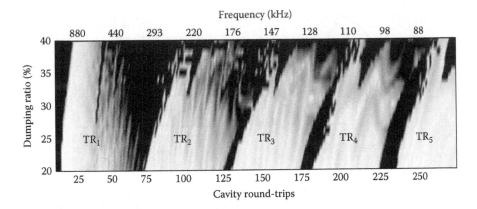

FIGURE 9.15
Stability regions of the system as predicted by simulation. Bright areas: low rms noise = high dynamical stability; Dark areas: poor dynamical stability. Upper horizontal axis: dumping frequency. Lower horizontal axis: number of round trips between dumping events at 22 MHz fundamental repetition frequency of the oscillator.

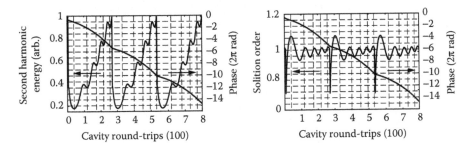

FIGURE 9.16
Left: calculated peak power evolution at a dumping rate of 80 kHz. Right: calculated transient of the soliton order. The pulse peak phase evolution is shown in both graphs. The dotted horizontal lines represent 2π-steps on the phase axes. The dotted vertical lines clarify the phase periodicity of the pulse. The distance between two dotted vertical lines indicates the number nR of cavity round trips during one phase period. The oscillations in the peak power and in the soliton order asymptotically approach the phase periodicity of the pulse.

high frequencies is caused by the additional losses of the cavity-dumping resulting in a significant drop in the average pulse energy and Q-switching instabilities. This can be compensated for by a stronger pump. The inward motion of the spectral sidebands during the transient (see Figure 9.10) can be well understood by the increasing intracavity power and the resulting change in the nonlinear phase shift. The movement of the spectral sidebands is accompanied by an oscillation of the pulse width, as shown in detail in Figure. 9.16. The figure shows both the transient peak power and the calculated soliton order

$$O_{\text{sol}} = \sqrt{\frac{\gamma A_0^2 \tau_{\text{sech}}^2}{|\beta_2|}}. \tag{9.18}$$

For the fundamental soliton, o_{sol} is equal to unity leading to the area theorem Equation 9.15. The black curve represents the peak phase evolution; dotted horizontal and vertical lines indicate a 2π phase step. The deviation from a linear phase is due to the change in pulse energy during one dumping cycle. Immediately after the dumping event, with strongly reduced energy, the pulse is too short to form a fundamental soliton, thus it sheds away energy into the continuum while broadening. The soliton order reacts with a damped oscillation around the steady state value of unity. The oscillation period asymptotically approaches the phase periodicity known from the fundamental soliton, as the soliton order gets close to unity.

In conclusion, the main pulse and spectral shaping mechanisms have been identified as a result of the solitary behavior of the pulse. Two distinct sets of Kelly sidebands have been observed and are due to the main two times-cales present in the system: the dumping period and the repetition period of the laser. Based on the better understanding of the underlying physical processes, design criteria for improved femtosecond cavity dumped laser systems could be deduced. For highest efficiency and pulse energy this leads to the following criterion:

$$f_{\text{dump}} \ll f_{\text{energy}}, \quad f_{\text{dump}} \ll f_{\text{soliton}} = \frac{1}{T_R n_R} \tag{9.19}$$

For the solitary mode-locked laser $f_{\text{dump}} \ll f_{\text{soliton}}$ represents a fundamental limitation to the accessible peak power P_{peak} of such a system, as the following relation can be derived from Equation 9.17:

$$f_{\text{soliton}} = \frac{1}{4\pi} \frac{\gamma}{T_R} P_{\text{peak}} \tag{9.20}$$

In order to maximize P_{peak} at a given magnitude of f_{soliton} it is necessary to reduce the SPM-parameter γ and increase the round-trip time T_R as much as possible. We applied these criteria to improve the performance of a Yb:KYW oscillator.

9.1.7 The Cavity-Dumped Yb:KYW Oscillator

A schematic setup of the laser system is shown in Figure 9.17. It was based on an AR-coated 1 mm-long Yb:KYW rod. The crystal was used in an n_g-cut geometry, such that the laser polarization was aligned along the n_p-axis (crystallographic b-axis). The pump diode was an internally collimated mul-tiemitter bar and capable of delivering 35 W at 981 nm. In the experiments 18 W were used of which 50% was absorbed under lasing conditions. The pump light was focused to a round spot with a diameter of approximately 300 μm by a combination of an achromatic and a cylindrical lens with a focal length of 30 mm and 100 mm, respectively (L1, L2). The modal size in the

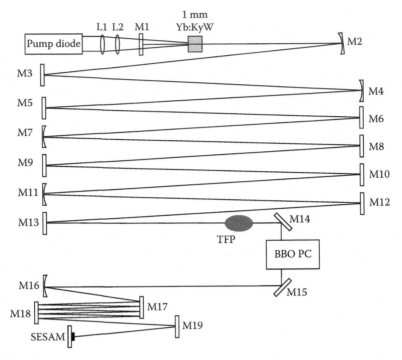

FIGURE 9.17

Setup of the Yb:KYW laser: L1, achromatic lens; L2, cylindrical lens; M1, dichroic pump mirror; M2, M4, M7, M11, and M16, curved mirrors; M14 and M15, high-reflecting mirrors; the rest, dispersive high-reflecting mirrors; other abbreviations as described in text.

active medium was estimated to be 240 μm defined by a curved mirror with 200 mm radius (M2). The cavity was stretched by four curved mirrors with radii of 1000 mm (M4), 3000 mm (M7, M11), and 1500 mm (M16) to a total length of 8.8 m, and a repetition frequency of f_{rep} = 17 MHz. We chose an electro-optical cavity dumper consisting of a 36 mm-long double beta barium borate (BBO) Pockels cell (PC) and a thin film polarizer (TFP) to avoid excess self-phase modulation. The high-voltage driving switch provided up to 1.54 kV at 1 MHz in a pulse short enough to suppress a postpulse with an extinction ratio of better than 1/2000. The laser is operated in the anomalous dispersion regime to enable soliton formation. The net dispersion of the laser cavity was measured to be –9200 fs² generated by dispersive mirrors, each having a dispersion of –500 fs². The soliton-like pulse was stabilized by a semiconductor saturable absorber mirror (SESAM) and the estimated spot-size on the SESAM was 750 μm. The small size of only 50 × 90 cm² makes the laser system very compact and easily portable.

With this setup we generate pulses with an energy of 1.35 μJ at a dumping frequency of 1 MHz. The typical spectral width is 3.3 nm full width at half maximum (FWHM) with a corresponding pulse length of 380 fs, measured by background free intensity autocorrelation. The time–bandwidth product of 0.34 suggests almost transform limited pulses, and the peak power is

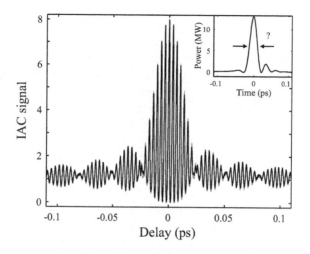

FIGURE 9.18

Power spectrum of the pulse after broadening in the LMA fiber. Black curve: measured spectrum; gray curve: simulation result. The inset depicts the typical spectrum of the laser.

3 MW. The inset in Figure 9.18 depicts the optical power spectrum. The spectral side peaks are the Kelly sidebands as discussed above [9] and result from the interference between the pulse and the continuum radiation generated by the periodic perturbance of the pulse due to the dumping process [9].

At certain dumping frequencies below 1 MHz and high dumping ratio we observed regimes of subharmonic behavior of both the pulse duration and the pulse energy [9]. In the stable regimes the laser shows an excellent noise performance with a pulse energy rms-value of below 0.5% over a time period of 12 h.

The generation of 1.35 μJ-pulses at a repetition rate of 1 MHz is a major breakthrough in the generation of high-energy pulses directly from a laser oscillator. Additionally, we will show that these pulses can be efficiently compressed in a simple fiber-prism compressor and peak powers exceeding 10 MW can be reached [30].

9.1.8 Pulse Compression

High peak power pulses with a duration of a few 100 fs show great potential for spectral broadening in large mode area fibers (LMA) and subsequent recompression in a prism sequence. Pulses of 12 MW have been obtained by combining a micro-structured fiber with a mode-locked thin disc laser [21]. In our case we used a 38 mm-long piece of ytterbium-doped double-clad fiber with an inner core radius of 29 μm and a numerical aperture of 0.054 for reasons of availability. Focusing was done with a 60 mm focal length plano-convex lens; for recollimation, we used a curved mirror with a radius of 50 mm. The prism sequence consisted of two quartz prisms with an apex distance of 96 cm and was used to compensate for the nonlinear chirp as well as for the fiber and lens dispersion. Pulses of 550 nJ could be launched without damaging the fiber, even when operating the system for hours. The pulse energy after the prisms was of 400 nJ and indicates an excellent coupling efficiency beyond 70%. The polarization after the fiber remained linear with a ratio of 1/200. Increasing the incident power substantially further led to fiber damage. The facet was not affected but the fiber bulk itself was. We believe that the destruction process is caused by catastrophic self-focusing, as a bright white spot becomes visible approximately 2 mm after the facet. The typical broadened spectrum of the pulse is depicted in Figure 9.18. The two prominent inner peaks are partly caused by the Kelly sidebands which are not affected by the nonlinearity. We characterized the pulses by means of an interferometric autocorrelation and the comparison to a numerical analysis of the process. The propagation of the typical laser pulse was simulated with the split-step Fourier method. In the model, second- and third-order dispersion, self-phase modulation, and self-steepening were included [25]. Stimulated Raman scattering was not significant in our particular case. Spectrum and autocorrelation from the simulation almost perfectly match the measured curves, as one can see in Figure 9.19. The side maxima of the autocorrelation are caused by the inevitable pre- and postpulses resulting from the shape of the broadened spectrum. With a launched pulse energy of 400 nJ we deduce from the simulation a pulse duration of 21 fs and a peak power of 13 MW [30]. The temporal pulse power profile is depicted in the inset of Figure 9.19.

Focusing the compressed high-peak-power pulses into a 3 mm-long sapphire plate we obtain a stable and smooth white light filament. An estimate

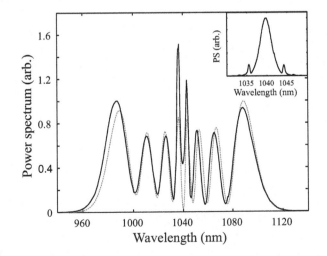

FIGURE 9.19
Autocorrelation signal of the compressed pulse (black curve) and simulated autocorrelation trace (gray curve). The inset shows the time dependence of the simulated pulse power with a pulse duration of $\tau = 21$ fs.

of the stability of the continuum is given by the rms-noise of only 1% behind an interference filter with a spectral bandwidth of 20 nm centered at 700 nm.

9.1.9 Conclusion

We reported on the generation of microjoule femtosecond pulses from a directly diode pumped cavity-dumped Yb:KYW oscillator. Based on the theoretical modeling and understanding of the main pulse shaping processes, peak powers exceeding 3 MW at a repetition frequency as high as 1 MHz have been obtained. With external fiber-prism compression down to 21 fs the peak power can be increased beyond 13 MW. This laser light source is well suited for many fields in science. It was applied successfully in multiphoton microscopy and material ablation. The very successful application in waveguide writing is reported in the second part of this contribution.

9.2 Optical Waveguide Writing

9.2.1 Introduction

Optical waveguides and related photonic devices, such as waveguide amplifiers and lasers, couplers, splitters, and interleavers, are finding increasing applications in optical communication systems. The integration on a single glass chip of several optical functions such as power splitting, wavelength multiplexing/demultiplexing, and gain can enable all-optical data processing and provide reliable and cheap devices for high-bandwidth optical fiber links, especially for Local Area Network (LAN) and Metropolitan Area Network (MAN) applications. Traditional waveguide manufacturing technologies include chemical vapor deposition with subsequent reactive ion etching (silica on silicon) [31], ion exchange [32] and sol-gel [33]. Though these techniques are well established, they present two main disadvantages: (1), they are multistep processes, including a photolithographic step limiting the flexibility of device fabrication and (2) they are essentially two-dimensional techniques, capable of producing only structures in planar geometry close to the surface of the sample.

Recently, a novel technique for the direct writing of waveguides and photonic circuits, exploiting refractive index modifications induced by focused femtosecond pulses, has emerged [34–38]. The basic physical mechanisms underlying this process can be outlined as follows: When a femtosecond pulse is tightly focused in a transparent material, a combination of nonlinear effects such as multiphoton absorption and avalanche ionization allows it to deposit energy is in a small volume around the focus [39,40]. The photogenerated hot electron plasma rapidly transfers its energy to the lattice, giving rise to high temperatures and pressures, and possibly leading to melting;

this localized absorption produces, by mechanisms still under investigation, an increase of refractive index over a micrometer-sized volume of the material. This photoinduced refractive index gradient allows one to produce, by moving the laser focus inside the glass substrate, a wide variety of devices, both active and passive.

Femtosecond micromachining shows, compared with traditional waveguide fabrication techniques, a number of distinct advantages:

1. It requires a simpler and less expensive device production equipment, avoiding clean-room facilities.

2. It enables rapid device prototyping, because the device pattern can be easily changed by simple software control, with significant cost reduction in comparison with standard techniques using photolithographic steps and requiring the production of a mask.

3. It is intrinsically a three-dimensional technique, because refractive index changes can be induced in any point in the bulk of the material within a given depth (100 μm to 1 mm) from the surface; this characteristic can be exploited to implement novel device functionalities, which are impossible with standard fabrication methods.

In this section we will review the current state of the art of femtosecond optical waveguide writing, with particular emphasis on the use of the diode-pumped cavity dumped Yb laser described in the previous section.

9.2.2 Absorption of Femtosecond Pulses in Transparent Materials

In order to understand the physical phenomena underlying optical waveguide writing by femtosecond laser pulses, it is necessary to analyze the absorption processes of intense laser pulses by transparent materials. Because these materials have, by definition, an energy gap greater than the laser photon energy, there is no linear absorption through interband transitions from valence to conduction band. At high intensities, however, absorption can take place through nonlinear phenomena, such as multiphoton, tunneling and avalanche ionization. Multiphoton ionization involves the simultaneous absorption of m photons of energy $h\nu$, where m is the minimum integer such that $mh\nu > E_g$, E_g being the band-gap energy of the dielectric material. This process is highly nonlinear with the laser intensity and can be described by the rate equation

$$\frac{dn}{dt} = \sigma_m I^m \tag{9.21}$$

where n is the electron density, I is the laser intensity, and σ_m is the cross section for the m-photon absorption process.

Tunneling ionization occurs when the very high electric field of the laser pulse lowers the Coulomb potential energy barrier and enables an electron to tunnel from the valence to the conduction band. At high intensities the multiphoton and tunneling ionization processes compete [41] and can be distinguished by the so-called Keldysh parameter, $\gamma = \omega(2m^*E_g)^{1/2}/eE$, where m^* and e are the effective mass and charge of the electron and E is the amplitude of the electric field oscillating at frequency ω. When $\gamma \gg 1$, multiphoton ionization dominates over tunneling.

To analyze the avalanche ionization process, let us suppose that an electron is free at the bottom of the conduction band of the material; if exposed to an intense light field, it can be accelerated and acquire kinetic energy. When its total energy exceeds the conduction band minimum by more than the bandgap energy, it can ionize another electron from the valence band, resulting in two electrons near the conduction band minimum. These electrons can be in turn accelerated by the electric field, causing an avalanche in which the free electron density grows exponentially; for sufficiently high densities, the dielectric becomes strongly absorbing. The avalanche ionization process can be described by the rate equation

$$\frac{dn}{dt} = \alpha I(t)n(t) \tag{9.22}$$

where α is the avalanche ionization rate.

Let us now try to understand the peculiar nonlinear mechanism by which femtosecond pulses are absorbed in transparent materials. If the dielectric is illuminated by a "long" pulse (with picosecond or nanosecond duration), the peak intensity is too low to allow multiphoton or tunneling ionization, even if the total pulse energy might be rather high. The only possible absorption mechanism is avalanche ionization starting from an initial "seed" of free electrons in the conduction band which, being the material an insulator, are due to impurities and dislocations within the focal volume of the laser pulse. As their number is subject to large fluctuations, the absorption process is erratic and poorly reproducible. With femtosecond laser pulses, on the other hand, the peak intensities are much higher, and multiphoton ionization becomes significant. When the intensity exceeds a given threshold, some free electrons are generated in the focal volume by multiphoton ionization. These electrons act as a *seed*, this time generated in a fully deterministic fashion, for the avalanche ionization process. It is therefore clear that only femtosecond pulses allow, by the unique combination of multiphoton and avalanche ionization, for the absorption of energy in a highly controlled and reproducible manner in a small volume inside the bulk of a transparent material.

9.2.3 Refractive Index Change

Though the process of nonlinear absorption of femtosecond pulses in dielectrics is well assessed, the physical mechanism by which it can induce permanent refractive index changes in glasses is not yet fully understood. Energy is deposited by the femtosecond pulses in a small volume of the bulk material, by the above described nonlinear mechanisms. If the absorbed energy is too high, catastrophic material damage occurs, including the formation of void-like structures; for a lower energy, there is a regime in which the material maintains its good optical quality, but there are permanent refractive index changes. Several mechanisms have been proposed to explain this index change, none of which seems to be fully general.

A first possible mechanism is color center formation: the femtosecond irradiation produces in the material a sufficient number of color centers which, while absorbing in the UV, modify the refractive index of the material at the wavelengths of interest by the Kramers–Kroenig mechanism. This has been a proposed mechanism for the refractive index change produced by deep-UV excitation of the Ge-doped silica fibers that results in fiber Bragg gratings. Characterization by electron spin resonance of glass substrates (fused silica and borosilicates) exposed to femtosecond pulses showed the presence of color centers; however, subsequent annealing strongly reduced the density of color centers, whereas the refractive index change was almost unaffected [42]. This indicates that color centers do not play a dominant role in refractive index modification.

An alternative mechanism is thermal; energy deposited by the laser melts the material in the focal volume, and the subsequent resolidification dynamics lead to density (and therefore refractive index) variations in the focal region [43]. Some glass types, such as fused silica, increase their density at higher temperatures; if they are rapidly cooled (quenched), the higher density (and therefore higher refractive index) structural arrangement is "frozen in." This hypothesis is confirmed by Raman measurements in fused silica, indicating the presence of this higher density structure of the glass after femtosecond irradiation [44]. However. in other glasses, such as Corning 0211, the density decreases with increasing temperature, yet waveguides are still formed by femtosecond irradiation. In this case, a possible mechanism responsible for waveguide formation is nonuniform cooling. After irradiation, the material melts out to a radius where the temperature equals melting temperature of the glass. Molten material just inside this maximum radius then quickly quenches and solidifies into a lower-density structural arrangement of the glass. This quenching continues radially inward as the glass continues to cool. Because there is no free surface that can expand to take up the extra volume occupied by the less dense glass formed by this quenching, the material near the focal region is put under pressure. As a result of this pressure, the material near the focal region solidifies into a higher density phase, leading to the higher refractive index at the core of the structures.

Finally, another possible mechanism is direct structural change induced by the femtosecond laser pulses, i.e., rearrangement of the network of chemical bonds in the glass matrix leading to a density increase.

In practical cases, all of the three mechanisms discussed above play a role in refractive index change, and it is difficult to disentangle their relative contributions. Finally, it is worth noting that in crystalline materials femtosecond laser irradiation generally produces a decrease in refractive index; this can be easily understood by considering that in a crystal the atoms are in the closest possible arrangement and that any change in the lattice order will lead to a lower density. Optical waveguides can nevertheless be created on the sides of the modified region, where stresses induce a refractive index increase [45,46].

9.2.4 State of The Art in Femtosecond Laser Waveguide Writing

Nonlinear absorption in glasses takes place for intensities around $1–5 \times 10^{13}$ W/cm^2 which, for a pulse duration of 100 fs, correspond to fluences of 1–5 J/cm^2. The pulse energy required to achieve such fluences depends on the focusing conditions: for "mild" focusing (1–3 μm beam waist), it is at the level of a few microjoules, and for extremely tight focusing, of the order of a half-wavelength (diffraction limited), it can be reduced to a few tens of nanojoules.

Two different regimes of femtosecond micromachining can be distinguished, depending on whether the pulse period is longer or shorter than the time required for heat to diffuse away from the focal volume: the "low-frequency" regime, in which material modification is produced by the individual pulses, and the "high-frequency" regime, in which cumulative effects take place. Because the heat diffusion time out of the absorption volume in the glass can be estimated at ≈1 μs, the transition between the two regimes takes place at frequencies around 1 MHz. Low-frequency systems typically use regeneratively amplified Ti:sapphire lasers (1–200 kHz repetition rate) [34–38], and high-frequency systems directly use Ti:sapphire oscillators with the cavity length stretched by a telescope or a multipass cell (5–20 MHz repetition rate) [47–53].

For the low-frequency regime, two different writing geometries are possible: longitudinal and transverse, in which the sample is translated, respectively, along and perpendicularly to the beam propagation direction (see Figure 9.20). In the longitudinal geometry, the waveguides are intrinsically symmetric, and their transverse size is determined by the focal spot size, making it possible to achieve fairly large diameters; however, the waveguide

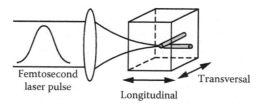

Femtosecond laser pulse

Longitudinal

Transversal

FIGURE 9.20
Waveguide writing geometries: longitudinal and transversal with respect to the femtosecond laser beam.

length is limited by the focal length of the focusing objective, and their quality is degraded by spherical aberrations, which depend on the depth of the focus inside the glass sample. The transverse geometry provides a much greater flexibility and allows one to write waveguides or waveguide structures of arbitrary length; it has, however, the disadvantage of producing a strong asymmetry in the waveguide cross section [54]. This asymmetry can be explained as follows: perpendicularly to the beam propagation direction, the waveguide size is given approximately by the beam focal diameter $2w_0$, whereas along the propagation direction, it is given by the confocal parameter $b = 2\pi w_0^2 / \lambda$. For focused diameters of the order of a few micrometers, this results in a large difference in waveguide sizes in the two directions. This asymmetry becomes particularly severe when the waveguide size is increased, as required for waveguiding at the optical communication wavelength of 1.5 μm, thus greatly reducing the efficiency of fiber butt coupling in conventional telecommunications setups. This problem can be overcome [55,56] by introducing a novel focusing geometry in which the femtosecond writing beam is astigmatically shaped by changing both the spot sizes in the tangential and sagittal planes and the relative positions of the beam waists. This shaping allows one to modify the interaction volume in such a way that the waveguide cross section can be made circular and with arbitrary size.

There are many studies on low-frequency optical waveguide writing with amplified Ti:Sapphire systems; such lasers are indeed commercial and are often already present in the laboratories for other kinds of experiments. The first demonstration of the possibility of writing waveguides in glasses by a femtosecond laser was given by Hirao's group [34–36], using an amplified Ti:Sapphire laser at 200 kHz. Subsequently, there were both fundamental investigations aimed at the study of the optical characteristics of the waveguides as a function of the writing conditions and glass substrate composition, and applied studies aimed at the fabrication of photonic devices. In the first class, we can include the papers studying the structural modifications that lead to refractive index variations in fused silica [34,38,41,44], and those exploring the possibility to write waveguides in different materials, such as doped silica, borosilicate, fluoride, chalcogenide, and doped phosphate glasses [35,57–60]. Among the demonstrated device functionalities, apart from straight waveguides, are splitters [38], directional couplers [47,61], Mach–Zehnder interferometers [62], and waveguide amplifiers [56,63]. Though some device architectures are similar to those used with standard waveguide fabrication methods, others exploit the unique 3D capabilities of femtosecond waveguide writing, such as a 1-to-3 coupler and three-dimensional waveguide arrays, recently fabricated in fused silica [64,65].

High-frequency micromachining is relatively new, because stretched-cavity Ti:sapphire oscillators with peak power sufficient for material modification have become available only recently. This writing regime offers several advantages: (1) simplification of the experimental setup, because the amplification stage is avoided, (2) much greater processing speeds, up to 20 mm/s, and (3) intrinsic symmetry of the waveguide transverse profile, determined by the isotropic heat diffusion, and the possibility of controlling the waveguide size by changing the writing speed, because of thermal accumulation effects.

However, the small available intensity range limits the process flexibility, and the need of tight focusing enables formation of structures only in close proximity to the surface, thus not allowing full exploitation of the three-dimensional capabilities of the process. Another problem of this technique is the strong dependence of the outcome on glass substrate composition. In fact, high quality optical waveguides can be created by high-frequency systems only in a few glass types (alkali-silicate glasses such as Corning 0211, Corning 0215, and Schott IOG10), while in other substrates (phosphate glasses and fused silica) irregular and void-like structures are formed [52]. The influence of the glass composition on waveguide formation is still poorly understood.

High-frequency lasers have been used to manufacture a variety of photonic devices such as directional couplers and unbalanced Mach–Zehnder interferometers, used as spectral filters [50]; recently, the three-dimensional capabilities of the technique have been exploited to create more complex devices, such as a three-waveguide directional coupler and a three-dimensional microring resonator [49].

9.2.5 Optical Waveguide Writing by Cavity-Dumped Yb Laser

Most of the experiments on femtosecond laser optical waveguide writing have used Ti:Sapphire laser systems. Quite recently, however, new Yb-based bulk and fiber lasers have been introduced, with medium-high repetition rate (200 kHz–1 MHz) and energy approaching the μJ level. Diode-pumping makes these systems particularly compact and efficient, which is important for industrial applications of the technique, requiring maintenance-free, turn-key operation. Here, we will report on optical waveguide writing experiments [66–68] using a femtosecond cavity-dumped Yb:glass oscillator described in the first part of this chapter.

The waveguide writing setup is very simple and is depicted schematically in Figure 9.21. The femtosecond pulses from the oscillator are suitably attenuated, then their polarization is rotated by a half-wave plate in order to be parallel to the translation direction; the beam is then deviated by a mirror to the vertical direction and focused by a microscope objective inside the glass sample. We tried different focusing objectives and demonstrated that the best results in terms of circular symmetry of the waveguide mode were obtained with a very high numerical aperture (N.A.) objective (100X oil-immersion Zeiss Plan-Apochromat, 1.4 N.A.); therefore, we will concentrate on the results obtained with such an objective. The waveguides are written inside the glass at a depth of 170 μm from the surface. This is the nominal depth to minimize the aberrations with this kind of objectives. However, the same objective provides uniform results at least for depths in a 100 μm range around the nominal one [53].

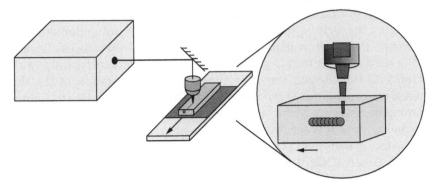

FIGURE 9.21
Waveguide writing setup: the femtosecond laser beam is focused by a microscope objective into the glass sample, which is suitably translated to obtain the desired optical circuit geometry.

The glass used is a commercial phosphate glass (QX, Kigre, Inc.) doped with 2% wt of Er_2O_3 and 4% wt of Yb_2O_3. Such doping levels have been optimized to obtain high gain per unit length in order to fabricate compact devices.

A transverse writing configuration is adopted, in which the sample is translated by motorized stages (M.511.DD, Physik Instrumente) in a direction perpendicular to the laser beam. Both end facets of the waveguides are polished after laser inscription.

In optical waveguide writing by femtosecond lasers, a strong dependence on the glass type and on the writing conditions (mainly repetition rate and pulse energy of the laser source) has been evidenced. On the other hand, the shape of the refractive index profile has a dominant impact on the waveguide properties. For this reason an accurate measurement of this profile is of the utmost importance. Such measurements often show quite unexpected features, with positive and negative refractive index peaks, that could not be predicted either from the writing parameters or from a near-field characterization of the guided mode but only from a direct measurement of the refractive index profile. Very few techniques have proved the capability of characterizing with high resolution, the refractive index profile of a femtosecond laser-written waveguide. Digital holography microscopy [52] and selective etching followed by atomic force microscopy [69] have been proposed, but both these techniques are destructive of the sample. We used a commercial refractive index profilometer (Rinck Elektronik) that is based on the refracted near-field technique [70]. This technique is nondestructive, applicable to almost any glass type and with high spatial resolution (0.5 μm) and sensitivity ($\Delta n = 10^{-5}$ relative and 10^{-4} absolute).

Figure 9.22 shows the measured refractive index profile for a waveguide written at 885 kHz repetition rate and 270 nJ, with 300 μm/sec translation speed. The actual spot size of the writing laser beam inside the sample is below 1 μm, due to the very tight focusing used. The size of the modified region ($\approx$20 μm) is much larger than the spot of the writing laser beam,

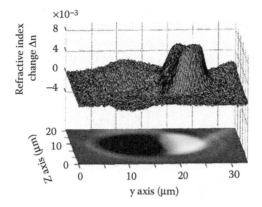

FIGURE 9.22
Refractive index profile of a femtosecond laser written waveguide measured by a refracted near-field profilometer.

indicating the occurrence of thermal diffusion. However, though the isotropy of thermal diffusion should give almost symmetric profiles, here we observe a clearly asymmetric profile consisting of two lobes, one positive and one negative. We believe that this asymmetry should not be ascribed to some misalignment of the focusing objective, because it is different from waveguide to waveguide and it changes by just varying the repetition rate or pulse energy. It seems to be related to the glass characteristics and requires further study to be fully understood. However, it is worth noting that when the product of the repetition rate with the pulse energy is about 250 mW, a rather reproducible profile shape is found. This refractive index profile shape is the most promising one for applications, as the positive part is single peaked with a very high index change. Although a certain asymmetry is present in the positive part of the refractive index profiles, it does not significantly affect the quality of the guided modes.

The key parameter for evaluating the waveguide performance is the insertion losses. The insertion losses of a waveguide are defined as the excess losses one measures in the transmission of a fiber when it is cut and a waveguide is inserted in the path. They consist of coupling losses and propagation losses. The former take into account the losses due to mismatch between the mode of the fiber and that of the waveguide and should be counted two times, one at the input and one at the output of the waveguide; the latter take into account the losses due to light scattering or absorption in the waveguide.

The insertion losses are measured by coupling the waveguide with two standard telecom fibers, with index matching fluid to minimize Fresnel losses, and by comparing the transmitted power to that measured with the same fibers spliced. The measurement was performed at the wavelength of 1600 nm, chosen because it falls outside the absorption band of the Er ions. Typical results were insertion losses around 1.1–1.2 dB for 22-mm-long waveguides.

In order to distinguish the contributions arising from coupling and propagation losses, we acquired the near-fields of the waveguide and fiber modes. The theoretical maximum coupling efficiency η was estimated from the overlapping integral of the field profiles in the waveguide (E_g) and the fiber (E_f), according to the formula

$$\eta = \frac{\left[\int E_g E_f^* dxdy \right]^2}{\int E_g E_g^* dxdy \int E_f E_f^* dxdy}$$
(9.23)

Figure 9.23 shows the near field of the single transverse mode supported at 1600 nm by a typical waveguide, acquired with a Vidicon Hamamatsu camera (Model C2400), together with the intensity profiles in the y and z directions. The measured intensity profiles of the coupling fiber at the same wavelength are also shown. The excellent mode matching between the two provides coupling losses as low as 0.1 dB. From the measured insertion losses it is thus possible to estimate an upper limit for the propagation losses of 0.4 dB/cm for the best waveguides. We expect that this figure can be further improved by a better control of the uniformity in the translation speed and by minimizing mechanical vibrations.

The substrate used for the waveguide inscription is erbium-doped and is intended to provide active devices at telecom wavelengths, where, in addition to insertion losses, optical gain is very important. A schematic of the

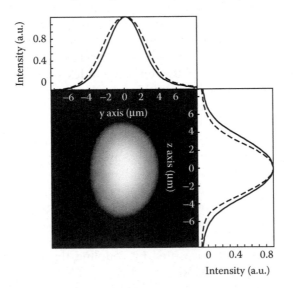

FIGURE 9.23
Near-field intensity of the guided mode of a femtosecond laser written waveguide. Side panels show the transverse cross sections of the waveguide (solid line) and fiber (dashed line) mode profiles.

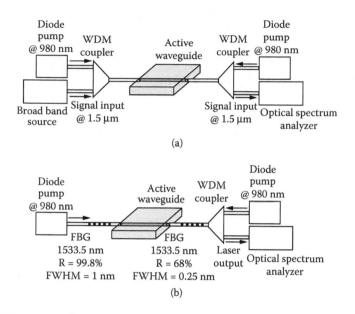

FIGURE 9.24
Schematic of the setup for (a) gain measurements and (b) laser action experiments in waveguides.

setup used for characterizing the optical gain is shown in Figure 9.24a. The 22-mm-long active waveguide is butt-coupled on both sides to standard telecom single mode fibers, using an index-matching fluid. A couple of 980-nm wavelength InGaAs laser diodes in a bipropagating pumping scheme supply up to 550 mW (300 mW + 250 mW) incident pump power through 980/1550 nm wavelength-division-multiplexers (WDMs). Signal radiation in the telecom C-Band (1530–1565 nm) is provided by a broadband source, properly attenuated to about –3 dBm over 100 nm bandwidth.

The small signal internal gain of the amplifier as a function of wavelength, measured with an accuracy of about 0.5 dB with a pump power level of 490 mW, is shown in Figure 9.25. An internal gain higher than 7 dB and 3 dB is measured, respectively, at 1535 nm and 1565 nm. Due to the very low insertion losses, this means that this waveguide is able to provide net gain in the whole C-Band, from 6dB at the peak to 2 dB at the edge. This waveguide can act as an optical amplifier or can be used as an active medium in a laser cavity.

The schematic of the laser setup is shown in Figure 9.24b. The cavity consists of two Fiber Bragg Gratings (FBGs) coupled with index matching fluid to the waveguide. This configuration, with respect to discrete mirrors or dielectric coatings, allows for efficient pumping with standard pigtailed diode lasers, simple change of the laser wavelength, and output coupling by substitution of the gratings and output power already coupled into a standard telecom fiber with no losses. Bidirectional diode pumping is used and output power is measured by an optical spectrum analyzer.

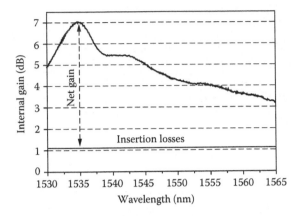

FIGURE 9.25
Internal gain, insertion losses, and net gain in the telecommunication C-band for a femtosecond laser-written waveguide.

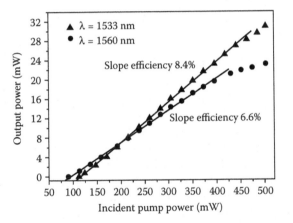

FIGURE 9.26
Output power of the waveguide laser as a function of incident pump power at two different wavelengths: 1533 nm (triangles), 1560 nm (circles). Interpolating lines for slope efficiency evaluation.

In Figure 9.26, the output power of the waveguide laser is reported with respect to the incident pump power, for an output coupling of 32% at a wavelength of 1533 nm. A pump power threshold of about 110 mW and a slope efficiency of 8.4% are found, with a maximum output power of 30 mW at 500 mW of pump power. For pump powers higher than 430 mW the output power shows a slight saturation. The slope efficiency is therefore calculated by fitting the experimental points before the saturation begins.

For telecom applications it would be extremely important to have a very compact laser source able to generate in parallel all the channels of a WDM transmission. The femtosecond laser writing technology is able to provide such device in a very easy and cheap way. Indeed, the waveguides we are presenting show a net gain in the whole C-Band and can, thus, be used to form a waveguide laser array emitting on the ITU grid in the whole wavelength range. To demonstrate this possibility, we made the waveguides lase also close to the C-Band edge (1560 nm). With an output coupling of 15%, we obtained a pump power threshold of 90 mW and a slope efficiency of 6.6%, with a maximum output power of 23 mW (see Figure 9.26). Again, for pump powers above 400 mW the laser characteristics deviate from linear thus the slope efficiency is calculated from points at lower pump power.

9.2.6 Conclusions

In conclusion, we have demonstrated the possibility to write high quality optical waveguides by means of the compact diode-pumped femtosecond oscillator described in the first part of the chapter. Such waveguides, realized on an erbium–ytterbium codoped phosphate glass, provide net gain and laser action in the whole C-Band with application to telecommunications. Femtosecond laser writing of optical waveguides and photonic devices is a still emerging field that has yet to exploit many of its potentialities, especially those involving three-dimensional structuring capabilities. Applications will span a wide range of fields from telecommunications to biophotonics.

References

1. M. Ramaswamy, M. Ulman, J. Paye, and J.G. Fujimoto, Cavity-dumped femtosecond Kerr-lens mode-locked Ti:Al$_2$O$_3$ laser, *Opt. Lett.*, Vol. 18, 1822–1824, 1993.
2. M.S. Pshenichnikov, W.P. de Boeij, and D.A. Wiersma, Generation of 13-fsec, 5-MW pulses from a cavity-dumped Ti:sapphire laser, *Opt. Lett.*, Vol. 19, 572–574, 1994.
3. G.N. Gibson, R. Klank, F. Gibson, and B.E. Bouma, Electro-optically cavity dumped ultrashort-pulse Ti:sapphire oscillator, *Opt. Lett.*, Vol. 21, 1055–1057, 1996.
4. A. Baltuska, Z.Wei, M. Pshenichnikov, D. Wiersma, and R. Szipöcs, All solid-state cavity dumped sub-5-fsec laser, *Appl. Phys. B*, Vol. 65, 175–188, 1997.
5. S. Schneider, A. Stockmann, and W. Schuesslbauer, Self-starting mode-locked cavity-dumped femtosecond Ti:sapphire laser employing a semiconductor saturable absorber mirror, *Opt. Eng.*, Vol. 6, 220–226, 2000.
6. A. Killi, U. Morgner, M.J. Lederer, and D. Kopf, Diode-pumped femtosecond laser oscillator with cavity dumping, *Opt. Lett.*, Vol. 29, 1288–1290, 2004.
7. A. Killi, J. Dörring, U. Morgner, M.J. Lederer, J. Frei, and D. Kopf, High speed electro-optical cavity dumping of mode-locked laser oscillators, *Opt. Express*, Vol. 13, 1916–1922, 2005.
8. C. Hönninger, R. Paschotta, F. Morier-Genoud, M. Moser, and U. Keller, Q-switching stability limits of continuous-wave passive mode locking, *J. Opt. Soc. Am. B*, Vol. 16, 46–56, 1999.

9. A. Killi and U. Morgner, Solitary pulse shaping dynamics in cavity-dumped laser oscillators, *Opt. Express*, Vol. 12, 3297–3307, 2004.
10. N.V. Kuleshov, A.A. Lagatsky, A.V. Podlipensky, V.P. Mikhailov, and G. Huber, Pulsed laser operation of Yb-doped KY(WO$_4$)$_2$ and KGd(WO$_4$)$_2$, *Opt. Lett.*, Vol. 22, 1317–1319, 1997.
11. P. Klopp, V. Petrov, U. Griebner, and G. Erbert, Passively mode-locked Yb:KYW laser pumped by a tapered diode laser, *Opt. Express*, Vol. 10, 108–113, 2002.
12. W.F. Krupke, Ytterbium solid-state lasers—the first decade, *IEEE J. Sel. Top Quantum Electron.*, Vol. 6, 1287–1296, 2000.
13. P. Lacovara, H.K. Choi, C.A. Wang, R.L. Aggarwal, and T.Y. Fan, Room temperature diode-pumped Yb:YAG laser, *Opt. Lett.*, Vol. 16, 1089–1091, 1991.
14. F. Druon, F. Balembois, P. Georges, A. Brun, A. Courjaud, C. Honninger, F. Salin, A. Aron, F. Mougel, G. Aka, and D. Vivien, Generation of 90-fsec pulses from a modelocked diode-pumped Yb^{3+}:Ca$_4$GdO(BO$_3$)($_3$) laser, *Opt. Lett.*, Vol. 25, 423–425, 2000.
15. F. Druon, S. Chenais, P. Raybaut, F. Balembois, P. Georges, R. Gaume, A. Aron, G. Aka, B. Viana, D. Vivien, A. Courjaud, C. Honninger, and F. Salin, Femtosecond laser oscillators with uncommon ytterbium-doped borate crystals, *J. De Phys. Iv*, Vol. 12, 351–353, 2002.
16. C. Hönninger, F. Morier-Genoud, M. Moser, U. Keller, L.R. Bovelli, and C. Harder, Efficient and tunable diode-pumped femtosecond Yb:glass lasers, *Opt. Lett.*, Vol. 23, 126–128, 1998.
17. C. Hönninger, R. Paschotta, M. Graf, F. Morier-Genoud, G. Zhang, M. Moser, S. Biswal, J. Nees, A. Braun, G. Mourou, I. Johannsen, A. Giesen, W. Seeber, and U. Keller, Ultra fast ytterbium-doped bulk lasers and laser amplifiers, *Appl. Phys. B*, Vol. 69, 3–17, 1999.
18. H. Liu, J. Nees, and G. Mourou, Diode-pumped Kerr-lens mode-locked Yb:KY(WO$_4$)$_2$ laser, *Opt. Lett.*, Vol. 26, 1723–1725, 2001.
19. G. Paunescu, J. Hein, and R. Sauerbrey, 100-fsec diode-pumped Yb:KGW modelocked laser, *Appl. Phys. B*, Vol. 79, 555–558, 2004.
20. E. Innerhofer, T. Sudmeyer, F. Brunner, R. Haring, A. Aschwanden, R. Paschotta, C. Honninger, M. Kumkar, and U. Keller, 60-W average power in 810-fsec pulses from a thin-disk Yb:YAG laser, *Opt. Lett.*, Vol. 28, 367–369, 2003.
21. T. Südmeyer, F. Brunner, E. Innerhofer, R. Paschotta, K. Furusawa, J.C. Baggett, T.M. Monro, D.J. Richardson, and U. Keller, Nonlinear femtosecond pulse compression at high average power levels by use of a large-mode-area holey fiber, *Opt. Lett.*, Vol. 28, 1951–1953, 2003.
22. S. Chenais, F. Druon, F. Balembois, P. Georges, A. Brenier, and G. Boulon, Diode-pumped Yb:GGG laser: comparison with Yb:YAG, *Opt. Mater.*, Vol. 22, 99–106, 2003.
23. M. Jacquemet, C. Jacquemet, N. Janel, F. Druon, F. Balembois, P. Georges, J. Petit, B. Viana, D. Vivien, and B. Ferrand, Efficient laser action of Yb:LSO and Yb:YSO oxyorthosilicates crystals under high-power diode-pumping, *Appl. Phys. B*, Vol. 80, 171–176, 2005.
24. H. Haus, Mode-locking of lasers, *IEEE J. Sel. Top. Quantum Electron.*, Vol. 7, 1173–1185, 2000.
25. G. Agrawal, *Nonlinear Fiber Optics*, Academic Press, San Diego, CA, 1995.
26. S.M. Kelly, Characteristic sideband instability of periodically amplified average soliton, *Electron. Lett.*, Vol. 28, 806–807, 1992.

27. N.J. Smith, K.J. Blow, and I. Andonovic, Sideband generation through perturbations to the average soliton model, *IEEE J. Lightwave Technol.*, Vol. 10, 1329–1333, 1992.

28. D.J. Jones, Y. Chen, H.A. Haus, and E.P. Ippen, Resonant sideband generation in stretched-pulse fiber lasers, *Opt. Lett.*, Vol. 23, 1535–1537, 1998.

29. T. Schibli, E. Thoen, F. Kartner, and E. Ippen, Suppression of Q-switched mode locking and break-up into multiple pulses by inverse saturable absorption, *Appl. Phys. B*, Vol. 70, 41–49, 2000.

30. A. Killi, A. Steinmann, U. Morgner, M. Lederer, D. Kopf, and C. Fallnich, High peak power pulses from a cavity-dumped Yb:KYW oscillator, *Opt. Lett.*, Vol. 30, 1811–1813, 2005.

31. K. Okamoto, *Fundamentals of Optical Waveguides*, Academic Press, New York, 2000.

32. R.V. Ramaswamy and R. Srivastava, Ion-exchanged glass waveguides: a review, *J. Lightwave Technol.*, Vol. 6, 984–1000, 1988.

33. X. Orignac, D. Barbier, X.M. Du, and R.M. Almeida, Fabrication and characterization of sol-gel planar waveguides doped with rare-earth ions, *Appl. Phys. Lett.*, Vol. 69, 895–897, 1996.

34. K.M. Davis, K. Miura, N. Sugimoto, and K. Hirao, Writing waveguides in glass with a femtosecond laser, *Opt. Lett.*, Vol. 21, 1729–1731, 1996.

35. K. Miura, J. Qiu, H. Inouye, T. Mitsuyu, and K. Hirao, Photo written optical waveguides in various glasses with ultrashort pulse laser, *Appl. Phys. Lett.*, Vol. 71, 3329–3331, 1997.

36. K. Hirao and K. Miura, Writing waveguides and gratings in silica and related materials by a femtosecond laser, *J. Non-Cryst. Solids*, Vol. 239, 91–95, 1998.

37. E.N. Glezer, M. Milosavljevic, L. Huang, R.J. Finlay, T.H. Her, J.P. Callan, and E. Mazur, Three-dimensional optical storage inside transparent materials, *Opt. Lett.*, Vol. 21, 2023–2025, 1996.

38. D. Homoelle, S. Wielandy, A.L. Gaeta, N.F. Borrelli, and C. Smith, Infrared photosensitivity in silica glasses exposed to femtosecond laser pulses, *Opt. Lett.*, Vol. 24, 1311–1313, 1999.

39. D. Du, X. Liu, G. Korn, J. Squier, and G. Mourou, Laser-induced breakdown by impact ionization in SiO_2 with pulse widths from 7 ns to 150 fsec, *Appl. Phys. Lett.*, Vol. 64, 3071–3073, 1994.

40. X. Liu, D. Du, and G. Mourou, Laser ablation and micromachining with ultrashort laser pulses, *IEEE J. Quantum Electron.*, Vol. QE33, 1706–1716, 1997.

41. S.S. Mao, F. Quere, S. Guizard, X. Mao, R.E. Russo, G. Petite, and P. Martin, Dynamics of femtosecond laser interactions with dielectrics, *Appl. Phys. A*, Vol. 79, 1695–1709, 2004.

42. A.M. Streltsov and N.F. Borrelli, Study of femtosecond-laser-written waveguides in glasses, *J. Opt. Soc. Am. B*, Vol. 19, 2496–2504, 2002.

43. C.B. Schaffer, J.F. Garcia, and E. Mazur, Bulk heating of transparent materials using a high repetition-rate femtosecond laser, *Appl. Phys. A*, Vol. 76, 351–354, 2003.

44. J.W. Chan, T. Huser, S. Risbud, and D.M. Krol, Structural changes in fused silica after exposure to focused femtosecond laser pulses, *Opt. Lett.*, Vol. 26, 1726–1728, 2001.

45. T. Gorelik, M. Will, S. Nolte, A. Tuennermann, and U. Glatzel, Transmission electron microscopy studies of femtosecond laser induced modifications in quartz, *Appl. Phys. A*, Vol. 76, 309–311, 2003.

46. V. Apostolopoulos, L. Laversenne, T. Colomb, C. Depeursinge, R.P. Salath, M. Pollnau, R. Osellame, G. Cerullo, and P. Laporta, Femtosecond irradiation induced refractive-index changes and channel waveguiding in bulk Ti^{3+}:sapphire, *Appl. Phys. Lett.*, Vol. 85, 1122–1124, 2004.

47. A.M. Streltsov and N.F. Borrelli, Fabrication and analysis of a directional coupler written in glass by nanojoule femtosecond laser pulses, *Opt. Lett.*, Vol. 26, 42–43, 2001.

48. C.B. Schaffer, A. Brodeur, J.F. Garcia, and E. Mazur, Micromachining bulk glass by use of femtosecond laser pulses with nanojoule energy, *Opt. Lett.*, Vol. 26, 93–95, 2001.

49. K. Minoshima, A.M. Kowalevicz, I. Hartl, E.P. Ippen, and J.G. Fujimoto, Photonic device fabrication in glass by use of nonlinear materials processing with a femtosecond laser oscillator, *Opt. Lett.*, Vol. 26, 1516–1518, 2001.

50. K. Minoshima, A.M. Kowalevicz, E.P. Ippen, and J.G. Fujimoto, Fabrication of coupled mode photonic devices in glass by nonlinear femtosecond laser materials processing, *Opt. Express*, Vol. 10, 645–652, 2002.

51. A.M. Kowalevicz, V. Sharma, E.P. Ippen, J.G. Fujimoto, and K. Minoshima, Three-dimensional photonic devices fabricated in glass by use of a femtosecond laser oscillator, *Opt. Lett.*, Vol. 30, 1060–1062, 2005.

52. R. Osellame, N. Chiodo, V. Maselli, A. Yin, M. Zavelani-Rossi, G. Cerullo, P. Laporta, L. Aiello, S. De Nicola, P. Ferraro, A. Finizio, and G. Pierattini, Optical properties of waveguides written by a 26 MHz stretched cavity Ti:sapphire femtosecond oscillator, *Opt. Express*, Vol. 13, 612–620, 2005.

53. R. Osellame, V. Maselli, N. Chiodo, D. Polli, R. Martinez Vazquez, R. Ramponi, G. Cerullo, Fabrication of 3D photonic devices at 1.55 μm wavelength by a femtosecond Ti:Sapphire oscillator, *Electron. Lett.*, Vol. 41, 315–316, 2005.

54. M. Will, S. Nolte, B.N. Chichkov, and A. Tuennermann, Optical properties of waveguides fabricated in fused silica by femtosecond laser pulses, *Appl. Opt.*, Vol. 41, 4360–4364, 2002.

55. G. Cerullo, R. Osellame, S. Taccheo, M. Marangoni, D. Polli, R. Ramponi, P. Laporta, and S. De Silvestri, Femtosecond micromachining of symmetric waveguides at 1.5 μm by astigmatic beam focusing, *Opt. Lett.*, Vol. 27, 1938–1941, 2002.

56. R. Osellame, S. Taccheo, M. Marangoni, R. Ramponi, P. Laporta, D. Polli, S. De Silvestri, and G. Cerullo, Femtosecond writing of active optical waveguides with astigmatically shaped beams, *J. Opt. Soc. Am. B*, Vol. 20, 1559–1567, 2003.

57. D. Ehrt, T. Kittel, M. Will, S. Nolte, and A. Tuennermann, Femtosecond-laser-writing in various glasses, *J. Non-Cryst. Solids*, Vol. 345-346, 332-337, 2004.

58. V.R. Bhardwaj, E. Simova, P.B. Corkum, D.M. Rayner, C. Hnatovsky, R.S. Taylor, B. Schreder, M. Kluge, and J. Zimmer, Femtosecond laser-induced refractive index modification in multicomponent glasses, *J. Appl. Phys.*, Vol. 97, 083102-1–9, 2005.

59. J.W. Chan, T.R. Huser, S.H. Risbud, J.S. Hayden, and D.M. Krol, Waveguide fabrication in phosphate glasses using femtosecond laser pulses, *Appl. Phys. Lett.*, Vol. 82, 2371–2373 2003.

60. R. Osellame, S. Taccheo, G. Cerullo, M. Marangoni, D. Polli, R. Ramponi, P. Laporta, and S. De Silvestri, Optical gain in Er-Yb doped waveguides fabricated by femtosecond laser pulses, *Electron. Lett.*, Vol. 38, 964–965, 2002.

61. W. Watanabe, T. Asano, K. Yamada, K. Itoh, and J. Nishii, Wavelength division with three-dimensional couplers fabricated by filamentation of femtosecond laser pulses, *Opt. Lett.*, Vol. 28, 2491–2493, 2003.

62. C. Florea, and K.A.Winick, Fabrication and characterization of photonic devices directly written in glass using femtosecond laser pulses, *J. Lightwave Technol.*, Vol. 21, 246–253, 2003.

63. Y. Sikorski, A.A. Said, P. Bado, R. Maynard, C. Florea, and K.A. Winick, Optical waveguide amplifier in Nd-doped glass written with near-IR femtosecond laser pulses, *Electron. Lett.*, Vol. 36, 226–227, 2000.

64. S. Nolte, M. Will, J. Burghoff, and A. Tuennermann, Femtosecond waveguide writing: a new avenue to three-dimensional integrated optics, *Appl. Phys. A*, Vol. 77, 109–111, 2003.

65. T. Pertsch, U. Peschel, F. Lederer, J. Burghoff, M. Will, S. Nolte, and A. Tuennermann, Discrete diffraction in two-dimensional arrays of coupled waveguides in silica, *Opt. Lett.*, Vol. 29, 468–470, 2004.

66. R. Osellame, N. Chiodo, G. Della Valle, S. Taccheo, R. Ramponi, G. Cerullo, A. Killi, U. Morgner, M. Lederer, and D. Kopf, Optical waveguide writing with a diode-pumped femtosecond oscillator, *Opt. Lett.*, Vol. 29, 1900–1902, 2004.

67. S. Taccheo, G. Della Valle, R. Osellame, G. Cerullo, N. Chiodo, P. Laporta, O. Svelto, A. Killi, U. Morgner, M. Lederer, and D. Kopf, Er:Yb-doped waveguide laser fabricated by femtosecond laser pulses, *Opt. Lett.*, Vol. 29, 2626–2628, 2004.

68. G. Della Valle, R. Osellame, N. Chiodo, S. Taccheo, G. Cerullo, P. Laporta, A. Killi, U. Morgner, M. Lederer, and D. Kopf, C-band waveguide amplifier produced by femtosecond laser writing, *Opt. Express*, Vol. 13, 5976–5982, 2005.

69. R.S. Taylor, C. Hnatovsky, E. Simova, D.M. Rayner, M. Mehandale, V.R. Bhardwaj, and P.B. Corkum, Ultra-high resolution index of refraction profiles of femtosecond laser modified silica structures, *Opt. Express*, Vol. 11, 775–781, 2003.

70. P. Oberson, B. Gisin, B. Huttner, and N. Gisin, Refracted nearfield measurements of refractive index and geometry of silica-on-silicon integrated optical waveguides, *Appl. Opt.*, Vol. 37, 7268–7272, 1998.

10

Solid-State Laser Technology for Optical Frequency Metrology

Oliver D. Mücke, Lia Matos, and Franz X. Kärtner

CONTENTS

10.1 Femtosecond-Laser-Based Frequency Combs for Optical Frequency Metrology

In the past decade, remarkable progress has been achieved in ultrashort pulse generation directly from the laser due to the advent of double-chirped mirror (DCM) technology for precision dispersion compensation. In standard 100 MHz Ti:sapphire lasers, phase coherent locking of approximately 2 million of longitudinal modes can be achieved, which results in pulses as short as 5 fs at a center wavelength of 800 nm (see Figure 10.1). The quest for ever shorter pulse durations was fuelled by the application of these pulses in ultrafast time-domain spectroscopy. In particular, with intense laser pulses containing less than two optical cycles underneath the electric field envelope, it is now possible to explore the regime of extreme nonlinear optics [1], in which light–matter interaction is directly governed by the time evolution of the laser electric field $E(t)$, which depends on the carrier-envelope (CE) phase ϕ_{CE}, i.e., the phase between the rapidly oscillating carrier wave and the electric field envelope (see Figure 10.2a). Thus, the CE phase can influence the outcome of nonlinear optics experiments. The influence of the CE phase has recently been observed, e.g., in multipath quantum interference effects [2,3], carrier-wave Rabi flopping [4,5], photoemission from a metal surface [6], above-threshold ionization [7], high-harmonic generation [8], and nonsequential double ionization [9]. Isolated attosecond soft-x-ray pulses generated by high-harmonic generation using few-cycle driving pulses also pave the way for attosecond science [10], and might furthermore be used to seed next generation x-ray free-electron lasers [11].

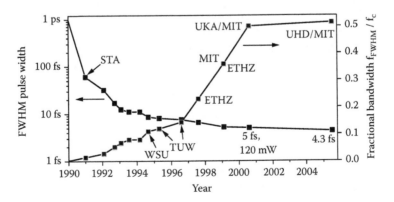

FIGURE 10.1
Evolution of the pulse duration and spectral bandwidth of ultrashort pulses from Ti:sapphire lasers achieved over the years. STA, University of St. Andrews; WSU, Washington State University; TUW, Technische Universität Wien; ETHZ, Eidgenössische Technische Hochschule Zürich; MIT, Massachusetts Institute of Technology; UKA, Universität Karlsruhe (TH); UHD, Universität Heidelberg.

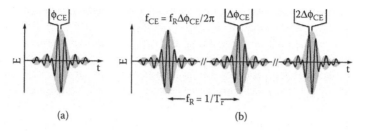

FIGURE 10.2

(a) Electric field vs. time (black line) of a single laser pulse containing only a few optical cycles of the carrier wave underneath the electric field envelope (i.e., inside the gray area). The carrier-envelope (CE) phase ϕ_{CE} measures the difference between the maximum of the electric field and the maximum of the field envelope. The pulse duration is defined as the full width at half maximum of the corresponding intensity profile. (b) Pulse train emitted by a femtosecond laser: The CE phase ϕ_{CE} changes from pulse to pulse by the amount $\Delta\phi_{CE}$ due to the difference between the group and phase velocities resulting from dispersive and nonlinear effects within the laser cavity. The CE frequency f_{CE} is smaller than the laser repetition frequency $f_R = 1/T_r$ with the cavity round-trip time T_r.

Today, as will be discussed in this chapter, frequency combs based on octave-spanning Ti:sapphire lasers also provide the basis for ultraprecise measurements of optical frequencies [12–14]. Modern high-precision spectroscopy allows the realization of standards for time, frequency, and length. In addition, these experiments enable precise tests of the fundamental laws of nature including quantum mechanics and general relativity, as well as the determination and search for possible time variations of fundamental constants [15,16]. In the past, optical frequency measurements required very large, expensive, inflexible, extremely sensitive and labor intensive experimental setups (so-called phase-coherent frequency chains), because no simple counting mechanism able to count optical frequencies (hundreds of THz) was known. These frequency chains were so labor intensive (see, e.g., Figure 1 in Reference 17) that only a few frequency chains have ever been implemented since their first demonstration [18] in 1973 [e.g., at the National Institute of Standards and Technology (NIST) in Boulder, CO, and the Physikalisch-Technische Bundesanstalt (PTB) in Braunschweig, Germany] [13].

Optical frequency metrology was revolutionized in 1999 by a new approach pioneered by the groups of Hänsch and Hall [19] at the Max Planck Institut für Quantenoptik (MPQ) in Garching, Germany, and at JILA in Boulder, CO. In the frequency domain, the pulse train emitted by a femtosecond mode-locked laser shown in Figure 10.2b corresponds to a frequency comb

$$\nu_m = mf_R + f_{CE}, \qquad (10.1)$$

which maps the two radio frequencies f_R and f_{CE} onto the optical frequencies ν_m via a large integer number $m = 10^5 - 10^6$. In Equation 10.1, f_R denotes the repetition frequency of the pulse train, f_{CE} is called carrier-envelope (CE) frequency. The femtosecond frequency comb thus allows one to establish a direct and rigid phase relationship between two microwave signals and the optical frequency comb of a mode-locked laser in a single step. By controlling the comb's two degrees of freedom f_R and f_{CE}, the optical frequency comb is fully controlled, and it can be used as a precise ruler in the frequency domain to measure optical frequencies by heterodyning an unknown optical frequency with the comb. Although the existence of the frequency comb was already recognized in early pioneering work on high-resolution two-photon spectroscopy using picosecond dye lasers [20–22], only the advent of broadband femtosecond mode-locked lasers opened the possibility to directly bridge THz frequency gaps and to fix the CE frequency f_{CE}, which needs to be known and stabilized to establish a stable frequency comb.

The experiments by the Hänsch group revealed that the frequency comb emitted by a mode-locked laser is remarkably equidistant. The comb uniformity (even after spectral broadening in a standard single-mode optical fiber) was confirmed at the level of 3×10^{-18} [23], and the comb line separation was found to agree with the laser repetition frequency within an accuracy of at least 6.0×10^{-16} [24]. These experiments guarantee the great potential of femtosecond frequency combs as clockworks for ultrahigh-precision frequency metrology.

The most straightforward approach to measure the CE frequency, the so-called ν-to-2ν self-referencing technique, necessitates a frequency comb with an octave bandwidth [25,26]. As can be seen from Equation 10.1, by heterodyning the frequency-doubled low-frequency components of the comb at frequency ν with the high-frequency components at frequency 2ν, the CE frequency f_{CE} can be determined. In 1999, the discovery of supercontinuum generation in a microstructure fiber by J. K. Ranka et al. [27] paved the way for the first implementation of the self-referenced optical frequency synthesizer [28,29]: the frequency comb of a 10–30-fs Ti:sapphire laser was first spectrally broadened in a microstructure fiber to cover one octave of bandwidth, and afterwards the CE frequency was detected and stabilized by ν-to-2ν self-referencing. With this self-referenced synthesizer based on a mode-locked laser, every spectroscopy laboratory is nowadays capable of measuring or synthesizing optical frequencies with unprecedented precision.

A particularly intriguing application of this technology are optical clocks [30–32]. To establish an optical clock, one needs to phase coherently derive the clock signal f_R from an optical frequency standard ν_S. A heterodyne beat signal f_b between the frequency standard ν_S and the frequency comb ν_m is used to establish a connection between ν_S and f_R, provided that f_{CE} is either independently stabilized or eliminated as an independent variable [30,31]. The stability and accuracy of frequency standards benefit from choosing as high transition frequencies as possible. The principal advantage of an atomic clock based on an optical transition over a microwave atomic clock is the much higher ($\sim 10^5$) operating frequency. This leads to a finer division of time and, thus, to a potentially higher stability. The frequency stability of an optical clock is most commonly characterized by the Allan deviation $\sigma_y(\tau)$, which gives the fractional frequency instability as a function of the averaging time τ (in seconds) [33,34]. It is anticipated that optical clocks can reach fractional frequency instabilities of about $10^{-16}\tau^{-1/2}$ [32–34], and Hertz-level accuracies have been reported [35].

Although microstructure fibers can easily broaden the femtosecond laser spectrum to bandwidths exceeding one octave, they have severe limitations with regard to long-term operation in optical clockworks as well as with homogeneous spectral coverage of the octave in the case of optical frequency synthesis. The CE phase noise added by amplitude-to-phase noise conversion in microstructure fibers has extensively been investigated [36,37], and broadband amplitude noise in the output light that might mask the heterodyne beats in the radio-frequency (RF) spectra may occur if too much broadening is necessary. For all these reasons, much experimental effort has been devoted to develop more reliable, more stable, and simpler optical clockworks without the need for spectral broadening in microstructure fibers.

In this chapter, we will discuss octave-spanning Ti:sapphire lasers that can be CE-phase stabilized by v-to-2v self-referencing without additional external broadening [38–42]. Meanwhile Ti:sapphire laser systems with repetition rates up to 1 GHz [43] were CE phase stabilized without the need for external broadening using the more complex 2v-to-3v self-referencing technique [44]. Alternatively, the CE frequency of few-cycle Ti:sapphire lasers can also be stabilized using the interference between spectral components generated by self-phase modulation (SPM) and second-harmonic generation (SHG) in thin ZnO crystals [45,46], and analogously, using the interference between SPM and difference-frequency generation (DFG) in a periodically-poled lithium niobate (PPLN) crystal (see Figure 10.3) [47].

Erbium fiber-laser-based frequency synthesizers also represent an attractive alternative for metrological applications with turnkey operation [48–50].

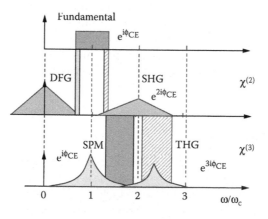

FIGURE 10.3
Origin of CE-phase sensitivity. The fundamental spectrum of an octave-spanning pulse centered at ω_c is shown at the top. Shown below are spectra generated by an instantaneous second- and third-order nonlinearity due to difference-frequency generation (DFG), second-harmonic generation (SHG), self-phase modulation (SPM), and third-harmonic generation (THG). The corresponding phases are indicated.

However, at present, the CE beat note of fiber lasers exhibits ~200 kHz linewidth in a 100 kHz resolution bandwidth [49], indicating increased CE phase fluctuations. The larger CE phase jitter might limit the usefulness of CE-phase stabilized fiber lasers for time-domain spectroscopy in the regime of extreme nonlinear optics. In optical frequency metrology, the larger CE beat linewidth implies decreased short-term stability and longer averaging times to obtain a desired stability.

This chapter is organized as follows: Section 10.2 reviews the laser dynamics and technology needed to achieve few-cycle laser pulses with octave-spanning spectra directly from a Ti:sapphire laser. In Section 10.3, the physical mechanism underlying CE phase control in prismless octave-spanning Ti:sapphire lasers via pump-power modulation is analyzed. Section 10.4 is devoted to a CE-phase-controlled 200 MHz octave-spanning Ti:sapphire laser. In section 10.5, we perform a complete noise analysis of the CE phase-lock loop and discuss its bandwidth limitations. Finally in Section 10.6, we will also discuss CE-phase independent optical clockworks based on sum/difference-frequency generation, which were recently employed as clockwork in a HeNe/CH_4 optical molecular clock [51,52].

10.2 Ultrabroadband Ti:Sapphire Lasers

The simplest and most robust approach to obtain a CE-phase stabilized train of pulses is to perform direct v-to-2v detection to the output of a laser oscillator. In this section, we will give a brief overview of the theory and technology needed to achieve octave-spanning spectra directly from the laser. We will start by briefly describing in Subsection 10.2.1 the pulse formation and dynamics in these lasers, followed by a discussion in Subsection 10.2.2 of the main technical challenges that must be overcome to enable the generation of ultrabroadband spectra. In Subsection 10.2.3, we will describe the experimental setup and results.

10.2.1 Laser Dynamics

Much insight has been obtained into the nonlinear dynamics of ultrashort pulse generation in general, and over the last thirty years a unified picture based on the master equation of mode locking has been developed. We will focus here on the description of Kerr-lens mode locking [53–56], which is the nonlinear mechanism that enables to generate pulses that are short enough to finally obtain octave-spanning spectra directly from the laser.

For passive mode locking to occur, a suitable saturable absorber mechanism is required. Depending on the ratio between saturable absorber recovery time and final pulse width, one may distinguish between different regimes of operation that result in different pulse formation processes. In the case of solid-state lasers, where the pulse energy is not large enough to saturate the gain medium in one single pass, gain saturation on the time scale of the pulse can be neglected. Therefore, a fast saturable absorber must be present which opens and closes the gain window immediately before and after the arrival of the pulse. This is the mode-locking principle known as fast saturable absorber mode locking [57]. Specifically, in Kerr-lens mode locking, the self-focusing which occurs in the gain medium due to the nonlinear Kerr effect, combined with a soft aperture created by the overlap of the laser beam with the gain profile creates a fast saturable absorption due to the ultrafast response of the Kerr effect, which is expected to be as fast as a few femtoseconds [58].

The dynamics of a laser mode locked with a fast saturable absorber in the limit of small changes per round trip is described by the master equation

$$T_r \frac{\partial}{\partial T} a = (g - \ell)a + \left(D_f + iD_2\right)\frac{\partial^2}{\partial t^2} a + (\gamma - i\delta)\,|\,a\,|^2\,a, \qquad (10.2)$$

where we have already factored out the carrier wave [59] and take into account the effects of gain g, loss l, finite gain or resonator bandwidth, fast saturable absorption, group-delay dispersion (GDD), and self-phase modulation (SPM) caused by the Kerr medium, which cannot be ignored when the pulse duration approaches the femtosecond regime. The complex amplitude $a \equiv a(T,t)$ is the slowly varying field envelope whose shape is investigated on two time scales: first, the global time T which is coarse-grained on the time scale of the round-trip time T_r, and second, the local time t which resolves the resulting pulse shape. $a(T,t)$ is normalized such that $|a(T,t)|^2$ is the instantaneous power and

$$W(T) = \int dt \,|a(T,t)|^2$$

the pulse energy at time T. D_f is the effective gain curvature, and the associated filtering action is represented by $D_f \frac{\partial^2}{\partial t^2} a$.

$$D_2 = \frac{d^2k}{df_c^2}\frac{L}{8\pi^2}$$

is the group-velocity dispersion (GVD) parameter for the cavity, i.e., the average dispersion per round trip in the cavity composed of the gain medium, the air path, and eventually including dispersion generated by reflections from dispersion compensating mirrors. γ is the self-amplitude modulation (SAM) coefficient. The Kerr coefficient is $\delta = (4\pi/\lambda_c)n_2L/A_{eff}$, where λ_c is the carrier wavelength, n_2 the nonlinear index, and A_{eff} the effective mode cross-sectional area. To simplify the dynamics, the gain is taken to follow adiabatically the intracavity pulse energy $g = g_0/(1 + W/W_{sat})$, with small-signal gain g_0 and saturation energy of the gain medium W_{sat}, as applicable for a gain medium with a relatively short relaxation time (which is the case for Ti:sapphire lasers). The bandwidth is assumed to be limited by a filter of bandwidth Ω_f resulting in $D_f = 1/\Omega_f^2$. Equation 10.2 has a simple steady-state solution [60]

$$a(t) = A_0 \mathrm{sech}^{(1-i\beta)}(t/\tau), \tag{10.3}$$

where β is the chirp parameter. Equation 10.3 describes the pulse evolution in a medium with continuously distributed dispersion and SPM [61].

However, as first pointed out by C. Spielmann et al. [62], for lasers generating pulses as short as 10 fs and below, large changes in the pulse width occur within one round trip through the resonator because the dispersive elements, positive and negative, as well as the SPM section are placed discretely, and the ordering of the elements of the resonator has a large impact on the resulting pulse width. This effect was first observed in a fiber laser and called stretched-pulse mode locking [63].

The spectral bandwidths of the octave-spanning Ti:sapphire lasers indicate that the pulse propagating in the cavity is as short as 5–6 fs. Therefore, we must consider the impact of the discrete action of dispersion and nonlinearity in the laser on the spectral shape of such laser pulses.

A mode-locked laser consists of a gain medium (i.e., the Ti:sapphire crystal) and dispersion balancing components, in our case DCMs. The system can be decomposed into the linear resonator arms and the nonlinear gain crystal, (Figure 10.4). To achieve ultrashort pulses, the dispersion-balancing components should produce close to zero net dispersion, while the dispersive elements individually produce significant group delay (GD) over the broad bandwidth of the laser pulse. A system with sufficient variation of dispersion can support nonlinear Bloch waves [64]. One can show that the Kerr nonlinearity produces a self-consistent nonlinear scattering potential that permits formation of a periodic solution with a simple phase factor in a system with zero net dispersion. It has been shown that nonlinear propagation along a dispersion-managed fiber near zero net GDD possesses a narrower spectrum in the segment of positive dispersion than in the segment of negative dispersion [64,65]. Thus, the effect of negative dispersion is greater than that of the positive dispersion and imparts to the pulse an effective net negative

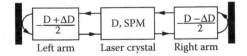

FIGURE 10.4
In analogy with dispersion-managed fiber transmission links, this schematic represents the order of the dispersive and nonlinear components in the resonator.

dispersion. This effective negative dispersion can balance the Kerr-induced phase, leading to steady-state pulses at zero net dispersion. This is true even if there is no nonlinearity in the negative dispersion segment. The pulses are analogous to solitons in that they are self-consistent solutions of the Hamiltonian (lossless) problem, but they are not secant hyperbolic in shape like the solution of Equation 10.2. Figure 10.5 shows a numerical simulation of such a self-consistent solution evolving through the resonator over one round trip.

The steady-state pulse formation can be understood in the following way. By symmetry the pulses are chirp-free in the middle of the dispersion cells. A chirp-free pulse starting in the center of the gain crystal, i.e., nonlinear segment, is spectrally broadened by the SPM and disperses in time due to the GVD, which generates a mostly linear chirp over the pulse. After the pulse leaves the crystal it experiences GVD in the arms of the laser resonator, which compresses the positively chirped pulse to its transform limit at the end of each arm, where an output coupler can be placed. Back propagation towards the crystal imposes a negative chirp, generating the time-reversed solution of the corresponding nonlinear Schrödinger equation (NLSE). Therefore, subsequent propagation in the nonlinear crystal compresses the pulse spectrally and temporally to its initial shape in the center of the crystal. The spectrum is narrower in the crystal than in the negative-dispersion sections, because it is negatively prechirped before it enters the SPM section, and spectral spreading occurs again only after the pulse has been compressed.

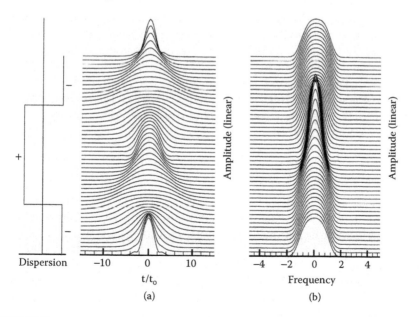

FIGURE 10.5
Numerical simulation of pulse shaping over one round trip in the resonator. (a) and (b) represent the pulse shape in the time and frequency domain, respectively. The dispersion map is shown on the left-hand side. SPM occurs only in the laser crystal, i.e., in the positive dispersion section.

In a laser with a linear cavity, for which the negative dispersion is equally distributed in both arms of the resonator, which is the case for the octave-spanning lasers described here, the pulse runs through the dispersion map twice per round trip. The pulse is short at each end of the cavity and, most importantly, the pulses are identical in both passes through the crystal, which exploits the KLM action twice per round trip [66], in contrast to an asymmetric dispersion distribution in the resonator arms, as is the case for lasers with intracavity prisms for dispersion compensation. Thus, a symmetric dispersion distribution may lead to an effective saturable absorption that is twice as strong as an asymmetric dispersion distribution, which results in substantially shorter pulses.

For gaining insight into the laser dynamics, a master equation approach describing the average pulse dynamics has been developed [61,67]. Because of the breathing of the pulse, the Kerr phase shift is produced by a pulse of varying amplitude and width as it circulates around the ring. The Kerr phase shift for a pulse of constant width, $\delta |a|^2$, has to be replaced by a phase profile that mimics the average shape of the pulse, weighted by its intensity. Therefore, the SPM action of Equation 10.2 is replaced by

$$\delta |a|^2 = \delta_0 |A_0|^2 \left(1 - \mu t^2 / \tau^2\right), \tag{10.4}$$

where A_0 is the pulse amplitude at the position of minimum width. Here, the Kerr phase profile is expanded to second order in t. The coefficient δ_0 and μ are evaluated variationally, the SAM action is similarly expanded. Finally, the net intracavity dispersion acting on average on the pulse is replaced by the effective dispersion D_{net} in the resonator within one round trip. Note that the effective dispersion D_{net} is different from the average linear dispersion in the cavity, because the pulse spectrum is less wide in the positive dispersion region than in the negative dispersion region, generating an effective net negative dispersion even at zero average linear dispersion. The master equation becomes

$$T_r \frac{\partial}{\partial T} a = (g - \ell)a + \left(D_f + iD_{net}\right)\frac{\partial^2}{\partial t^2} a + (\gamma_0 - i\delta_0)|A_0|^2 \left(1 - \mu t^2 / \tau^2\right)a. \tag{10.5}$$

This equation has Gaussian pulse solutions. Because the approximations made in arriving at Equation 10.5 are not applicable to the wings of the pulse, the wings are not in fact Gaussian. However, in few-cycle lasers, the wings are not very well described by the simple parabolic gain profile either, rather the mirrors and output couplers are what form a hard spectral filter for the pulse. As we will see in this case, the saturable absorber action is strong enough to stabilize generation of spectrum beyond the high reflectivity bandwidth of the output coupler and the output spectrum shows strong wings due to the enhanced output coupling. This will become evident when we show the experimental results in Subsection 10.2.3. The master equation 10.5

is a patchwork; it is not an ordinary differential equation. The coefficients in the equation depend on the pulse solution and eventually have to be found iteratively. Nevertheless, the equation accounts for the pulse shaping in the system in an analytic fashion. The pulse is shaped by the interplay between GVD and SPM. However, this pulse is not stable in the presence of gain filtering or a finite cavity bandwidth. Stabilization is achieved by saturable absorber action due to KLM that favors the pulse and suppresses background radiation that can benefit from the peak of the gain.

10.2.2 Technical Challenges in Generating Octave-Spanning Spectra

To make use of the full potential of the dispersion-managed soliton pulse formation, several technical challenges need to be overcome. We will highlight here the three most important aspects to be considered in the design of such ultrabroadband lasers.

The first important aspect is to design the laser cavity in such a way as to maximize the artificial saturable absorber action in the laser crystal. As mentioned earlier, fast saturable absorption is obtained by combining the self-focusing due to the Kerr effect in the crystal with a soft aperture caused by the proper selection of the pump mode size in the crystal. In this case, the change in the beam waist enhances the overlap between the laser mode and the pump mode, and therefore, enhances the gain of the self-focused laser mode. As a consequence, because the pulse peak sees higher gain/lower losses than the wings, the pulse shortens in each round trip. The response time of the nonresonant Kerr effect is on the order of a few femtoseconds, allowing the nonlinear index of refraction to follow the pulse almost instantaneously, enhancing the KLM action.

The proper resonator design for optimum KLM is a complex, time-consuming problem. In order to map out the complete parameter range of the laser system, a complete spatiotemporal evolution of the pulse needs to be done numerically, until steady state is reached, which is a challenging task even with today's computers. However, a resonator design based on ABCD matrices or q-parameter analysis, where the KLM is modeled by an intensity-dependent lens, already gives a good starting point for the laser construction [68–71]. The final optimization is then done by the experimentalist by varying the cavity parameters in order to achieve the strongest KLM, i.e., broadest spectrum.

The second essential element in the generation of octave-spanning spectra is a mirror technology that supports an octave-spanning spectrum with high reflectivity and custom-designed dispersion properties. For this purpose we invented the so-called double-chirped mirror pairs (see Figure 10.6 and Figure 10.7). DCMs have been developed to enable precise dispersion control and high reflectivity simultaneously over a fractional bandwidth of as much as $\Delta f/f_c = 0.4$ [72,73], where f_c is the center frequency of the pulse. DCMs work on the principle of chirped mirrors, where the Bragg wavelength of the mirror pairs is continuously chirped to generate a wavelength-dependent

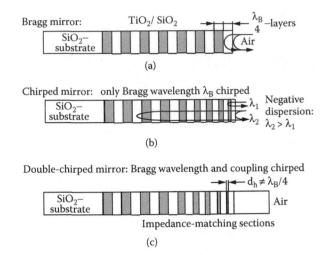

(a) Standard Bragg mirror, (b) simple chirped mirror, (c) double-chirped mirror (DCM) with matching sections to avoid residual reflections causing undesired oscillations in the GD and GDD of the mirror.

FIGURE 10.6
(a) Standard Bragg mirror, (b) simple chirped mirror, (c) double-chirped mirror (DCM) with matching sections to avoid residual reflections causing undesired oscillations in the GD and GDD of the mirror.

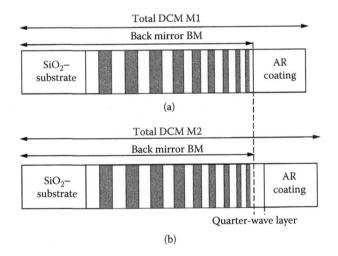

FIGURE 10.7
DCM pair consisting of DCMs M1 (a) and M2 (b). M1 can be decomposed into a double-chirped back-mirror BM matched to a medium with the index of the topmost layer. In M2, a layer with a quarter-wave thickness at the center frequency of the mirror and an index equivalent to the topmost layer of BM is inserted between the back mirror and the AR coating. The new mirror containing BM and the quarter-wave layer can be reoptimized to achieve the same phase as M1 with an additional phase shift over the whole octave of bandwidth.

penetration depth into the mirror. However, chirping of the Bragg wavelength is not enough to accomplish this. In addition to the Bragg wavelength, impedance matching has to be ensured such that the waves do not encounter spurious reflections before the turning point in the mirror is reached. Hence, the Bragg wavelength and the impedance matching has to be tuned over the operational wavelength of the mirror, which is emphasized by the name double-chirped mirror [73]. Most importantly, to avoid reflections at the surface of the mirror to the air, a high quality antireflection (AR) coating is necessary. To avoid spurious reflections during propagation of the grating structure, the grating is switched on adiabatically by increasing the thickness of the high index layer continuously from small values to a quarter-wave thickness (Figure 10.6c). However, to extend this concept to cover one full octave, i.e., $\Delta f / f_c = 0.66$, the requirements on the quality of the AR coating of such mirrors become impossible to realize and even after reoptimization of the mirror, a small amount of impedance mismatch is still present leading to undesired large oscillations in the GDD.

The high reflectivity range of the back mirror can easily be extended to one octave, simply by chirping slow enough and having a sufficient number of layer pairs. However, the smoothness of the resulting GDD strongly depends on the quality of matching provided by the AR coating.

The reflections occurring at the AR coating, similar to those in Gires–Tournois interferometers (GTIs), add up coherently when multiple reflections on chirped mirrors occur inside the laser over one round trip, leading to pre- and postpulses if the mode-locking mechanism is not strong enough to suppress them sufficiently. Experimental results indicate that a residual reflection in the AR coating of $r < 0.01$ and smaller, depending on the number of reflections per round trip, is required so that the pre- and postpulses are sufficiently suppressed. This corresponds to an AR coating with less than 10^{-4} residual power reflectivity, which, for a bandwidth approaching one octave, is no longer possible [74]. A way out of this limitation is offered by the observation that a coherent subtraction of the pre- and postpulses to the first order in r is possible by reflections on a mirror pair M1 and M2 (Figure 10.7). A series of two reflections on mirror M1 and on a similar mirror M2 with an additional phase shift of π between the AR coating and the back mirror leads to coherent subtraction of the first-order GTI effects.

Figure 10.8 shows in the top graph, the designed reflectivity of both mirrors of the pair in high resolution, taking into account the absorption in the layers. The graph below shows the reflectivity of the mirror, which has in addition high transmission between 510–550 nm for pumping of the Ti:sapphire crystal. Each mirror consists of 40 layer pairs of SiO_2 and TiO_2 fabricated using ion-beam sputtering [75,76]. Both mirror reflectivities cover more than one octave of bandwidth from 580 to 1200 nm or 250 to 517 THz, with an average reflectivity of about 99.9% including the absorption in the layers. In addition, the mirror dispersion corrects for the second- and higher-order dispersion of all intracavity elements. The choice for the lower wavelength boundary in dispersion compensation is determined and limited by the pump window of Ti:sapphire. The oscillations in the group delay of each mirror are about 10 times larger than those of high-quality DCMs

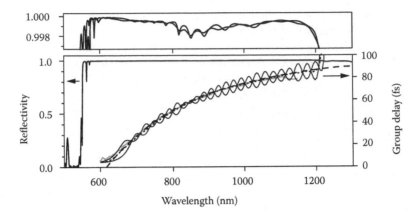

FIGURE 10.8
Reflectivity (left scale) of the type I DCMs shown as thick solid line. The group delay design is given by the thin dashed line. The individual group delays (right scale) of type I and II DCMs are shown as thin lines and its average as a thick dotted line, which is almost identical to the design goal over the wavelength range of interest from 650-1200 nm. The group delay, measured using white light interferometry, is shown as the thick gray line from 600-1100 nm. Beyond 1100 nm the sensitivity of the Si detector limited the measurement.

covering 350 nm of bandwidth [66]. However, in the average group delay of both mirrors, the oscillations are ideally suppressed due to cancellation by more than a factor of ten. Therefore, the effective residual reflectivity of the mirror pair covering one octave, r^2, is even smaller than that of conventional DCMs. Because of slight fabrication errors the oscillations in the group delay still do not precisely cancel. Close to 900 nm and 1000 nm especially, deviations from the design goal of the order of 1–2 fs occur, which will lead to observable spectral features in the spectral output of the lasers described in Subsection 10.2.3.

The last key element in the laser design is the output coupling mirror. In order for an ultrashort pulse to build up in the laser cavity, the output coupler should have a large enough reflectivity bandwidth so as not to impose significant spectral filtering on the pulse. However, to perform direct v-to-2v detection to the laser, the wings of the spectrum need to be enhanced by increased output coupling relative to the center of the spectrum. Therefore, the reflectivity curve of the output coupler has to be chosen to balance these two effects. To illustrate the importance of this effect, we have measured the intracavity and external spectrum as well as output coupler transmission curves for two different output couplers OC1 and OC2 (mirrors with 1% transmission over the center wavelength range and different bandwidth), which can be seen in Figure 10.9. In the case of OC1, the output coupler is a strong spectral filter in the laser, which limits the spectrum that can be generated to a bandwidth similar to that of the transmission curve. Although there is some enhancement of the spectral wings, it is not sufficient to achieve enough power at the v and 2v points. In contrast, in the case of OC2 the spectral filter is relatively weak which allows a broader spectrum to be generated inside the cavity, and properly enhanced to achieve significant power at the spectral wings.

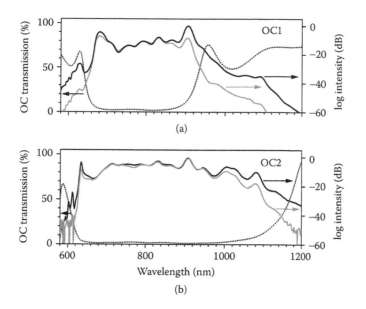

FIGURE 10.9

Output spectra (black curves) and intracavity spectra (gray curves) generated by a 80 MHz Ti:sapphire laser using (a) a broadband output coupler OC1, and (b) an ultrabroadband output coupler OC2. The output coupler transmission (black dotted curves) are shown for comparison.

Unfortunately, engineering a broadband reflectivity OC with no penalties to the GDD is not an easy task. The best results have been obtained so far by employing a simple high-index contrast quarter-wave dielectric stack made out of $ZnSe/MgF_2$ layers with 1% transmission at the center wavelength. The bandwidth is related to the index contrast of the high- and low-index layers. With such structure, we have designed output couplers that produce 1% transmission from 650 nm to 1100 nm, as can be seen in Figure 10.11. The experimentally observed spectra (Figure 10.11) show the generation of octave-spanning spectra directly from the laser using these OCs.

10.2.3 Experimental Setup and Results

The octave-spanning lasers demonstrated here consist of astigmatism-compensated cavities (Figure 10.10). They have a 2-mm long Ti:sapphire crystal with an absorption of $\alpha = 7$ cm^{-1} at 532 nm and are pumped by diode-pumped, frequency-doubled $Nd:YVO_4$ lasers, either Millennia X*s* by Spectra-Physics or Verdi-V6 by Coherent. In the 80-MHz repetition rate version, the radius of curvature (ROC) of the folding mirrors is 10 cm, and the pump lens has a 60 mm focal length; in the 200 MHz version, the corresponding values are 7.5 cm and 50 mm, respectively. All mirrors in the cavity, except for the end mirrors, are type I (gray) and type II (black) DCMs that generate smooth group-delay dispersion when used together in pairs, as shown in Figure 10.8. One cavity-end mirror is a silver mirror, the other one is the broadband

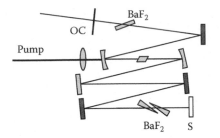

FIGURE 10.10
Scheme of the octave-spanning prismless Ti:sapphire laser. The Z-folded cavity is astigmatically compensated. Twelve bounces on the DCM pairs [DCM type I (gray) and type II (black)] provide smooth and broadband compensation of the dispersion of the laser crystal, the BaF_2 plate and wedges, and the air within the laser cavity. The BaF_2 wedges are used for dispersion fine-tuning. S, silver end mirror; OC, output coupling mirror.

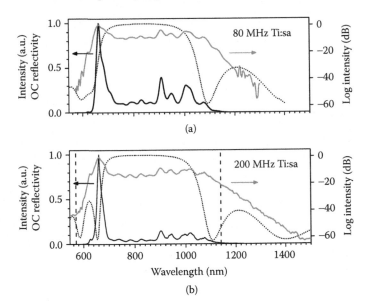

FIGURE 10.11
(a) Output spectrum of the 80 MHz Ti:sapphire laser on a linear (black curve) and on a logarithmic scale (gray curve). The reflectivity of the $ZnSe/MgF_2$ output coupler (dotted curve) is shown for comparison. (b) Same for the 200 MHz Ti:sapphire laser. The wavelengths 570 and 1140 nm used for v-to-2v self-referencing are indicated by two dashed lines. The Fourier limit of the pulse spectrum is 3.6 fs.

$ZnSe/MgF_2$ output coupler (OC) with 1% transmission in the center. In one round trip of the laser pulse through the cavity, the 12 bounces on DCMs generate the precise negative dispersion required to compensate for the positive second- and up to sixth-order dispersion caused by the laser crystal, the air path in the cavity, and the BaF_2 plate and wedges used to fine-tune the dispersion. We used BaF_2 for dispersion compensation because it has the lowest ratio of third- to second-order dispersion in the wavelength range from 600–1200 nm and the slope of the dispersion of BaF_2 is nearly identical to that of air. This allows us to scale the cavity length and repetition rate without changing the overall intracavity dispersion. To achieve mode-locked operation, it is usually necessary to reduce the amount of BaF_2 inside the laser cavity (i.e., by withdrawing one of the wedges). The broadest spectrum can be achieved by optimizing the insertion of the BaF_2 wedge. The spectral width of the laser depends critically on the dispersion balance, but with the prism-less lasers adjusting the dispersion does not significantly change the cavity alignment. In contrast to prism-compensated cavities, a slight misalignment of the resonator does not affect intracavity dispersion, and therefore, it is possible to operate the laser in octave-spanning mode for as much as a full day without interruption. Figure 10.11 shows the spectrum under broadband operation of a 80 MHz and a 200 MHz laser [41,42], the corresponding average output powers are ~120 mW and ~270 mW, respectively. The octave is reached at a spectral density about 25 dB below the average power level. The same plot also shows the OC reflectivity curve. As already described, the detailed shape of this reflectivity curve is also a determining factor for the width of the output spectrum, because it significantly enhances the spectral wings and simultaneously allows for spectral buildup inside the cavity.

Because the dispersion of 0.5 mm of BaF_2 is similar to that of 1 m of air, the cavity can be scaled up to GHz repetition rates by removing air path and correspondingly adding BaF_2 to maintain the proper dispersion balancing. This is not possible in lasers with intracavity prisms because a minimum distance between the prisms is required to provide negative dispersion.

10.3 Carrier-Envelope Phase Control

Active control of the CE frequency of mode-locked lasers is a prerequisite for many applications in both time and frequency domain. For octave-spanning lasers which do not use intracavity prisms for dispersion compensation [41,42], this control can be done by utilizing the response of f_{CE} to intracavity power, i.e., by controlling the intracavity pulse energy via modulation of the pump power.

Detailed studies of the intensity-related f_{CE} dynamics have been performed in Ti:sapphire mode-locked lasers which employ external fiber broadening for the generation of octave-spanning spectra and f_{CE} control [77]. A connection between the dependence of the CE frequency, the laser repetition rate and the center frequency of the pulse spectrum on intracavity pulse energy was found. The experiments showed that the CE phase control coefficient $C_{fPp} = \Delta f_{CE} / \Delta P_p$, where ΔP_p is a change in pump power, could change signs, consistent with the shifting of the spectrum with pump power also changing signs. In order to obtain optimum conditions for f_{CE} control, the dependence of C_{fP_p} on intracavity power, and therefore, on pump power is of prime importance.

In octave-spanning lasers the CE phase dynamics is much simpler [78]. This is due to the fact that the center frequency of the pulse has no observable change with pump power, because the pulse spectrum completely fills up the available bandwidth and is in fact limited by the bandwidth of the output couplers available. Then a change of the CE frequency due to changes of the center frequency of the pulse is absent; the CE frequency responds only to changes in the intracavity pulse energy and the concomitant changes in phase and group velocity related to the nonlinear refractive index. This response has a simple linear behavior, provided care is taken to work in a power region where the laser is operating in a unique single pulse regime, i.e., no continuous-wave (cw) component or multiple pulses are present.

In this section, we summarize the linear and nonlinear effects in the laser cavity that may lead to a CE phase shift per round trip $\Delta\phi_{CE}$ and, therefore, contribute to the CE frequency via

$$f_{CE} = \frac{\Delta\phi_{CE}}{2\pi T_r}. \tag{10.6}$$

If we assume that the laser operates at carrier frequency f_c, then the complex carrier wave of the pulse is given by

$$e^{i2\pi f_c(t-z/v_p)}, \tag{10.7}$$

where v_p is the phase velocity of the carrier wave in the cavity. In the absence of nonlinearities, the phase velocity is simply the ratio between frequency and wavenumber due to the linear refractive index of the media in the cavity, i.e., $v_p = v_p(f_c) = 2\pi f_c/k(f_c)$. The envelope of a pulse that builds up in the cavity due to the mode-locking process will travel at the group velocity due to the presence of the linear media given by $v_g = v_g(f_c) = 2\pi[dk(f_c)/df_c]^{-1}$. Therefore, after one round trip of the pulse over a distance $2L$, which takes the time $T_r = 2L/v_g$, we obtain from Equation 10.7 that the linear contribution to the CE phase shift caused by the difference between phase and group velocities is

$$\Delta\phi_{CE} = 2\pi f_c\left(1 - \frac{v_g(f_c)}{v_p(f_c)}\right)T_r \tag{10.8}$$

and for the subsequent CE frequency, we obtain

$$f_{CE} = f_c\left(1 - \frac{v_g(f_c)}{v_p(f_c)}\right). \tag{10.9}$$

In a dispersive medium, group and phase velocities depend on the carrier frequency. Therefore, if the carrier frequency shifts as a function of the intracavity pulse energy, the linear CE frequency becomes energy- and pump power-dependent as found in Reference 77.

In a mode-locked laser there are also nonlinear processes at work that may directly lead to an energy-dependent CE frequency. There are many effects that may contribute to such a shift. Here we rederive briefly the effects due to the intensity-dependent refractive index as discussed by Haus and Ippen [79] for the case of a laser with strong soliton-like pulse shaping, which can be evaluated analytically using soliton perturbation theory. We then argue that the same analysis holds for the general case where steady-state pulse formation is different from conventional soliton pulse shaping.

We start our analysis with a master equation of the form shown in Equation 10.2. Strictly speaking, as already discussed in section 10.2.1, Equation 10.2 applies only to a laser with small changes in pulse shape within one round trip. Obviously this is not the case for few-cycle laser pulses where the pulse formation is governed by dispersion-managed mode locking [64]. Nevertheless we want to understand this propagation equation as an effective equation of motion for the laser, where some of the parameters need to be determined self-consistently [59].

Let us assume that the laser operates in the negative GVD regime, where a conventional soliton-like pulse forms, and that it is stabilized by the effective saturable absorber action against the filtering effects. Then the steady-state pulse solution is close to a fundamental soliton, i.e., a symmetric sech-shaped pulse that acquires an energy-dependent nonlinear phase shift per round trip due to the nonlinear index

$$a(T,t) = A_0 \text{sech}\left(t / \tau\right) e^{-i\phi_s T / T_r}, \tag{10.10}$$

see Reference 79. The nonlinear soliton phase shift per round trip is

$$\phi_s = \frac{1}{2}\delta A_0^2. \tag{10.11}$$

A more careful treatment of the influence of the Kerr effect on the pulse propagation, especially for few-cycle pulses, needs to take the self-steepening of the pulse into account, i.e., the variation of the index during an optical cycle, by adding to the master equation (Equation 10.2) the term [80]

$$L_{\text{pert}} = -\frac{\delta}{\omega_c}\frac{\partial}{\partial t}\left(|a(T,t)|^2 a(T,t)\right). \tag{10.12}$$

We emphasize that this term is a consequence of the Kerr effect and is not related to soliton propagation. It can be viewed as a perturbation to the master equation (Equation 10.2). For pulses with τ much longer than an optical cycle, this self-steepening term is unimportant in pulse shaping, because it is on the order of $1/\omega_c\tau \ll 1$. However, this term is always of importance when the phase shifts acquired by the pulse during propagation are considered. Haus and Ippen found, by using soliton perturbation theory based on the eigensolutions of the unperturbed linearized Schrödinger equation, analytic expressions for the changes in phase and group velocity. The nonlinear phase shift per round trip of the soliton adds an additional phase shift to the pulse in each round trip.

If the term in Equation 10.12 is applied to a real and symmetric waveform, it generates an odd waveform. An odd waveform added as a perturbation to the symmetric waveform of the steady-state pulse leads, to first order, to a temporal shift of the steady-state pulse. For a soliton-like steady-state solution this timing shift can be evaluated with soliton perturbation theory, i.e., using the basis functions of the linearized operator, which results in a timing shift [79,81]

$$T_r \left.\frac{\partial \Delta t(T)}{\partial T}\right|_{\text{self-steep}} = \Delta\left(\frac{1}{v_g}\right) = \frac{\delta}{\omega_c} A_0^2 = \frac{2\phi_s}{\omega_c}. \tag{10.13}$$

In total, the compound effect of self-phase modulation, self-steepening, and linear dispersion on the pulse results in a CE frequency of

$$f_{CE} = \frac{f_R}{2\pi}\Delta\phi_{CE} = -f_R\frac{\phi_s}{2\pi} + f_c\left.\frac{\partial}{\partial T}\Delta t(T)\right|_{\text{self-steep}} + f_c\left(1 - \frac{v_g(f_c)}{v_p(f_c)}\right)$$

$$= -\frac{f_R}{4\pi}\delta A_0^2 + 2\frac{f_R}{4\pi}\delta A_0^2 + f_c\left(1 - \frac{v_g(f_c)}{v_p(f_c)}\right). \tag{10.14}$$

As the above expression shows, the term arising from the group delay change due to self-steepening is twice as large and of opposite sign compared with the one due to self-phase modulation. In total we obtain

$$f_{CE} = \frac{f_R}{4\pi}\delta A_0^2 + f_c\left(1 - \frac{v_g(f_c)}{v_p(f_c)}\right). \tag{10.15}$$

We emphasize that soliton perturbation theory was used in this derivation only for analytical evaluation of timing shifts. If the pulse shaping in the laser is not governed by conventional soliton formation but rather by dispersion-managed soliton dynamics [64] or a saturable absorber, the fundamental physics stays the same. If the steady-state solution has a real and symmetric component, the self-steepening term converts this component via the derivative into a real and odd term, which is to first order a timing shift in the autonomous dynamics of the free running mode-locked laser. Another mechanism that leads to a timing shift is, for example, the action of a slow saturable absorber, which absorbs only the front of the pulse. So care needs to be taken to include all relevant effects when a given laser system is analyzed.

The derivation above shows that the group velocity change due to self-steepening of the pulse leads to a change in sign of the energy-dependent contribution at fixed center wavelength of the pulse. We checked this prediction by observing the CE frequency shift in a 200 MHz repetition rate octave-spanning Ti:sapphire laser, which is discussed in detail in Section 10.4. The identification, which of the peaks in the RF spectrum corresponds to the CE frequency, can be done by inserting BaF_2 material in the laser and observing which peak moves up in frequency (adding dispersion causes v_g/v_p to decrease, thus increasing the magnitude of the second term in Equation 10.15). Variation of the pump power and observation that the same peak also moves up in frequency confirms the predictions of Equation 10.15. Figure 10.12a shows the CE frequency shift as a function of pump power. As Equation 10.15 predicts, the CE frequency shift follows linearly the soliton phase shift ϕ_s and therefore the pump power over the range where the intracavity laser power (or pulse energy) depends linearly on the pump power. This is the case as long the laser operates in a unique single pulse regime. For higher pump powers, cw background radiation breaks through and the theory on which Equation 10.15 is based no longer holds because the intracavity power is now divided between two components, the pulse and the cw solution (Figure 10.12b). The observed turning point is not inherent to the pulse dynamics itself but is due to the appearance of a cw component in the spectrum, as is easy to see from Figure 10.12. After the cw component is present, any increase in pump power enhances the cw component and most likely decreases the pulse energy, causing f_{CE} to eventually shift in the opposite direction. This is a consequence of the fact that the Kerr lens mode-locking (KLM) action does not increase indefinitely, i.e., there is an upper value for the pulse energy above which a further increase in pump

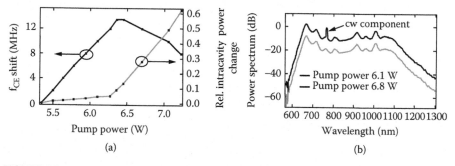

FIGURE 10.12

(a) CE frequency shift (black) and relative intracavity power change (gray) as function of pump power change. Both curves make evident the presence of a cw breakthrough for pump powers above 6.3 W, which is confirmed by the appearance of a cw component in the optical spectrum shown in (b). For clarity, the spectra are vertically offset by 10 dB.

power will either contribute to cw breakthrough or to multiple pulses. There-fore, care must be taken to operate at an optimum pump power level which is significantly below this threshold value for single pulse instabilities. From the data shown in Figure 10.12, the pump power to CE frequency conversion coefficient for the 200 MHz lasers is $C_{fP_p} = 11$ MHz/W. Figure 12a also shows the relative change of the intracavity pulse energy as a function of the same variation in pump power. The appearance of a cw component is also explic-itly indicated in this measurement by the abrupt change in the observed slope. The shallow slope of the change in the average power in pulsed operation as compared to the change in power in cw operation is an indica-tion of the strength of the saturable absorption and the bandwidth limitation of the laser. From this data, we can infer a relatively weak response in the change of the pulse energy in mode-locked operation and therefore a corre-spondingly weak response in the f_{CE} change, which in fact will be confirmed in the transfer function analysis discussed in Subsection 10.5.1.

Now, we can compare quantitatively the measured shift of f_{CE} in Figure 10.12a, with the theoretically derived result from soliton perturbation theory. It turns out that the measurement and theory agree very well, despite the fact that the laser dynamics differs from the ideal conventional soliton oper-ation regime. The conversion coefficient

$$C_{fP_{\text{intra}}} = \Delta f_{CE} \, / \, \Delta P_{\text{intra}}, \tag{10.16}$$

where ΔP_{intra} is the intracavity power in mode-locked operation, can now be determined in terms of known cavity parameters under the assumption of a fixed pulse width

$$\frac{\Delta f_{CE}}{\Delta P_{\text{intra}}} = \frac{Ln_2}{4\lambda_c A_{\text{eff}} \tau}, \tag{10.17}$$

where $L = 4$ mm is the path length per round trip through the Ti:sapphire crystal, $n_2 = 3 \times 10^{-20}$ m^2/W is the nonlinear index of refraction for Ti:sap-phire, $\lambda_c = 800$ nm is the carrier wavelength, $A_{\text{eff}} = \pi w_0^2$ ($w_0 = 16\,\mu$m) is the mode cross-sectional area, $\tau = \tau_{\text{FWHM}}/1.76$ with a pulse width of $\tau_{\text{FWHM}} = 5$ fs, and $P_{\text{intra}} = 12$ W is the intracavity power. This expression gives, for a 5% change in intracavity power, a corresponding change in f_{CE} of 9.6 MHz, which agrees well with the results shown in Figure 10.12a.

Despite this surprisingly good agreement, one has to be aware that the spot size and other parameters are rough estimates, which may easily change depending on cavity alignment. Also the pulse width is not constant in the crystal but rather stretching and compressing by more than a factor of two. Nevertheless, the experimental observations in Figure 10.12 agree well with the above theoretical estimate obtained from conventional soliton propagation.

10.4 Carrier-Envelope Phase-Controlled 200 MHz Octave-Spanning Ti:Sapphire Lasers

Scaling the repetition rate of Ti:sapphire lasers to higher values brings along many advantages for optical frequency metrology, ultrafast time-domain spectroscopy, and other applications. In optical frequency metrology, higher repetition rate lasers yield a larger power per comb line leading to larger signal-to-noise ratios (SNRs) of the measured heterodyning beat signals. Moreover, for repetition rates above ~150 MHz, the individual comb lines can conveniently be resolved using commercial wavemeters. In ultrafast time-domain spectroscopy, higher repetition rate lasers enable shorter data acquisition times and improved SNRs. In this section, we discuss CE phase stabilization results for a 200 MHz octave-spanning Ti:sapphire laser as described in Subsection 10.2.3.

The 200 MHz octave-spanning Ti:sapphire laser (Figure 10.13) is pumped by focusing ~6.5 W (measured in front of the acousto-optic modulator (AOM)) of 532 nm light emitted by a Coherent Verdi-V6 pump laser into the gain crystal using a 50 mm focal length lens, and it emits an average output power of ~270 mW.

Our v-to-2v self-referencing setup represents a major improvement over previous setups: in them, to generate a CE beat note with sufficient SNR for phase locking (~30 dB in 100 kHz resolution bandwidth), the short- and long-wavelength portions of the laser spectrum were spatially separated using a dichroic mirror in a Mach–Zehnder type interferometer [28] or a prism in a prism-based interferometer [40]. The long-wavelength portion was frequency doubled in a second-harmonic generation (SHG) crystal, and the short-wavelength fundamental light and the SHG light were recombined again. Although the wavelength components, that interfere with each other

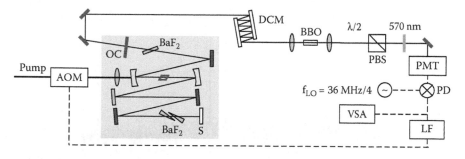

FIGURE 10.13
CE-phase stabilized 200 MHz octave-spanning Ti:sapphire laser. The femtosecond laser itself (located inside the gray area) has a compact 20 cm × 30 cm footprint. AOM, acousto-optical modulator; S, silver end mirror; OC, output coupling mirror; PBS, polarizing beam splitter cube; PMT, photomultiplier tube; PD, digital phase detector; LF, loop filter; VSA, vector signal analyzer. The CE frequency is phase locked to 36 MHz.

generating the CE beat note, can conveniently be overlapped in space and time, these interferometer setups tend to be bulky and alignment sensitive. Even more important, mechanical fluctuations in the interferometer introduce additional CE phase noise in the frequency range up to ~10 kHz. In our setup (Figure 10.13), in contrast, the time delay between the 570 and 1140 nm spectral components used for ν-to-2ν self-referencing is produced by 10 bounces on DCMs, which is intrinsically more stable than the above-mentioned interferometric setups. After the DCM-based delay line, the Ti:sapphire output is focused onto a 2 mm-thick BBO crystal cut for type I SHG at 1160 nm. We measured an SHG conversion efficiency on the order of 10^{-3}. The emitted SHG light and the orthogonally polarized fundamental light are projected onto a common axis using a half-waveplate and a polarizing beam splitter cube. Then it is spectrally filtered using a 10 nm wide interference filter centered at 570 nm, spatially filtered using a ~1mm diameter aperture, and finally detected using a photomultiplier tube (Hamamatsu H6780-20).

In the RF power spectrum shown in Figure 10.14, we observe a peak at the CE frequency with a SNR of ~35 dB in a 100 kHz resolution bandwidth. This SNR is sufficient for direct and routine CE phase stabilization. Phase locking is achieved by a phase-lock loop (PLL, see Section 10.5) in feeding an error signal back to an AOM placed into the pump beam which regulates the pump power and thus changes the CE frequency [13]. A bandpass filter is used to select the CE beat signal at 36 MHz. This signal is amplified, divided by four in frequency to enhance the locking range of the PLL, and compared with a reference frequency supplied by a signal generator using a digital phase detector. The output signal is amplified in the loop filter, which in our case is a proportional–integral (PI) controller, and fed back to the AOM, closing the loop. The output of the digital phase detector is proportional to the remaining jitter between the CE phase evolution and the local oscillator reduced by the division ratio of four.

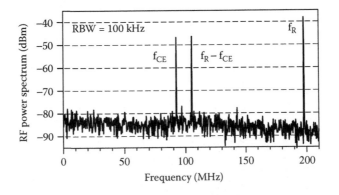

FIGURE 10.14
Radio-frequency power spectrum of fundamental and frequency-doubled light transmitted through a 10 nm wide interference filter centered at 570 nm, resolution bandwidth (RBW) is 100 kHz. The peak at the CE frequency f_{CE} exhibits a signal-to-noise ratio of ~35 dB, sufficient for direct and routine CE phase stabilization.

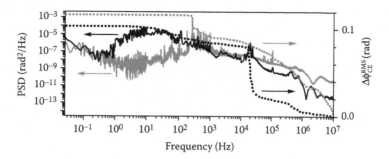

FIGURE 10.15
Power spectral density (PSD) of the CE phase fluctuations $S_{\phi_{CE}}$ (gray and black curves) and accumulated RMS CE phase error $\Delta\phi_{CE}^{RMS}$ (gray and black dotted curves) measured with a digital phase detector and analog mixer, respectively.

The power spectral density (PSD) of the CE phase fluctuations $S_{\phi_{CE}}$ measured with a vector signal analyzer (VSA) at the output of the digital phase detector and properly rescaled by the division factor is shown in Figure 10.15. The accumulated (root-mean-square) CE phase error $\Delta\phi_{CE}^{RMS}$ can be obtained from $S_{\phi_{CE}}$ by integration over frequency according to

$$\Delta\phi_{CE}^{RMS} = \sqrt{-2\int_{10\,MHz}^{f} S_{\phi_{CE}}(f')\,df'} \tag{10.18}$$

resulting in a value of 0.103 rad (integrated from 2.5 mHz to 10 MHz), equivalent to 44 attosecond CE phase jitter at 800 nm.

Because a digital phase detector requires a low pass filter (~1.9 MHz in our case) for operation, any signal with a higher frequency than the low-pass filter cutoff will be attenuated, yielding a CE phase error which does not represent the true CE phase error of the system. Fundamentally this makes a separate out-of-loop measurement [37] consisting of a separate SHG process, CE beat detection, and phase comparison necessary. The Allan deviation of an SHG process using a nonlinear crystal was measured to be on the order of 10^{-16} for an averaging time of 1 s [82], thus it is reasonable to assume no significant contribution of the SHG process to the CE phase noise. Furthermore, our monolithic v-to-2v self-referencing setup employing a DCM-based delay line is not expected to introduce additional CE phase noise either, in contrast to the commonly used Mach–Zehnder type or prism-based interferometers. Hence, assuming that our v-to-2v self-referencing setup and the photomultiplier detection truthfully reflect the CE phase dynamics, only a phase detector with a high enough intermediate frequency (IF) bandwidth is necessary to measure the true CE phase error within the loop.

In a second measurement, also depicted in Figure 10.15, we therefore replaced the digital phase detector with an analog mixer. This measurement yielded a slightly higher value for the accumulated CE phase error of 0.117 rad, equivalent to 50 attosecond CE phase jitter at 800 nm. Both measurements are in good agreement with each other considering that they were taken on different days with different loop-filter settings. These results reflect the elaborate acoustic vibration isolation and shielding against environmental perturbations (e.g., air currents) as well as the effectiveness of our PI control loop up to ~20 kHz. At present, the bandwidth of our PI control loop is limited by the AOM used to modulate the pump power. By using an electro-optic modulator, the CE phase fluctuations are expected to be even further suppressed in the future.

10.5 Noise Analysis of Carrier-Envelope Frequency-Stabilized Lasers

In 2005, we have built two 200 MHz octave-spanning frequency combs (OSFCs). One is pumped by a single-longitudinal-mode (slm) $Nd:YVO_4$ pump laser (Verdi-V10, Coherent) and the other one by a multi-longitudinal-mode (mlm) $Nd:YVO_4$ pump laser (Millennia Xs, Spectra-Physics). Figure 10.16 depicts the measured relative intensity noise (RIN) for the mlm pump laser and slm pump laser. The mlm pump laser shows significantly higher RIN in the high-frequency range, whereas the slm pump laser has higher RIN at very low frequencies. Recently, S. Witte et al. [83] also characterized the influence of RIN of these pump lasers on the residual CE phase noise for a 10-fs Ti:sapphire laser, employing chirped mirrors for intracavity dispersion compensation and external spectral broadening in a microstructure fiber. However, no rigorous noise analysis has been performed so far. The purpose of this section is to elucidate the impact of the RIN of different pump lasers on the finally achievable CE phase noise and how the feedback mechanism and the design of the feedback loop employed impacts residual CE phase noise [78].

Figure 10.17 shows the corresponding spectrally resolved and integrated CE phase error measured for the two 200 MHz OSFCs. In agreement with

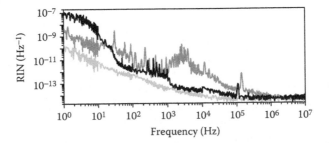

FIGURE 10.16

Relative intensity noise (RIN) of a Coherent Verdi-V6 (black curve) and a Spectra-Physics Millennia Xs (gray curve). The measurement noise floor is given by the light gray curve.

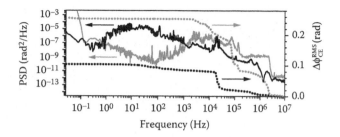

FIGURE 10.17

Comparison of the CE phase noise of a self-referenced 200 MHz Ti:sapphire frequency comb pumped by a Coherent Verdi-V6 (black solid and black dotted curves, data already shown in Figure 10.15) and by a Spectra-Physics Millennia X*s* (gray solid and gray dotted curves).

the data measured by S. Witte et al., the high-frequency CE phase noise is found to be larger for the mlm pump laser. The CE phase fluctuations at lower frequencies, which are smaller for the mlm pump laser, are strongly suppressed by the large PI control loop gain and, therefore, do not contribute significantly to the residual CE phase noise. The residual CE phase fluctuations of the OSFC pumped by the mlm pump laser amount to 0.257 rad, compared to only 0.117 rad if the slm laser is used. First of all, it is surprising that the mlm pumped system is only a factor of 2.2 worse than the slm pumped system, despite the fact that the high-frequency noise of the mlm pump is worse. As we will see, this is so because the feedback gain is large below 100 kHz. Obviously the system pumped by the mlm pump laser could do equally well if the feedback-loop bandwidth could be extended by one order of magnitude. The reasons why the high-frequency noise of the mlm pump cannot be further suppressed will be elaborated further in the following feedback analysis.

From a control systems point of view, the f_{CE}-stabilized laser is a PLL [84], where the voltage-controlled oscillator (VCO) is the CE-frequency controlled OSFC, which is the block indicated by the dashed box in Figure 10.18. When the laser is turned on, the CE frequency f_{CE} is determined by the cavity parameters and alignment, equivalent to the center frequency of oscillation of the VCO in a PLL. A voltage applied to the AOM driver changes this frequency by an amount proportional to the equivalent VCO gain of the system. The model depicted in Figure 10.18 includes all the electronic components used in the stabilization (phase detector, AOM, and loop filter), whose transfer characteristics are easy to measure and to describe by analytic expressions. Assuming an instantaneous response of the CE frequency to pump power via a constant C_{fP_p}, one is not able to reproduce the measured CE phase noise spectrum. Therefore, the impact of the frequency response of the OSFC system must be taken into account in the analysis, which was done by considering the transfer function between the intracavity laser power (or pulse energy) and the pump power via the laser gain dynamics.

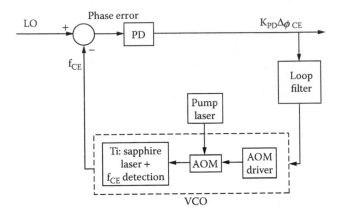

FIGURE 10.18
Block diagram of the phase-lock loop (PLL) composed of the f_{CE}-stabilized laser. The voltage-controlled oscillator (VCO) is depicted in the dashed box.

10.5.1   Transfer Function Representation for the Pulse Energy versus Pump-Power Dynamics

The starting point for derivation of the transfer function are the laser rate equations for pulse energy and gain, which can be derived from the master equation (Equation 10.2) by proper elimination of the remaining degrees of freedom in the mode-locked laser, as has been derived, for example, in the case of soliton lasers mode locked by slow saturable absorbers [85,86]. One can write

$$T_r \frac{dE}{dT} = \left(g - l - q(E)\right)E \tag{10.19}$$

$$T_r \frac{dg}{dT} = -\frac{g - g_0}{\tau_L / T_r} - g\frac{E}{E_{sat}}, \tag{10.20}$$

where we have used

T_r = cavity round-trip time
τ_L = upper state lifetime
l = total nonsaturable loss
$q(E) \equiv q_{ml}(E)$ = effective energy-dependent saturable absorber and filter loss (for more details, see Reference 86)
g_0 = small-signal gain, which is proportional to pump power
E_{sat} = saturation energy of the gain medium.

Many assumptions have been made when using these equations to describe the energy and gain dynamics. For example, possible frequency shifts and back action of the background radiation onto the energy and gain dynamics, i.e., the details of the pulse-shaping mechanism, are neglected. We have

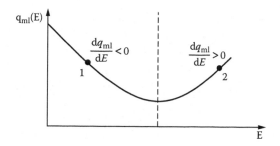

FIGURE 10.19
Mode locking related energy-dependent loss $q_{ml}(E)$. Stable mode locking occurs at operating point 2 where $dq_{ml}/dE > 0$.

denoted the time coordinate as capital T to emphasize the fact that this dynamics occurs on the time scale of many cavity round trips. This is possible for solid-state laser gain media because the interaction cross section is small and therefore the gain saturates with average power rather than with a single passage of the pulse through the gain medium [85]. Note that the most important term in the rate equations is the mode locking related energy-dependent loss $q_{ml}(E)$, which comprises the loss during saturation of the absorber as well as additional losses due to the bandwidth limitations of the system that increase with additional spectral broadening or increasing intracavity pulse energy [86]. A typical characteristic dependence of q_{ml} on intracavity pulse energy E is shown in Figure 10.19. To derive a transfer function for the laser, we linearize Equation 10.19 and Equation 10.20 around the steady-state operating point, denoted by the subscripts:

$$E = E_s + \Delta E, \ g = g_s + \Delta g, \ g_0 = g_{0s} + \Delta g_0,$$

$$q(E) = q(E_s + \Delta E) \approx q(E_s) + \left.\frac{\partial q}{\partial E}\right|_{E=E_s} \Delta E \equiv q_s + \frac{\partial q}{\partial E_s} \Delta E$$

to get the set of linearized equations

$$T_r \frac{d\Delta E}{dT} = -\frac{\partial q}{\partial E_s} E_s \Delta E + E_s \Delta g \qquad (10.21)$$

$$T_r \frac{d\Delta g}{dT} = -\frac{T_r}{\tau_{stim}} \Delta g - \frac{g_s}{E_{sat}} \Delta E + \frac{T_r}{\tau_L} \Delta g_0, \qquad (10.22)$$

where τ_{stim} is the stimulated lifetime given by $\tau_L(1 + \tau_L F_s/T_r E_{sat})^{-1}$. By taking the Laplace transform of the above equations, it is straightforward to derive a *pump power* to *pulse energy* transfer function, by writing g_0 as $K_{g0}P_p$. Defining the pump parameter $r = 1 + \tau_L E_s/T_r E_{sat}$, which indicates how many times the laser operates above threshold, we arrive at

$$\Delta \tilde{E} = \frac{P_s K_0 / \tau_L}{s^2 + \left(\dfrac{r}{\tau_L} + \dfrac{\partial q}{\partial E_s} P_s \right) s + P_s \dfrac{\partial q}{\partial E_s} \dfrac{r}{\tau_L} + \dfrac{r-1}{\tau_L \tau_p} \left(1 + \dfrac{q_s}{l} \right)} \Delta \tilde{P}_p, \qquad (10.23)$$

where $\tau_p = T_r/l$ is the photon decay time due to the linear cavity losses in cw operation. So far we have considered the mode-locked case, but a similar relation can be obtained for the intracavity power in cw operation simply by setting the saturable absorber terms in Equation 10.23 to zero. However, the laser parameters, like mode cross section in the gain medium likely change values when the laser changes from cw operation to mode-locked operation. But because we have no way of measuring them in mode-locked operation, we have estimated them based on our knowledge of laser parameters in cw operation and used those to examine the effect of the inclusion of the saturable absorber into the transfer function. Figure 10.20 shows the pump power to intracavity pulse energy transfer functions, in amplitude and phase, for cw and mode-locked operation for different values of the term $(\partial q/\partial E_s)P_s$. It is obvious that the mode locking of the laser drastically changes the transfer characteristic between pump power and intracavity power due to the pulse operation, which introduces the term $(\partial q/\partial E_s)P_s$ into the rate equation (Equation 10.21). Depending on its sign this term enhances (for $(\partial q/\partial E_s)P_s < 0$ it may lead to Q-switching when large enough) or strongly damps (for $(\partial q/\partial E_s)P_s > 0$) intracavity energy fluctuations. In cw operation its absence usually leads to pronounced relaxation oscillations (Figure 10.20). As we can see, the stronger the effective inverse saturable absorption, the more damped become the relaxation oscillations in the laser, and the weaker becomes the response at all frequencies. This result has to be expected, because the stronger the inverse saturable absorption, the more clamped become the pulse energy and the average power.

The model is verified by measuring the transfer function of the laser in cw and mode-locked operation. The measurement is performed by using a network analyzer as shown in Figure 10.21. Care has to be taken to assure

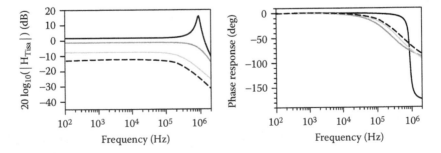

FIGURE 10.20
Calculated amplitude and phase response of intracavity power with pump power for cw operation (black) and for different values of saturable absorption: $(\partial q/\partial E_s)P_s = 10r/\tau_L$ (gray), $(\partial q/\partial E_s)P_s = 50r/\tau_L$ (light gray), and $(\partial q/\partial E_s)P_s = 150r/\tau_L$ (dashed).

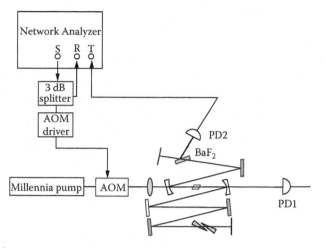

FIGURE 10.21
Schematic of the transfer function measurement setup. In order to measure only the contribution from the laser itself, a first calibration measurement is performed with PD1 measuring the AOM and AOM driver response (PD1 and PD2 are the same photo detector). The laser transfer function is measured by detecting with PD2 the reflection from one intracavity BaF$_2$ plate and subtracting the AOM and AOM driver response.

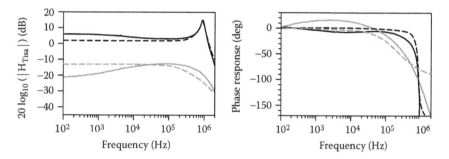

FIGURE 10.22
Measured (black and gray curves) and modeled (black and gray dashed curves) amplitude and phase response of OSFC laser, in cw and mode-locked operation.

that no cw component is present in mode-locked operation during the measurements. The results are shown in Figure 10.22. Also shown in the same plots are the amplitude and phase of the transfer function given by Equation 10.23. For the cw case, $q(E) = 0$. The parameter r was measured to be 3.22, and the intracavity loss l, which determines the value of τ_p, was determined by matching the relaxation oscillation frequency in the model with the measured result. The value obtained was $l = 0.22$. K_{go} was then calculated using the relationship $g_0 = K_{go}P_p = rl = 0.15$. As can be seen in Figure 10.22, the model describes well the gain dynamics in cw operation. For the mode-locked case, we included the effect of the saturable absorber and modified the other parameters in such a way as to match the measured transfer function as closely as possible, especially for frequencies beyond 10 kHz, where the impact on the final noise calculation is most pronounced. This was achieved by setting $(\partial q/\partial E_s)P_s = 150\,r/\tau_L$, $q_s = l$, $l = 0.17$, $r = 3.5$, and $K_{go} = 0.23$. Figure 10.22 shows that the approximations made in the model do not fully describe the system, i.e., we neglect the interaction with the continuum, which is an infinite-dimensional system. In the measurements, a significant change in the amplitude response from low to high frequencies, in the 1–100 kHz range, is observed in mode-locked operation, which is an indication of additional slow processes occurring in the mode-locked laser that are absent in the cw laser. When pushing the laser to octave-spanning operation, i.e., for the shortest pulse and widest spectrum, one is always pushing the laser towards its stability boundary resulting in modes that approach zero damping time, i.e., dynamics with time constants of many round trips. These modes, which are neglected in the analysis, are most likely responsible for the deviations in the low-frequency range of the laser when mode locked. Nevertheless, the model gives good qualitative and quantitative description of the laser dynamics and the transfer function mimics the global behavior of the measured transfer function while still being simple. It confirms our observations on the strength of the saturable absorption in such systems, explicit in the measurements shown in Figure 10.12a. As will be shown in the next section, the inclusion of this transfer function in the noise analysis is essential in deriving the correct noise behavior of the system.

10.5.2 Determination of the Carrier-Envelope Phase Error

To calculate the CE phase noise spectrum of the OSFC, a linear noise analysis is performed. The block diagram in Figure 10.23 shows the closed-loop system. The input noise source, characterized by the PSD of the pump noise $S_p(s)$, which is the RIN multiplied by the square of the pump power, is converted to the CE phase noise spectral density $S_\phi(s)$ in the laser and is partially suppressed in the feedback loop. The feedback path consists of the phase detector, the loop filter and the AOM, with transfer functions denoted by $H_{PD}(s)$, $H_{LF}(s)$, and $H_{AOM}(s)$, respectively. Table 10.1 shows the corresponding analytic expressions. The loop filter consists of a simple PI controller with time constants τ_1, τ_2, and τ_3, and the AOM is, up to a small drop in amplitude for higher frequencies, equivalent to a delay line with a propagation delay given by the time it takes for the acoustic wave to travel from the

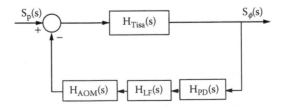

FIGURE 10.23
Block diagram describing the addition of intensity noise to the CE phase-lock loop.

TABLE 10.1

Transfer functions of the Ti:sapphire laser, phase detector, loop filter, and AOM

$H_{\text{Tisa}}(s)$	$\dfrac{P_s K_0 / \tau_L}{s^2 + \left(\dfrac{r}{\tau_L} + \dfrac{\partial q}{\partial E_s} P_s\right) s + P_s \dfrac{\partial q}{\partial E_s} \dfrac{r}{\tau_L} + \dfrac{r-1}{\tau_L \tau_p}\left(1 + \dfrac{q_s}{l}\right)} \times \dfrac{C_{f P_{\text{intra}}}}{s T_r}$
$H_{\text{PD}}(s)$	K_{PD}
$H_{\text{LF}}(s)$	$\dfrac{1 + \tau_2 s}{\tau_1 s} \dfrac{1}{\left(1 + \tau_3 s\right)^3}$
$H_{\text{AOM}}(s)$	$e^{-s\tau_{\text{delay}}}$

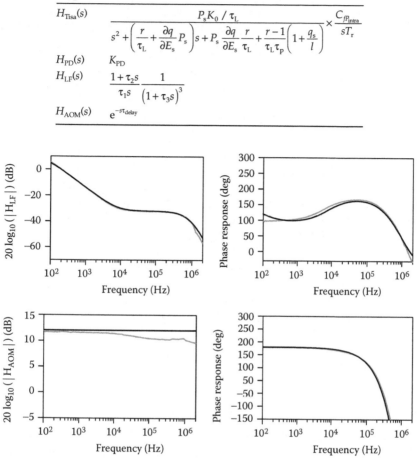

FIGURE 10.24

Measured (gray) and calculated (black) transfer functions for loop filter (top) and AOM (bottom). Amplitude response is on the left and phase response on the right.

piezoelectric transducer to the optical beam. We set this delay to 1.73 μs to match the measured transfer function. The finite bandwidth of the AOM is not taken into account here because the measurements show that the magnitude of the AOM transfer function is approximately constant out to 1 MHz. The calculated and measured transfer functions for the loop filter and the AOM are shown in Figure 10.24. We consider the phase detector (Analog Devices AD9901) transfer function to be flat, based on its datasheet.

In the noise analysis we consider the case of the OSFC pumped by the multi-longitudinal-mode pump laser because of the increased high-frequency noise in such systems as discussed earlier in Section 10.5. Given the RIN measurement in Figure 10.16 as the input noise source, we calculate $S_\phi(s)$ using the transfer functions in Table 10.1 to derive the closed-loop transfer function that describes the conversion of pump noise $S_p(s)$ to CE phase noise $S_\phi(s)$

$$H_{CL}(s) = \frac{H_{Tisa}(s)}{1 + H_{Tisa}(s)H_{PD}(s)H_{LF}(s)H_{AOM}(s)}. \tag{10.24}$$

Then, $S_\phi(s)$ is obtained by multiplying the intensity noise spectrum of the pump laser $S_p(s)$ with the square modulus of $H_{CL}(s)$ given by Expression 10.24

$$S_\phi(s) = |H_{CL}(s)|^2 S_p(s). \tag{10.25}$$

The calculated and measured CE phase noise is shown in Figure 10.25. Already given in the previous Subsection 10.5.1, the values for the parameters used in $H_{Tisa}(s)$ are those which closely match the measured and calculated OSFC transfer function in mode-locked operation in the range between 10 kHz and 1 MHz. When only the pump noise is used as a noise source, we found that the calculated and measured $S_\phi(s)$ showed good agreement up to 200 kHz, beyond which point the measurement showed enhanced noise. This was determined to be due to electronic noise at the output of the phase detector, which was then included as an additional noise source in the analysis to get a good fit beyond 200 kHz. We added this white noise source S_{PD} in the loop, which converts to a final contribution to $S_\phi(s)$ by the closed-loop transfer function

$$H_{CL}^{wn}(s) = \frac{1}{1 + H_{Tisa}(s)H_{PD}(s)H_{LF}(s)H_{AOM}(s)} \tag{10.26}$$

where the magnitude of the white noise source S_{PD} was estimated based on the measurement. Improved noise performance could be obtained by increasing

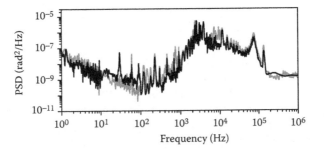

FIGURE 10.25
CE phase noise spectrum of the Millennia-X*s* pumped OSFC. The black curve is the calculated spectrum, the gray curve shows the measurement.

the closed-loop bandwidth, which is currently about 100 kHz. This bandwidth is dictated by the phase margin of the feedback loop, which according to the Nyquist theorem may run unstable when the gain is larger than 1 while the phase approaches 180° [87]. Note that the VCO integrates a frequency deviation into a phase deviation causing the feedback loop to start off with a −90° phase, as can be seen in Figure 10.26. When the bandwidth of the gain medium is approached, additional phase accumulates from the gain dynamics and the time delay from the AOM (see Figure 10.13), which renders the system unstable if the loop gain is not properly reduced. Of course, for a reasonably stable system and optimum operation of the feedback loop, large enough gain and phase margins are necessary. This imposes a limitation to the maximum loop gain because the jointly added phase from $H_{AOM}(s)$ and $H_{Tisa}(s)$ reduces the phase margin.

However, now that the different phase contributions to the feedback loop are well understood, the noise suppression at high frequencies could be improved by custom design of the control electronics. For example, by adding a lead-lag compensator [87], the limitation arising from the phase delays of some of the components could be reduced. Also, replacing the AOM by an electro-optic modulator or intracavity loss modulation could further the contribution to the open-loop phase from the time delay of the AOM.

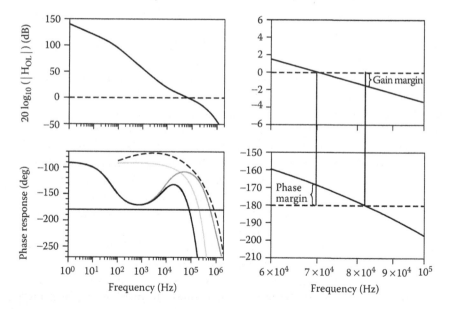

FIGURE 10.26
Amplitude (top) and phase (bottom) of open-loop transfer function of the CE phase-lock loop (black curve) corresponding to Figure 10.23. The contribution from the loop filter (gray), AOM (light gray), and laser dynamics (dashed) to the open-loop phase are also shown. The zoomed plots on the right-hand side confirm that the PLL satisfies the Nyquist stability criterion (gain margin 2 dB, phase margin 12°).

10.6 Carrier-Envelope Phase-Independent Optical Clockwork for the HeNe/CH$_4$ Optical Molecular Clock

An intriguing alternative to optical clockworks based on octave-spanning Ti:sapphire lasers are CE phase independent optical clockworks, which have less demanding bandwidth requirements and allow extending frequency combs into the infrared (IR) spectral region. In 2002, A. Baltuška et al. [88]. have shown that the phase relationship between pump, signal, and idler pulses in a parametric interaction enables the generation of CE phase-independent idler pulses. Similarly, by using difference-frequency generation (DFG) between two different spectral portions $v_m = mf_R + f_{CE}$ and $v_n = nf_R + f_{CE}$ (with m, n integers) of the same frequency comb, a CE frequency-independent DFG comb in the infrared spectral region $v_m - v_n = (mf_R + f_{CE}) - (nf_R + f_{CE})$ $= (m - n)f_R$ with excellent accuracy and stability can be generated [89]. Furthermore, a CE frequency independent DFG comb tuned to 800 nm might also pave the way to an all-optical CE phase stabilization scheme by injection locking [90]. Equivalent to the DFG approach, sum-frequency generation (SFG) between an IR optical frequency standard v_S and the low-frequency portion of a frequency comb $nf_R + f_{CE}$ yields an SFG comb $v_S + nf_R + f_{CE}$ that can be tuned to spectrally overlap with the high-frequency portion $mf_R + f_{CE}$ of the original frequency comb. The heterodyne beat $v_S - (m - n)f_R$ results from the coherent superposition of many corresponding comb lines and does not depend on the CE frequency f_{CE}, thus allowing the construction of a CE frequency-independent optical clockwork. A similar scheme to eliminate the CE frequency by orthogonalizing error signals derived from two optical heterodyne beats of the comb against a reference laser v_S and its second harmonic $2v_S$ was already used earlier [31]: If the two optical beats are expressed as $f_{b1} = v_S - (mf_R + f_{CE})$ and $f_{b2} = 2v_S - (2mf_R + f_{CE})$, then the difference between the two beats gives $f_{b2} - f_{b1} = v_S - mf_R$, permitting control of f_R from v_S without stabilizing f_{CE}. In another experiment, in order to avoid the use of microstructure fibers or the need for special octave-spanning lasers, A. Amy-Klein et al. [91] phase locked two modes of the same frequency comb to two cw diode lasers. The difference frequency between the two cw lasers was phase locked to a stable IR reference. This technique is insensitive to f_{CE}, but involves numerous phase locks and lasers.

In collaboration with the groups of Jun Ye (JILA in Boulder, CO, U.S.A.) and Mikhail A. Gubin (Lebedev Physical Institute in Moscow, Russia), our group implemented a CE frequency independent optical clockwork for the HeNe/CH$_4$ optical clock based on SFG/DFG in a periodically-poled lithium niobate (PPLN) crystal [51,52]. Although we first demonstrated the SFG based CE phase independent clockwork [51], we restrict our discussion here to the DFG based optical clockwork, which eventually was employed in the HeNe/CH$_4$ optical molecular clock [52].

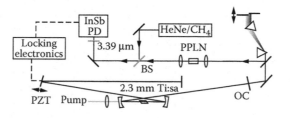

FIGURE 10.27

Experimental setup of the HeNe/CH₄ optical molecular clock. The heterodyne beat between the cw output of a methane-stabilized helium neon laser at 3.39 μm and the CE frequency independent comb obtained by difference-frequency generation in a periodically poled lithium niobate (PPLN) crystal is stabilized via feedback to a piston-mode piezoelectric transducer on which the fold mirror of the Ti:sapphire laser is mounted. Thus, the repetition frequency f_R is phase coherently derived from the frequency of the HeNe/CH₄ laser. OC, output coupling mirror; BS, beam splitter; PD, photodiode; PZT, piezoelectric transducer.

The experimental setup of the HeNe/CH₄ optical molecular clock is shown in Figure 10.27. The three key ingredients arc (i) a custom-designed Ti:sapphire frequency comb, (ii) a transportable double-mode methane-stabilized helium neon laser [92–94] at 3.39 μm serving as IR frequency standard, and (iii) a difference-frequency generation setup using a PPLN crystal.

The X-folded Ti:sapphire laser operates at a repetition frequency f_R of 78 MHz and emits an average output power of typically 150 mW. It employs a 2.3-mm thick Ti:sapphire gain crystal and 7 bounces on DCMs for dispersion compensation. The two concave DCMs, both with a focal length of 50 mm, produce a beam waist inside the gain crystal, which is oriented at Brewster's angle of ~60°. Approximately 6.5 W of 532 nm pump light emitted by a frequency-doubled Nd:YVO₄ laser is focused into the gain crystal using an achromatic doublet lens (76.2 mm effective focal length). The output spectrum depicted in Figure 10.28a is spectrally shaped by means of a custom-designed narrowband output coupler consisting of five pairs of SiO₂/TiO₂ layers. The resulting transmission (not shown here) is about 0.5% in the center of the output coupler at about 800 nm, it increases strongly to 5% at the designated wavelengths of 685 nm and 866 nm [51]. Due to these transmission characteristics of the output coupler, the output spectrum exhibits spectral peaks at 670 nm and 834 nm (see Figure 10.28a), which have a frequency difference corresponding to 3.39 μm wavelength. A third peak at ~902 nm is not used in the experiment. Approximately 12.3 mW of average power is contained within the 7 nm FWHM bandwidth of the 670 nm peak, while a 10 nm wide spectral region centered at 831 nm has 7.8 mW of average power.

The Ti:sapphire output is focused into a periodically-poled lithium niobate (PPLN) crystal using a 40 mm focal length calcium fluoride lens. This 5-mm long PPLN crystal has a quasi-phase-matching period of 16.2 μm and is heated to 130°C. The resulting phase-matching bandwidth is ~0.75 THz. The output light is collimated with a 38 mm calcium fluoride lens. For efficient DFG inside the PPLN, the Ti:sapphire beam traverses a prism-based delay

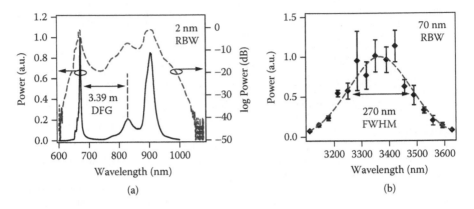

FIGURE 10.28
(a) Output spectrum of the custom-designed 78 MHz Ti:sapphire laser with strong peaks at 670 and 834 nm. (b) CE-frequency-independent IR comb at 3.39 μm generated by difference-frequency mixing of the Ti:sapphire comb in a PPLN crystal. RBW, resolution bandwidth.

line (consisting of two SF10 prisms) before the PPLN for fine-adjusting the temporal overlap of the 670 and 834 nm spectral components. The resultant CE frequency independent IR comb passes through a 50 nm FWHM band-width interference filter centered at 3.39 μm. We measured ~10 μW of average power transmitted through the filter, corresponding to nearly 600 pW per mode of the IR comb. With an IR monochromator [95] and lock-in detection, we observed that the IR comb has a bandwidth of ~270 nm (7 THz) centered at ~3.4 μm (88.5 THz) as shown in Figure 10.28b.

The optical frequency standard is a compact, transportable, double-mode helium neon laser stabilized to resonances of the $F_2^{(2)}$ ($P(7)v_3$) transition in methane (HWHM ~200 kHz, hyperfine structure is unresolved) at 3.39 μm. Saturated absorption and saturated dispersion resonances are both used to phase lock the reference laser frequency, providing optical radiation with a linewidth < 100 Hz. This type of reference HeNe laser is being routinely incorporated in transportable HeNe/CH$_4$ systems with telescopic beam expanders for providing resolution of the hyperfine structure of methane and frequency repeatability of 2×10^{-13} over several years [92]. Measurements with a frequency chain indicate an instability of $< 4 \times 10^{-13}$ at 1 s, limited by a hydrogen maser. A direct optical comparison of two versions of the system (one was a nontransportable, resolved hyperfine-structure version) has a demonstrated Allan deviation of $\sim 2 \times 10^{-14}$ at 1 s. A second HeNe laser (heterodyne laser) is phase locked to the reference laser with a fixed ~600 kHz frequency offset, and provides the optical field we use to stabilize our IR comb. An amplifier HeNe tube is used to increase the power from 300 μW to ~1 mW.

After the cw HeNe beam and the IR comb are combined on a 60/40 beam splitter, they pass through a 50 nm FWHM interference filter centered at 3.39 μm. A 1.8 cm focal length calcium fluoride lens focuses the beams onto a liquid-nitrogen-cooled InSb photodiode with a 250 μm diameter active area.

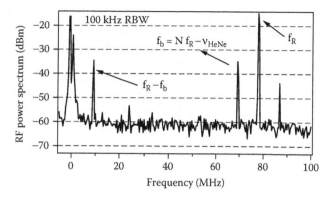

FIGURE 10.29
Heterodyne beat signal f_b between the HeNe/CH$_4$ laser and a neighboring DFG comb line exhibiting a signal-to-noise ratio of > 25 dB in a 100 kHz RBW.

A heterodyne beat between the cw HeNe beam at ν_{HeNe} and the Nth line of the IR comb is detected and amplified. As shown in Figure 10.29, the beat frequency $f_b = Nf_R - \nu_{HeNe}$ has a SNR of 25 dB in a 100 kHz bandwidth. It is subsequently phase locked via an RF tracking oscillator, which consists of a voltage-controlled oscillator and a digital phase-lock loop, to a 70 MHz synthesized signal (with a negligible noise contribution) derived from a stable cesium reference. Control of f_b is achieved by feedback of the PLL error signal to a piston-mode piezoelectric transducer on which the fold mirror of the Ti:sapphire laser is mounted. Now the RF frequency f_R is directly expressed in terms of the optical frequency ν_{HeNe} as $f_R = (\nu_{HeNe} + 70 \text{ MHz})/N$. The optical clock signals f_R are independently detected using visible light from the second port of the beam splitter.

The stability of the RF clock signal is compared to both a hydrogen maser and the repetition frequency of a second Ti:sapphire laser operating as part of an optical clock based on a molecular iodine transition [31]. For comparison with the maser, the tenth harmonic of f_R is compared to a 780 MHz signal derived from NIST's (National Institute of Standards and Technology in Boulder, CO) ST-22 hydrogen maser and transferred to JILA via the Boulder Research and Administrative Network (BRAN) single-mode fiber [96]. The I$_2$ clock consists of a second Ti:sapphire laser with a 100 MHz repetition rate. Its frequency comb is stabilized by locking f_{ce} using ν-to-2ν self-referencing, and by phase locking one comb line near 1064 nm to a Nd:YAG laser which is stabilized to molecular iodine [31]. The seventh harmonic of this I$_2$-referenced Ti:sapphire laser's f_R is compared to the ninth harmonic of the CH$_4$-referenced Ti:sapphire laser's f_R at ~702 MHz. We deliberately introduce an offset frequency of 10 kHz between the two stabilized f_R harmonics. A double-balanced mixer detects this 10 kHz difference frequency that is subsequently filtered and counted [97] with a 1 s gate time in order to determine the Allan deviation $\sigma_y(\tau)$ of the HeNe/CH$_4$ clock as compared to the hydrogen maser or the I$_2$ clock.

Figure 10.30a shows representative frequency counting data for both comparisons. Denoting the comparison frequency (780 MHz for the maser, 702 MHz for the I_2 clock) as f_c, frequency fluctuations Δf_c are normalized to f_c for fair evaluation of the two stability measurements. Parasitic back reflections from surfaces inside the HeNe resonator shift the centers of saturated absorption and saturated dispersion resonances of the reference system [98]. As the position of the reflecting surface moves due to temperature changes, both a slow drift and a slow oscillation of ν_{HeNe} are observed. Therefore, a quadratic drift of Δf_c tightly correlated with recorded temperature changes of the HeNe system (monitored independently with the laser's intermode beat frequency), has been subtracted from the raw data. However, an oscillation with period > 20 s is still present. Figure 10.30b shows the Allan deviations determined from various frequency-counting records. Measurements against the maser (filled diamonds) are limited by the maser's intrinsic short-term instability of ~3×10^{-13} at 1 s. Data run I (bow ties) against the I_2 clock used free-space photodiodes for detection of both clocks' repetition frequencies, and data run II (circles) used fiber-coupled detectors which exhibit greater phase noise and amplitude-to-phase noise conversion. The noisy plateau in the Allan deviation for data run II for $\tau < 20$ s strongly suggests that amplitude-to-phase noise conversion in the detection process limits our instability for short time scales [99]. However, for time scales 20 s < τ < 100 s, data run II exhibits comparable stability to data run I, even though they were taken under very different conditions of temperature fluctuations in the laboratory. We therefore conclude that the instability of our HeNe/CH_4 clock as compared to the I_2 clock is ~1.2×10^{-13} at 1 s, averaging down as $\tau^{-1/2}$ for $\tau < 100$ s, limited by excess noise in the photodetection process [99].

In order to characterize the phase noise of our HeNe/CH_4 clock signal, we examine the 10 kHz signal derived from comparison with the I_2 clock (using

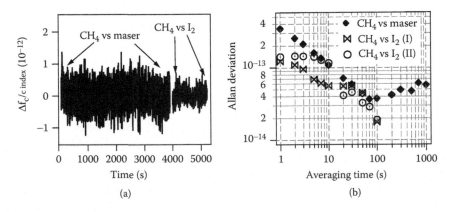

FIGURE 10.30
Comparison of the HeNe/CH_4 clock signal against the NIST ST-22 hydrogen maser and the iodine clock signal. (a) Fractional frequency fluctuations $\Delta f_c/f_c$ for 1 s counter gate time. (b) Allan deviations for comparison with the hydrogen maser (diamonds), the iodine clock with fiber-coupled detection of f_R (circles), and the iodine clock with free-space-coupled detection of f_R (bow ties).

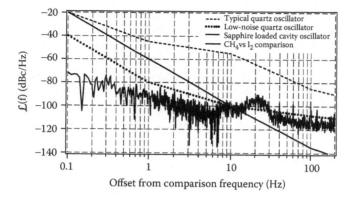

FIGURE 10.31

Single-sideband (SSB) phase noise of the HeNe/CH$_4$ clock signal when compared with the iodine clock signal. Additional curves show the approximate SSB phase noise of extremely low-phase-noise microwave sources for comparison. All data were scaled for a 1 GHz carrier for ease of comparison.

free-space detection of the clock signals) with a fast-Fourier-transform (FFT) spectrum analyzer. These measurements represent an upper limit to the single-sideband (SSB) phase noise, as they are sensitive to amplitude noise as well. Figure 10.31 shows our measurement results after scaling the phase noise up to a 1 GHz carrier frequency for ease of comparison against other oscillators. We include curves representing the typical SSB phase noise of other extremely stable microwave sources [100]. Less than 10 Hz from the carrier, our SSB phase noise is superior to some of the lowest phase-noise microwave sources available, achieving –93 dBc/Hz at 1 Hz offset. Note that if the phase was driven by white noise sources, one would expect a 20 dB per decade roll-off of the SSB phase noise. Independent measurements of the phase noise of our RF amplifiers indicate that, below 10 Hz, they are limiting the SSB phase noise of our system.

Without the need for stabilization of f_{CE}, and coupled with the use of a portable highly stable HeNe/CH$_4$ optical frequency standard, our setup represents a compact, reliable optical clock with high stability and exceedingly low phase noise, which can in principle be operated for long (> 24 h) periods of time. We also note the applicability of our scheme to absolute frequency measurements throughout the near- and mid-IR spectral range.

10.7 Conclusions

In this chapter we have shown that, besides their well-known importance for ultrafast time-domain spectroscopy, few-cycle Ti:sapphire lasers yield highest-quality frequency combs for optical frequency metrology. We reviewed the pulse formation and dynamics in these lasers, discussed the main technical challenges that must be overcome to enable the generation of ultrabroadband spectra, and presented octave-spanning Ti:sapphire lasers which can directly be carrier-envelope phase stabilized by v-to-2v self-referencing with residual carrier-envelope phase jitter of only 50 asec. We furthermore performed a complete noise analysis of the carrier-envelope phase-lock loop and discussed its bandwidth limitations. This system represents a very attractive clockwork for ultrastable and ultraprecise optical clocks based on optical transitions in trapped single ions and neutral atoms. An intriguing alternative to optical clockworks based on octave-spanning Ti:sapphire lasers are carrier-envelope phase independent optical clockworks based on difference-frequency generation, which have less demanding bandwidth requirements and allow to extend frequency combs into the IR spectral region. Finally we have presented a HeNe/CH_4 optical molecular clock employing such a carrier-envelope phase-independent clockwork and demonstrated the synthesis of ultrastable microwave signals from this optical frequency reference.

Acknowledgments

We gratefully acknowledge contributions to this work from R. Ell, A. Winter, J. Kim, J. R. Birge, J. Chen, O. Kuzucu, T. R. Schibli, F. N. C. Wong, D. Kleppner, and E. P. Ippen (M.I.T.), S. M. Foreman, A. Marian, D. J. Jones, L.-S. Ma, J. L. Hall, and J. Ye (JILA), E. A. Petrukhin, and M. A. Gubin (P. N. Lebedev Physical Institute, Moscow), and financial support from ONR under contract N00014-02-1-0717 and AFOSR under contract FA9550-04-1-0011. In addition, O. D. Mücke gratefully acknowledges support from the Alexander von Humboldt Foundation.

References

1. Brabec, T. and Frausz, F., Intense few-cycle laser fields: frontiers of nonlinear optics, *Rev. Mod. Phys.* 72, 545, 2000.
2. Fortier, T.M., Roos, P.A., Jones, D.J., Cundiff, S.T., Bhat, R.D.R., and Sipe, J.E., Carrier-envelope phase-controlled quantum interference of injected photocurrents in semiconductors, *Phys. Rev. Lett.* 92, 147403, 2004.
3. Marian, A., Stowe, M.C., Lawall, J.R., Felinto, D., and Ye, J., United time-frequency spectroscopy for dynamics and global structure, *Science* 306, 2063, 2005.
4. Mücke, O.D., Tritschler, T., Wegener, M., Morgner, U., and Kärtner, F.X., Role of the carrier-envelope offset phase of few-cycle pulses in nonperturbative resonant nonlinear optics, *Phys. Rev. Lett.* 89, 127401, 2002.

5. Mücke, O.D., Tritschler, T., Wegener, M., Morgner, U., Kärtner, F.X., Khitrova, G., and Gibbs, H.M., Carrier-wave Rabi flopping: role of the carrier-envelope phase, *Opt. Lett.* 29, 2160, 2004.

6. Apolonski, A., Dombi, P., Paulus, G.G., Kakehata, M., Holzwarth, R., Udem, T., Lemell, C., Torizuka, K., Burgdörfer, J., Hänsch, T.W., and Krausz, F., Observation of light-phase-sensitive photoemission from a metal, *Phys. Rev. Lett.* 92, 073902, 2004.

7. Paulus, G.G., Lindner, F., Walther, H., Baltuška, A., Goulielmakis, E., Lezius, M., and Krausz, F., Measurement of the phase of few-cycle laser pulses, *Phys. Rev. Lett.* 91, 253004, 2003.

8. Baltuška, A., Udem, T., Uiberacker, M., Hentschel, M., Goulielmakis, E., Gohle, C., Holzwarth, R., Yakovlev, V.S., Scrinzi, A., Hänsch, T.W., and Krausz, F., Attosecond control of electronic processes by intense light fields, *Nature* 421, 611, 2003.

9. Liu, X., Rottke, H., Eremina, E., Sandner, W., Goulielmakis, E., Keeffe, K.O., Lezius, M., Krausz, F., Lindner, F., Schätzel, M.G., Paulus, G.G., and Walther, H., Nonsequential double ionization at the single-optical-cycle limit, *Phys. Rev. Lett.* 93, 263001, 2004.

10. Drescher, M., Hentschel, M., Kienberger, R., Uiberacker, M., Yakovlev, V., Scrinzi, A., Westerwalbesloh, T., Kleineberg, U., Heinzmann, U., and Krausz, F., Time-resolved atomic inner-shell spectroscopy, *Nature* 419, 803, 2002.

11. Graves, W.S, Farkhondeh, M., Kärtner, F.X., Milner, R., Tschalaer, C., van der Laan, J.B., Wang, F., Zolfaghari, A., Zwart, T., Fawley, W.M., and Moncton, D.E., X-ray laser seeding for short pulses and narrow bandwidth, *Proc. 2003 Part. Accel. Conf.* 2, 959, 2003.

12. Udem, T., Holzwarth, R., and Hänsch, T.W., Optical frequency metrology, *Nature* 416, 233, 2002.

13. Cundiff, S.T., Ye, J., and Hall, J.L., Optical frequency synthesis based on mode-locked lasers, *Rev. Sci. Instrum.* 72, 3749, 2001.

14. Hollberg, L., Diddams, S., Bartels, A., Fortier, T., and Kim, K., The measurement of optical frequencies, *Metrologia* 42, S105, 2005.

15. Bize, S., Diddams, S.A., Tanaka, U., Tanner, C.E., Oskay, W.H., Drullinger, R.E., Parker, T.E., Heavner, T.P., Jefferts, S.R., Hollberg, L., Itano, W.M., and Bergquist, J.C., Testing the stability of fundamental constants with the $^{199}Hg^+$ single-ion optical clock, *Phys. Rev. Lett.* 90, 150802, 2003.

16. Fischer, M., Kolachevsky, N., Zimmermann, M., Holzwarth, R., Udem, T., Hänsch, T.W., Abgrall, M., Grünert, J., Maksimovic, I., Bize, S., Marion, H., Pereira Dos Santos, F., Lemonde, P., Santarelli, G., Laurent, P., Clairon, A., Salomon, C., Haas, M., Jentschura, U.D., and Keitel, C.H., New limits on the drift of fundamental constants from laboratory measurements, *Phys. Rev. Lett.* 92, 230802, 2004.

17. Schnatz, H., Lipphardt, B., Helmcke, J., Riehle, F., and Zinner, G., First phase-coherent frequency measurement of visible radiation, *Phys. Rev. Lett.* 76, 18, 1996.

18. Evenson, K.M., Wells, J.S., Petersen, F.R., Danielson, B.L., and Day, G.W., Accurate frequencies of molecular transitions used in laser stabilization: the 3.39-μm transition in CH_4 and the 9.33- and 10.18-μm transitions in CO_2, *Appl. Phys. Lett.* 22, 192, 1973.

19. "For their contributions to the development of laser-based precision spectroscopy, including the optical frequency comb technique," J. L. Hall and T. W. Hänsch were awarded the Nobel Prize in Physics 2005 (together with R. J. Glauber).

20. Teets, R., Eckstein, J., and Hänsch, T.W., Coherent two-photon excitation by multiple light pulses, *Phys. Rev. Lett.* 38, 760, 1977.

21. Baklanov, E.V. and Chebotaev, V.P., Two-photon absorption of ultrashort pulses in a gas, *Sov. J. Quantum Electron.* 7, 1252, 1977.

22. Eckstein, J.N., Ferguson, A.I., and Hänsch, T.W., High-resolution two-photon spectroscopy with picosecond light pulses, *Phys. Rev. Lett.* 40, 847, 1978.

23. Holzwarth, R., Measuring the Frequency of Light Using Femtosecond Laser Pulses, Ph.D. thesis, Ludwig-Maximilian-Universität, München, 2001.

24. Udem, T., Reichert, J., Holzwarth, R., and Hänsch, T.W., Accurate measurement of large optical frequency differences with a mode-locked laser, *Opt. Lett.* 24, 881, 1999.

25. Reichert, J., Holzwarth, R., Udem, T., and Hänsch, T.W., Measuring the frequency of light with mode-locked lasers, *Opt. Commun.* 172, 59, 1999.

26. Telle, H.R., Steinmeyer, G., Dunlop, A.E., Stenger, J., Sutter, D.H., and Keller, U., Carrier-envelope offset phase control: a novel concept for absolute frequency measurement and ultrashort pulse generation, *Appl. Phys. B* 69, 327, 1999.

27. Ranka, J.K., Windeler, R.S., and Stentz, A.J., Visible continuum generation in air-silica microstructure optical fibers with anomalous dispersion at 800 nm, *Opt. Lett.* 25, 25, 2000.

28. Jones, D.J., Diddams, S.A., Ranka, J.K., Stentz, A., Windeler, R.S., Hall, J.L., and Cundiff, S.T., Carrier-envelope phase control of femtosecond modelocked lasers and direct optical frequency synthesis, *Science* 288, 635, 2000.

29. Holzwarth, R., Udem, T., Hänsch, T.W., Knight, J.C., Wadsworth, W.J., and Russell, P.St.J., Optical frequency synthesizer for precision spectroscopy, *Phys. Rev. Lett.* 85, 2264, 2000.

30. Diddams, S.A., Udem, T., Bergquist, J.C., Curtis, E.A., Drullinger, R.E., Hollberg, L., Itano, W.M., Lee, W.D., Oates, C.W., Vogel, K.R., and Wineland, D.J., An optical clock based on a single trapped [199]Hg+ ion, *Science* 293, 825, 2001.

31. Ye, J., Ma, L.S., and Hall, J.L., Molecular iodine clock, *Phys. Rev. Lett.* 87, 270801, 2001.

32. Wilpers, G., Binnewies, T., Degenhardt, C., Sterr, U., Helmcke, J., and Riehle, F., Optical clock with ultracold neutral atoms, *Phys. Rev. Lett.* 89, 230801, 2002.

33. Hollberg, L., Oates, C.W., Curtis, E.A., Ivanov, E.N., Diddams, S.A., Udem, T., Robinson, H.G., Bergquist, J.C., Rafac, R.J., Itano, W.M., Drullinger, R.E., and Wineland, D.J., Optical frequency standards and measurements, *IEEE J. Quantum Electron.* 37, 1502, 2001.

34. Gill, P., Optical frequency standards, *Metrologia* 42, S125, 2005.

35. Margolis, H.S., Barwood, G.P., Huang, G., Klein, H.A., Lea, S.N., Szymaniec, K., and Gill, P., Hertz-level measurement of the optical clock frequency in a single [88]Sr+ ion, *Science* 306, 1355, 2004.

36. Fortier, T.M., Ye, J., Cundiff, S.T., and Windeler, R.S., Nonlinear phase noise generated in air-silica microstructure fiber and its effect on carrier-envelope phase, *Opt. Lett.* 27, 445, 2002.

37. Fortier, T.M., Jones, D.J., Ye, J., Cundiff, S.T., and Windeler, R.S., Long-term carrier-envelope phase coherence, *Opt. Lett.* 27, 1436, 2002.

38. Morgner, U., Ell, R., Metzler, G., Schibli, T.R., Kärtner, F.X., Fujimoto, J.G., Haus, H.A., and Ippen, E.P., Nonlinear optics with phase-controlled pulses in the sub-two-cycle regime, *Phys. Rev. Lett.* 86, 5462, 2001.

39. Ell, R., Morgner, U., Kärtner, F.X., Fujimoto, J.G., Ippen, E.P., Scheuer, V., Angelow, G., and Tschudi, T., Generation of 5 fsec pulses and octave-spanning spectra directly from a Ti:sapphire laser, *Opt. Lett.* 26, 373, 2001.

40. Fortier, T.M., Jones, D.J., and Cundiff, S.T., Phase stabilization of an octave-spanning Ti:sapphire Laser, *Opt. Lett.* 28, 2198, 2003.

41. Matos, L., Kleppner, D., Kuzucu, O., Schibli, T.R., Kim, J., Ippen, E.P., and Kaertner, F. X., Direct frequency comb generation from an octave-spanning, prism-less Ti:sapphire laser, *Opt. Lett.* 28, 2198, 2004.

42. Mücke, O.D., Ell, R., Winter, A., Kim, J., Birge, J.R., Matos, L., and Kärtner, F.X., Self-referenced 200 MHz octave-spanning Ti:sapphire laser with 50 attosecond carrier-envelope phase jitter, *Opt. Express* 13, 5163, 2005.

43. Bartels, A. and Kurz, H., Generation of a broadband continuum by a Ti:sapphire femtosecond oscillator with a 1-GHz repetition rate, *Opt. Lett.* 27, 1839, 2002.

44. Ramond, T.M., Diddams, S.A., Hollberg, L., and Bartels, A., Phase-coherent link from optical to microwave frequencies by means of the broadband continuum from a 1-GHz Ti:sapphire femtosecond oscillator, *Opt. Lett.* 27, 1842, 2002.

45. Mücke, O.D., Tritschler, T., Wegener, M., Morgner, U., and Kärtner, F.X., Determining the carrier-envelope offset frequency of 5-fs pulses using extreme nonlinear optics in ZnO, *Opt. Lett.* 27, 2127, 2002.

46. Tritschler, T., Hof, K.D., Klein, M.W., and Wegener, M., Variation of the carrier-envelope phase of few-cycle laser pulses owing to the Gouy phase: a solid-state-based measurement, *Opt. Lett.* 30, 753, 2005.

47. Fuji, T., Rauschenberger, J., Apolonski, A., Yakovlev, V.S., Tempea, G., Udem, T., Gohle, C., Hänsch, T.W., Lehnert, W., Scherer, M., and Krausz, F., Monolithic carrier-envelope phase-stabilization scheme, *Opt. Lett.* 30, 332, 2005.

48. Washburn, B.R., Diddams, S.A., Newbury, N.R., Nicholson, J.W., Yan, M.F., and Jørgensen, C.G., Phase-locked, erbium-fiber-laser-based frequency comb in the near infrared, *Opt. Lett.* 29, 250, 2004.

49. Schibli, T.R., Minoshima, K., Hong, F.-L., Inaba, H., Onae, A., Matsumoto, H., Hartl, I., and Fermann, M.E., Frequency metrology with a turnkey all-fiber system, *Opt. Lett.* 29, 2467, 2004.

50. Kubina, P., Adel, P., Adler, F., Grosche, G., Hänsch, T.W., Holzwarth, R., Leitenstorfer, A., Lipphardt, B., and Schnatz, H., Long term comparison of two fiber based frequency comb systems, *Opt. Express* 13, 904, 2005.

51. Mücke, O.D., Kuzucu, O., Wong, F.N.C., Ippen, E.P., Kärtner, F.X., Foreman, S.M., Jones, D.J., Ma, L.-S., Hall, J.L., and Ye, J., Experimental implementation of optical clockwork without carrier-envelope phase control, *Opt. Lett.* 29, 2806, 2004.

52. Foreman, S.M., Marian, A., Ye, J., Petrukhin, E.A., Gubin, M.A., Mücke, O.D., Wong, F.N.C., Ippen, E.P., and Kärtner, F.X., Demonstration of a HeNe/CH$_4$-based optical molecular clock, *Opt. Lett.* 30, 570, 2005.

53. Spence, D.E., Kean, P.N., and Sibbett, W., 60-fsec pulse generation from a self-mode-locked Ti:sapphire laser, *Opt. Lett.* 16, 42, 1991.

54. Negus, D.K., Spinelli, L., Goldblatt, N., and Feugnet, G., Sub-100 femtosecond pulse generation by Kerr lens modelocking in Ti:sapphire, in *Advanced Solid-State Lasers*, Dubé, G. and Chase, L., Optical Society of America, Washington, D.C., 1991, p. 120.

55. Salin, F., Squier, J., and Piché, M., Modelocking of Ti:Al$_2$O$_3$ lasers and self-focusing: a Gaussian approximation, *Opt. Lett.* 16, 1674, 1991.

56. Piché, M., Beam reshaping and self-mode-locking in nonlinear laser resonators, *Opt. Commun.* 86, 156, 1991.

57. Haus, H.A., Theory of mode locking with a fast saturable absorber, *J. Appl. Phys.* 64, 3049, 1975.

58. Haus, H.A., Fujimoto, J.G., and Ippen, E.P., Analytic theory of additive pulse and Kerr lens mode locking, *IEEE J. Quantum Electron.* 28, 2086, 1992.
59. Haus, H. A., Mode-locking of lasers, *IEEE J. Sel. Top. Quantum Electron.* 6, 1173, 2000.
60. Martinez, O.E., Fork, R.L., and Gordon, J.P., Theory of passively modelocked lasers including self-phase modulation and group-velocity dispersion, *Opt. Lett.* 9, 156, 1984.
61. Kärtner, F.X., Morgner, U., Schibli, T.R., Ell, R., Haus, H.A., Fujimoto, J.G., and Ippen, E.P., Few-cycle pulses directly from a laser, in *Few-Cycle Laser Pulse Generation and Its Applications*, Topics in Applied Physics, Vol. 95, Ed. Kärtner, F.X., Springer-Verlag, New York, 2004, p. 73.
62. Spielmann, C., Curley, P.F., Brabec, T., and Krausz, F., Ultrabroadband femtosecond lasers, *IEEE J. Quantum Electron.* 30, 1100, 1994.
63. Tamura, K., Ippen, E.P., Haus, H.A., and Nelson, L.E., 77-fs pulse generation from a stretched-pulse modelocked all-fiber ring laser, *Opt. Lett.* 18, 1080, 1993.
64. Chen, Y., Kärtner, F.X., Morgner, U., Cho, S.H., Haus, H.A., Fujimoto, J.G., and Ippen, E.P., Dispersion managed mode-locking, *J. Opt. Soc. Am. B* 16, 1999, 1999.
65. Nijhof, J.H.B., Doran, N.J., Forysiak, W., and Knox, F.M., Dispersion managed solitons, *Electron. Lett.* 23, 1726, 1997.
66. Morgner, U., Kärtner, F.X., Cho, S.H., Haus, H.A., Fujimoto, J.G., Ippen, E.P., Scheuer, V., Angelow, G., and Tschudi, T., Sub-two cycle pulses from a Kerr-lens modelocked Ti:sapphire laser, *Opt. Lett.* 24, 411, 1999.
67. Haus, H.A., Tamura, K., Nelson, L.E., and Ippen, E.P., Stretched-pulse additive pulse mode-locking in fiber ring lasers: theory and experiment, *IEEE J. Quantum Electron.* 31, 591, 1995.
68. Magni, V., Cerullo, G., De Silvestri, S., and Monguzzi, A., Astigmatism in Gaussian-beam self-focusing and in resonators for Kerr-lens mode locking, *J. Opt. Soc. Am. B* 12, 476, 1995.
69. Penzkofer, A., Wittmann, M., Lorenz, M., Siegert, E., and MacNamara, S., Kerr lens effects in a folded-cavity four-mirror linear resonator, *Opt. Quantum Electron.* 28, 423, 1996.
70. Herrmann, J., Theory of Kerr-lens mode locking: role of self-focusing and radially varying gain, *J. Opt. Soc. Am. B* 11, 498, 1994.
71. Grace, E.J., Ritsataki, A., French, P.M.W., and New, G.H.C., New optimization criteria for slit-apertured and gain-apertured KLM all-solid-state lasers, *Opt. Commun.* 183, 249, 2000.
72. Kärtner, F.X., Matuschek, N., Schibli, T., Keller, U., Haus, H.A., Heine, C., Morf, R., Scheuer, V., Tilsch, M., and Tschudi, T., Design and fabrication of double-chirped mirrors, *Opt. Lett.* 22, 831, 1997.
73. Matuschek, N., Kärtner, F.X., and Keller, U., Theory of double-chirped mirrors, *IEEE J. Sel. Top. Quantum Electron.* 4, 197, 1998.
74. Dobrowolski, J.A., Tikhonravov, A.V., Trubetskov, M.K., Sullivan, B.T., and Verly, P.G., Optimal single-band normal-incidence antireflection coatings, *Appl. Opt.* 35, 644, 1996.
75. Tilsch, M., Scheuer, V., Staub, J., and Tschudi, T., Direct optical monitoring instrument with a double detection system for the control of multilayer systems from the visible to the near infrared., *SPIE Conf. Proc.* 2253, 414, 1994.
76. Scheuer, V., Tilsch, M., and Tschudi, T., Reduction of absorption losses in ion beam sputter deposition of optical coatings for the visible and near infrared, *SPIE Conf. Proc.* 2253, 445, 1994.

77. Holman, K.W., Jones, R.J., Marian, A., Cundiff, S.T., and Ye, J., Detailed studies and control of intensity-related dynamics of femtosecond frequency combs from mode-locked Ti:sapphire lasers, *IEEE J. Sel. Top. Quantum Electron.* 9, 1018, 2003.

78. Matos, L., Mücke, O.D., Chen, J., and Kärtner, F.X., Carrier-envelope phase dynamics and noise analysis in octave-spanning Ti:sapphire lasers, *Opt. Express* 14, 2497, 2006.

79. Haus, H.A. and Ippen, E.P., Group velocity of solitons, *Opt. Lett.* 26, 1654, 2001.

80. Agrawal, G.P., *Nonlinear Fiber Optics*, 3rd ed., Academic Press, San Diego, CA, 2001.

81. Haus, H.A. and Lai, Y., Quantum theory of soliton squeezing: a linearized approach, *J. Opt. Soc. Am. B* 7, 386, 1990.

82. Stenger, J., Schnatz, H., Tamm, C., and Telle, H.R., Ultraprecise measurement of optical frequency ratios, *Phys. Rev. Lett.* 88, 073601, 2002.

83. Witte, S., Zinkstok, R.T., Hogervorst, W., and Eikema, K.S.E., Control and precise measurement of carrier-envelope phase dynamics, *Appl. Phys. B* 78, 5, 2004.

84. Gardner, F.M., *Phaselock Techniques*, 3rd ed., John Wiley & Sons, Hoboken, NJ, 2005.

85. Kärtner, F.X., Brovelli, L.R., Kopf, D., Kamp, M., Calasso, I., and Keller, U., Control of solid state laser dynamics by semiconductor devices, *Opt. Eng.* 34, 2024, 1995.

86. Kärtner, F.X., Jung, I.D., and Keller, U., Soliton mode-locking with saturable absorbers, *IEEE J. Sel. Top. Quantum Electron.* 2, 540, 1996.

87. Franklin, G.F., Powell, J.D., and Emami-Naeini, A., *Feedback Control of Dynamic Systems*, 3rd ed., Addison-Wesley, Reading, MA, 1994.

88. Baltuška, A., Fuji, T., and Kobayashi, T., Controlling the carrier-envelope phase of ultrashort light pulses with optical parametric amplifiers, *Phys. Rev. Lett.* 88, 133901, 2002.

89. Zimmermann, M., Gohle, C., Holzwarth, R., Udem, T., and Hänsch, T.W., Optical clockwork with an offset-free difference-frequency comb: accuracy of sum- and difference-frequency generation, *Opt. Lett.* 29, 310, 2004.

90. Fuji, T., Apolonski, A., and Krausz, F., Self-stabilization of carrier-envelope offset phase by use of difference-frequency generation, *Opt. Lett.* 29, 632, 2004.

91. Amy-Klein, A., Goncharov, A., Daussy, C., Grain, C., Lopez, O., Santarelli, G., and Chardonnet, C., Absolute frequency measurement in the 28-THz spectral region with a femtosecond laser comb and a long-distance optical link to a primary standard, *Appl. Phys. B* 78, 25, 2004.

92. Gubin, M.A., Tyurikov, D.A., Shelkovnikov, A.S., Koval'chuk, E.V., Kramer, G., and Lipphardt, B., Transportable He-Ne/CH4 optical frequency standard and absolute measurements of its frequency, *IEEE J. Quantum Electron.* 31, 2177, 1995.

93. Basov, N.G. and Gubin, M.A., Quantum frequency standards, *IEEE J. Sel. Top. Quantum Electron.* 6, 857, 2000.

94. Gubin, M.A. and Protsenko, E.D., Laser frequency standards based on saturated-dispersion lines in methane, *Quantum Electron. (Moscow)* 27, 1048, 1997.

95. Foreman, S.M., Jones, D.J., and Ye, J., Flexible and rapidly configurable femtosecond pulse generation in the mid-IR, *Opt. Lett.* 28, 370, 2003.

96. Ye, J., Peng, J.-L., Jones, R.J., Holman, K.W., Hall, J.L., Jones, D.J., Diddams, S.A., Kitching, J., Bize, S., Bergquist, J.C., Hollberg, L.W., Robertsson, L., and Ma, L.-S., Delivery of high-stability optical and microwave frequency standards over an optical fiber network, *J. Opt. Soc. Am. B* 20, 1459, 2003.

97. All synthesizers and frequency counters used in the experiment were referenced to a local cesium clock.

98. Krylova, D.D., Shelkovnikov, A.S., Petrukhin, E.A., and Gubin, M.A., Effect of intracavity back reflections on the accuracy and stability of optical frequency standards, *Quantum Electron. (Moscow)* 34, 554, 2004.

99. Ivanov, E.N., Diddams, S.A., and Hollberg, L., Analysis of noise mechanisms limiting the frequency stability of microwave signals generated with a femtosecond laser, *IEEE J. Sel. Top. Quantum Electron.* 9, 1059, 2003.

100. Manufacturer's data for: Poseidon Scientific Instruments SLCO; Frequency Electronics, Inc. FE-102A-100 and FE-205A.

11

Solid-State Ultrafast Optical Parametric Amplifiers

Giulio Cerullo and Cristian Manzoni

CONTENTS

11.1 Introduction

Ultrafast optical science is a rapidly expanding field based on the capability, offered by lasers, to generate ultrashort light pulses, with duration ≤ 100 fs (see Chapter 6 to Chapter 10). On the one hand such short pulses enable the observation of very fast dynamical processes in atoms, molecules, and solids [1]; on the other hand, their very high intensities enable the study of new regimes in light-matter interaction [2]. In a classical "pump–probe" experiment, the system is resonantly excited by a "pump" pulse, and its subsequent evolution is monitored by the pump-induced absorption change of a delayed "probe" pulse. Our capability to observe fast dynamical processes is limited by the duration of the light pulses available to us, thus calling for the generation of shorter and shorter pulses; however, the need to excite a system on resonance and probe optical transitions occurring at different photon energies requires frequency tunability of both pump and probe pulses. Therefore, frequency tunability is as important an issue as short pulse duration for an ultrafast optical spectroscopy system.

The first sources of femtosecond light pulses were based on the dye laser [3–5] (i.e., "liquid laser") technology. While in some cases very short pulses could be obtained, the peak powers were limited by the low saturation intensities of the dye molecules, and the presence of dye jets made the systems very sensitive to alignment, preventing long term stability; partial frequency tunability could be achieved in a cumbersome way by changing the laser dye.

The 1990s have witnessed a revolution in ultrafast laser technology, in which dye laser systems have been completely replaced by solid state ones. Three main innovations have been responsible for this evolution: (1) the availability of high optical grade Ti:sapphire crystals [6], (2) the discovery of Kerr-lens mode locking (KLM) [7], and (3) the introduction of the chirped-pulse amplification (CPA) technique [8,9]. Ti:sapphire is a laser material with excellent thermal properties and exceptionally broad-gain bandwidth in the near-IR (from 600–1000 nm). KLM is a powerful passive mode-locking technique which allows generating from Ti:sapphire oscillators pulses with sub-100-fs duration and energy in the nJ range; CPA enables to boost the energy from an oscillator by some six orders of magnitude, up to the mJ level. Typical figures of commercial Ti:sapphire laser systems are pulse energy >1 mJ, pulse duration <100 fs, and repetition rate >1 kHz. However, the frequency tunability of such laser sources is limited in a narrow range around the peak gain wavelength of 800 nm or 400 nm, obtained by second harmonic generation (SHG). Despite this limitation, the very high peak powers of these sources, in excess of 10 GW, enable exploitation of the second-order nonlinear optical effect known as optical parametric amplification (OPA) to extend their tuning range [10]. The principle of OPA is quite simple: in a suitable second-order nonlinear crystal, a high- frequency and high-intensity beam (the *pump beam*, at frequency ω_3) amplifies a lower-frequency, lower-intensity beam (the *signal beam*, at frequency ω_2); in addition a third beam (the *idler beam*, at frequency ω_1) is generated. The OPA process can be given a simple corpuscular interpretation: a photon at frequency ω_3 is absorbed by a virtual level of the material and two photons at frequencies ω_2 and ω_1 are emitted. In this interaction both energy conservation

$$\hbar\omega_3 = \hbar\omega_1 + \hbar\omega_2 \tag{11.1}$$

and momentum conservation

$$\hbar k_3 = \hbar k_1 + \hbar k_2 \tag{11.2}$$

must be fulfilled. The signal frequency can in principle vary from $\omega_3/2$ (the so-called *degeneracy* condition) to ω_3 and, correspondingly, the idler varies from $\omega_3/2$ to 0 (at degeneracy signal and idler coincide). To summarize, an OPA transfers energy from an high intensity, fixed frequency pump beam, to a low intensity, variable frequency signal beam; in this process a third beam, the idler, is also generated. The OPA acts as a broadband optical amplifier with continuously tunable center frequency, which can amplify by many orders of magnitude a suitably generated weak signal beam (the so-called *seed beam*).

In this chapter we will discuss the state of the art of femtosecond OPAs including the most recent developments [11]. In Section 11.2 we will briefly review the theory of optical parametric amplification and point out the features relevant to ultrashort pulses, deriving some design criteria. In Section 11.3 we will outline the classical designs of femtosecond OPA systems, covering the near-IR, visible and mid-IR spectral ranges. In Section 11.4 we will show how OPAs have the capability of generating pulses much shorter than the pump pulses, exploiting the ultrabroad gain bandwidths available in certain configurations. In Section 11.5 we will describe another unique feature of OPAs, namely, the capability to generate, under suitable conditions, pulses with a constant carrier-envelope phase. Finally, in Section 11.6 we will show how OPAs, thanks to their very high gain and low thermal loading, may provide an alternative route to the generation of ultrahigh peak power pulses, exceeding the PW level.

11.2 Theory of Optical Parametric Amplification

In order to highlight the main factors influencing the optical parametric amplification process with ultrashort pulses, we will give a brief theoretical treatment [12,13]. We start with a linearly polarized light pulse with carrier frequency ω, envelope $A(z,t)$ and plane wavefront, propagating in the z direction

$$E(z,t) = A\left(z,t\right) \exp\left[j\left(\omega t - kz\right)\right] + c.c. \tag{11.3}$$

in a medium with a nonlinear polarization

$$P_{NL}(z,t) = \varepsilon_0 d_{eff} E^2(z,t) \tag{11.4}$$

Here, d_{eff} is the so-called effective nonlinear optical coefficient, which is linked through the tensor $\chi^{(2)}$ to the propagation direction and polarization of the field.

If the propagation equation

$$\frac{\partial^2 E(z,t)}{\partial z^2} - \mu_0 \frac{\partial^2 D(z,t)}{\partial t^2} = \mu_0 \frac{\partial^2 P_{NL}(z,t)}{\partial t^2} \tag{11.5}$$

is manipulated considering linear dispersion and the *slowing varying envelope approximation* (SVEA), one gets [13]

$$\left[\frac{\partial A}{\partial z} + \frac{\partial k}{\partial \omega}\frac{\partial A}{\partial t} + \frac{1}{2j}\frac{\partial^2 k}{\partial \omega^2}\frac{\partial^2 A}{\partial t^2} \right](-2jk)\exp[j(\omega t - kz)] = \mu_0 \frac{\partial^2 P_{NL}(z,t)}{\partial t^2} \tag{11.6}$$

where μ_0 is the magnetic permeability of vacuum. By defining the group velocity

$$v_g = \frac{\partial \omega}{\partial k}$$

and the group velocity dispersion (GVD)

$$D = \frac{\partial^2 k}{\partial \omega^2},$$

Equation 11.6 can be recast in the form:

$$\left[\frac{\partial A}{\partial z} + \frac{1}{v_g}\frac{\partial A}{\partial t} + \frac{1}{2j}D\frac{\partial^2 A}{\partial t^2} \right](-2jk)\exp[j(\omega t - kz)] = \mu_0 \frac{\partial^2 P_{NL}(z,t)}{\partial t^2} \tag{11.7}$$

This equation shows that the pulse envelope propagates through the medium with the group velocity v_g, whereas GVD causes a change of its shape; the nonlinear polarization acts as a source term driving its amplitude variations.

Let us now consider the propagation in a second-order nonlinear medium of three collinear linearly polarized waves at frequencies ω_1, ω_2, and ω_3, with $\omega_1 < \omega_2 < \omega_3$ and $\omega_1 + \omega_2 = \omega_3$. The polarizations of the three waves are not necessarily parallel. The resulting electric field can be written as:

$$E(z,t) = A_1(z,t)\exp\left[j(\omega_1 t - k_1 z)\right] + A_2(z,t)\exp\left[j(\omega_2 t - k_2 z)\right]$$
$$+ A_3(z,t)\exp\left[j(\omega_3 t - k_3 z)\right] + c.c. \tag{11.8}$$

and the nonlinear second-order polarization becomes

$$P_{NL}(z,t) = 2\varepsilon_0 d_{eff} A_2{}^*(z,t)A_3(z,t)\exp j[\omega_1 t - (k_3 - k_2)z] +$$
$$+ 2\varepsilon_0 d_{eff} A_1{}^*(z,t)A_3(z,t)\exp j[\omega_2 t - (k_3 - k_1)z] + \tag{11.9}$$
$$+ 2\varepsilon_0 d_{eff} A_1(z,t)A_2(z,t)\exp j[\omega_3 t - (k_1 + k_2)z] + c.c.$$

where we have rejected the constant terms (optical rectification) and the oscillating components at frequencies different from ω_1, ω_2, and ω_3.

Equation 11.9 shows that (1) the nonlinear polarization at one frequency is proportional to the product of the electric fields at the other two frequencies, so that the waves become nonlinearly coupled, and (2) the wavevector of the nonlinear polarization at a given frequency does not coincide with that of the wave.

The forcing term $\partial^2 P_{NL}/\partial t^2$ can be easily calculated and simplified applying the SVEA, thus getting:

$$\frac{\partial^2 P_{NL}}{\partial t^2} = -2\varepsilon_0 d_{eff}\omega_1^2 A_2{}^*(z,t)A_3(z,t)\exp j[\omega_1 t - (k_3 - k_2)z] +$$

$$-2\varepsilon_0 d_{eff}\omega_2^2 A_1{}^*(z,t)A_3(z,t)\exp j[\omega_2 t - (k_3 - k_1)z] + \qquad (11.10)$$

$$-2\varepsilon_0 d_{eff}\omega_3^2 A_1(z,t)A_2(z,t)\exp j[\omega_3 t - (k_1 + k_2)z] + c.c.$$

By inserting (10) into (7) and rejecting the complex conjugated, one derives the three coupled nonlinear equations:

$$\frac{\partial A_1}{\partial z} + \frac{1}{v_{g1}}\frac{\partial A_1}{\partial t} + \frac{1}{2j}D_1\frac{\partial^2 A_1}{\partial t^2} = -\frac{jd_{eff}\omega_1}{c_0 n_1}A_2{}^* A_3\exp\left[-j\Delta kz\right] \quad (11.11a)$$

$$\frac{\partial A_2}{\partial z} + \frac{1}{v_{g2}}\frac{\partial A_2}{\partial t} + \frac{1}{2j}D_2\frac{\partial^2 A_2}{\partial t^2} = -\frac{jd_{eff}\omega_2}{c_0 n_2}A_1{}^* A_3\exp\left[-j\Delta kz\right] \quad (11.11b)$$

$$\frac{\partial A_3}{\partial z} + \frac{1}{v_{g3}}\frac{\partial A_3}{\partial t} + \frac{1}{2j}D_3\frac{\partial^2 A_3}{\partial t^2} = -\frac{jd_{eff}\omega_3}{c_0 n_3}A_1 A_2\exp\left[j\Delta kz\right] \quad (11.11c)$$

where $\Delta k = k_3 - k_2 - k_1$ is the so-called wave vector mismatch. Equation 11.11 show that the three fields exchange energy through the nonlinear polarization; if two fields are injected into the medium, a third one is generated by the nonlinear interaction. According to the initial condition, two main phenomena can arise: sum-frequency generation (SFG) and difference-frequency generation (DFG). In SFG, two fields at frequencies ω_1 and ω_2 interact to produce a field at the frequency $\omega_3 = \omega_1 + \omega_2$ (SHG is just a particular case in which $\omega_1 = \omega_2$). In DFG, two fields at frequencies ω_3 and ω_2 interact; the field at ω_3 loses energy in favor of the field at ω_2 and of the newly generated frequency $\omega_1 = \omega_3 - \omega_2$ (optical rectification is the limiting case when $\omega_3 = \omega_2$).

OPA is a mechanism similar to DFG, except for the strength of the interacting fields: DFG arises when the fields at ω_2 and ω_3 have comparable intensities, whereas OPA occurs when the field at ω_2 is much weaker. In OPA, therefore, an intense beam at ω_3 (pump beam) generates a beam at $\omega1$ (idler beam) and transfers energy to the beam at ω_2 (signal beam), thereby amplifying it.

To capture the main features of the OPA process, we will simplify Equation 11.11 by considering monochromatic waves

$$\left(\frac{\partial A(z,t)}{\partial t}=0\right):$$

$$\frac{dA_1}{dz}=-\frac{jd_{eff}\omega_1}{c_0\,n_1}A_2{}^*A_3\exp[-j\Delta kz] \tag{11.12a}$$

$$\frac{dA_2}{dz}=-\frac{jd_{eff}\omega_2}{c_0\,n_2}A_1{}^*A_3\exp[-j\Delta kz] \tag{11.12b}$$

$$\frac{dA_3}{dz}=-\frac{jd_{eff}\omega_3}{c_0\,n_3}A_1A_2\exp[-j\Delta kz] \tag{11.12c}$$

These equations can be easily solved if one neglects the depletion of the pump beam ($A_3 \cong$ cost.) and assumes an initial signal intensity A_{20} (seed beam) and no initial idler beam ($A_{10}=0$). One obtains, after propagation for a length L of nonlinear material:

$$I_2\left(L\right)=I_{20}\left\{1+\left[\frac{\Gamma}{g}\text{senh}\left(gL\right)\right]^2\right\} \tag{11.13a}$$

$$I_1\left(L\right)=I_{20}\frac{\omega_1}{\omega_2}\left[\frac{\Gamma}{g}\text{senh}\left(gL\right)\right]^2 \tag{11.13b}$$

where Γ and g are defined as follows:

$$\Gamma^2=\frac{\omega_1\omega_2 d_{eff}^2\left|A_3\right|^2}{n_1 n_2 c_0^2}=\frac{2\omega_1\omega_2 d_{eff}^2 I_3}{n_1 n_2 n_3 \varepsilon_0 c_0^3}\quad\text{and}\quad g^2=\Gamma^2-\left(\frac{\Delta k}{2}\right)^2 \tag{11.14}$$

In the case of large gain ($gL\gg1$), Equation 11.13 simplifies to

$$I_2\left(L\right)=\frac{1}{4}\left(\frac{\Gamma}{g}\right)^2 I_{20}\exp\left(2gL\right) \tag{11.15a}$$

$$I_1\left(L\right)=\frac{\omega_1}{4\omega_2}\left(\frac{\Gamma}{g}\right)^2 I_{20}\exp\left(2gL\right) \tag{11.15b}$$

When $\Delta k = 0$, $g = \Gamma$ and expressions (15) simplify into:

$$I_2(L) = \frac{1}{4} I_{20} \exp(2gL) \tag{11.16a}$$

$$I_1(L) = I_{20} \frac{\omega_1}{\omega_2} \exp(2gL) \tag{11.16b}$$

Equation 11.15 and Equation 11.16 show that, for a given small-signal gain, both signal and idler intensities grow exponentially (within the no-depletion approximation) with crystal length L and allow to define a parametric gain as

$$G = \frac{I_2(L)}{I_{20}} = \frac{1}{4}\left(\frac{\Gamma}{g}\right)^2 \exp(2gL) \tag{11.17}$$

Note that this exponential growth is qualitatively different from the quadratic growth observed in other second-order processes like SFG and SHG and shows that the OPA behaves like a real amplifier. With respect to a classical optical amplifier based on population inversion in an atomic or molecular transition, however, an OPA has three important differences: (1) it does not have any energy storage capability, i.e., the gain is present only during the pump pulse; (2) the gain center frequency is not fixed but can be continuously adjusted by varying the phase- matching condition; and (3) the gain bandwidth is not limited by the linewidth of the transition but rather by the possibility of satisfying the phase-matching condition over a broad range of frequencies.

Let us now consider the factors influencing the parametric gain G:

1. $G \propto \exp(g)$ strongly depends on the parameter g, which is maximum when $\Delta k = 0$ (phase-matching condition). G rapidly decreases for nonzero values of Δk, suggesting that phase-matching is a key condition to be fulfilled in order to get significant amplification from the nonlinear material. In the following we will focus on the case $\Delta k = 0$, leading to $g = \Gamma$.

2. $G \propto \exp(d_{eff})$ depends exponentially on the second-order nonlinear optical coefficient of the crystal d_{eff}; one should therefore select the crystal with the highest d_{eff}. There are, however, other considerations leading to the choice of the crystal, such as phase-matching range, dispersive properties, availability, and optical damage threshold.

3. $G \propto \exp\left(\sqrt{I_p}\right)$ scales as the exponential of the square root of the pump intensity. This indicates the suitability of ultrashort pulses for OPAs, due to their high peak powers. One should try to use the highest possible pump intensity before the onset of other nonlinear optical phenomena such as self-focusing, self-phase modulation and beam breakup. In order to be able to use high pump intensities, it is important to have a spatially clean beam profile, without hot spots.

4. $G \propto \exp(L)$ scales as the exponential of the crystal length. This gain is strikingly different from the efficiency of SHG, which scales as the square of the crystal length. During the nonlinear interaction the idler beam is generated at the difference frequency between pump and signal. The idler then gets amplified, and a beam at the signal frequency and phase is generated, thus adding power to the amplified seed. This mechanism acts as positive feedback, giving rise to an exponential gain of the signal during its propagation. We will see that with ultra-short light pulses, the optimum crystal length has to be chosen considering the durations and group velocities of the interacting pulses.

5. $G \propto \exp\left(\sqrt{\omega_1 \omega_2}\right)$ scales as the exponential of the square root of the product of signal and idler frequencies. This seems to indicate an advantage to use high pump frequencies. However, we will see that with ultrashort pulses this advantage is often offset by the larger difference in group velocities of the interacting pulses.

Let us consider two practical examples of parametric gain calculation; both refer to the crystal β-barium borate (BBO), which is the most widely used. For an OPA pumped at $\lambda_3 = 0.8$ μm and a signal wavelength $\lambda_2 = 1.2$ μm, employing a 3-mm-thick BBO crystal and pump intensity $I_3 = 40$ GW/cm², we obtain G = 2 × 10⁴; for an OPA pumped at $\lambda_3 = 0.4$ μm and a signal wavelength $\lambda_2 = 0.6$ μm, employing a 3-mm-thick BBO crystal and pump intensity $I_3 = 40$ GW/cm², we obtain G = 6 × 10⁵. Note the advantage of using shorter pump wavelengths.

It is worthwhile mentioning that the phase-matching condition $k_3 = k_2 + k_1$ and the relationship between the frequencies, $\omega_3 = \omega_2 + \omega_1$, can be rewritten as

$$\hbar k_3 = \hbar k_2 + \hbar k_1 \tag{11.18a}$$

$$\hbar \omega_3 = \hbar \omega_2 + \hbar \omega_1 \tag{11.18b}$$

which represent momentum and energy conservation for the interacting pump, signal, and idler photons in the corpuscular interpretation of parametric processes. This interpretation allows extension of the phase-matching condition to a noncollinear geometry by simply using the vector equation:

$$\hbar \mathbf{k}_3 = \hbar \mathbf{k}_1 + \hbar \mathbf{k}_2 \tag{11.19}$$

Let us now discuss how to obtain the collinear phase-matching condition $\Delta k = 0$ necessary for high gain. The condition $k_3 = k_1 + k_2$ is equivalent to

$$n_3 = \frac{\omega_1 n_1 + \omega_2 n_2}{\omega_3} \tag{11.20}$$

It is easy to show that (20) is normally not satisfied in isotropic materials because of normal dispersion; in the limiting case $\omega_1 = \omega_2 = \omega$, $\omega_3 = 2\omega$ (SHG), it would correspond to n(2ω) = n(ω). In birefringent crystals, which are, anyway, those used for OPAs because, due to the lack of inversion symmetry, they have a second-order nonlinear optical coefficient, one can exploit the polarization dependence of refractive index. In particular, one can choose for the high-frequency pump field the polarization direction corresponding to the lower refractive index. One can then select the propagation direction in the crystal in order to exactly satisfy (20) for a given set of wavelengths.

We consider the case, common in femtosecond OPAs, of negative uniaxial birefringent crystals ($n_e < n_o$); according to the previous discussion, the pump beam has extraordinary polarization. If both signal and idler beam have ordinary polarization, we talk about type I phase matching. If either the signal or the idler beams have extraordinary polarization, we talk about type II phase matching [14]. Both types of phase matching have their specific advantages and can be used depending on the case. Let us consider type I phase matching in a negative uniaxial crystal (such as for example BBO). In this case Equation 11.20 becomes:

$$n_{e3}{}^* = \frac{\omega_1 n_{o1} + \omega_2 n_{o2}}{\omega_3} \tag{11.21}$$

which allows us to compute $n_{e3}{}^*$, i.e., the refractive index at the pump frequency required for phase matching. The refractive index of the extraordinary beam depends on the angle θ between the wave vector and the optical axis according to the:

$$\frac{1}{n_e^2(\theta)} = \frac{\sin^2(\theta)}{n_e^2} + \frac{\cos^2(\theta)}{n_o^2} \tag{11.22}$$

where n_e is the principal extraordinary refractive index. Putting together Equation 11.21 and Equation 11.22, one obtains the phase-matching angle as

$$\theta_m = \arcsen\left[\frac{n_{e3}}{n_{e3}{}^*}\sqrt{\frac{n_{o3}^2 - n_{e3}{}^{*2}}{n_{o3}^2 - n_{e3}^2}}\right] \tag{11.23}$$

In this case, the idler is efficiently generated with ordinary polarization, whereas extraordinary components are negligible as they are phase-mismatched. For type II phase matching, Equation 11.21 cannot be solved explicitly, only approximate solutions can be found with numerical methods [14].

Figure 11.1 shows the phase-matching angle as a function of wavelength for BBO type I and II OPAs at the pump wavelengths 0.8 µm. Note that, in general, the phase-matching angle has a less pronounced wavelength dependence for type I with respect to type II phase matching.

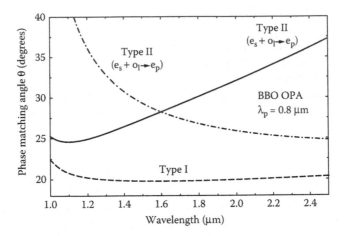

FIGURE 11.1
Phase-matching angle as a function of wavelength for a BBO OPA pumped at 0.8 μm for type I phase matching (dashed line), type II phase matching (signal extraordinary, dash-dotted line) and type II phase matching (idler extraordinary, solid line).

So far we have discussed the case of monochromatic waves $(\partial A(z,t)/\partial t = 0)$; we will now see how the previous considerations can be extended to OPAs with ultrashort pulses. In this case one needs to go back to Equation 11.11, containing the time derivatives of the pulse envelopes. As previously discussed, the second derivatives describe broadening of the individual pulses within the nonlinear crystal due to dispersion and can be neglected for reasons that will appear clearer later. By transforming to a frame of reference that is moving with the group velocity of the pump pulse $(\tau = t\text{-}z/v_{g3})$ we obtain the equations

$$\frac{\partial A_1}{\partial z} + \left(\frac{1}{v_{g1}} - \frac{1}{v_{g3}} \right) \frac{\partial A_1}{\partial \tau} = -\frac{jd_{eff}\omega_1}{c_0\,n_1} A_2{}^* A_3 \exp\left[-j\Delta k z\right] \qquad (11.24a)$$

$$\frac{\partial A_2}{\partial z} + \left(\frac{1}{v_{g2}} - \frac{1}{v_{g3}} \right) \frac{\partial A_2}{\partial \tau} = -\frac{jd_{eff}\omega_2}{c_0\,n_2} A_1{}^* A_3 \exp\left[-j\Delta k z\right] \qquad (11.24b)$$

$$\frac{\partial A_3}{\partial z} = -\frac{jd_{eff}\omega_3}{c_0\,n_3} A_1 A_2 \exp\left[j\Delta k z\right] \qquad (11.24c)$$

These equations show that the main limiting factor in parametric interaction with ultrashort pulses is the difference in group velocities, also called group velocity mismatch (GVM), between the interacting pulses. With respect to the pump pulse, which is fixed in this frame of reference, the signal and idler pulses move with different velocities and after a while separate temporally, thus stopping the parametric interaction. The relative speeds v_{grel}

of signal and idler pulses with respect to the pump pulse are given, according to Equation 11.24, by

$$\frac{1}{v_{grel\,j}} = \frac{1}{v_{gj}} - \frac{1}{v_{g3}} \qquad j = 1,2 \tag{11.25}$$

One can define pulse splitting length as the propagation length after which the signal (or the idler) pulse temporally separates from the pump pulse in the absence of gain; this represents the length over which parametric interaction takes place, i.e., the maximum useful crystal length. It can be expressed as

$$l_{j3} = \tau\, v_{grel\,j} \qquad j = 1,2 \tag{11.26}$$

where τ is the pump pulse duration. Note that the pulse splitting length becomes shorter for decreasing pulse duration and for increasing GVM. GVM depends on the crystal type, pump wavelength and type of phase matching. Figure 11. 2 and Figure 11.3 show examples of GVM curves for a BBO OPA pumped by 0.8 μm (Figure 11.2) and 0.4 μm (Figure 11.3) pulses, respectively. Note that, due to greater dispersion values in the visible, GVM is in general larger in this wavelength range.

We have shown that the GVM between pump and signal/idler limits the maximum useful length of the nonlinear crystal. We will now consider which factors determine the amplification bandwidth. Ideally, one would like to have a broadband amplifier, i.e., an amplifier which, for a fixed pump frequency ω_3, provides a more or less constant gain over a broad range of signal frequencies. Practically, however, the phase-matching condition can be satisfied only for a given set of frequencies (ω_1, ω_2, and ω_3). If the pump frequency

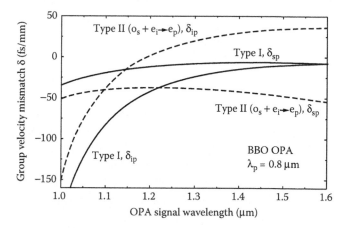

FIGURE 11.2
Pump-signal (δ_{sp}) and pump-idler (δ_{ip}) group velocity mismatch curves for a BBO OPA pumped at 0.8 μm for type I phase matching (solid lines) and type II phase matching (idler extraordinary, dashed lines).

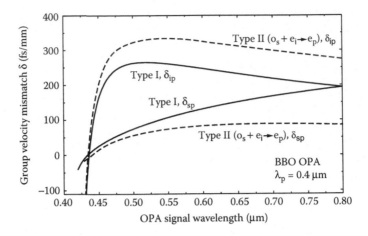

FIGURE 11.3
Pump-signal (δ_{sp}) and pump-idler (δ_{ip}) group velocity mismatch curves for a BBO OPA pumped at 0.4 μm for type I phase matching (solid lines) and type II phase matching (idler extraordinary, dashed lines).

is fixed at ω_3 and the signal frequency changes to $\omega_2+\Delta\omega$, then by energy conservation the idler frequency changes to $\omega_1-\Delta\omega$. The ensuing wave vector mismatch can be approximated to the first order as

$$\Delta k \cong -\frac{\partial k_2}{\partial \omega_2}\Delta\omega + \frac{\partial k_1}{\partial \omega_1}\Delta\omega = \left(\frac{1}{v_{g1}} - \frac{1}{v_{g2}}\right)\Delta\omega \qquad (11.27)$$

The full width at half maximum parametric gain bandwidth can be calculated from Equation 11.17, within the large gain approximation, as:

$$\Delta v = \frac{2(\ln 2)^{1/2}}{\pi}\left(\frac{g}{L}\right)^{1/2}\frac{1}{\left|\dfrac{1}{v_{g2}} - \dfrac{1}{v_{g1}}\right|} \qquad (11.28)$$

Equation 11.28 shows that the gain bandwidth is inversely proportional to the GVM between signal and idler and has only a square root dependence on small-signal gain and crystal length. For the case when $v_{g1}= v_{g2}$, Equation 11.28 loses validity and Equation 11.27 must be expanded to the second order, giving

$$\Delta v = \frac{2(\ln 2)^{1/4}}{\pi}\left(\frac{g}{L}\right)^{1/4}\frac{1}{\left|D_1 + D_2\right|} \qquad (11.29)$$

Figure 11.4 and Figure 11.5 show typical plots of phase-matching bandwidths for BBO OPAs, pumped at 0.8 μm (Figure 11.4) and 0.4 μm (Figure

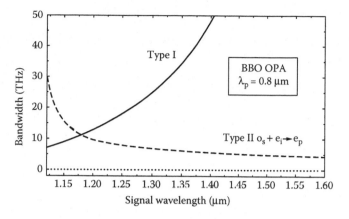

FIGURE 11.4
Phase-matching bandwidth for a BBO OPA pumped at 0.8 μm for type I phase matching (solid line) and type II phase matching (idler extraordinary, dashed line). Crystal length is 4 mm and pump intensity 50 GW/cm².

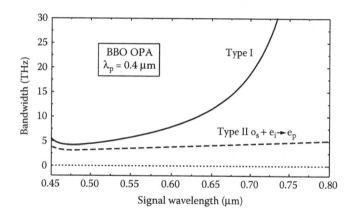

FIGURE 11.5
Phase-matching bandwidth for a BBO OPA pumped at 0.4 μm for type I phase matching (solid line) and type II phase matching (idler extraordinary, dashed line). Crystal length is 2 mm and pump intensity 100 GW/cm².

11.5), respectively. We see a remarkable difference between type I and II phase matching: for type II interaction, the bandwidth is smaller than in type I and stays more or less constant over the tuning range, while for type I interaction, as previously said, the bandwidth increases as the OPA approaches degeneracy. These features can be exploited for different applications: type I phase matching is used to achieve the shortest pulses, whereas type II phase matching allows us to obtain relatively narrow bandwidths over broad tuning ranges, which are required for many spectroscopic investigations.

11.3 Optical Parametric Amplifiers from the Visible to the Mid-IR

A general scheme of an ultrafast OPA is presented in Figure 11.6. The system is powered by energetic femtosecond pulses, typically coming from an amplified Ti:sapphire laser at 800 nm. A fraction of the beam is split and used to generate the seed beam. Then the pump beam (which may be optionally frequency doubled) and the seed, after their timing has been adjusted by a delay line, interact in a first amplification stage. It is possible to further amplify the signal in a second stage (power amplifier), using a previously split fraction of the pump beam. The two-stage approach has two advantages: (1) it allows compensation from the GVM arising between pump and signal beams in the first stage, and (2) it enables to adjust the pump intensity, and thus the parametric gain, separately in the two stages. In particular, the power amplifier requires a much lower gain. At the OPA output, after the pump has been spectrally filtered, both signal and idler beams are available. In some cases, it is necessary to use a pulse compressor to restore the transform-limited duration of the pulses.

In the following we will first briefly describe the seed generation techniques, then give examples of OPA designs in the near-IR, visible, and mid-IR frequency ranges.

11.3.1 Seed Generation

The first stage of any OPA is the seed generator, i.e., a stage producing the initial photons at the signal wavelength. As the seed beam is at a different wavelength from the pump beam, a nonlinear process is required. The two

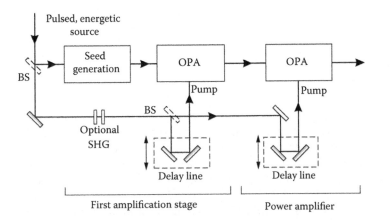

FIGURE 11.6
Scheme of an ultrafast optical parametric amplifier. BS, beam splitter; OPA, optical parametric amplification stage.

main techniques employed for seed generation are parametric superfluorescence and supercontinuum generation.

Parametric superfluorescence [15] is parametric amplification of vacuum noise and can also be thought as two-photon emission from a virtual level excited by the pump field. In practice it is achieved by pumping a suitable second-order nonlinear crystal (often of the same kind as those employed in the later OPA stages); amplification of vacuum noise occurs at those wavelengths for which the phase-matching condition is satisfied. Disadvantages of this technique are the poor spatial quality of the generated seed beam and its large fluctuations (inherent in a process which is starting from noise); for these reasons, it is nowadays seldom used in OPAs.

Supercontinuum (white light) generation [16] is a phenomenon occurring when an intense light pulse is focused inside a transparent material, such as fused silica or sapphire. It can be understood as a result of self-phase-modulation (SPM) of the pulse. When an intense laser beam impinges on a material of thickness L, its refractive index becomes a linear function of the intensity (Kerr effect):

$$n(t) = n_0 + n_2\, I(t) \tag{11.30}$$

so that the phase shift experienced by a quasi-monochromatic pulse of carrier frequency ω_0 is

$$\phi(t) = \omega_0 t - \frac{\omega_0}{c_0} n(t) L = \omega_0 t - \frac{\omega_0}{c_0} n_0 L - \frac{\omega_0}{c_0} n_2 L\, I(t) \tag{11.31}$$

and the instantaneous frequency becomes

$$\omega(t) = \frac{d\phi(t)}{dt} = \omega_0 - \frac{\omega_0 n_2 L}{c_0} \frac{dI}{dt} = \omega_0 \left(1 - \frac{n_2 L}{c_0} \frac{dI}{dt} \right). \tag{11.32}$$

Equation 11.32 suggests that the instantaneous carrier frequency of the pulse changes in time; in particular, during the leading edge of the pulse ($dI/dt > 0$), it is shifted to the red, whereas during the trailing edge ($dI/dt < 0$) it is shifted to the blue, leading to a dramatic broadening of the spectrum. Note that this is an oversimplified description of the phenomenon; in fact, other phenomena such as spatial self-focusing, temporal self-steepening, and space time-focusing play a role (see [17] and [18] for a detailed description).

Practically, white light generation is achieved by focusing 800-nm, 100-fs pulses with energy from 1 to 3 μJ into a sapphire plate, with thickness ranging between 1 and 3 mm. Sapphire is chosen because of excellent thermal conductivity and low UV absorption, preventing long-term degradation. The white light extends throughout the visible (down to $\approx 0.4\ \mu m$) and the near-IR (up to $\approx 1.5\ \mu m$), with an energy of approximately 10 pJ per nm of bandwidth; it displays very high pulse-to-pulse stability and excellent spatial beam quality.

11.3.2   OPAs in Near-IR

The simplest way of obtaining OPAs tunable in the near-IR is by pumping with the fundamental wavelength (800 nm) of an amplified Ti:sapphire laser [19–26]. They have the following advantages: (1) high available pump energies (up to the mJ-level) and (2) low pump-signal and pump-idler GVM values, allowing the use of long nonlinear crystals and the obtainment of high gains. These advantages are partially offset by a lower gain for the parametric interaction in the near-IR compared to the visible because of lower signal and idler frequencies. Tunability is limited by the losses due to absorption of the idler wave in the nonlinear crystal. This is typically a problem at wavelengths longer than ≈ 3 μm. Therefore, the signal beam is tunable from degeneracy (1.6 μm) to 1.1 μm whereas the idler beam tunes from 1.6 to 3 μm. This leaves a "hole" in the tuning range from 0.8 to 1.1 μm.

A typical setup for a near-IR OPA is shown in Figure 11.7 [27]. It is pumped by a standard CPA Ti:sapphire laser generating 500-μJ, 50-fs pulses at 1 kHz repetition rate. A small fraction of the pump (≈2 μJ) is used to generate white light in a 2-mm-thick sapphire plate; 50 μJ of the pump are used to amplify a near-IR wavelength in the continuum in a preamplifier stage consisting of a 3-mm-thick BBO crystal cut for type II phase matching (θ = 26°, φ = 0°). Wavelength tuning is achieved by tilting the crystal, thus changing the phase-matching condition. Typical signal energies after the preamplifier are up to 6 μJ. The power amplifier stage consists of an identical BBO crystal pumped by 450 μJ; in this case signal energies up to 200 μJ are generated. The signal beam is tunable from 1.1 to 1.6 μm and the idler up to 2.8 μm; pulsewidth ranges from 30 to 50 fs according to wavelength.

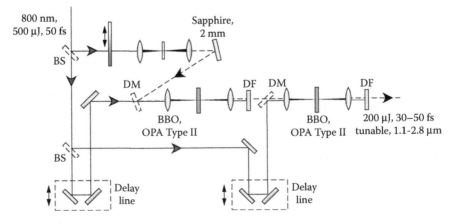

FIGURE 11.7
Scheme of a near-IR OPA. BS, beam splitter; DF, dichroic filter; DM, dichroic mirror.

11.3.3 OPAs in the Visible

The generation of femtosecond pulses tunable in the visible is important for a variety of spectroscopic applications in physics, chemistry, and biology. A straightforward way of achieving tunable visible pulses consists in frequency- doubling the output of an 800-nm-pumped near-IR OPA [26]; however, as absorption of the idler in the nonlinear crystal prevents signal amplification for wavelengths shorter than $\approx$1100 nm, the SH would be tunable down to only 550 nm, leaving a substantial part of the visible range uncovered. Pumping with the SH of an amplified Ti:sapphire laser around 400 nm [28–33], the signal can be tuned through most of the visible range, from $\approx$450 nm to degeneracy (800 nm). Correspondingly, the idler tunes from 800 nm to 2.5 μm; this fills the gap in the tuning range left by near-IR OPAs. Visible OPAs in general produce lower energies than near-IR ones because of the lower pump energy available from a frequency-doubled pump. Furthermore, GVM is much larger in the visible range, preventing the use of long nonlinear crystals. This disadvantage is partially compensated for by the larger gain for parametric interaction in the visible. Also in visible OPAs, the most popular nonlinear material is BBO. Type II phase matching provides gain bandwidths that are narrower and stay essentially constant over the tuning range, which may be beneficial for some spectroscopic applications [30]. Using type I phase matching, the amplified pulse bandwidth strongly depends on signal wavelength, increasing in the red as degeneracy is approached. The collinear interaction geometry limits the available phase-matching bandwidth. A possible solution to this problem consists of changing the phase-matching angle in subsequent amplification stages, so as to amplify at each pass a different spectral region of the white light seed; this technique makes it possible to generate 30 fs pulses tunable throughout the visible [32]. A more effective solution to the problem, which consists in using noncollinear interaction geometry, will be discussed in Section 11.4.

11.3.4 OPAs in Mid-IR

The mid-IR spectral region (3-10 μm) is spectroscopically very interesting because it covers the vibrational transitions in molecules and the intersubband transitions in low-dimensional semiconductors; the generation of ultrashort pulses tunable in this frequency range is therefore of great interest to chemists and physicists. Although several approaches have been proposed, we will report on the most widely used one, based on DFG between signal and idler pulses generated by a near-IR OPA pumped at 800 nm [34–36]. A typical experimental setup is shown in Figure 11.8. The signal and idler pulses are generated by a type II near-IR OPA and thus have perpendicular polarization, as required in the DFG process. The two pulses, with a combined energy of $\approx$ 200 μJ, are separated by a dichroic mirror, reflecting the idler and transmitting the signal, and recombined by an identical mirror. In this way their relative delay can be adjusted. The two collinear and

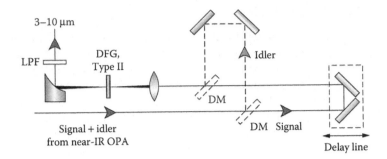

FIGURE 11.8
Scheme of a mid-IR pulse generation stage. DM, dichroic mirrors; LPF, long-pass filter.

temporally overlapped pulses are then focused on a 1-mm-thick AgGaS$_2$ crystal, cut for type II phase matching ($\theta = 40°$); the mid-IR pulses are separated from the residual near-IR by a long-pass filter and collimated by an off-axis paraboloid. By tuning the OPA and simultaneously readjusting the phase-matching angle of the DFG crystal, the mid-IR pulses can be tuned from 3 to 10 μm with energies in excess of 1 μJ; pulsewidths down to 70 fs, corresponding to just a few cycles of the mid-IR electric field, are obtained.

11.4 Ultrabroadband Optical Parametric Amplifiers

OPAs provide an easy way of tuning over a broad range the frequency of an otherwise fixed femtosecond laser system, and this is their main spectroscopic application. On the other hand, they can be broadband optical amplifiers, and thus can be used to dramatically shorten, by more than an order of magnitude, the duration of the pump pulse. One can, therefore, start with a femtosecond system producing relatively long pulses (100–200 fs) and use an OPA to shorten their duration to the sub-10-fs regime. In this section we will describe such ultrabroadband OPAs based on noncollinear phase matching.

We have seen in Section 11.2 that the gain bandwidth of an OPA is determined by the GVM between signal and idler pulses. For 400-nm-pumped, visible OPAs, the idler is in the NIR and moves at a significantly larger group velocity with respect to the signal, due to the lower refractive index, resulting in narrow gain bandwidths. One solution to this problem is to adopt a noncollinear interaction geometry (see Figure 11.9a) in which pump, signal, and idler propagate along different directions. It is then possible to choose the angle Ω between signal and idler wave vectors is such way that the projection of the idler group velocity along the signal direction equals the signal group velocity:

$$v_{g2} = v_{g1} \cos\Omega. \tag{11.33}$$

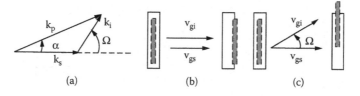

FIGURE 11.9
(a) Schematic of a non-collinear interaction geometry; (b) representation of signal and idler pulses in the case of collinear interaction; and (c) same as (b), for non-collinear interaction.

This effect is shown pictorially in Figure 11.9; for a collinear geometry (Figure 11.9b), signal and idler moving with different group velocities get quickly separated, giving rise to pulse lengthening and bandwidth reduction, whereas in the noncollinear case (Figure 11.9c) the two pulses manage to stay temporally overlapped. Note that Equation 11.33 can be satisfied only if $v_{g2} > v_{g1}$; this is, however, always the case if the signal is in the visible range and type I phase matching in a negative uniaxial crystals, where both signal and idler see the ordinary refractive index, is used. Equation 11.33 allows determining the signal-idler angle Ω required for broadband phase matching; from a practical point of view, it is more useful to know the pump-signal angle α, which is given by

$$\alpha = \arcsin\left(\frac{1 - v_{g2}^2/v_{g1}^2}{1 + 2\,v_{g2}n_2\omega_2/v_{g1}n_1\omega_1 + n_2^2\omega_2^2/n_1^2\omega_1^2}\right)^{1/2} \tag{11.34}$$

Note that in the noncollinear geometry, the phase matching condition is a vector equation

$$k_3 = k_1 + k_2 \tag{11.35}$$

which, projected in the directions parallel and perpendicular to the signal wave vector and according to the geometry sketched in Figure 11.9a, becomes

$$k_3 \cos\alpha = k_1 \cos\Omega + k_2 \tag{11.36a}$$

$$k_3 \sin\alpha = k_1 \sin\Omega. \tag{11.36b}$$

Note that in a practical situation the pump-signal angle α is fixed while the signal idler angle Ω adjusts itself, according to Equation 11.36b, to satisfy the phase-matching condition; so, the idler is emitted at a different angle for each wavelength, i.e., is angularly dispersed.

As an example, in a type I BBO OPA pumped at 400 nm, for a signal wavelength of 600 nm broadband noncollinear phase matching is achieved, according to Equation 11.34, for $\alpha = 3.7°$. To better illustrate the effect of noncollinear

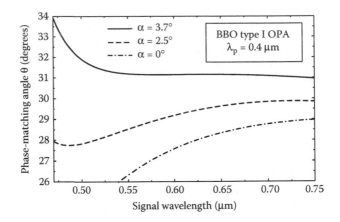

FIGURE 11.10

Phase-matching angle as a function of signal wavelength for a non-collinear type I BBO OPA pumped at 0.4 μm, as a function of pump-signal angle α.

phase matching, in Figure 11.10 we plot, for the above described OPA, the phase-matching angle θ_m as a function of signal wavelength for different values of α. For a collinear configuration ($\alpha = 0°$), θ_m shows a strong dependence on the signal wavelength so that, for a fixed crystal orientation, phase-matching can be achieved only over a narrow signal frequency range. By going to a noncollinear configuration and increasing α, the wavelength dependence of θ_m becomes progressively weaker until, for the optimum value $\alpha = 3.7°$, a single crystal orientation ($\theta \cong 31°$) allows us to achieve simultaneously phase matching over an ultrabroad bandwidth, extending from 0.5 to 0.75 μm.

This favorable property of noncollinear geometry for broadband parametric amplification was first recognized by Gale et al. [37–39] and was exploited to build broadband optical parametric oscillators generating pulses as short as 13 fs; more recently, the same concept was extended by several research groups to OPAs seeded by the white-light continuum [40–49]. The noncollinear OPA (NOPA) has nowadays become a workhorse for ultrafast spectroscopy with the highest temporal resolution.

In the following we will describe the NOPA design used by our research group [41,43,46,47], the schematic of which is shown in Figure 11.11. The system starts with a conventional CPA Ti:sapphire laser (Clark-MXR, Model CPA-1) generating 150-fs, 800-nm pulses at 1 kHz with energy up to 500 μJ. The system has enough energy for simultaneously pumping up to three NOPAs. A fraction of the beam is used to generate the pump pulses at 400 nm by SHG in a 1-mm-thick lithium triborate crystal; they have energy up to 50 μJ, and their duration is slightly lengthened to $\approx$180 fs by GVM during the SHG process. Another small fraction of the beam, with energy of approximately 2 μJ, is focused into a 1-mm-thick sapphire plate to generate the seed pulses; by carefully controlling the energy incident on the plate (using a variable-optical-density attenuator) and the position of the plate around the

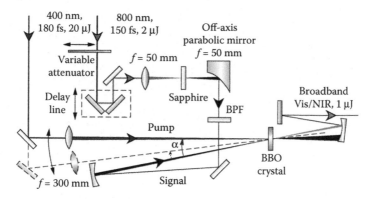

FIGURE 11.11
Scheme of a non-collinear OPA pumped at 0.4 μm BPF, bandpass filter.

focus, a highly stable single-filament white light continuum is generated. The group delay of the visible portion of the white light, measured with the technique of optical Kerr gating, is small and fairly linear with frequency, with group-delay dispersion (GDD) ranging from 75 fs^2 at 700 nm to 115 fs^2 at 500 nm; the measured chirp matches fairly well the GDD introduced by the sapphire plate. This can be understood by recalling that, as explained in Section 11.3.1, most of the visible frequency components are generated in a short time interval during the leading edge of the pulse, around the point of maximum steepness. To avoid the introduction of additional chirp, only reflective optics are employed to guide the white light to the amplification stage. Parametric gain is achieved in a 1-mm-thick BBO crystal, cut at $\theta = 31°$, using a single-pass configuration to increase the gain bandwidth. The chosen crystal length is close to the pulse-splitting length for signal and pump in the wavelength range of interest. To minimize the effects of self-focusing, we position the BBO crystal beyond the focus of the pump beam. In that position the pump spot size is approximately 120 μm, corresponding to an intensity of 120 GW/cm^2; at higher intensities distortions and beam breakup are observed. The white-light seed is imaged by a spherical mirror (Figure 11.11) in the BBO crystal, with a spot size matching that of the pump beam.

When it is illuminated by the pump pulse and aligned perpendicularly to the pump beam, the BBO crystal emits a strong off-axis parametric superfluorescence in the visible in a cone with an apex angle of 6.15° (corresponding to an angle of 3.7° inside the crystal); this is the direction for which the group velocities of signal and idler are matched and therefore the gain bandwidth is maximized. We carefully adjust the pump-signal angle to match the cone apex angle, giving us a visual aid in order to optimize the system for broadband phase-matching. In these conditions, for optimum pump–seed delay, an ultrabroad gain bandwidth that extends over most of the visible is observed. A typical amplified pulse spectrum shown in Figure 11.12a: It displays a FWHM of 180 THz. The amplified pulses have energy of approximately 2 μJ, peak-to-peak fluctuations of less than 7% and maintain

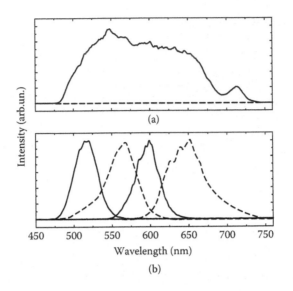

FIGURE 11.12
(a) NOPA spectrum under optimum alignment conditions (pump-signal angle α = 3.7°); (b) spectra generated by the NOPA with α = 2°, demonstrating continuous tunability.

a good TEM_{00} beam quality. Higher energies, up to ≈ 10 μJ, can be extracted by a second pass in the BBO crystal. After the gain stage, the amplified pulses are collimated by a spherical mirror and sent to the compressor.

The NOPA generates pulses with very broad bandwidths and thus potentially very short; in order to obtain the minimum pulsewidth compatible with their bandwidth (the so-called transform-limited duration), one needs to accurately control their spectral phase. In the following we will briefly outline the basics of pulse compression: let us consider an ultrashort pulse with frequency spectrum:

$$E(\omega) = A(\omega) \exp\left[j\phi(\omega)\right] \qquad (11.37)$$

where $A(\omega)$ is the spectral amplitude and $\phi(\omega)$ is the spectral phase. By a Taylor expansion of the spectral phase around the central frequency ω_0 (carrier frequency) one obtains

$$\phi(\omega) = \phi(\omega_0) + \phi'(\omega_0)(\omega - \omega_0) + \frac{1}{2}\phi''(\omega_0)(\omega - \omega_0)^2$$
$$+ \frac{1}{6}\phi'''(\omega_0)(\omega - \omega_0)^3 + ... \qquad (11.38)$$

By differentiating ϕ with respect to ω one obtains the group delay, representing the transit time for a quasi-monochromatic wavepacket through the system:

$$\tau_g(\omega) = \frac{d\phi}{d\omega} = \tau_{g0} + GDD(\omega - \omega_0) + \frac{1}{2}TOD(\omega - \omega_0)^2 + \dots \quad (11.39)$$

where $\tau_{g0} = \phi'(\omega_0)$ is the group delay of the carrier frequency, $GDD = \phi''(\omega_0)$ is known as second-order dispersion (or group-delay dispersion), and $TOD = \phi'''(\omega_0)$ is known as third-order dispersion. In order to obtain a transform-limited pulse, the group delay should be constant with frequency, so that the different frequency components of the pulse arrive simultaneously. Propagation in a transparent material gives, in the visible spectral range, a "positive" dispersion ($GDD > 0$, $TOD > 0$), corresponding to the blue colors traveling more slowly than the red ones. In order to compensate for it, a compressor, i.e., an optical system providing "negative" dispersion (blue colors traveling faster than the red ones), is needed. Simple compressors can be obtained by double-passing prism pairs or grating pairs; in this case, one can adjust the parameters (prisms or gratings distance) in order to compensate for the second- order dispersion of the material:

$$GDD_{material} + GDD_{compressor} = 0 \quad (11.40)$$

This correction is sufficient for pulses with moderately broad bandwidths, as TOD cannot be controlled independently [40–42]. For ultrabroadband pulses one can use prism-grating sequences in order to be able to correct simultaneously second and third order dispersion [44,45]; alternatively, one can use compressors with custom-tailored dispersion characteristics, which can accurately correct to high orders the phase distortions in the system. Such compressors are either chirped dielectric mirrors, in which both GDD and TOD can be independently controlled or adaptive systems based on deformable mirrors. Using ultrabroadband double-chirped mirrors, pulses as short as 5.7 fs have been generated from NOPAs [47], whereas the pulse-width could be pushed down to 3.9 fs using deformable mirrors [48].

Although some applications benefit from the ultrabroad bandwidths generated by the visible NOPA under optimum alignment, in other cases it is necessary to be more frequency selective and reduce the NOPA bandwidth, generating pulses which are still very short (15–20 fs) but have center frequencies tunable across the visible range. Two possible strategies to reach this goal are: (1) increasing the chirp of the white light before the amplification stage and (2) detuning the pump-signal angle α from the optimum value. In the first case the spectral components of the white light seed are properly delayed so that only few components are temporally superposed to the 180-fs pump and get amplified. This can be easily obtained by adding a glass block on the seed path; its thickness changes the group delay of the various colors and influences the bandwidth of the seed superposed to and amplified by the pump. Tunability, on the contrary, is obtained adjusting the seed-pump delay only, thanks to the broad acceptance bandwidth of the amplifying crystal. The disadvantage of this technique lies in the need to compensate for the chirp added by the glass, which increases when thicker glasses are chosen.

In order to avoid any correction in the compression stage, a second approach can be adopted: since, as already explained through Figure 11.10, the amplified bandwidth depends on the angle α between pump and signal, one can choose to set the two beams at a suitable angle, according to the desired bandwidth. For $\alpha \approx 2.5°$, for instance, the bandwidth acceptance of the amplifier allows pulses to get about 20 fs long after compression; in this case tunability can be achieved by tilting the crystal towards the desired θ_m. Figure 11.12b shows a series of spectra acquired in this configuration for different values of θ_m, demonstrating the continuous tunability of the setup. The bandwidth can be changed without any consequence on the pulse chirp, thus avoiding any changes in the compressor configuration.

The tunable visible NOPA can be easily configured to generate pulses in the near-IR spectral region; the only difference lays in the stage of signal generation and filtering. In order to increase the signal spectral energy density in the IR, white light is generated in a 2-mm-thick sapphire plate and then filtered with a 1-mm-thick Schott long-pass filter. For center wavelengths near 900 nm, pulse compression is necessary to compensate for the chirp introduced by self-phase modulation and propagation in the sapphire plate: since no suitable chirped mirrors are available for the purpose, a prism-pair compression stage can be adopted. Two fused silica prisms cut at Brewster angle and kept at a distance of 21 cm give excellent results. From 900 nm to 1200 nm pulses from 12 fs to 20 fs can be obtained, and the compression can be adjusted so that measured pulses are only 15% longer than the transform-limited expected duration. For longer wavelengths dispersion gets negligible since GDD approaches 0, and the duration of the amplified pulses goes from 30 fs to 45 fs, approaching the transform limit also without any compression stage.

11.5 Self-Phase-Stabilized Optical Parametric Amplifiers

Ultrabroadband light pulses containing only a few optical carrier cycles under their envelope are currently produced by several methods, including direct generation from a mode-locked oscillator [50], spectral broadening in a guiding medium [51] and optical parametric amplification (see previous section). For such short pulses, the maximum amplitude of the electric field varies significantly between consecutive optical half-cycles, so that it becomes important to control the evolution of the electric field underneath the pulse envelope. Mathematically, the electric field of an ultrashort pulse can be written as

$$E(t) = A(t) \cos(\omega_c t + \varphi) \qquad (41)$$

where $A(t)$ is the pulse envelope with its maximum at $t = 0$, ω_c is the carrier frequency and φ is the carrier-envelope phase (CEP). If $\varphi = 0$, a maximum of the electric field corresponds to the peak of the pulse envelope (cosine pulse), whereas if $\varphi = \pi/2$, the electric at the peak of the pulse envelope is zero (sine pulse). Control of the CEP becomes important for extreme non-linear optics experiments which are sensitive to the electric field rather than the intensity of the pulse. Examples of such effects in the non-resonant case, requiring high pulse energies, are above-threshold ionization [52] and high-harmonic generation [53], for which CEP-control is a prerequisite to the production of attosecond pulses [54]. Resonant CEP-sensitive phenomena, requiring much lower pulse energies, are for example sideband interference during carrier-wave Rabi flopping [55], multiphoton photoemission from metal surfaces [56] and quantum interference between one and two-photon absorption pathways in semiconductors [57].

A standard mode-locked laser oscillator produces a pulse train in which the CEP changes from pulse to pulse, due to the difference between phase and group velocities during one roundtrip. The CEP slippage $\Delta\varphi$, in addition, is not constant in time, but drifts due to fluctuations in the cavity parameters. The goal of producing CEP-stabilized pulses has been achieved by first measuring the CEP slippage $\Delta\varphi$ and stabilizing it using active electronic feedback [58,59]; by subsequent picking and amplification of selected pulses, a train of pulses with constant CEP is achieved [53].

A completely different method was recently proposed by Baltuska et al. [60], who recognized that in a white-light seeded OPA in which the pump and the seed are derived from the same source, the idler is automatically phase stabilized. The physical mechanisms underlying this effect are basically three: (1) a white light continuum generated by SPM maintains the same value of CEP as the driving pulse; (2) the OPA process preserves the phase of the seed pulse, carrying it to the signal wave; (3) in an OPA, the idler is produced by DFG between pump and signal, so that in this nonlinear process the CEPs of the two pulses add up with different signs. Let us now consider the case in which both pump and seed are derived from the same pulse: in this situation their CEPs are equal and thus cancel in the idler beam, leading to passive, all-optical phase stabilization. If we consider previously described OPA configurations, those pumped by the FW satisfy this condition, because the 800-nm beam is used both for seed generation and as a pump; on the other hand, the visible NOPA does not satisfy this condition, because the pump is the SH and carries CEP 2φ while the seed is generated by the FW and carries CEP φ. To solve this problem, Baltuska et al. built a NOPA in which the white light seed is generated by the SH instead of the FW, in a CaF_2 plate; in this case, both pump and signal carry a CEP 2φ and the idler is self phase stabilized. Since the idler of the NOPA spans over an octave of bandwidth, its self-frequency -stabilization can be simply verified by a nonlinear spectral interferometer (NLSI) [61–63]. The pioneering work by Baltuska et al. suffers from the disadvantage that the self-phase-stabilized pulses obtained from the NOPA idler have a strong angular dispersion. This

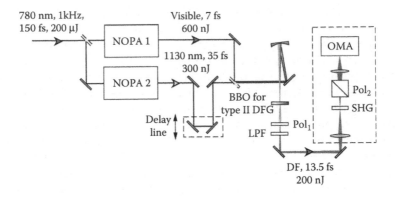

FIGURE 11.13
Scheme of the experimental setup for DFG of self-phase stabilized pulses: Pol_1, near-IR sheet polarizer; LPF, long-pass filter; SHG, 2-mm-thick BBO crystal; Pol_2, cube polarizer; OMA, optical multichannel analyzer.

was overcome in a recent work by our group, in which collinear DFG was performed between pulses generated by two synchronized NOPAs sharing the same CEP: an ultrabroadband visible and a narrowband near-IR NOPA. The obtained self-phase-stabilized pulses are free from angular dispersion, have 200-nJ energy and ultrabroad bandwidth spanning over an octave, from 800 to 1700 nm [64]; tuning of the difference frequency (DF) spectrum is possible by changing the near-IR carrier frequency.

The experimental setup is shown in Figure 11.13. An amplified Ti:sapphire of the type described in Section 11.4 drives simultaneously two NOPAs. Both NOPAs are pumped by the SH, seeded by white light and use BBO crystals cut for type I phase-matching ($\theta = 31°$, $\varphi = 0°$). The first is a standard visible NOPA, already described in Section 11.4, employing the pump-seed angle (3.7°) required for broadband visible amplification. It generates pulses with up to 2 μJ energy and spectrum extending from 490 to 720 nm, compressed to nearly transform-limited sub-10-fs duration by 10 bounces onto ultra-broadband chirped mirrors. The second NOPA is configured to operate in the near-IR by selecting the near-IR portion of the white-light seed with a long-pass filter and reducing the pump-seed angle to $\approx 2.4°$. Upon suitably tilting the BBO crystal, pulses centered at 1130 nm with 50 nm bandwidth and up to 1 μJ energy are produced. Due to the low dispersion in the near-IR, nearly transform-limited 30 fs pulsewidths are obtained without pulse compression.

The pulses produced by the two NOPAs are synchronized by a delay line, combined by a dichroic beam splitter and collinearly focused on a 400-μm-thick BBO crystal cut for DFG with type II phase-matching ($\theta = 32°$, $\varphi = 30°$): both pulses from the NOPAs have extraordinary polarization, while the DF pulse is generated with ordinary polarization. Type II DFG provides a broad acceptance bandwidth for the high-frequency beam and a much narrower one for the low-frequency beam. This is well suited to our experimental

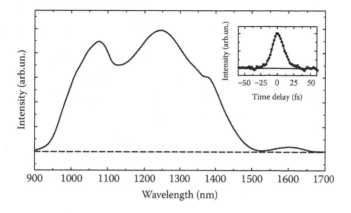

FIGURE 11.14
Spectrum of the DF pulse. Inset, non-collinear autocorrelation of DF pulses.

configuration with an ultrabroadband visible and a narrowband near-IR beam: numerical simulations show that, even with a 400-μm-thick BBO crystal, it should be possible to down-convert the full visible bandwidth to the near-IR. The collinear configuration avoids angular dispersion of the DF beam but poses the problem of its separation from the generating beams, one of which spectrally overlaps with the DF: the type II configuration enables to select the DF light by a near-IR polarizer together with a long-pass filter.

A typical DF spectrum is shown in Figure 11.14: it extends from 900 to 1700 nm, which are the responsivity limits of the InGaAs spectrometer used for this measurement. Use of a silicon spectrometer enables to detect a tail of the DF signal extending to wavelengths lower than 800 nm; therefore the DF spectrum spans, albeit with its tails, over an octave of bandwidth. With incident energies of 600 nJ and 300 nJ from the visible and near-IR NOPAs, respectively, typical measured energy of the DF pulses is 200 nJ, in excellent agreement with numerical simulations. A Fourier transform of the spectrum in Figure 11.14 yields a 7.6-fs pulsewidth: note that, given the 3.8-fs carrier period, this corresponds to about two cycles under the pulse envelope. Preliminary pulse characterization using a non-collinear autocorrelator yielded a 13.5-fs pulsewidth without the use of any compressor (see inset in Figure 11.14).

To investigate self-phase-stabilization of the DF pulses, we set up a NLSI [65]. The DF pulses are focused on a 2-mm frequency-doubling BBO crystal and the spatially overlapped FW and SH are directed to the silicon spectrometer through a polarizer. Self-phase-modulation in the BBO crystal slightly broadens the FW spectrum, so that it is possible to achieve spectral overlap between its blue edge and the red edge of the SH spectrum (see Figure 11.15a). Suitably adjusting their intensities by rotating the polarizer, an interference pattern is observed; Figure 11.15a shows one such pattern obtained by averaging over 1000 laser shots. The appearance of a stable fringe pattern upon averaging is a clear proof of the shot-to-shot stability of the CEP due to the

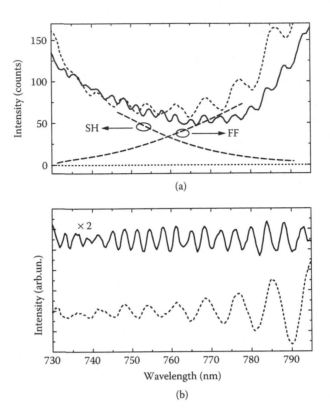

FIGURE 11.15
(a) Spectral interference patterns between the FF and the SH of the DF pulses as generated (dashed curve) and with a 2-mm-thick SF10 plate on the beam path (solid curve). The dotted lines are a guide to the eye. (b) Oscillatory components of interferograms.

self-phase-stabilization mechanism of DFG; the reduced fringe visibility results from a combination of imperfect mode matching between the interfering beams, leakage in the polarizer and residual CEP fluctuations due to environmental noise. By taking a Fourier transform of the oscillatory component of the interferogram (shown in Figure 11.15b) one obtains a 210-fs delay between the interfering pulses (see Figure 11.16); this is attributed to the residual chirp between the frequency components of the DF pulse and to the group delay (GD) introduced by the focusing lens and the BBO crystal. To confirm that the observed fringes stem from FW-SH interference, we inserted a 2-mm-thick SF10 plate in front of the DF beam. According to the dispersive properties of SF10, this plate should introduce a GD of about 275 fs between the 760 nm and the 1520 nm components of the DF. The fringes showed a reduced period (solid lines in Figure 11.15b) corresponding to an increase of the FF-SH delay by 313 fs (solid line in Figure 11.16), in excellent agreement with the expected value. Note that all the configurations discussed in this section and based on the DFG process do not require a system with an actively CEP-stabilized oscillator.

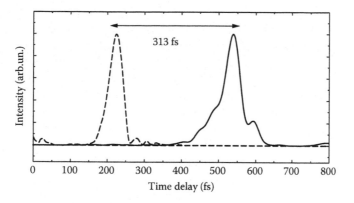

FIGURE 11.16
Fourier transforms of the oscillatory components of the interferograms in Figure 11.15(b).

11.6 Optical Parametric Chirped-Pulse Amplification

There is a great interest in the generation of ever increasing laser peak powers and focused intensities, for a number of current and potential applications. Using the CPA technique, peak powers in excess of 1 PW [66] and intensities greater than 10^{21} W/cm^2 have been demonstrated; to scale this performance to even higher levels, a number of issues must be faced in conventional CPA systems. Since the energy levels are approaching the damage threshold of the compressor gratings, significant increases of the peak power can be only achieved by shortening the pulse duration, which in turn requires an increase of the gain bandwidth. For strongly driven amplifying media, however, the phenomenon of gain narrowing reduces the available bandwidth. In addition, the prepulse due to amplified spontaneous emission (ASE) spoils the temporal pulse contrast, and the large linear and nonlinear phases accumulated in the long paths through the amplifying media prevents transform-limited pulse recompression and diffraction-limited focusing. A novel high power amplification scheme, solving most of these problems, was recently proposed by Ross et al., based on the seminal work by Dubietis et al. [67], and termed "optical parametric chirped-pulse amplification" (OPCPA). In this scheme parametric gain is achieved by coupling a quasi-monochromatic high energy pump field (such as, for example, a picosecond or nanosecond pulse generated by a neodymium laser) to a chirped, low energy broadband seed field in a nonlinear crystal [68–70]. If the seed pulse is sufficiently stretched, good energy extraction from the pump field can be achieved, and subsequent recompression makes it possible to reach very high peak powers. The OPCPA concept has some very important advantages with respect to the standard CPA:

1. The parametric amplification process, in the non-collinear geometry, can provide gain bandwidths well in excess of those achievable with conventional amplifiers and could sustain pulse spectra corresponding to a transform-limited duration of ≈5 fs. High energies are possible by using large nonlinear crystals, such as KDP, which can be grown to sizes of tens of centimeters. These crystals should be capable of withstanding pump energies of hundreds of joules.

2. The OPCPA has the capability of providing a high gain in a relatively short path; for example, a lithium triborate (LBO) crystal pumped by 0.5 ns pulses at 0.526 μm, at intensities below the damage threshold, can have a gain coefficient of 12 cm^{-1}. This short path length allows a compact, tabletop amplifier setup and also minimizes the linear and nonlinear phase distortions and ensures an excellent temporal and spatial quality of the pulses.

3. In OPCPA amplification occurs only during the pump pulse, so that the ASE and the consequent prepulse pedestal are greatly reduced.

4. In a standard amplifier, even for good energy extraction, there is always some thermal loading, due to the quantum defect (difference between energies of the pump and emitted photons, usually absorbed by the material in nonradiative decay processes); this becomes very relevant for high energy amplifiers, often requiring cryogenic cooling. In OPCPA, on the other hand, no fraction of the pump photon energy is deposited in the medium, because it is transformed in the sum of the signal and idler photon energies; so thermal loading effects, apart from parasitic absorption, are completely absent, greatly reducing spatial aberration effects on the beams.

The OPCPA concept is very promising and has been already implemented in a proof-of-principle experiment to generate TW-level pulses [70]; in a low repetition rate systems pumped by a Nd:glass laser chain, pulses with energy up to 570 mJ with 115 fs duration, corresponding to a peak power of 3.67 TW, have been obtained [71]. More recently, the ultrabroad gain bandwidth obtainable from OPCPAs has been exploited to generate 8-mJ pulses with 100-THz bandwidth [72], subsequently recompressed to 5-mJ, 10-fs pulses. The system was pumped by a frequency-doubled, amplified Nd:YAG laser producing 60-mJ, 60-ps pulses at 20 Hz repetition rate and 532 nm wavelength, and seeded by a broadband Ti:sapphire oscillator; parametric gain was achieved in two 4-mm type I BBO crystals. In addition, since OPAs preserve the CEP of the seed beam, it is possible to achieve phase-stable amplification of ultrashort pulses; recently, 11.8-fs, 120-μJ pulses at 1 kHz have been obtained [73,74].

Future developments include the use of OPCPAs pumped by large-frame high energy lasers, in order to generate PW-class systems. As an example, the design for an OPCPA pumped by the high-power iodine laser Asterix IV is given in Ref. [75]. The Asterix IV pump laser delivers at the fundamental wavelength (1.315 μm) up to 1.2 kJ of energy in pulses with a duration of 500 ps; the beam can be efficiently frequency tripled to generate over 500 J of energy at 0.438 μm. The 10-fs seed pulses are generated by a Ti:sapphire oscillator and then stretched to several hundreds picoseconds; parametric amplification takes place in three stages, using a non-collinear interaction geometry, and telescopes increase the beam size after each stage. The first two stages employ LBO because of its high nonlinear coefficient and broad amplification bandwidth, while the last stage uses KDP because this nonlinear crystal can be grown in the large sizes ($\approx$ 30 cm) required to keep the fluence below the damage threshold. After the compressor, energies of 100 J with pulse duration of 20 fs are expected, corresponding to a peak power of 5 PW and to a focused intensity of 10^{23} W/cm^2.

11.7 Conclusions

We have seen in this chapter that ultrafast OPAs are a mature technology, making it possible to extend considerably the tuning range of femtosecond Ti:sapphire laser systems. Several OPA designs have become standard and are commercially available. In particular, near-IR OPAs (pumped by the FW of a Ti:sapphire laser) offer tunability from 1.1 to 2.5 μm with several tens of μJ energy, while visible OPAs (pumped by the SH of a Ti:sapphire laser) are tunable from 0.45 to 2.5 μm with somewhat lower energies. Typical pulsewidths obtainable from these systems are in the 50-200 fs range, depending on the specific design and the pump pulse duration. SFG and DFG techniques allow extending their tunability from the UV range to the mid-IR, out to 12 μm.

In addition to their standard application as continuously frequency tunable optical amplifiers, OPAs have some other remarkable properties that make them useful in ultrafast optical science:

1. Under some conditions, in particular in a non-collinear interaction geometry, OPAs offer very broad gain bandwidths, thus enabling the generation of very short pulses, down to a few optical cycles. Such pulse durations are much shorter than that of the driving pulse, so that OPAs act as effective pulse compressors.

2. The DFG process which takes place in the OPA leads to a subtraction of the carrier-envelope phases of pump and signal pulses. Therefore, if pump and signal carry the same CEP, the idler pulses are automatically self-phase-stabilized. This passive, all-optical self-phase-stabilization technique, in conjunction with the broad phase-matching bandwidths of OPAs, offers an interesting route to the generation of few-optical-cycles with precisely controlled electric fields.

3. OPAs also have the capability of providing very high gains over broad bandwidths: these characteristics, together with the relatively short material path, negligible thermal loading of the crystals, low linear and nonlinear phase distortions and low levels of amplified spontaneous emission, make them very attractive candidates for large-scale, high peak power amplifiers. The OPCPA concept holds promise to increase the peak powers available from lasers well above the current limit of 1 PW.

References

1. A.H. Zewail, Femtochemistry: atomic-scale dynamics of the chemical bond, *J. Phys. Chem. A*, Vol. 104, 5660–5694, 2000.
2. T. Brabec and F. Krausz, Intense few-cycle laser fields: Frontiers of nonlinear optics, *Rev. Mod. Phys.*, Vol. 72, 545–591, 2000.
3. R.L. Fork, B.I. Greene, and C.V. Shank, Generation of optical pulses shorter than 0.1 psec by colliding pulse mode locking, *Appl. Phys. Lett.*, Vol. 38, 671–672, 1981.
4. W.H. Knox, M.C. Downer, R.L. Fork, and C.V. Shank, Amplified femtosecond optical pulses and continuum generation at 5-kHz repetition rate, *Opt. Lett.*, Vol. 9, 552–554, 1984.
5. R.L. Fork, C.H. Brito Cruz, P.C. Becker, and C.V. Shank, Compression of optical pulses to six femtoseconds by using cubic phase compensation, *Opt. Lett.*, Vol. 12, 483–485, 1986.
6. P.F. Moulton, Spectroscopic and laser characteristics of $Ti:Al_2O_3$, *J. Opt. Soc. Am. B*, Vol. 3, 125–133, 1986.
7. D.E. Spence, P.N. Kean, and W. Sibbett, 60-fsec pulse generation from a self-mode-locked Ti:sapphire laser, *Opt. Lett.*, Vol. 16, 42–44, 1991.
8. D. Strickland and G. Mourou, Compression of amplified chirped optical pulses, *Opt. Commun.*, Vol. 56, 219–221,1985.
9. S. Backus, C.G. Durfee, III, M.M. Murnane, and H.C. Kapteyn, High power ultrafast lasers, *Rev. Sci. Instrum.*, Vol. 69, 1207–1223, 1998.
10. J.A. Giordmaine and R.C. Miller, Tunable coherent parametric oscillation in $LiNbO_3$ at optical frequencies, *Phys. Rev. Lett.*, Vol. 14, 973–976, 1965.
11. G. Cerullo and S. De Silvestri, Ultrafast optical parametric amplifiers, *Rev. Sci. Instrum.*, Vol. 74, 1–18, 2003.
12. Y.R. Shen, *The Principles of Nonlinear Optics*, John Wiley & Sons, New York, 1984.
13. R.W. Boyd, *Nonlinear Optics*, 2nd ed., Academic Press, Boston, MA, 2003.
14. V.G. Dmitriev, G.G. Gurzadyan, and D.N. Nikogosyan, *Handbook of Nonlinear Optical Crystals*, Springer-Verlag, Berlin, 1991.
15. S.E. Harris, M.K. Oshman, and R.L. Byer, Observation of tunable optical parametric fluorescence, *Phys. Rev. Lett.*, Vol. 18, 732–734, 1967.
16. R.R. Alfano, Ed., *The Supercontinuum Laser Source*, Springer-Verlag, New York, 1989.
17. J.K. Ranka and A.L. Gaeta, Breakdown of the slowly varying envelope approximation in the self-focusing of ultrashort pulses, *Opt. Lett.*, Vol. 23, 534–536, 1998.
18. A.L. Gaeta, Catastrophic collapse of ultrashort pulses, *Phys. Rev. Lett.*, Vol. 84, 3582–3585, 2000.

19. W. Joosen, P. Agostini, G. Petite, J.P. Chambaret, A. Antonetti, Broadband femtosecond infrared parametric amplification in β-BaB$_2$O$_4$, *Opt. Lett.*, Vol. 17, 133–135, 1992.

20. G.P. Banfi, P. Di Trapani, R. Danielius, A. Piskarkas, R. Righini, I. Sa'nta, Tunable femtosecond pulses close to the transform limit from traveling-wave parametric conversion, *Opt. Lett.*, Vol. 18, 1547–1549, 1993.

21. G.P. Banfi, R. Danielius, A. Piskarkas, P. Di Trapani, P. Foggi, R. Righini, Femtosecond traveling-wave parametric generation with lithium triborate, *Opt Lett.*, Vol. 18, 1633–1635, 1993.

22. F. Seifert, V. Petrov, F. Noack, Sub-100-fs optical parametric generator pumped by a high-repetition-rate Ti:sapphire regenerative amplifier system, *Opt Lett.*, Vol. 19, 837–839, 1994.

23. V. Petrov, F. Seifert, F. Noack, High repetition rate traveling wave optical parametric generator producing nearly bandwidth limited 50 fs infrared light pulses, *Appl. Phys. Lett.*, Vol. 65, 268–270, 1994.

24. M. Nisoli, S. De Silvestri, V. Magni, O. Svelto, R. Danielius, A. Piskarkas, G. Valiulis, A. Varavinicius, Highly efficient parametric conversion of femtosecond Ti:sapphire laser pulses at 1 kHz, *Opt. Lett.*, Vol. 19, 1973–1975, 1994.

25. M. Nisoli, S. Stagira, S. De Silvestri, O. Svelto, G. Valiulis, A. Varavinicius, Parametric generation of high-energy 14.5-fs light pulses at 1.5 μm, *Opt. Lett.*, Vol. 23, 630–632, 1994.

26. V.V. Yakovlev, B. Kohler, K.R. Wilson, Broadly tunable 30-fs pulses produced by optical parametric amplification, *Opt. Lett.*, Vol. 19, 2000–2002, 1994.

27. K.R. Wilson and V.V. Yakovlev, Ultrafast rainbow: tunable ultrashort pulses from a solid-state kilohertz system, *J. Opt. Soc. Am. B*, Vol. 14, 444–449, 1997.

28. M.K. Reed, M.K. Steiner-Shepard, D.K. Negus, Widely tunable femtosecond optical parametric amplifier at 250 kHz with a Ti:sapphire regenerative amplifier, *Opt. Lett.*, Vol. 19, 1855–1857, 1994.

29. M.K. Reed, M.S. Armas, M.K. Steiner-Shepard, D.K. Negus, 30-fs pulses tunable across the visible with a 100-kHz Ti:sapphire regenerative amplifier, *Opt. Lett.*, Vol. 20, 605–607, 1995.

30. S.R. Greenfield and M.R. Wasielewski, Near-transform-limited visible and near-IR femtosecond pulses from optical parametric amplification using Type II β-barium borate, *Opt. Lett.*, Vol. 20, 1394–1396, 1995.

31. P. Di Trapani, A. Andreoni, C. Solcia, G.P. Banfi, R. Danielius, A. Piskarskas, P. Foggi, Powerful sub-100-fs pulses broadly tunable in the visible from a blue-pumped parametric generator and amplifier, *J. Opt. Soc. Am. B*, Vol. 14, 1245–1248, 1997.

32. T.S. Sosnowski, P.B. Stephens, and T.B. Norris, Production of 30-fs pulses tunable throughout the visible spectral region by a new technique in optical parametric amplification, *Opt. Lett.*, Vol. 21, 140–142, 1996.

33. K.S. Wong, Z.R. Qui, H. Wang, and G.K.L. Wong, Efficient visible femtosecond optical parametric generator and amplifier using tilted pulse-front pumping, *Opt. Lett.*, Vol. 22, 898–900, 1997.

34. F. Seifert, V. Petrov, M. Woerner, Solid-state laser system for the generation of midinfrared femtosecond pulses tunable from 3.3 to 10 μm, *Opt. Lett.*, Vol. 19, 2009–2011, 1994.

35. M.K. Reed and M.K. Steiner-Shepard, Tunable infrared generation using a femtosecond 250 kHz Ti:sapphire regenerative amplifier, *IEEE J. Quantum Electron.*, Vol. 32, 1273–1277, 1996.

36. B. Golubovic and M.K. Reed, All-solid-state generation of 100-kHz tunable mid-infrared 50-fs pulses in type I and type II AgGaS$_2$, *Opt. Lett.*, Vol. 23, 1760–1762, 1998.

37. T.J. Driscoll, G.M. Gale, and F. Hache, Ti:sapphire second-harmonic-pumped visible range femtosecond optical parametric oscillator, *Opt. Commun.*, Vol. 110, 638–644, 1994.

38. G.M. Gale, M. Cavallari, T.J. Driscoll, and F. Hache, Sub-20-fs tunable pulses in the visible from an 82-MHz optical parametric oscillator, *Opt. Lett.*, Vol. 20, 1562–1564, 1995.

39. G.M. Gale, M. Cavallari, and F. Hache, Femtosecond visible optical parametric oscillator, *J. Opt. Soc. Am. B*, Vol. 15, 702–714, 1998.

40. T. Wilhelm, J. Piel, E. Riedle, Sub-20-fs pulses tunable across the visible from a blue-pumped single-pass noncollinear parametric converter, *Opt. Lett.*, Vol. 22, 1494–1496, 1997.

41. G. Cerullo, M. Nisoli, S. De Silvestri, Generation of 11 fs pulses tunable across the visible by optical parametric amplification, *Appl. Phys. Lett.*, Vol. 71, 3616–3618, 1997

42. A. Shirakawa and T. Kobayashi, Noncollinearly phase-matched femtosecond optical parametric amplification with a 2000 cm^{-1} bandwidth, *Appl. Phys. Lett.*, Vol. 72, 147–149, 1998.

43. G. Cerullo, M. Nisoli, S. Stagira, S. De Silvestri, Sub-8-fs pulses from an ultra-broadband optical parametric amplifier in the visible, *Opt. Lett.*, Vol. 23, 1283–1285, 1998.

44. A. Shirakawa, I. Sakane, T. Kobayashi, Pulse-front-matched optical parametric amplification for sub-10-fs pulse generation tunable in the visible and near infrared, *Opt. Lett.*, Vol. 23, 1292–1294, 1998.

45. A. Shirakawa, I. Sakane, M. Takasaka, T. Kobayashi, Sub-5-fs visible pulse generation by pulse-front-matched noncollinear optical parametric amplification, *Appl. Phys. Lett.*, Vol. 74, 2668–2670, 1999.

46. G. Cerullo, M. Nisoli, S. Stagira, S. De Silvestri, G. Tempea, F. Krausz, K. Ferencz, Mirror-dispersion-controlled sub-10-fs optical parametric amplifier tunable in the visible, *Opt. Lett.*, Vol. 24, 1529–1531, 1999.

47. M. Zavelani-Rossi, G. Cerullo, S. De Silvestri, L. Gallmann, N. Matuschek, G. Steinmeyer, U. Keller, G. Angelow, V. Scheuer, T. Tschudi, Pulse compression over 170-THz bandwidth in the visible using only chirped mirrors, *Opt. Lett.*, Vol. 26, 1155–1157, 2001.

48. A. Baltuska, T. Fuji, and T. Kobayashi, Visible pulse compression to 4 fs by optical parametric amplification and programmable dispersion control, *Opt. Lett.*, Vol. 27, 306–308, 2002.

49. M.R. Armstrong, P. Plachta, E.A. Ponomarev, and R.J.D. Miller, Versatile 7-fs optical parametric pulse generation and compression by use of adaptive optics, *Opt. Lett.*, Vol. 26, 1152–1154, 2001.

50. R. Ell, U. Morgner, F.X. Kärtner, J.G. Fujimoto, E.P. Ippen, V. Scheuer, G. Angelow, T. Tschudi, M.J. Lederer, A. Boiko, B. Luther-Davies, Generation of 5-fs pulses and octave-spanning spectra directly from a Ti:sapphire laser, *Opt. Lett.*, Vol. 26, 373–375, 2001.

51. B. Schenkel, J. Biegert, U. Keller, C. Vozzi, M. Nisoli, G. Sansone, S. Stagira, S. De Silvestri, and O. Svelto, Generation of 3.8-fs pulses from adaptive compression of a cascaded hollow fiber supercontinuum, *Opt. Lett.*, Vol. 28, 1987–1989, 2003.

52. G.G. Paulus, F. Grabson, H. Walther, P. Villoresi, M. Nisoli, S. Stagira, E. Priori, and S. De Silvestri, Absolute-phase phenomena in photoionization with few-cycle laser pulses, *Nature*, Vol. 414, 182–184, 2001.

53. A. Baltuska, T. Udem, M. Uiberacker, M. Hentschel, E. Goulielmakis, C. Gohle, R. Holzwarth, V.S. Yakovlev, A. Scrinzi, T.W. Hänsch, and F. Krausz, Attosecond control of electronic processes by intense light fields, *Nature*, Vol. 421, 611–615, 2003.

54. M. Hentschel, R. Kienberger, C. Spielmann, G.A. Reider, N. Milosevic, T. Brabec, P. Corkum, U. Heinzmann, M. Drescher, and F. Krausz, Attosecond metrology, *Nature*, Vol. 414, 509–513, 2001.

55. O.D. Mücke, T. Trischler, M. Wegener, U. Morgner, F.X. Kärtner, Role of the carrier-envelope offset phase of few-cycle pulses in nonperturbative resonant nonlinear optics, *Phys. Rev. Lett.*, Vol. 89, 127401, 2002.

56. A. Apolonski, P. Dombi, G.G. Paulus, M. Kakehata, R. Holzwarth, T. Udem, C. Lemell, K. Torizuka, J.Burgdörfer, T. Hänsch, and F. Krausz, Observation of light-phase-sensitive photoemission from a metal, *Phys. Rev. Lett.*, Vol. 92, 073902, 2004.

57. T.M. Fortier, P.A. Roos, D.J. Jones, S.T. Cundiff, R.D.R. Bhat, J.E. Sipe, Carrier-envelope phase-controlled quantum interference of injected photocurrents in semiconductors, *Phys. Rev. Lett.*, Vol. 92, 147403, 2004.

58. D.J. Jones, S.A. Diddams, J.K. Ranka, A. Stentz, R.S. Windeler, J.L. Hall, and S.T. Cundiff, Carrier-envelope phase control of femtosecond mode-locked lasers and direct optical frequency synthesis, *Science*, Vol. 288, 635–639, 2000.

59. A. Apolonski, A. Poppe, G. Tempea, C. Spielmann, T. Udem, R. Holzwarth, T. W. Hänsch, and F. Krausz, Controlling the phase evolution of few-cycle light pulses, *Phys. Rev. Lett.*, Vol. 85, 740–743, 2000.

60. A. Baltuska, T. Fuji, and T. Kobayashi, Controlling the carrier-envelope phase of ultrashort light pulses with optical parametric amplifiers, *Phys. Rev. Lett.*, Vol. 88, 133901, 2002.

61. M. Kakehata, H. Takada, Y. Kobayashi, K. Torizuka, Y. Fujihara, T. Homma, and H. Takahashi, Single-shot measurement of carrier-envelope phase changes by spectral interferometry, *Opt. Lett.*, Vol. 26, 1436–1438, 2001.

62. S. Adachi, P. Kumbhakar, and T. Kobayashi, Quasi-monocyclic near-infrared pulses with a stabilized carrier-envelope phase characterized by noncollinear cross-correlation frequency-resolved optical gating, *Opt. Lett.*, Vol. 29, 1150–1152, 2004.

63. X. Fang and T. Kobayashi, Self-stabilization of the carrier-envelope phase of an optical parametric amplifier verified with a photonic crystal fiber, *Opt. Lett.*, Vol. 29, 1282–1284, 2004.

64. C. Manzoni, G. Cerullo, S. De Silvestri, Ultrabroadband self-phase-stabilized pulses by difference-frequency generation, *Opt. Lett.*, Vol. 29, 2668–2670, 2004.

65. A. Baltuska, T. Fuji, and T. Kobayashi, Self-referencing of the carrier-envelope slip in a 6-fs visible parametric amplifier, *Opt. Lett.*, Vol. 27, 1241–1243, 2002.

66. M.D. Perry, D. Pennington, B.C. Stuart, G. Tietbohl, J.A. Britten, C. Brown, S. Herman, B. Golick, M. Kartz, J. Miller, H.T. Powell, M. Vergino, and V. Yanovsky, Petawatt laser pulses, *Opt. Lett.*, Vol. 24, 160–162, 1999.

67. A. Dubietis, G. Jonusauskas, and A. Piskarskas, Powerful femtosecond pulse generation by chirped and stretched pulse parametric amplification in BBO crystal, *Opt. Commun.*, Vol. 88, 437–440, 1992.

68. I.N. Ross, P. Matousek, M. Towrie, A.J. Langley, and J.L. Collier, The prospects for ultrashort pulse duration and ultrahigh intensity using optical parametric chirped pulse amplifiers, *Opt. Commun.*, Vol. 144, 125–133, 1997.

69. J. Collier, C. Hernandez-Gomez, I.N. Ross, P. Matousek, C.N. Danson, J. Walczak, Evaluation of an ultrabroadband high-gain amplification technique for chirped pulse amplification facilities, *Appl. Opt.*, Vol. 38, 7486–7493, 1999.

70. I.N. Ross, J. Collier, P. Matousek, C.N. Danson, N. Neely, R.M. Allott, D.A. Pepler, C. Hernandez-Gomez, and K. Osvay, Generation of terawatt pulses by use of optical parametric chirped pulse amplification, *Appl. Opt.*, Vol. 39, 2422–2427, 2000.

71. X. Yang, Z. Xu, Y. Leng, H. Lu, L. Lin, Z. Zhang, R. Li, W. Zhang, D. Yin, and B. Tang, Multiterawatt laser system based on optical parametric chirped pulse amplification, *Opt. Lett.*, Vol. 27, 1135–1137, 2002.

72. N. Ishii, L. Turi, V. S. Yakovlev, T. Fuji, F. Krausz, A. Baltuska, R. Butkus, G. Veitas, V. Smilgevicius, R. Danielius, and A. Piskarskas, Multimillijoule chirped parametric amplification of few-cycle pulses, *Opt. Lett.*, Vol. 30, 567–569, 2005.

73. C.P. Hauri, P. Schlup, G. Arisholm, J. Biegert, and U. Keller, Phase-preserving chirped-pulse optical parametric amplification to 17.3 fs directly from a Ti:sapphire oscillator, *Opt. Lett.*, Vol. 29, 1369–1371, 2004.

74. R. Th. Zinkstok, S. Witte, W. Hogervosrt, and K.S.E. Eikema, High-power parametric amplification of 11.8-fs laser pulses with carrier-envelope phase control, *Opt. Lett.*, Vol. 30, 78–80, 2005.

75. P. Matousek, B. Rus, and I.N. Ross, Design of a multi-petawatt optical parametric chirped-pulse amplifier for the iodine laser ASTERIX IV, *IEEE J. Quantum Electron.*, Vol. 36, 158–163, 2000.

12

Noise of Solid-State Lasers

Rüdiger Paschotta, Harald R. Telle, and Ursula Keller

CONTENTS

12.1 Introduction

Laser noise has been of high interest for many years, both in the context of fundamental physics and for a variety of laser applications. Simple types of lasers, particularly single-frequency lasers, constitute cases of strong interest in quantum optics which have been studied in great detail. The obtained understanding has also brought significant benefits for the further development of lasers with favorable noise properties, as required for many applications. For example, lasers are nowadays used in fiber-optic communication systems the data rates of which are fundamentally limited by noise issues. An important part of the noise in such a system comes from the laser source in the transmitter, another one usually from fiber amplifiers, where similar effects occur. Noise is similarly important for various kinds of metrology applications, including ultraprecise interferometric length measurements (e.g., for gravitational wave detection), spectroscopic measurements, and optical clocks surpassing cesium clocks in performance, to name only a few examples. Over quite a few decades, such types of applications have continued to act as a driver for the development of laser sources with still better noise properties, and of improved measurement techniques to monitor the success of such attempts. A decent understanding of how laser noise is generated does not only help to improve the performance of particular devices but also to realize fundamental limitations in order to correctly judge the future potential of different types of lasers for low-noise performance.

This chapter is meant to be an introduction into the area of laser noise. We start with the essential mathematical basics, also explaining some details of the notation, which often cause confusion in the literature. We then first focus on single-frequency lasers, which have been studied in most detail and deserve significant attention for various reasons, including the fact that their understanding also greatly facilitates the understanding of more complicated situations — for example, in mode-locked lasers. The following sections are then devoted to multimode continuous-wave lasers, Q-switched lasers, and mode-locked lasers. Each section begins with a discussion of the basic physics, and most of them end with an overview on experimental methods for measurements. The emphasis is on single-frequency lasers and mode-locked lasers.

12.2 Mathematical Basics

In this section, we very briefly recall a few conceptual and mathematical basics for the description of noise, as required in later sections. However, throughout this book chapter we will limit the mathematical discussion as far as possible and put more emphasis on conceptual aspects.

One typically starts with the measurement of time-dependent fluctuating quantities such as a laser output power, optical phase, or the temporal positions of laser pulses. The given examples show that these fluctuating quantities can be either continuously varying over time or occur at discrete temporal positions as defined, for example, by the temporal positions of pulses emitted by a mode-locked laser. In either case, the data are normally thought of being available for an infinite span of time, although this is obviously not possible in practice. This aspect has some mathematical implications, as is discussed later on in this section.

The time-varying quantities themselves may be directly of interest — for example, when they carry important measurement information or when they are needed in some compensation schemes. In many cases, however, one is more interested in statistical descriptions based on time and/or ensemble-averaged data. In the time domain, this is typically done with auto- or cross-correlation functions. For example, the autocorrelation function corresponding to the laser power P is

$$R_P(\tau) = \left\langle P(t)P(t+\tau) \right\rangle \tag{12.1}$$

where the angle brackets indicate temporal and/or ensemble averaging (which deliver the same results for ergodic systems).

More frequently used are spectral descriptions in the form of power densities. For a spectral quantity which exists only in a finite span of time, a frequency-dependent power density $S(f)$ is easily defined via the squared modulus of the Fourier-transformed data. For quantities with an infinite span of time, however, the Fourier integrals usually do not converge. A laser power, for example, will fluctuate around some mean value, which can not be integrated over an infinite span. In this situation, one may in principle use a modified definition of power density, which involves a limit process:

$$S_P(f) \equiv \lim_{T \to \infty} \frac{1}{T} \left| \int_{-T/2}^{+T/2} P(t)\, e^{-i2\pi f t} \mathrm{d}t \right|^2 \tag{12.2}$$

with units of W^2/Hz. Here, the Fourier integral is restricted to a time interval of width T, then divided by this width, and the limit for $T \to \infty$ is taken. This definition is conceptually simple, but often awkward to use directly. Therefore, one usually applies the Wiener–Khinchin theorem which gives a connection to the autocorrelation function:

$$S_P(f) = \int_{-\infty}^{+\infty} R_P(\tau)\, e^{i2\pi f \tau} \, \mathrm{d}\tau \tag{12.3}$$

This power density is two-sided — that is, defined for positive as well as negative frequencies, even though the negative-frequency part does not carry additional information for real quantities such as a laser power. In the following we will normally use one-sided power densities (as used in the engineering disciplines), which are two times larger but defined only for positive frequencies, apart from the point $\omega = 0$, which must be handled with care in integrations. In any case, the autocorrelation function can be retrieved with an inverse Fourier transform

$$R_P(\tau) = \int S_P(f)\, e^{-i2\pi f\tau}\, df \tag{12.4}$$

where the frequency range must be chosen according to the use of one- or two-sided power densities. As a special case, one may calculate a variance according to

$$\sigma_P^2 \equiv \left\langle P^2 \right\rangle = R_P(0) = \int S_P(f)\, df \ . \tag{12.5}$$

The description of phase noise involves some additional mathematical aspects. It can be described with a power density $S_\varphi(f)$ of the optical phase φ, or alternatively by a power density $S_\nu(f)$ of the instantaneous optical frequency, which is related to the phase according to

$$\nu = \frac{1}{2\pi}\frac{d\varphi}{dt} \ . \tag{12.6}$$

(Note that for clarity we use the symbol ν for optical frequencies and f for noise frequencies.) From Equation 12.6, one can infer that

$$S_\nu(f) = f^2 S_\varphi(f) \tag{12.7}$$

(with units of Hz^2/Hz), which shows that, for example, white frequency noise corresponds to $S_\varphi(f) \propto f^{-2}$.

Phase noise is often retrieved from the spectrum of the electric field of a light beam. However, the relation between the power densities of the optical phase and the electric field are not trivial. In cases where the excursions of the optical phase are limited to a small range with a width well below, for example, 1 rad, the field spectrum has a delta function at the mean optical frequency plus noise sidebands with a power density which is directly proportional to $S_\varphi(f)$. The linewidth of the field spectrum is then zero. In other cases, where the optical phase undergoes a random walk which involves an unbounded long-term drift, one obtains a field spectrum with nonzero width and a nontrivial relation of the power densities (see Equation (7) and (8) in Reference 1):

$$S_E(\Delta\nu) = \int_{-\infty}^{+\infty} \exp\left[-\int_0^{+\infty} S_\varphi(f)\,(1-\cos(2\pi f\tau))\, df\right] \exp(2\pi i\, \Delta\nu\, \tau)\, d\tau \tag{12.8}$$

where $S_E(\Delta v)$ is the power density of the electric field, expressed as a function of the offset Δv from the mean optical frequency. $S_E(\Delta v)$ is a two-sided and $S_\varphi(f)$ is a one-sided power density.

For the simple case of a one-sided $S_\varphi(f) = C/f^2$, $S_E(\Delta v)$ has a Lorentzian shape with a full width at half maximum $\Delta v_{FWHM} = \pi C$.

12.3   Single-Frequency Lasers

12.3.1   Quantum Noise Limits

Concerning the noise properties, the simplest possible laser system is that of a single-frequency laser — that is, of a laser which operates in a continuous-wave fashion on a single cavity mode. Even if the gain bandwidth is much larger than the frequency spacing of the cavity modes, single-frequency operation can be achieved by using an appropriate intracavity filter. Any effects of mode competition, which would complicate the dynamics, are then excluded. Basically, we have to deal with only two different kinds of noise: intensity noise, as measured, for example, with a photodiode, and phase noise, which is relevant in situations involving interference or precise frequency measurements.

There is a very important difference between the characteristics of intensity and phase noise. Gain saturation provides a "restoring" force for the output power, which always pulls the output power toward the steady-state solution, even under the influence of noise. On the other hand, the optical phase has no restoring force: the laser has no phase (or time) reference, so that the phase undergoes a random walk, which can lead to ever increasing phase deviations compared to the noiseless case. (This situation is well known for any free-running oscillator.) The consequence of this is a finite emission linewidth. In contrast to that, a quantum-limited beam in a so-called coherent state [2] also has some intensity and phase noise but can have a zero linewidth: there is no unbounded drift of the optical phase. The corresponding intensity noise is called *shot noise*. This was originally interpreted as a result of the detection process, not the light field itself, as a constant probability for detecting a photon in a certain time interval already produces this shot noise. However, it was later realized that shot noise actually results from the quantum fluctuations of the light field, and a rather convincing support for this interpretation came with the demonstration of so-called amplitude-squeezed light [3,4], which can cause the detected intensity noise to fall below the shot noise level.

The noise of the output of a free-running single-frequency laser can approach both the shot noise level and quantum-limited phase noise at high enough noise frequencies but usually does not so at low frequencies. In the simplest situation, only quantum noise affects the laser operation. Such noise has two basic origins: the laser gain medium and any optical elements introducing cavity losses. The noise impact of losses can be illustrated with the random nature of the removal of part of the photons, whereas the effect of laser gain is explained with spontaneous emission into the cavity mode. In fact, it is known [5] that any phase-insensitive optical amplifier with a given gain has to add at least the amount of noise, which an ideal four-level laser gain medium adds due to spontaneous emission — even though the physical origin of this noise may be different. In the case of a (nondegenerate) parametric amplifier; for example, the dominant noise influence results from the vacuum fluctuations entering the idler port.

In a laser with quantum noise influences only, the circulating field acquires fluctuations in each cavity round trip due to the quantum noise associated with the gain and the cavity losses. If the latter are dominated by the output coupler mirror, these can be understood as the effect of vacuum fluctuations which can get into the cavity through the output coupler. As the fluctuations corresponding to different round trips are statistically independent, their variances add up, and the variance, for example, of the optical phase grows linearly with time. In Appendix 1, we show that this leads to a finite linewidth of the laser output which is given by the Schawlow–Townes formula [6] (here presented in a slightly modified form)

$$\Delta\nu_{\text{laser}} = \frac{h\nu\, l_{\text{tot}} T_{\text{oc}}}{4\pi\, T_R^2\, P_{\text{out}}} = \frac{h\nu\, l_{\text{tot}}}{4\pi\, T_R^2\, P_{\text{int}}} \tag{12.9}$$

for the FWHM (full width at half maximum) laser linewidth, where $h\nu$ is the photon energy, l_{tot} denotes the total power losses per cavity round trip, T_{oc} the output coupler transmission, T_R the cavity round-trip time, P_{out} the laser output power, and P_{int} the intracavity power. The formula is exactly equivalent to the original Schawlow–Townes formula if there are no parasitic losses (i.e., $l_{\text{tot}} = T_{\text{oc}}$). In that case, for a given output power the linewidth scales with the square of the output coupler transmission, as a higher transmission means both higher laser gain (introducing stronger fluctuations) and a lower intracavity power (making quantum effects relatively stronger). The linewidth also scales with the inverse square of the round-trip time, which means that long laser cavities are potentially better in terms of linewidth — provided that classical noise (see below) can be well suppressed.

The derivation of the Schawlow–Townes formula considers the spontaneous emission noise, but does not refer to any noise sources within the gain medium which are associated with pump noise (which is usually at the shot noise level or above) and dipole fluctuation noise. It is possible to treat these noise sources as well, using, for example, a semiclassical technique based on Langevin equations. Such a model as, for example, used in Reference 7 through Reference 9 contains the full laser dynamics and allows for calculation of both the quantum-limited intensity and phase noise. In the following, we briefly review the results obtained for an example case, namely, a

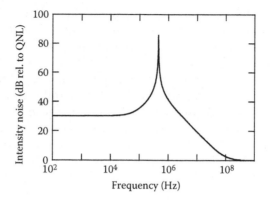

FIGURE 12.1
Calculated intensity noise (in dB relative to the QNL) of a diode-pumped solid-state laser. The pump noise is 30 dB above the QNL.

compact diode-pumped Nd:YAG laser with 100 mW output power at 1064 nm, 5% output coupler transmission, 1% parasitic losses and a round-trip time of 133 ps = 1/(7.5 GHz). The pump source is a diode laser, the intensity noise of which is assumed to be white and 30 dB above the shot noise limit. Figure 12.1 shows the calculated intensity noise relative to the shot noise level (with all calculations based on Reference 7), which would correspond to the one-sided power density

$$S_{P,sn}(f) = 2h\nu \cdot \overline{P} \, , \tag{12.10}$$

with $\overline{P}$ being the average power. The corresponding relative intensity noise (RIN) at the shot noise level has the power density

$$S_{RIN,sn}(f) = \frac{2h\nu}{\overline{P}} \, . \tag{12.11}$$

At low noise frequencies, the intensity noise of the laser of Figure 12.1 is ≈30 dB above the shot noise limit, as the power fluctuations of the pump are directly translated into output power fluctuations. The peak at 445 kHz is due to relaxation oscillations [10], which are excited mainly by vacuum fluctuations, despite the significant pump excess noise. For frequencies above this peak, the intensity noise is still dominated by vacuum fluctuations from the gain medium, until the noise level finally reaches the shot noise level above a few hundred MHz. In the semiclassical picture, the noise in this high-frequency regime results mainly from vacuum fluctuations that enter the output beam by reflection at the output coupler. This mechanism illustrates the fundamental nature of the shot noise level, which in this regime can not even be strongly affected by any dynamics inside the laser cavity. Here, the combination of the transmitted intracavity field with the (dominant) vacuum noise approximates a coherent state, as mentioned above. The noise of the intracavity field can be below the shot noise level at high frequencies, basically because the light field is circulating in the cavity and experiences only slight modifications of its fluctuations within a few cavity round trips (assuming small cavity losses). This intracavity field, however, is not directly accessible.

A higher output power of, for example, 1 W would significantly modify the spectral density in Figure 12.1 only around the relaxation oscillation peak, which would become reduced in height but becomes broader. On the other hand, a longer cavity would reduce the relaxation oscillation frequency and thus allow reaching the shot noise limit at a lower frequency; for this reason, a laser with long cavity can be superior over a compact monolithic laser for measurements where the shot noise limit must be reached at not too high frequencies. The use of a shot-noise-limited pump source (with otherwise same parameters as in Figure 12.1) would lead to Figure 12.2, with the low-frequency noise reduced to a level which is much closer to the shot noise, whereas a significant influence of excess noise from quantum fluctuations in the gain medium is then seen on both sides of the relaxation oscillation peak.

The same kind of model also allows calculating the phase noise power density of the output beam (see, e.g., Equation (15) in Reference 9). Compared to the Schawlow–Townes result, there is only one deviation: for very high noise frequencies, the phase noise power density becomes constant (as for a coherent state), i.e., does not continue to fall in proportion to f^{-2}. This effect is again due to vacuum fluctuations, which are reflected at the output coupler mirror and cause the high-frequency noise to stay at the fundamental quantum noise limit. The linewidth remains unchanged, as it is determined by low-frequency noise only. Note that in the simple case considered here the phase noise does not exhibit a peak related to the relaxation oscillations, as the dynamics of intensity and phase noise are completely decoupled.

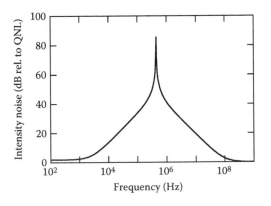

FIGURE 12.2
Same as Figure 12.1 but with a shot-noise-limited pump source.

12.3.2 Other Noise Sources

So far, we have considered only noise from quantum fluctuations and from excess noise of the pump source. However, real lasers also exhibit additional noise sources, as well as coupling mechanisms, which can also increase various kinds of noise.

Particularly in the context of phase noise, mirror vibrations can be important unless a very stable cavity setup is used. First considering only cavity length changes but no tilts (which would lead to misalignment), mirror vibrations can be expected to have basically no effect on the intensity noise but a strong effect on the phase noise. A change δL of the round-trip length (corresponding to twice the cavity length change in the case of a linear cavity) will change the optical phase by $\delta\varphi = k\delta L = 2\pi\delta L/\lambda$ (with $k = 2\pi/\lambda$) in each round trip. This means that the phase errors are integrated over subsequent round trips. This can be described with the differential equation

$$T_R \frac{\mathrm{d}}{\mathrm{d}t}\delta\varphi = k \cdot \delta L \qquad (12.12)$$

where $\delta\varphi$ now describes the phase evolution over many cavity round trips and T_R the cavity round-trip home. In the frequency domain, we have

$$\delta\varphi(f) = \frac{k}{i2\pi f \, T_R}\delta L(f) \; . \qquad (12.13)$$

From this we can conclude that random cavity length changes with a power density $S_{\delta L}(f)$ lead to a phase noise power density

$$S_\varphi(f) = \left(\frac{k}{2\pi f \, T_R}\right)^2 S_{\delta L}(f) \; . \qquad (12.14)$$

The integrating nature of the effect results in the f^{-2} dependence of the coupling factor, which shows that particularly the low-frequency phase noise is sensitive to mirror vibrations. (In other words, changes of the cavity length lead to changes of optical frequency, which is basically the derivative of the phase.) The high frequency phase noise is much less affected, both because of the smaller coupling factor and the decay of the power density of the mechanical vibrations above the mechanical resonance frequencies.

The effect of mirror tilts is much more complicated to describe, because these lead to cavity misalignment. How exactly a laser will respond to that, depends on the cavity design (determining how the caused beam offset depends on the position in the cavity) as well as on factors like the pump distribution and thermal lensing in the gain medium. It is easy to see that both intensity and phase noise can be affected by such tilts, whereas a general quantitative description would appear to be very difficult.

Another source of noise is related to thermal fluctuations, which can occur in the gain medium mainly as a result of pump power fluctuations. These may mainly translate into additional phase noise, particularly at low frequencies but can also cause drifts of the maximum of the gain spectrum which might affect the laser wavelength, possibly even trigger mode hops.

Finally, there are effects which provide a coupling between intensity and phase noise. This is well known for semiconductor lasers, where fluctuations of the carrier density affect the refractive index and thus cause additional phase noise. Quantitatively, this coupling can be described with Henry's linewidth enhancement factor α [11], which leads to a phase change $\delta\varphi = \alpha\delta g/2$ when the light passes the gain medium while there is a fluctuation δg of the intensity gain. In a laser, white spontaneous emission noise causes white noise of the carrier density (at noise frequencies well below the relaxation oscillation frequency) and thus phase noise with a power density proportional to f^{-2}, which contributes to the linewidth. In effect, the coupling increases the linewidth by a factor $1 + \alpha^2$, as derived in Reference 11 and in a different way in Appendix 2. For semiconductor lasers, $1 + \alpha^2$ can easily be between 10 and 100. For solid-state lasers based on ion-doped crystals or glasses, such a factor is usually neglected, although this may not always be fully justified. In particular for quasi-three-level lasers (including semiconductor lasers), but also for wavelength-tuned lasers, the shape of the gain curve can be quite asymmetric around the laser wavelength, and Kramers–Krönig relations, result in a nonvanishing dependence of the refractive index at the laser wavelength on the excitation of the medium. To complicate things further, allowed transitions in the ultraviolet spectral region can give additional Kramers–Krönig type contributions [12] even for four-level laser gain media with laser operation at the maximum of a symmetric line.

Figure 12.3 shows the phase noise power density for a linewidth enhancement factor of 3, again relative to the quantum noise limit (corresponding to a coherent state), calculated from Equation (15) in Reference 9. Here, the low-frequency phase noise and thus the linewidth are increased due to the intensity/phase coupling. The most pronounced effect is a peak at the relaxation oscillation frequency.

An even more direct coupling of intensity and phase noise is in principle also possible via the Kerr effect: the refractive index can be increased for high optical intensities. However, this effect is very weak in continuous-wave lasers: typical phase shifts are in the region of microradians per round-trip, and the fluctuations of these phase shifts (caused by intensity fluctuations) even smaller. Therefore, other effects as discussed above are usually dominating.

Semiconductor lasers differ from ion-doped solid-state lasers not only in terms of the above mentioned coupling of intensity and phase noise. More importantly, their general parameters are very different. In particular, the laser transition (typically an interband transition) has a very large oscillator strength, whereas ion-doped solid-state lasers a usually emitting on so-called forbidden transitions with very low oscillator strength. As a consequence,

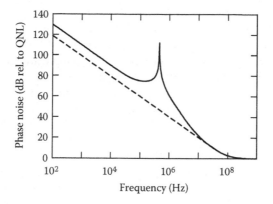

FIGURE 12.3
Solid curve: phase noise for the same parameters as in Figure 12.1 but with $\alpha = 3$. Dashed curve: same for $\alpha = 0$.

semiconductor lasers have a much stronger coupling between light field and excitation of the gain medium, and a much lower upper-state lifetime. This leads to a much higher relaxation oscillation frequency, which is often in the gigahertz range, and to a strong damping of relaxation oscillations [11,13–16]. There is a wide range of noise frequencies where excess noise in the pump current can in principle directly affect the laser noise. However, the pump noise itself can be rather weak when a suitable diode driver is used, so that the intensity noise does not need to be large, at least in single transverse mode diodes where mode hopping does not occur.

12.3.3   Noise Measurements

12.3.3.1   Measurement of Relative Intensity Noise

The relative intensity noise of a laser is usually measured with a photodiode. A sensitivity high enough to detect the shot noise can often be achieved but may be difficult in some cases, for example, if the requirement of a multi-GHz detection bandwidth enforces measures which degrade the sensitivity. A high photocurrent is desirable in order to achieve a strong signal, which has to compete mainly with thermal and other excess noise in the detection electronics. However, the photocurrent per unit detection area is limited by the onset of detector saturation (which tends to suppress the apparent noise), and large-area devices tend to have a reduced detection bandwidth. In the visible spectral region, medium-sized PIN silicon photodiodes such as the model Hamamatsu G5832-02 with bandwidths in the MHz range are widely used because of their demonstrated ability to handle fairly large power levels (with a linear response above 100 mW optical power), as well as their high surface uniformity (with a 2% loss of sensitivity at half the distance from the diode center) [17,18]. For longer wavelengths (e.g., 1.5 μm), InGaAs detectors are commonly used; these are rather expensive but offer a much better compromise between bandwidth and photocurrent than germanium-based devices, for example. Velocity-matched devices [19] even allow bandwidths of tens of gigahertz combined with photocurrents of tens of milliamperes.

For the electronic processing, one often has to separate the DC part of the photocurrent from the noise part. The calibration may then require a separate measurement, for example, involving a mechanical chopper or acousto–optic modulator, producing a well-defined power modulation.

12.3.3.2 Measurement of Phase Noise

The measurement of phase noise always requires some kind of phase reference. Different phase noise measurement schemes can rely on different kinds of phase references and can be grouped as follows:

1. Schemes based on *passive devices* such as filters or resonators, which can provide a direct conversion from the frequency of the DUT (device under test) to a power.

2. *Interferometric interrogation schemes,* where the instantaneous phase of the DUT is measured via interference with a reference signal which itself is derived from the output of the DUT, typically with some time delay or temporal averaging. This is the basis of self-heterodyning methods.

3. *Heterodyne schemes* with a reference signal generated by an auxiliary oscillator of superior frequency stability. Alternatively, one may employ two similar optical oscillators. Assuming uncorrelated noise processes with same power density for both lasers, the measured noise power densities are two times those of the single devices.

In the simplest case — the passive device schemes, sometimes named direct or homodyne detection — frequency fluctuations are converted into power fluctuations by means of the frequency-dependent transmission of the discriminator. The slope of a Fabry–Pérot interferometer (FPI) or an absorption line of some molecules can be employed as such a discriminator, in such cases yielding a Lorentzian-like line shape. If the oscillator or the filter is tuned such that the mean frequency of the oscillator is at the slope, preferably near the inflection point, the power transmitted by the filter varies, to first order, linearly with the frequency of the signal as $P(v - v_s) \cong P(v_s) + k_d(v - v_s)$ where k_d is the slope of the filter at v_s. A photodetector behind the filter converts the power fluctuations into fluctuations of the photocurrent, which can be characterized by means of an electronic spectrum analyzer. Modern spectrum analyzers show a quantity directly related to the spectral density of the fluctuations of the signal. To obtain the power density of the frequency fluctuations in Hz^2/Hz, the slope k_d of the frequency discriminator has to be determined.

The application of this method requires that the contributions of other noise sources do not affect the measurement. However, there may be contributions from fluctuations of the center frequency of the filter or from fluctuations of the signal amplitude. The latter contribution may be eliminated by stabilizing the amplitude of the input or by normalizing the transmitted signal based on the separately measured amplitude fluctuations. The unwanted response of such a slope discriminator to intensity fluctuations can be substantially reduced with a symmetric setup [20] which is well known from RF (radio frequency) techniques. As a fundamental limitation, the transfer function from frequency to photocurrent fluctuations drops with f^{-2} at noise frequencies above the half-width of a Lorentzian shaped filter, yielding a f^{-4} dependence of the frequency noise detectivity. However, a resonance width as narrow as possible is highly desirable in order to obtain optimum detectivity at low noise frequencies.

An improved high-frequency performance, namely a f^{-2} dependence of the detectivity, is available using the interferometric interrogation methods of the second group listed. Here, the instantaneous field incident on an optical resonator is compared with the field already stored in the resonator. Both field components are superimposed on the resonator input mirror, where the input field is partly reflected and where, on the other hand, a small amount of the intra-cavity field is leaking out in backward direction. In principle, an FM demodulation at the slope of the on-resonance notch of this reflected signal would give the desired improved high-frequency response. However, the contrast of such spectral notches might be deteriorated, for example, due to imperfect transverse mode matching or out-of-band background signals. Thus, essentially two methods have been established to improve the contrast, for instance, to reduce the AM sensitivity. They rely either on phase modulation (PM) (Pound-Drever-Hall scheme [21,22]) or on polarization (Hänsch-Couillaud [23]). In the first case, the resonance is "seen" only by the carrier but not by the PM sidebands if the modulation frequency is chosen sufficiently high. Hence, the interference of both field components as detected at the modulation frequency provides the desired information about DUT frequency fluctuations. The Hänsch–Couillaud scheme, on the other hand, employs an intracavity loss [23] or birefringent [20] element in order to distinguish between the instantaneous field and the stored reference field while background signals are suppressed.

In both examples discussed so far, the reference signal was derived from the DUT output and stored in an optical resonator. A similar storing effect can be achieved using a long optical fiber in one arm of a highly unbalanced interferometer. In the domain of RF techniques, such a setup is commonly referred to as delay-time discriminator. For ease of data processing, one usually inserts a frequency shifter (e.g., an AOM = acousto-optic modulator) into one of the interferometer arms. The beat note between both field components at the interferometer output is then detected with a photodiode and characterized with an RF spectrum analyzer. If the delay time of such a self-heterodyning setup is chosen larger than the coherence time of the laser emission, the fluctuations of both optical fields become uncorrelated, and the output spectrum becomes a self-convolution of the laser output spectrum.

This situation is the simplest concerning data processing, and it corresponds to the heterodyning of similar but independent lasers, that is, to the third group listed. For highly coherent laser oscillators, however, a fiber-optic delay with a length of the order of 100 kilometers or more may be required, which is fairly impractical due to cost and fiber losses. Therefore, measurements with such long delay times are often carried out with amplifying recirculating fiber loops [24], where a fiber piece of moderate length can be passed many times.

One can also work with delay lengths below the coherence length of the laser emission. In this case, the power density of the interferometer output is comprised of a delta function peak at the drive frequency of the AOM, superimposed on a pedestal, the shape of which depends on the noise spectrum of the DUT [25]. In the simplest case, where the fiber delay τ_d is very short compared to the coherence time τ_c and where the frequency noise power density is constant $S_v(f) = S_{v0}$ (corresponding to white frequency noise), the power density of the interferometer output corresponds to a delta function at the AOM frequency and a sinc^2-shaped pedestal

$$S_{ped}(\Delta v) \propto \frac{\sin^2(\pi \, \Delta v \, \tau_d)}{(\pi \, \Delta v \, \tau_d)^2} \, . \tag{12.15}$$

Here, the approximation of small modulation index

$$R_\varphi(0) \ll 1 \, \text{rad}^2 \tag{12.16}$$

holds, and the pedestal can be normalized to the total noise power, yielding

$$L(\Delta v) = S_{ped}(\Delta v) / R_\varphi(0)$$

as measured in units of dBc/Hz. Then, the value of S_{v0} can be inferred from the ratio between the delta function power and the maximum of the sinc^2-function:

$$S_{v0} = \frac{2L(\Delta v = 0)}{(2\pi\tau_d)^2} \tag{12.17}$$

12.4 Multimode Continuous-Wave Lasers

Many continuous-wave lasers are not operating on a single cavity mode but rather on several or even many modes simultaneously. In such a multimode regime, there is a competition by the different cavity modes that are using the same gain medium. The competition may be strong in the sense that saturation of the gain by one mode can reduce the gain of other modes to about the same extent, or weak in the sense that these cross-saturation effects are weak. A reduction of the mode competition is called inhomogeneous gain saturation and may occur as a result of microscopic details of the gain medium (e.g., the existence of different lattice sites of laser-active ions or different velocity groups in the case of a gas laser), but also as a consequence of spatial hole burning in a linear cavity. The latter effect results from the standing-wave pattern in the gain medium, which has essentially two consequences: the gain is saturated mostly around the antinodes of the pattern, and the mode "sees" the excitation of the gain medium mostly in these regions.

Whatever the exact situation of mode competition is, this competition can easily lead to significant additional noise (called *mode partition noise*), because it often does not lead to a stable situation with a more or less constant distribution of power between the modes. Even if it does for some time, temperature changes in the gain medium or other influences can disturb the mode pattern, and the result can be a significant redistribution of power between the modes, which is usually also associated with changes of the overall power. This is not surprising; consider, for example, a mode which originally was below the laser threshold but suddenly gets some more gain due to a change of the pattern of lasing modes. The power in that mode will rise exponentially and will reach its equilibrium (if there is one) only in an oscillatory manner. Therefore, there is not only the mode beating as such, but also an increased low-frequency intensity noise that can be of fairly broadband nature due to the complicated nonlinear dynamics of the mode powers. It is known that the mode partition noise tends to be highest for lasers operating on just a few cavity modes with comparable powers in these modes [26–28]. Even if most of the power is carried by a single mode, a small fraction of the power in other modes can cause significant noise [29], with a power density which can be orders of magnitude higher than that, for example, of a high-power laser diode operating on many modes.

For a laser running only on fundamental (usually Gaussian) transverse modes, the mode-beating noise contains only high frequencies around the mode spacing and integer multiples thereof. A significantly more complicated situation arises for lasers with transverse multimode operation, where many more mode frequencies and thus also many more beat frequencies, including those with much lower frequencies, occur. Therefore, one can expect lasers with a diffraction-limited beam quality to exhibit lower intensity noise.

Mode competition (and the related dynamics) is essentially responsible for the excess noise of high-power diode lasers, as used, for example, for pumping solid-state lasers. When the total output power is collected on the detector, the obtained intensity noise is typically several tens of dB above the shot-noise limit. Even much higher noise can be obtained when collecting only a part of the spatial pattern, because the noise in the different spatial modes can be highly anticorrelated.

Phase noise is, of course, more difficult to specify in multimode lasers: there is not a well-defined phase of the overall field. One can, however, consider the phases of the different modes, which also must be expected to undergo complicated nonlinear dynamics.

A common situation is that a solid-state laser is pumped with a high-power semiconductor laser diode, with the latter exhibiting significant mode partition noise. Even if the total power of the semiconductor laser is fed into the solid-state laser (to avoid the above-mentioned detrimental effect of spatial filtering), the resulting intensity noise of the solid-state laser can be significant, because the degree of pump absorption can significantly vary across the spectrum of the pump diode, so that the absorbed pump power fluctuates more than the total incident pump power.

12.5 Q-Switched Lasers

12.5.1 General Remarks

In a similar fashion as in a continuous-wave laser, operation on multiple cavity modes can lead to mode beating and associated intensity noise. The mode beating periods often is on the order of one tenth of the pulse duration, corresponding to a pulse duration approximately one order of magnitude above the cavity round-trip time. Note that significant mode beating occurs even in situations where one mode carries most of the power. A special aspect of the situation in a Q-switched laser is that there is often too little time for establishing clean single-frequency operation during the build-up phase of the Q-switched pulse, even if one mode has some gain advantages over all others. For that reason, it may take special precautions to suppress mode beating. Possible measures include strong spectral filtering in the cavity, the use of a short cavity (leading to a large mode spacing), injection seeding with an additional single-frequency laser, and feedback-stabilized single-frequency prelasing [30].

In contrast to pulse generation by mode locking (see Section 12.6), each pulse in a Q-switched laser is normally built up from noise (typically from fluorescence in the gain medium), unless some mechanism for injection seeding is used. A consequence of this is that even for single-frequency operation the relative optical phase of subsequent pulses is entirely random. The build-up time and thus the pulse energy and temporal position can also be influenced by the noise in the initial period of pulse buildup.

12.5.2 Active vs. Passive Q Switching

In the following we briefly discuss some aspects which depend on whether an active or passive Q-switching mechanism is used. It will become apparent that such differences are caused not only by possible noise in an active Q switch or its driver electronics. In fact, the most fundamental difference results from the fact that the start of the pulse formation process is differently influenced by pump noise.

Consider an actively Q-switched laser first. Here, the pulse formation process is triggered by an external event, which is not influenced by the current energy content of the gain medium. Still, the temporal position of the generated pulse depends on the energy stored in the gain medium: the more energy there is, the higher the laser gain in the pulse-build-up phase, and the earlier the pulse maximum will be reached. Therefore, in a situation with a fixed repetition rate of the active Q switch, pump noise can cause bounded fluctuations of the pulse energy, the pulse duration, and the temporal pulse position.

In contrast, pulse formation in a passively Q-switched laser (with a saturable absorber in the cavity) is triggered not by an external signal but rather at the time where the energy stored in the gain medium becomes sufficient to generate a positive net gain per cavity round trip. Assuming constant cavity losses and an always fully recovered saturable absorber, this occurs always for the same level of stored energy, independent of the pump power. In principle, the pump power can still have an influence during the pulse buildup time, but this time is often so short that the pump influence is negligible. Therefore, the pulse energy and the pulse duration are then nearly independent of the pump power, and changes of the pump power only affect the pulse repetition rate, as observed, for example, in passively Q-switched microchip lasers [31]. This can cause reduced noise in the pulse energy at the expense of higher timing jitter. Note, however, that this does not fully hold for cases with incomplete absorber recovery between the pulses.

12.6 Mode-Locked Lasers

12.6.1 Types of Noise

Compared to continuous-wave lasers, mode-locked lasers exhibit many more degrees of freedom which may be affected by noise, and different kinds of noise can be relevant in different situations. For getting some overview, we first list possible types of noise in such lasers:

- Intensity noise can be understood as noise of the average power, and this is related to noise of the pulse energy and repetition rate.

- Absolute timing jitter is the deviation of the pulse positions (defined by the times with maximum optical power or better as some "center of gravity") from the "ticks" of a hypothetical ideal clock. Timing jitter can also be specified relative to some other oscillator, for example, relative to some other laser or to an electronic oscillator which is used to actively mode-lock the laser under test. In such cases, the term *residual jitter* is sometimes used.

- All involved cavity modes exhibit phase noise, which is partially correlated between the modes.

- The carrier-envelope offset phase [32] is related to the phases of the cavity modes.

- Other parameters of the pulses can also exhibit noise, in particular the pulse duration, center frequency, chirp, etc.

In contrast to noise in continuous-wave lasers, pulse parameters — similar to energy, temporal position, duration, etc. — are not defined on a continuous time scale but only on a discrete temporal grid. A consequence of this is that noise spectra can be specified only up to half the pulse repetition rate, corresponding to the Nyquist frequency. A mathematically more subtle point is that the optical phase (or amplitude) of a mode is actually well defined only over time scales well above the pulse period, as required to spectrally resolve the line. However, this is normally not a problem, as one usually considers the phase noise at lower frequencies.

In principle, all kinds of noise of the output of a mode-locked laser are determined by the specification of the amplitude and phase noise of all the lines in the optical spectrum, plus the correlations between these fluctuations. Therefore, one may see quantities like timing jitter or intensity noise as projections of the overall noise on some smaller (typically one-dimensional) spaces. In practice, one is mostly dealing with such projections, which often embrace everything relevant for a particular experiment. For example, the phase noise in a single line largely determines the beat frequency noise with a (noiseless) single-frequency reference signal.

12.6.2   Coupling Effects

Before considering the most important types of noise, we discuss a number of effects which couple different kinds of noise in a mode-locked laser with each other. We do not claim to present a complete picture — which is hardly possible, given that a great variety of coupling effects can occur in different situations — but rather try to discuss the most important mechanisms and to create a reasonable understanding of what consequences such coupling mechanisms can have.

The basic idea is to consider the noise in different dynamical variables describing the pulse properties and to discuss in which ways different kinds of noise can affect each other. As explained in Section 12.6.1, these dynamical variables actually correspond to projections of the overall noise in a mode-locked laser, which could alternatively be described as amplitude and phase noise of all lines in the spectrum. However, the picture used in the following is natural because it involves experimentally very relevant variables. The variables considered in the following are the pulse energy E_p, the optical phase φ_{opt} of the central line, the pulse duration τ_p, the temporal position Δt, and the deviation Δv_c of the center frequency (defined as a "center of gravity") from its average value. We supplement these variables by the gain g of the laser medium, which is directly related to the energy stored in the gain medium (at least if homogeneously broadened media are considered and spatial mode profiles are disregarded). Figure 12.4 illustrates the interactions between these variables, which we briefly discuss in the following text. Note that some analytical models have been described (see, for example, Reference 33) which work with differential equations for some limited set of such dynamical variables.

As is well known, the evolution of pulse energy and gain is closely coupled by the effects of laser amplification and gain saturation: any deviation of the gain from the net loss in the cavity leads to changes of the pulse energy, whereas larger pulse energies tend to reduce the gain. Both effects are typically of integrating nature; for example, gain saturation in a mode-locked solid-state laser (where the pulse energy is typically far below the saturation energy) affects the temporal derivative of the gain, rather than its current value. The resulting dynamics lead to the well-known relaxation oscillations. Note, however, that these dynamics can be coupled to other variables; for example, the gain can also depend on the spectral width of the pulses.

The pulse energy can then couple to the pulse duration and other variables. For example, a shorter pulse has a higher peak power, leading to a lower loss on a fast saturable absorber (but not on a slow absorber) [34]. In lasers with a soliton circulating in the cavity, pulse duration and pulse energy are coupled to each other by the interplay of dispersion and nonlinearity. The optical phase can be affected by intensity fluctuations (and thus by fluctuations of pulse energy or duration) via the Kerr nonlinearity and also by gain fluctuations due to refractive index changes as described by Kramers–Krönig relations.

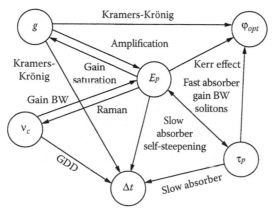

FIGURE 12.4
Illustration of the interactions of various pulse parameters and the laser gain.

The timing (temporal position) of pulses can be affected by the temporal delay caused on a slow saturable absorber, which depends on pulse energy and duration [35]. The timing can also be modified via the self-steepening effect [36] (which is important for very short pulses), by center frequency fluctuations if there is dispersion [33], and by gain fluctuations via Kramers–Krönig effects. The center frequency can affect the pulse energy via the frequency-dependent laser gain and/or cavity losses, and can itself be influenced by pulse energy fluctuations if the Raman self-frequency shift [37] of pulses is relevant (for example, in some fiber lasers).

It is apparent that the situation can be rather complex, even though not all possible coupling effects will usually be simultaneously relevant in any concrete situation. The simplest consequence of such coupling effects is that noise of one quantity is transferred into noise of some other quantity. However, strong bidirectional coupling can also introduce complicated nonlinear dynamics, and even in a numerical model as presented, for example, in Reference 38 the interpretation of the overall dynamics may not always be simple.

12.6.3   Intensity Noise

At a first glance, one may expect that the intensity noise of a passively mode-locked laser is basically the same as for a single-mode continuous-wave laser as discussed in Section 12.2. However, at least for passively mode-locked lasers this is not true because the saturable absorber can significantly alter the gain dynamics, and in fact usually does so in practical situations. The key point is that the cavity losses become dependent on the pulse energy (for a slow absorber) or the peak power (for a fast absorber) [34]; usually, they are reduced for higher pulse energies. In effect, the damping of the relaxation oscillations is reduced, or the relaxation oscillations can even become undamped — typically, for pump powers below a certain well defined threshold value. In the latter case, the continuous-wave mode-locked regime becomes unstable [39,40]. The resulting regime with strong oscillations of the pulse energy is called *Q-switched mode locking*. This regime can be rather noisy with large fluctuations of maximum pulse energy, pulse duration, optical phase, etc., if the pulse energy between subsequent bunches of pulses gets very low. However, Q-switched mode loading can also be very stable.

Even if this regime is avoided, the reduced damping of the relaxation oscillations increases the intensity noise around the relaxation oscillation frequency, whereas the width of the relaxation oscillation peak is reduced [41]. Transfer function measurements with a sinusoidally modulated pump power have demonstrated the increased peak height and decreased peak width [42], and the mentioned effect on the noise is clearly to be expected. On the other hand, the impact on noise at frequencies far below or far above the relaxation oscillation frequency should not too strongly deviate from that of a continuous-wave laser.

The intensity noise of any laser can, of course, be strongly increased by excess noise of its pump source. This is particularly the case for lasers with gain media which have a very short upper-state lifetime, such as semiconductor lasers or dye lasers. For a discussion in the context of a passively mode-locked dye laser, see Reference 43.

12.6.4 Timing Jitter

Most noise measurements on mode-locked lasers are focused on timing jitter, which is relevant for many applications, such as data transmission or optical sampling measurements. Therefore, we also focus on timing jitter in this book chapter but also present some results on optical phase noise in Subsection 12.6.6. The current section is devoted to the mathematics and physics of timing jitter, whereas the section following it addresses jitter measurements.

Before going into details of pulse generation, we briefly discuss how timing jitter is usually specified. Conceptually, one can consider the timing errors of the pulses emitted by a mode-locked laser, as obtained by comparing their temporal positions with the ideal temporal positions for a noiseless pulse source ($\rightarrow$ absolute jitter), or with some real source which will itself have some noise ($\rightarrow$ relative jitter). One can specify the statistical properties of timing noise by the power spectral density $S_{\Delta t}(f)$ of the timing errors. It is actually quite common to replace the timing error Δt with a *timing phase error* $\varphi_t = 2\pi f_{rep}\Delta t$, where f_{rep} is the pulse repetition rate, and the corresponding power density is

$$S_{\varphi t}(f) = \left(2\pi f_{rep}\right)^2 S_{\Delta t}(f) . \tag{12.18}$$

This picture results from considering the emitted pulse train as a (usually highly anharmonic) oscillation of the optical power, which in the noiseless case can be seen as consisting of a sinusoidal signal at f_{rep} and integer harmonics. The optical field oscillation (sometimes called the *carrier oscillation*) is disregarded in this picture, which basically describes the output of a photodiode illuminated with the pulse train. Note that using the term *phase noise* in this case can provoke confusion with *optical phase noise*, so we suggest using the term *timing phase noise*, instead.

A rather different method of specifying timing noise of oscillators, which is often used in metrology, is the Allan variance [44,45]. It is a generally agreed [46] means to characterize frequency fluctuations in time domain. The so-called two-sample variance without dead time is commonly referred to as $\sigma_y^2(2,\tau,\tau)$ or $\sigma_y^2(\tau)$ and is defined as

$$\sigma_y^2(\tau) \equiv \frac{1}{2} < (\bar{y}_{k+1} - \bar{y}_k)^2 > \tag{12.19}$$

with

$$\bar{y}_k = \frac{1}{\tau} \int\limits_{t_k}^{t_k+\tau} y(t) \, dt \text{ and } y \equiv \frac{\Delta v}{v_0}. \tag{12.20}$$

Here, the frequency deviations are normalized with the mean frequency v_0 and averaged over subsequent time intervals of width $\tau = t_{k+1} - t_k$, where t_k are equidistant time values with an integer index k and a spacing τ. The Allan variance and its square root, sometimes called the Allan (standard) deviation, are based on differences of adjacent (time-averaged) frequency values rather than on frequency differences from the mean value, as is the "conventional" standard deviation. Plots usually show the Allan variance as a function of the averaging time τ, and the shape of such curves can be used to retrieve information on the noise processes.

Alternatively, the Allan variance can be determined from the normalized phase or timing deviation $x(t)$. For a given measurement interval τ it follows that

$$\overline{y}_k = \frac{\overline{x}_{k+1} - \overline{x}_k}{\tau}$$

which after insertion gives

$$\sigma_y^2(\tau) = \frac{1}{2\tau^2} \left\langle (\overline{x}_{k+2} - 2\overline{x}_{k+1} + \overline{x}_k)^2 \right\rangle . \tag{12.21}$$

The Allan variance can be calculated from the frequency noise density according to [47]

$$\sigma_y^2(\tau) = \frac{2}{v_0^2} \int_0^\infty S_v(f) \frac{\sin^4(\pi f \tau)}{(\pi f \tau)^2}\, df , \tag{12.22}$$

while it is not possible to unambiguously retrieve the frequency noise spectrum from the Allan variance [48].

Equation 12.22 can be simplified in certain cases [47]. White frequency noise $S_v(f) = S_0$, for example, results in $\sigma_y^2(\tau) = S_0 / 2\tau$, whereas so-called flicker frequency noise $S_v(f) = a/f$ leads to an Allan variance which is independent of the measuring time: $\sigma_y^2(\tau) = 2a\ln(2)$. A linear frequency drift $y(t) = bt$, which is typical for frequency standards, results in an Allan variance that increases with the squared measuring time: $\sigma_y^2(\tau) = (b\tau)^2 / 2$.

We now turn to the timing jitter of pulse trains from mode-locked lasers. Here, we need to distinguish between actively and passively mode-locked lasers. The fundamental difference between those laser types is that an actively mode-locked laser has a timing reference connected to its modulator, which introduces a kind of restoring force for the relative timing error between both signals. Passively mode-locked lasers, on the other hand, have no timing reference (unless they are equipped with some active timing stabilization), so that the timing errors can undergo an unbounded drift. In the noise spectrum, this kind of random walk leads to a divergence for zero noise frequency, similar to the one for the optical phase noise. However, the distinction is somewhat blurred by the existence of regenerative active mode-locking schemes, where the modulator frequency is automatically adjusted to match the cavity round-trip frequency, so that cavity length drifts play the same role as in passively mode-locked lasers and the (absolute) timing error displays the same kind of low-frequency divergence. The same is actually even true for "ordinary" actively mode-locked lasers if one considers that the RF signal of the modulator driver has a diverging phase noise, but the divergence does not occur for the *relative* jitter between pulse train and modulator driver signal.

There have been a number of publications presenting theoretical results for the timing jitter of actively and/or passively mode-locked lasers. In this chapter, we focus on three kinds of models, which are based on somewhat different approaches. The first one is the model by H. A. Haus and A. Mecozzi [33], which has been developed on the basis of soliton perturbation theory and the assumption of a fast saturable absorber. Here, it is assumed that the pulse dynamics are basically determined by soliton effects, with additional influences from quantum noise which are treated as weak perturbations. The circulating pulse is described with four dynamical variables, namely pulse energy, optical phase, temporal position, and spectral position (i.e., offset of the center frequency from its mean value). Of course, this is already a simplification, as additional degrees of freedom such as chirps or pulse shape distortions are not included. Also, the pulse duration is not rigidly coupled to the pulse energy even for the case of soliton pulses; only in the stationary case, the soliton dynamics lead to a constant product of pulse energy and pulse duration. In addition, the description of the gain dynamics is rather crude: there is no additional dynamical variable for the laser gain, so that relaxation oscillation dynamics can not be described. However, an extension in this direction is possible and has indeed been reported [49].

The quantum noise can then be projected onto the dynamical variables of the model, leading to Langevin-type equations, and analytical results for the noise spectra corresponding to these variables have been obtained. As a brief review of the results, we focus on the central result concerning timing jitter, which has two contributions. The first one is the direct impact of quantum noise on the pulse position, i.e., on the timing phase error, which causes a random walk of this variable, corresponding to a power spectral density proportional to f^{-2} and to white frequency noise. The second contribution comes from fluctuations of the optical center frequency, which are bounded due to the limited laser gain bandwidth (which always tends to pull the spectrum back to its equilibrium position) and couple to the pulse timing via dispersion — that is, the dependence of the group velocity on the center frequency. The integrating nature of this coupling leads to an additional f^{-2} dependence, so that this contribution from the center frequency is proportional to f^{-2} at low frequencies (where the center frequency noise is basically white) and proportional to f^{-4} at higher frequencies, where, however, finally the direct contribution of quantum noise on the pulse timing dominates.

Although the Haus–Mecozzi model has been developed for passively mode-locked lasers, it can relatively easily be modified for active mode locking [49,50], as far as soliton shaping effects still stay dominant. Technically, this involves including an additional term in the dynamical equation of the temporal position, representing a "restoring force" as generated by the active modelocker. The result is that the power spectral density of the timing noise levels off at $f = 0$, eliminating the mentioned divergence.

Of course, the Haus–Mecozzi model is not valid for mode-locked lasers that are not based on soliton pulse shaping. As such lasers are also very important, a numerical model has recently been developed [38] that requires only a significantly less restricting set of assumptions. Here, the pulse is represented by an array of complex amplitudes, which resemble the pulse shape in the time or frequency domain. A wide range of effects acting on the pulse can be described in the time or frequency domain, and fast Fourier transforms allow to switch between domains as required. Thus, the model allows us to treat basically all kinds of mode-locked lasers, regardless of their mode-locking mechanism (active or passive, fast or slow absorber), gain dynamics, and additional coupling effects.

The first situation investigated with this numerical model was one of a soliton mode-locked laser, where the Haus–Mecozzi model is applicable. After applying some trivial corrections to the latter (see the erratum [51] and Reference 38), perfect agreement was obtained, as had been expected. An initially quite surprising result was that the agreement persisted in cases without any dispersion and Kerr nonlinearity, i.e., without soliton shaping effects, where the assumptions of the Haus–Mecozzi model are clearly violated. This lead to the insight that the noise properties (including the timing jitter) are not directly affected by the detailed pulse-shaping processes (e.g., soliton effects or saturable absorber action): the noise influences come from quantum effects (spontaneous emission), and their impact on the pulse parameters depends only slightly on the pulse shape (e.g., sech^2 or Gaussian) but not on the shaping mechanism itself. The latter only influences the coupling of different kinds of noise. For example, the presence of dispersion (as is required, e.g., for soliton mode locking) leads to the above mentioned coupling of center frequency noise to timing noise.

This new insight triggered the development of a new analytical model [52] which does not reduce the dynamics to the interplay of a few dynamical parameters and emphasizes less the role of the pulse-shaping mechanism. In this model, the above mentioned direct effect of quantum noise on the pulse timing can be explained as follows. The quantum noise added to the pulse during one cavity round trip can be described by adding uncorrelated noise amplitudes with equal noise variance to all samples (complex amplitudes) in the time domain. The total power in each sample can be increased or reduced, depending on the relative phase of original amplitude and noise contribution. In effect, the pulse position (defined as the "center of gravity") experiences some random change while the pulse shape distortions are subsequently dampened by the pulse shaping mechanism. Even though on average this timing change is zero, the variances of the timing errors for subsequent cavity round trips simply add up (because the corresponding noise influences are statistically independent), which leads to a linear growth of variance with time. The latter corresponds to the f^{-2} dependence of the timing noise power spectral density. The indirect effect of quantum noise via the optical center frequency can be described similarly, and the results are exactly identical to those of the (corrected) Haus–Mecozzi model. So we see that the Haus–Mecozzi results are much more general than the underlying assumptions, and this is proven by the new model which does not need these assumptions, in particular not those on the pulse-shaping mechanism.

As a consequence, we can now apply such analytical results to a much wider range of lasers. However, we have to keep in mind that a great variety of coupling effects (as discussed in Section 12.6.2) can occur. In any particular case, one has to check which coupling effects can occur and analyze their importance. (Note that the importance of particular coupling effects can greatly vary between different types of mode-locked lasers, which can have totally different values of pulse duration, repetition rate, peak power, etc., and different components in the cavity.) The relevant effects then have to be included in the equations. The challenge is to make sure that all relevant effects have been identified and correctly included. In this situation, it can be very helpful to employ the above mentioned numerical model at least for validation purposes, as the description of particular interactions (e.g., with a slow saturable absorber) tends to be simpler than the analysis of all resulting coupling effects. This kind of sanity check has been done in Reference 52 for some simple cases. More complicated situations of practical interest will still have to be analyzed.

12.6.5 Jitter Measurements

For the measurement of the timing jitter of mode-locked lasers, a great variety of techniques is available, which strongly differ in terms of the utilized phase (or timing) reference and also in terms of performance and applicability in different situations.

A frequently used technique for jitter measurements is based on the direct spectral analysis of a photodiode signal with a microwave spectrum analyzer [53], which displays sharp peaks of the power spectral density at integer multiples of the repetition rate f_{rep} and noise sidebands around them, which contain information on timing jitter. A simplified analysis, using the approximation of small timing excursions (compared to the pulse period) and ignoring various correlations between different fluctuating quantities, leads to the simple result that the intensity noise causes equal contributions to the sidebands of all orders, whereas the contributions of the timing phase noise scale with the square of the sideband order. Thus, one can in principle easily separate the contributions of intensity and timing phase noise from each other, if one has data from at least two different sidebands.

The simplicity of this method, both in terms of experimental setup and data analysis, makes it very attractive. However, one should be aware of a number of limitations. Fundamental limitations arise from large phase excursions in passively mode-locked lasers and from the possibility of the above mentioned correlations [54–56]. An important remark concerns the fact that the effectively used phase reference is the local oscillator of the spectrum analyzer, which itself of course exhibits phase noise. It turns out that the phase noise of state-of-the-art spectrum analyzers can easily exceed the timing phase noise of certain types of mode-locked lasers, as has been demonstrated, for example, with passively mode-locked miniature Er:Yb:glass lasers [57,58]. Other technical issues can arise from the difficulty to detect high enough harmonics for multi-GHz lasers and from possible AM/PM conversion (i.e., conversion from intensity to phase noise) in the photodetector [59,60]. Note also that the lowest detectable noise frequency is in the order of three times the minimum resolution bandwidth of the spectrum analyzer. Finally, note that noise data obtained with electronic spectral analyzers require certain corrections related to envelope detection, logarithmic averaging, and the effective noise bandwidth of the RF filter. Such corrections typically amount to adding ≈2 dB to the measured data, resulting in ≈16% higher power densities. Other errors can result from inadequate detector modes or signal power levels. (For example, see Agilent Application Note 150 for details.)

Another class of jitter measurement techniques is based on phase detectors. Typically, a microwave mixer is used as phase detector. Its inputs are the photodiode signal and a sinusoidal reference signal from a stable electronic oscillator. The relative phase of these must be adjusted to be $\pi/4$ on average. As long as the phase deviations from this condition are small enough (well below 1 rad), the average voltage output of the mixer is proportional to the phase deviation. For larger phase differences, of course, the method is not suitable. Therefore, it is applicable to cases where the phase excursion remains small for all times, as is typically the case. Examples would be for the residual jitter in actively mode-locked lasers [60–62], for timing-stabilized passively mode-locked lasers [63–66], and for synchronously pumped lasers [67] but not for free-running passively mode-locked lasers.

Technical difficulties arise from parasitic effects such as mixer offsets, which lead to incomplete suppression of amplitude noise, and excess noise of electronic components. Such problems can be addressed with chopper schemes [60] and/or with correlation methods [68,69], allowing suppression of excess noise from electronic components like mixers and amplifiers to a high degree [70]. Other methods involve the conversion of the microwave reference into an optical amplitude or phase modulation [71]. Another issue, of course, is the stability of the electronic oscillator. For high sensitivity, one may employ an ultra low-noise sapphire-loaded cavity oscillator [72] which, however, is rather expensive, or replace this with a second mode-locked laser, which should either be superior in terms of noise or have noise properties similar to those of the first laser. With such a scheme and the combination of two phase detectors, operating on different harmonics of the photodiode signal, timing stabilization and measurement have become possible with relative rms jitter values in the order of only 1 fs [73].

A very high (subfemtosecond) sensitivity for timing jitter detection can be obtained with optical phase detectors [74] when dealing with very short femtosecond pulses. Here, the inputs are two laser beams carrying femtosecond pulse trains, and the phase detector is a nonlinear crystal or (for better intensity noise suppression) a combination of two nonlinear crystals [75]. Note that such a phase detector can work only for timing errors in a very narrow range in the order of the pulse duration; that is, not for free-running lasers. For short pulses, however, the achieved sensitivity is extremely high and can go beyond the limits set by shot noise for all photodiode-based methods.

Recently, a novel method for relative timing jitter measurement has been demonstrated [57], which uses microwave mixers but not as phase detectors. It is very sensitive and much more versatile than phase detector methods. The principle setup for comparing the timing of two lasers is shown in Figure 12.5. The outputs of the two lasers are detected with fast photodetectors and mixed with the signal from a tunable electronic oscillator to obtain mixing products at much lower frequencies of, for example, 200 kHz. These are then simultaneously recorded with a two-channel digital storage oscilloscope (or a sampling card in a PC). Both mixing products are actually affected by phase noise of the electronic reference oscillator, which may be stronger than the phase noise arising from the timing noise of the lasers. However, with suitable Fourier transform techniques one can extract the *relative* phase of the two recorded data traces, and this relative phase reflects the timing difference of the lasers, with no influence of the phase noise of the electronic oscillator. It has been shown that the system phase noise resulting from thermal and digitizing noise is almost white, so that particularly at low noise frequencies the sensitivity of this method is much higher than for other methods which are affected by phase noise from a reference oscillator. Figure 12.6 shows data from measurements on two 10-GHz passively mode-locked Er:Yb:glass

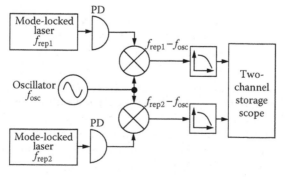

FIGURE 12.5
Setup for relative timing jitter measurements. PD = photodiode.

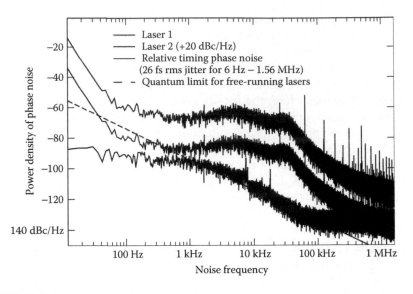

FIGURE 12.6

Relative timing noise spectrum as measured with a setup according to Figure 12.5 with two timing-stabilized passively mode-locked Er:Yb:glass lasers. See the text for explanations.

lasers (see Reference 57). The two upper noise traces indicate the noise spectrum corresponding to the two recorded channels, which are essentially equal and dominated by the electronic oscillator. (One of the curves has been raised by 20 dBc/Hz in order to distinguish the otherwise overlapping curves.) The lowest noise trace shows the relative timing noise as obtained from the phase differences. It is close to the quantum limit of the free-running lasers (dashed line) at about 1–100 kHz is lower than that for low noise frequencies (due to the timing stabilization) and hits the noise floor above ≈100 kHz. The noise floor is at ≈136 dBc/Hz, set by noise of the used sampling card, but in other cases it may be limited by the photodetectors. In the reported experiments, traveling-wave photodiodes have been used that can handle high photocurrents and thus allow for a particularly low noise floor.

Note that such a time domain method allows characterization of noise at very low frequencies, limited only by the available sampling memory. For low noise frequencies, another time domain method has been discussed in Reference 76.

Instead of an external phase reference, one may derive the phase reference from the input signal itself — pretty much in analogy to methods discussed in Section 12.3.3.2 in the context of optical phase noise. The direct analogue of a self-heterodyning setup in the electronic domain is rarely used despite its potential sensitivity. In principle, one could use a photodiode for detection of the repetition signal and a delay-time discriminator comprising a long RF delay line. However, the short pulse durations of today's mode-locked lasers cannot be exploited with such systems due to the limited bandwidth of photodetectors and RF cables. A better approach is therefore to delay the pulse train of the DUT in a long fiber and to compare the arrival times of the input pulses with those of the delayed pulses, using an optical intensity correlator — for example, a nonlinear crystal or a photodetector based on two-photon absorption. Such systems are inherently fast if the overall group velocity dispersion of the delay line is kept low, using a proper combination of different fibers. Output pulse durations well below 1 ps have been reported for delays of many kilometers [77]. Thus, the discriminator slope of a fiber-based timing discriminator can be about two orders of magnitude steeper than that of a microwave delay line, so that a superior frequency noise sensitivity can be achieved.

12.6.6 Optical Phase Noise and Carrier-Envelope Offset Noise

In the case of a mode-locked laser, the definition of the optical phase is not a completely straight-forward issue. One is tempted to consider that phase as a property of each single pulse, similar to the pulse energy, duration, etc. This picture, however, leads to some complications — for example, the question of which temporal position to take the phase value: at a constant position (even if the pulse position drifts) or at the pulse maximum (which itself may not always be very well defined)? The relevance of the optical phase for beat measurements actually suggests a different definition: the spectrum of a regular pulse train consists of an equidistant comb of lines, and one can take one of these lines (typically the central one) to define an optical phase. As each of these lines is similar to the output of a single-frequency laser, this does not introduce conceptual problems and at the same time leads exactly to that quantity which is relevant when, for example, a beat note between that line and the output of a single-frequency laser is measured. There is only one remark to be made: one requires a measurement time of several pulse periods to measure the phase of an individual line, so this kind of optical phase is, strictly speaking, defined only with a bandwidth which is some fraction of the pulse repetition rate; this usually doesn't matter. So we basically have a multitude of phase values, corresponding to all lines in the spectrum of the laser output. Even if one is interested only in a single beat note, it is worthwhile to consider the full frequency comb, as this has implications which in recent years have turned out to be very important for frequency metrology. One can exploit the exact equidistance of the lines (at least in a noiseless case, where all lines have zero width) and write the frequency of the line with the integer index j as [32,78]

$$\nu_j = \nu_{ceo} + j \cdot f_{rep} \qquad (12.23)$$

with the so-called carrier-envelope offset (CEO) frequency ν_{ceo} and the repetition rate f_{rep}. Now we consider the situation with noise. In principle, all lines are subject to noise, which is not completely correlated. In the long run, however, the mode-locking mechanism does not allow the pulse to fall apart (for example, by broadening indefinitely), so that the phases of all lines are prevented from drifting arbitrarily. For an actively mode-locked laser, there is only one degree of freedom which is able to do an unbounded drift: the phases of all lines can drift simultaneously by the same amount; that is, the differences between them are bounded. From this one can conclude that the linewidths of all lines must be equal, irrespective of their optical powers. For the case that only quantum noise acts in the laser, it has been shown [79] that this common linewidth equals the Schawlow–Townes linewidth (Equation 12.9), evaluated with the *total* power of all lines — that is, the laser's average power. This holds despite the fact that the high-frequency (short-term) noise is only weakly correlated: the linewidth is determined by the diverging low-frequency noise not by the high-frequency noise.

The situation is somewhat different for a passively mode-locked laser, which has one more degree of freedom for an unbounded drift; the pulse timing, which has no "restoring force" in that case and is related to phase, changes proportionally to the deviation of the frequency from the center frequency of the comb. This means that the slope of the optical phases can undergo an unbounded drift, which affects the wings of the spectrum but not the center. As a result of this, one obtains somewhat increased linewidth for the lines in the wings of the optical spectrum. More important, the optical phase can be coupled to other variables by various effects; for example, it can be shown (and demonstrated with a numerical model as described above) that a slow saturable absorber can strongly couple center frequency fluctuations to those of the optical phase. As a result, the optical phase noise can be orders of magnitude stronger than that at the Schawlow–Townes level. This, however, remains to be carefully checked in experiments.

Although one can characterize the steady state of the frequency comb with only two parameters (ν_{ceo} and f_{rep}), a similar decomposition for the noise is somewhat problematic. It does work for the low noise frequencies, as explained above, but not for high frequencies. Even for low noise frequencies, there are practical problems, namely the difficulty to determine the CEO phase without picking up uncontrolled phase drifts in the measurement setup. We also note that the CEO frequency is a quantity which is not directly measured but rather extrapolated from optical frequencies, so one should recognize the optical frequencies as being closer to the physical reality. The overall noise in a mode-locked laser is therefore perhaps more naturally seen as the result of noise in all lines of the comb, rather than as being caused by a superposition of common phase noise with carrier envelope offset noise. Notwithstanding these thoughts, currently used schemes for the stabilization of frequency combs usually involve the detection and stabilization of the CEO frequency [32,78,80], and to achieve this one requires control parameters which can influence the pulse repetition rate and the CEO frequency. Currently, different experimental approaches are under investigation with the goal of generating stabilized frequency combs with minimum noise, as required for high-precision frequency metrology.

We can conclude that the area of optical phase noise and noise of the carrier-envelope offset frequency of mode-locked lasers is far from being fully explored. More in-depth theoretical studies are to be done and, experimentally, the field probably has the potential for significant further development.

Acknowledgments

The authors thank the all-solid-state laser group at ETH Zürich for useful discussions, in particular Simon Zeller and Adrian Schlatter for comments on the manuscript. Work at ETH on laser noise has been funded by the Hasler Stiftung and the Swiss innovation promotion agency KTI/CTI (Commission for Technology and Innovation).

Appendix 1: Derivation of Schawlow–Townes Formula

The Schawlow–Townes formula for the linewidth of a laser with quantum noise influences only has been presented by Schawlow and Townes in 1958 (see Reference 6). This reference, however, does not contain a derivation, but only referred to a result for thermal noise influences in masers and simply replaced $k_B T$ with $h\nu$. In the following, we give a derivation with a detailed physical explanation, which is instructive and confirms a very important result.

Assume a four-level single-frequency laser without amplitude/phase coupling. The output coupler transmission is T_{oc}, and there can be additional (parasitic) cavity losses l_{par}, so that the total cavity losses are $l_{tot} = T_{oc} + l_{par}$. On average, the gain must balance the losses: $g = l_{tot}$.

We describe the circulating field with a complex amplitude A, normalized so that the intracavity power is $P_{int} = |A|^2$.

During each round-trip, the gain medium adds a fluctuating amplitude δA where each of the quadrature components has the variance

$$\sigma^2 = \frac{h\nu}{4 \cdot T_R} g \tag{12.24}$$

with T_R being the round-trip time. This amount of noise is consistent, for example, with the results of Reference 5 and with a 3-dB noise figure of a high-gain laser amplifier. Its effect is that the optical phase changes by

$$\delta\varphi = \frac{\delta A_q}{A} \tag{12.25}$$

where δA_q is the quadrature component perpendicular to A in the complex plane. The cavity losses contribute another noise amplitude proportional to l_{tot}. Therefore, in total the variance of the phase is increased by

$$\delta\sigma_\varphi^2 = \frac{h\nu}{4 \cdot T_R \cdot |A|^2} g + \frac{h\nu}{4 \cdot T_R \cdot |A|^2} l_{tot} = \frac{h\nu}{2 \cdot T_R \cdot P_{int}} l_{tot} \qquad (12.26)$$

per round trip, where we have used $g = l_{tot}$, because the gain must balance the cavity losses. As each round trip contributes statistically independent phase fluctuations, the phase variance grows with time according to

$$\sigma_\varphi^2(t) \equiv \left\langle \left(\varphi(t) - \varphi(0)\right)^2 \right\rangle = \frac{h\nu}{2 \cdot T_R \cdot P_{int}} l_{tot} \frac{t}{T_R} = \frac{h\nu \, l_{tot}}{2 \cdot T_R^2 \cdot P_{int}} t. \qquad (12.27)$$

Such a linear growth of variance is consistent with a two-sided power density

$$S_\varphi(f) = \frac{C}{f^2} \qquad (12.28)$$

where

$$C = \frac{1}{(2\pi)^2} \frac{h\nu \, l_{tot}}{2 \cdot T_R^2 \cdot P_{int}}. \qquad (12.29)$$

(One can verify this by calculating the phase variance for a given power density proportional to f^{-2}, which leads to an integral that can be analytically solved and is proportional to the time t.) One can further show that the power density of Equation 12.28 corresponds to a Lorentzian field spectrum with

$$L(\Delta f) = \frac{C}{(\pi C)^2 + (\Delta f)^2}, \qquad (12.30)$$

which approaches the phase noise power density $S_\varphi(f)$ for large frequency offsets Δf. It is normalized for a unity integral and leads to the FWHM field linewidth

$$\Delta\nu = 2\pi C = \frac{h\nu \, l_{tot}}{4\pi \cdot T_R^2 \cdot P_{int}} = \frac{h\nu \, l_{tot} T_{oc}}{4\pi \, T_R^2 \, P_{out}} \qquad (12.31)$$

(see also the end of Section 12.2). For $l_{tot} = T_{oc}$ (i.e., without parasitic cavity losses), this result is exactly equivalent to the original Schawlow–Townes formula [6]

$$\Delta v = \frac{4\pi\, hv\left(\Delta v_{\mathrm{c}}\right)^{2}}{P_{\mathrm{out}}} \tag{12.32}$$

where

$$\Delta v_{\mathrm{c}} = \frac{1}{4\pi}\frac{l_{\mathrm{tot}}}{T_{\mathrm{R}}} \tag{12.33}$$

is the *half width* of the resonances of the cavity (with unpumped gain medium).

Appendix 2: Derivation of Linewidth Formula for Lasers with Amplitude/Phase Coupling

We use the same notation as in Appendix 1. Any fluctuation δA in the amplitude quadrature of the intracavity field of the laser amounts to a change

$$\delta P_{\mathrm{int}} = 2A\,\delta A = \left(\frac{2}{\sqrt{P_{\mathrm{int}}}}\,\delta A\right) P_{\mathrm{int}} \tag{12.34}$$

of the intracavity optical power. Written in this way, one can consider the addition of the fluctuation amplitude as a kind of gain, which for low frequencies must be compensated by the opposite change of the laser gain, caused by gain saturation:

$$\delta g = -\frac{2}{\sqrt{P_{\mathrm{int}}}}\,\delta A \tag{12.35}$$

The optical amplitudes have white noise with the two-sided power density

$$S_{\delta A}(f) = \frac{hv}{2}\,g \tag{12.36}$$

if the fluctuations from gain medium and losses are added. Therefore, we have

$$S_{\delta g}(f) = \frac{4}{P_{\mathrm{int}}}S_{\delta A}(f) = \frac{2hv}{P_{\mathrm{int}}}\,g, \tag{12.37}$$

which indicates white noise in the gain. For the instantaneous optical frequency v_{opt} and the optical phase φ we have the dynamical equation

$$2\pi\,\delta v_{\mathrm{opt}} = \frac{d}{dt}\varphi = \frac{\alpha}{2T_{\mathrm{R}}}\,\delta g, \tag{12.38}$$

because the optical phase changes by $\alpha\delta g/2$ per round trip where α is the linewidth enhancement factor. From this we obtain the power density

$$S_{\nu,\text{opt}}(f) = \frac{\alpha^2}{16\pi^2 T_R^2}\, S_{\delta g}(f) = \frac{\alpha^2\, h\nu\, g}{8\pi^2 T_R^2 P_{\text{int}}} \tag{12.39}$$

of the optical frequency and

$$S_\varphi(f) = S_{\nu,\text{opt}}(f)\,/\,f^2 = \frac{\alpha^2\, h\nu\, g}{8\pi^2 T_R^2 P_{\text{int}}}\, f^{-2} \tag{12.40}$$

of the optical phase, from which we obtain the contribution to the linewidth (in the same way as in Appendix 1) as

$$\Delta\nu = 2\pi\, \frac{\alpha^2\, h\nu\, g}{8\pi^2 T_R^2 P_{\text{int}}} + \frac{h\nu\, g}{4\pi\, T_R^2\, P_{\text{int}}} = \alpha^2\, \frac{h\nu\, l_{\text{tot}} T_{\text{oc}}}{4\pi\, T_R^2\, P_{\text{out}}}\,. \tag{12.41}$$

Finally, one has to add this contribution to that one of the direct influence of quantum noise on the phase fluctuations, as both fluctuations are from different quadrature components and thus statistically independent. In total, one obtains

$$\Delta\nu = 2\pi\, \frac{\alpha^2\, h\nu\, g}{8\pi^2 T_R^2 P_{\text{int}}} + \frac{h\nu\, g}{4\pi\, T_R^2\, P_{\text{int}}} = \left(1 + \alpha^2\right) \frac{h\nu\, l_{\text{tot}} T_{\text{oc}}}{4\pi\, T_R^2\, P_{\text{out}}}\,. \tag{12.42}$$

References

1. B. Daino, P. Spano, and S. Piazolla, Phase noise and spectral line shape in semiconductor lasers, *IEEE J. Quant. Electron.*, Vol. 19, 266–270, 1983.
2. R.J. Glauber, Coherent and incoherent states of the radiation field, *Phys. Rev.*, Vol. 131, 2766, 1963.
3. D.J. Walls, Squeezed states of light, *Nature*, Vol. 306, 141–146, 1983.
4. R.E. Slusher, L.W. Hollberg, B. Yurke, J.C. Mertz, and J.F. Valley, Observation of squeezed states generated by four wave mixing in an optical cavity, *Phys. Rev. Lett.*, Vol. 55, 2409–2412, 1985.
5. C.M. Caves, Quantum limits on noise in linear amplifiers, *Phys. Rev. D*, Vol. 26, 1817, 1982.
6. A.L. Schawlow and C.H. Townes, Infrared and optical masers, *Phys. Rev.*, Vol. 112, 1940–1949, 1958.
7. C.C. Harb, T.C. Ralph, E.H. Huntington, D.E. McClelland, and H.-A. Bachor, Intensity-noise dependence of Nd:YAG lasers on their diode-laser pump source, *J. Opt. Soc. Am. B*, Vol. 14, 2936–2945, 1997.
8. B.C. Buchler, E.H. Huntington, C.C. Harb, and T.C. Ralph, Feedback control of laser intensity noise, *Phys. Rev. A*, Vol. 57, 1286–1294, 1998.

9. T.C. Ralph, C.C. Harb, and H.-A. Bachor, Intensity noise of injection-locked lasers: quantum theory using a linearized input-output method, *Phys. Rev. A*, Vol. 54, 4359–4369, 1996.

10. K.J. Weingarten, B. Braun, and U. Keller, In-situ small-signal gain of solid-state lasers determined from relaxation oscillation frequency measurements, *Opt. Lett.*, Vol. 19, 1140–1142, 1994.

11. C.H. Henry, Theory of the linewidth of semiconductor lasers, *IEEE J. Quantum Electron.*, Vol. 18, 259–264, 1982.

12. J.W. Arkwright, P. Elango, G.R. Atkins, T. Whitbread, and M.J.F. Digonnet, Experimental and theoretical analysis of the resonant nonlinearity in ytterbium-doped fiber, *J. Lightwave Technol.*, Vol. 16, 798–806, 1998.

13. R. Lang and K. Kobayashi, External optical feedback effects on semiconductor injection laser properties, *IEEE J. Quant. Electron.*, Vol. 16, 347–351, 1980.

14. J. Poizat and P. Grangier, Quantum noise of laser diodes, *J. Mod. Opt.*, Vol. 47, 2841–2856, 2000.

15. Y. Yamamoto, AM and FM quantum noise in semiconductor lasers, *IEEE J. Quantum Electron.*, Vol. 19, 34–46, 1983.

16. K. Vahala and A. Yariv, Semiclassical theory of noise in semiconductor lasers, *IEEE J. Quant. Electron.*, Vol. 19, 1096–1101, 1983.

17. R. Rollins, D. Ottaway, M. Zucker, R. Weiss, and R. Abbott, Solid-state laser intensity stabilization at the 10^{-8} level, *Opt. Lett*, Vol. 29, 1876–1878, 2004.

18. R.P. Scott, C. Langrock, and B.H. Kolner, High-dynamic-range laser amplitude and phase noise measurement techniques, *IEEE J. Sel. Top. Quantum Electron.*, Vol. 7, 641–655, 2001.

19. L.Y. Lin, M.C. Wu, T. Itoh, T.A. Vang, R.E. Muller, D.J. Sivco, and A.Y. Cho, Velocity-matched distributed photodetectors with high saturation power and large bandwidth, *IEEE Photon. Technol. Lett.*, Vol. 8, 1376–1378, 1996.

20. H.R. Telle, Narrow linewidth laser diodes with broad continuous tuning range, *Appl. Phys. B*, Vol. 49, 217–226, 1989.

21. R.W.P. Drever, J.L. Hall, F.V. Kowalski, J. Hough, G.M. Ford, A.J. Monley, and H. Ward, Laser phase and frequency stabilization using an optical resonator, *Appl. Phys. B*, Vol. 31, 97–105, 1983.

22. R.V. Pound, Electronic frequency stabilization of micowave oscillators, *Rev. Sci. Instrum.*, Vol. 17, 490–505, 1946.

23. T.W. Hänsch and B. Couillaud, Laser frequency stabilization by polarization spectroscopy of a reflecting reference cavity, *Opt. Commun.*, Vol. 35, 441–444, 1980.

24. F.K. Fatami, J.W. Lou, and T.F. Carruthers, Frequency comb linewidth of an actively mode-locked fiber laser, *Opt. Lett*, Vol. 29, pp. 944–946, 2004.

25. L.E. Richter, H.I. Mandelberg, M.S. Kruger, and P.A. McGrath, Linewidth determination from self-heterodyne measurements with subcoherence delay times, *IEEE J. Quant. Electron.*, Vol. 22, 2070–2074, 1986.

26. H. Jackel and G. Guekos, High frequency intensity noise spectra of axial mode groups in the radiation from cw GaAlAs diode lasers, *Opt. Quantum. Electron.*, Vol. 9, 233–239, 1977.

27. M. Ohtsu, Y. Otsuka, and Y. Teramachi, Precise measurement and computer simulations of mode-hopping phenomena in semiconductor lasers, *Appl. Phys. Lett*, Vol. 46, 108–110, 1984.

28. P.L. Liu, Photon statistics and mode partition noise of semiconductor lasers, in *Coherence, Amplification and Quantum Effects in Semiconductor Lasers*, Y. Yamamoto, Ed., New York: John Wiley & Sons, 1991, pp. 411–459.

29. C.H. Henry, P.S. Henry, and M. Lax, Partition fluctuations in nearly single-longitudinal-mode lasers, *J. Lightwave Technol.*, Vol. 2, 209–216, 1984.

30. C. Bollig, W.A. Clarkson, and D.C. Hanna, Stable high-repetition-rate single-frequency Q-switched operation by feedback suppression of relaxation oscillations, *Opt. Lett.*, Vol. 20, 1383–1385, 1995.

31. G.J. Spühler, R. Paschotta, R. Fluck, B. Braun, M. Moser, G. Zhang, E. Gini, and U. Keller, Experimentally confirmed design guidelines for passively Q-switched microchip lasers using semiconductor saturable absorbers, *J. Opt. Soc. Am. B*, Vol. 16, 376–388, 1999.

32. H.R. Telle, G. Steinmeyer, A.E. Dunlop, J. Stenger, D.H. Sutter, and U. Keller, Carrier-envelope offset phase control: A novel concept for absolute optical frequency measurement and ultrashort pulse generation, *Appl. Phys. B*, Vol. 69, 327–332, 1999.

33. H.A. Haus and A. Mecozzi, Noise of mode-locked lasers, *IEEE J. Quantum Electron.*, Vol. 29, 983–996, 1993.

34. U. Keller, Ultrafast solid-state lasers, *Prog. Opt.*, Vol. 46, 1–115, 2004.

35. R. Paschotta and U. Keller, Passive mode locking with slow saturable absorbers, *Appl. Phys. B*, Vol. 73, 653–662, 2001.

36. N. Tzoar and M. Jain, Self-phase modulation in long-geometry optical waveguides, *Phys. Rev. A*, Vol. 23, 1266–1270, 1981.

37. J.P. Gordon, Theory of the soliton self-frequency shift, *Opt. Lett.*, Vol. 11, 662–664, 1986.

38. R. Paschotta, Noise of mode-locked lasers. Part I: numerical model, *Appl. Phys. B*, Vol. 79, 153–162, 2004.

39. F.X. Kärtner, L.R. Brovelli, D. Kopf, M. Kamp, I. Calasso, and U. Keller, Control of solid-state laser dynamics by semiconductor devices, *Opt. Eng.*, Vol. 34, 2024–2036, 1995.

40. C. Hänninger, R. Paschotta, F. Morier-Genoud, M. Moser, and U. Keller, Q-switching stability limits of continuous-wave passive mode locking, *J. Opt. Soc. Am. B*, Vol. 16, 46–56, 1999.

41. U. Keller, Ultrafast all-solid-state laser technology, *Appl. Phys. B*, Vol. 58, 347–363, 1994.

42. A. Schlatter, S.C. Zeller, R. Grange, R. Paschotta, and U. Keller, Pulse energy dynamics of passively mode-locked solid-state lasers above the Q-switching threshold, *J. Opt. Soc. Am. B*, Vol. 21, 1469–1478, 2004.

43. M.C. Nuss, U. Keller, G.T. Harvey, M.S. Heutmaker, and P.R. Smith, Dramatic reduction of the amplitude noise of colliding-pulse-mode locking dye laser, *Opt. Lett.*, Vol. 15, 1026–1028, 1990.

44. D.W. Allan, Statistics of atomic frequency standards, *Proc. IEEE*, Vol. 54, 221–230, 1966.

45. J.A. Barnes, A.R. Chi, L.S. Cutler, D.J. Healey, D.B. Leeson, T.E. McGunigal, J.A. Mullan, W.L. Smith, R.L. Sydnor, R.F.C. Vessot, and G.M.R. Winkler, Characterization of frequency stability, *IEEE Trans. Instrum. Meas.*, Vol. IM-20, 105–120, 1971.

46. J. Rutman, Characterization of phase and frequency instabilities in precision frequency source: fifteen years of progress, *Proc. IEEE*, Vol. 66, 1048–1075, 1978.

47. P. Lesage and C. Audoin, Characterization and measurement of time and frequency stability, *Radio Sci.*, Vol. 14, 521–539, 1979.

48. C.A. Greenhall, Spectral ambiguity of Allan Variance, *IEEE Trans. Instrum. Meas.*, Vol. 47, 623–627, 1998.

49. L.A. Jiang, M.E. Grein, H.A. Haus, and E.P. Ippen, Noise of mode-locked semiconductor lasers, *IEEE J. Sel. Top. Quantum Electron.*, Vol. 7, 159–167, 2001.

50. M.E. Grein, H.A. Haus, Y. Chen, and E.P. Ippen, Quantum-limited timing jitter in actively mode locked lasers, *IEEE J. Quant. Electron.*, Vol. 40, 1458–1470, 2004.

51. H.A. Haus and A. Mecozzi, Correction to "Noise of mode-locked lasers," *IEEE J. Quantum Electron.*, Vol. 30, 1966, 1994.

52. R. Paschotta, Noise of mode-locked lasers. Part II: timing jitter and other fluctuations, *Appl. Phys. B*, Vol. 79, pp. 163–173, 2004.

53. D.v.d. Linde, Characterization of the noise in continuously operating mode-locked lasers, *Appl. Phys. B*, Vol. 39, 201–217, 1986.

54. I.G. Fuss, An interpretation of the spectral measurement of optical pulse train noise, *IEEE J. Quantum Electron.*, Vol. 30, 2707–2710, 1994.

55. D. Eliyahu, R.A. Salvatore, and A. Yariv, Noise characterization of a pulse train generated by actively mode-locked lasers, *J. Opt. Soc. Am. B*, Vol. 13, 1619–1626, 1996.

56. D. Eliyahu, R.A. Salvatore, and A. Yariv, Effect of noise on the power spectrum of passively mode-locked lasers, *J. Opt. Soc. Am. B*, Vol. 14, 167–174, 1997.

57. R. Paschotta, B. Rudin, A. Schlatter, G.J. Spühler, L. Krainer, N. Haverkamp, H.R. Telle, and U. Keller, Relative timing jitter measurements with an indirect phase comparison method, *Appl. Phys. B*, Vol. 80, 185–192, 2005.

58. A. Schlatter, B. Rudin, S.C. Zeller, R. Paschotta, G.J. Spühler, L. Krainer, N. Haverkamp, H.R. Telle, and U. Keller, Nearly quantum noise limited timing jitter from miniature Er: Yb: glass lasers, to appear in *Opt. Lett.*, June 15, 2005.

59. M.J.W. Rodwell, K.J. Weingarten, D.M. Bloom, T. Baer, and B.H. Kolner, Reduction in the timing fluctuations in a modelocked Nd:YAG laser by electronic feedback, *Opt. Lett.*, Vol. 11, 638–640, 1986.

60. M.J.W. Rodwell, D.M. Bloom, and K.J. Weingarten, Subpicosecond laser timing stabilization, *IEEE J. Quantum Electron.*, Vol. 25, 817–827, 1989.

61. T.R. Clark, T.F. Carruthers, P.J. Matthews, and I.N. Duling, III, Phase noise measurements of ultrastable 10 GHz harmonically mode locked fiber laser, *Electron. Lett.*, Vol. 35, 720–721, 1999.

62. L.A. Jiang, M.E. Grein, E.P. Ippen, C. McNeilage, J. Searls, and H. Yokoyama, Quantum-limited noise performance of a mode-locked laser diode, *Opt. Lett.*, Vol. 27, 49–51, 2002.

63. D.J. Jones, K.W. Holman, M. Notcutt, J. Ye, J. Chadalia, L.A. Jiang, E.P. Ippen, and H. Yokoyama, Ultralow-jitter, 1550-nm mode-locked semiconductor laser synchronized to a visible optical frequency standard, *Opt. Lett*, Vol. 28, 813–815, 2003.

64. J.B. Schlager, B.E. Callicoatt, R.P. Mirin, N.A. Sanford, D.J. Jones, and J. Ye, Passively mode-locked glass waveguide laser with 14-fs timing jitter, *Opt. Lett*, Vol. 28, 2411–2413, 2003.

65. U. Keller, C.E. Soccolich, G. Sucha, M.N. Islam, and M. Wegener, Noise characterization of femtosecond color center lasers, *Opt. Lett.*, Vol. 15, 974–976, 1990.

66. G.T. Harvey, M.S. Heutmaker, P.R. Smith, M.C. Nuss, U. Keller, and J.A. Valdmanis, Timing jitter and pump-induced amplitude modulation in the colliding-pulse mode-locked (CPM) laser, *IEEE J. Quantum Electron.*, Vol. 27, 295–301, 1991.

67. U. Keller, K.D. Li, M.J.W. Rodwell, and D.M. Bloom, Noise characterization of femtosecond fiber Raman soliton lasers, *IEEE J. Quantum Electron.*, Vol. 25, 280–288, 1989.

68. F.L. Walls, S.R. Stein, J.E. Gray, and D.J. Glaze, Design considerations in state-of-the-art signal processing and phase noise measurement systems, *Proc. 30th Annu. Symp. Freq. Control*, pp. 269–274, 1976.

69. E. Rubiola and V. Giordano, Corelation-based phase noise measurements, *Rev. Sci. Instrum.*, Vol. 71, 3085–3091, 2000.

70. E.N. Ivanov, S.A. Diddams, and L. Hollberg, Experimental study of noise properties of a Ti: sapphire femtosecond laser, *IEEE Trans. Ultrasonics Ferroelectr. Freq Control*, Vol. 45, 355–360, 2003.

71. P.W. Juodawlkis, J.C. Twichell, J.L. Wasserman, G.E. Betts, and R.C. Williamson, Measurement of mode-locked laser timing jitter by use of phase-encoded optical sampling, *Opt. Lett*, Vol. 26, 289–291, 2001.

72. E.N. Ivanov, M.E. Tobar, and R.A. Woode, Applications of interferometric signal processing to phase-noise reduction in microwave oscillators, *IEEE Trans. Microwave Theory Technol.*, Vol. 46, 1537–1545, 1998.

73. R.K. Shelton, S.M. Foreman, L.-S. Ma, J.L. Hall, H.C. Kapteyn, M.M. Murnane, M. Notcutt, and J. Ye, Subfemtosecond timing jiter between two independent, actively synchronized, mode-locked lasers, *Opt. Lett.*, Vol. 27, 312–314, 2002.

74. R. Schimpe, Intensity noise associated with the lasing mode of a (GaAl)As diode laser, *IEEE J. Quant. Electron.*, Vol. 19, 895–897, 1983.

75. T.R. Schibli, J. Kim, O. Kuzucu, J.T. Gopinath, S.N. Tandon, G.S. Petrich, L.A. Kolodziejski, J.G. Fujimoto, E.P. Ippen, and F.X. Kärtner, Attosecond active synchronization of passively mode-locked lasers by balanced cross correlation, *Opt. Lett.*, Vol. 28, 947–949, 2003.

76. H. Tsuchida, Time-interval analysis of laser-pulse-timing fluctuations, *Opt. Lett.*, Vol. 24, 1434–1436, 1999.

77. C.-C. Chang and A.M. Weiner, Fiber transmission for sub-500-fs pulses using a dispersion-compensating fiber, *IEEE J. Quant. Electron.*, Vol. 33, 1455–1464, 1997.

78. F.W. Helbing, G. Steinmeyer, and U. Keller, Carrier-envelope offset phase-locking with attosecond timing jitter, *IEEE J. Sel. Top. Quantum Electron.*, Vol. 9, 1030–1040, 2003.

79. P.-T. Ho, Phase and amplitude fluctuations in a mode-locked laser, *IEEE J. Quant. Electron.*, Vol. 21, 1806–1813, 1985.

80. S.T. Cundiff and J. Ye, Colloquium: femtosecond optical frequency combs, *Rev. Mod. Phys.*, Vol. 75, 325–342, 2003.

Index